Insects and forests

The role and diversity of insects in the forest environment

Roger Dajoz

Professor at the Muséum National d'Histoire Naturelle, Paris.

Translated from the French by
G.-M. de Rougemont

Intercept Ltd

11, rue Lavoisier
F-75384 Paris cedex 08

LONDRES - PARIS - NEW YORK

DANGER
PHOTOCOPYING
KILLS BOOKS

© **TECHNIQUE & DOCUMENTATION, 2000**
ISBN : 2-7430-0331-6
Originally published as : "Les insectes et la forêt"
by Technique et Documentation
ISBN : 2-7430-0254-9

© **INTERCEPT LTD – 2000**

ISBN : 1-898298-68-8

Contents

Chapter 3 ▰▰▰ THE ROLE OF BIOTIC FACTORS

Chapter 4. ▰▰▰ DIVERSITY AND ABUNDANCE OF INSECTS IN FOREST

Chapter 5 ▰▰▰ INSECTS AND THE FOREST ECOSYSTEM

Chapter 6 ■■■■■ FOREST INSECT PESTS

Chapter 7 ■■■■■ CANOPY INSECTS

Chapter 8 ■■■ DEFOLIATING LEPIDOPTERA

Chapter 9 ■■■ PROCESSIONARY CATERPILLARS

Chapter 10 ■■■ OTHER DEFOLIATORS: SAWFLIES, BEETLES, FLIES

Chapter 11 ▬▬▬ SAP-SUCKERS: SCALE INSECTS, APHIDS AND BUGS

Chapter 12 ▬▬▬ GALLS AND GALL INSECTS

Chapter 13 ▬▬▬ INSECTS ON FLOWERS, FRUITS AND SEEDS

Chapter 14 ■■■ WOOD AND ITS USE BY XYLOPHAGOUS ORGANISMS

Chapter 15 ■■■ BARK BEETLES (SCOLYTIDAE) AND THEIR ASSOCIATED FAUNA

Chapter 16 ◼◼ **SAPROXYLIC INSECTS**

Chapter 17 ▬▬ **WOOD DECAY AND INSECT SUCCESSIONS**

Chapter 18 ▬▬ **TWO SPECIAL HABITATS: TREE HOLES
                        AND FUNGI**

Chapter 19 ▬▬ INSECTS OF FOREST SOILS

Chapter 20 ▬▬ NORTH AMERICAN FOREST INSECTS

Chapter 21 ▬▬ TROPICAL RAIN FOREST INSECTS

Introduction

Forest entomology has a long history. Linnaeus himself recommended steeping felled tree trunks in water to deter infestation by wood-boring insects. This principle is still applied today, when tree trunks stored in forests after felling are periodically sprayed with water. European forest entomology was born in Germany, a country in which extensive stands of conifers had for many years suffered considerable damage, leading to research into the biology and systematics of the insects that caused it. Julius Theodor Christian Ratzeburg devoted himself entirely to forest entomology; he is the author of a monumental work, *Die Forstinsekten*, which was published in three volumes between 1837 and 1844, and of *Die Ichneumonen der Forstinsekten*, also in three volumes, published between 1844 and 1852. In *Die Forstinsekten* great importance was given to scolytids, whose role was already known. At about the same time Edouard Perris, who lived in the Landes in Southwest France, a true "land of the maritime pine", wrote an *Histoire des insectes du pin maritime* of which only the part devoted to Coleoptera was published, in 1863. In this work, which is mainly devoted to the biology of all insects without applied considerations, Perris reveals himself as the first experimenter by cutting trees at different seasons in order to describe the faunal successions. He also demonstrated the effects of the physiological state of a tree on its colonisation by insects, and foresaw the part played by attractive substances emitted by sick trees.

After Ratzeburg's death in 1871 studies in systematics and biology increased. Karl Escherich's monumental work *Die Forstinsekten Europas* was published in four volumes between 1913 and 1940. *Traité d'entomologie forestière* by Auguste Barbey appeared in 1925, and *Die Fichtenkäfer Finnlands* by the Finn U. Saalas came out between 1917 and 1923.

In North America studies in forest entomology began at an early date, understandably, given the extent and economic value of forests there. In Canada forest pests were mentioned from the very beginning of colonisation, as is shown by contemporary references to the presence of spruce bud worm in forests in the East of the country. Recent research by R.F. Morris on populations of this species gave rise to the use of mortality tables in the study of the dynamics of forest defoliators, and have been applied in Europe to the larch bud moth, the pine sawfly *Diprion pini*, and the pine processionary caterpillar. Many researchers in the United States have devoted themselves to forest entomology, including A.S. Packard, *Insects injurious to forest and shade trees*, 1881; S.A. Graham, *Principles of forest entomology*, 1929; K. Graham, *Concepts of forest entomology*, 1963; and Furniss & Carolin, *Western forest insects*, 1977. Biological control of *Lymantria dispar* and *Euproctis chrysorrhea*, which were accidentally introduced to the United States, was the origin of the work of Howard and Fiske (1912) on the predators of these species and their roles as control agents.

For many years forest entomology was almost exclusively devoted to the study of pest species in increasingly attered forests and to research into methods of control. There are many publications devoted to this aspect of entomology, one of the most recent being the annual revue called "La Santé des Forêts" published by the Ministry of Agriculture and Fisheries in France, which contains recent advances concerning various pest species, both old and recent.

A different aspect of forest entomology is developing today. This consists of studying the biology and ecology of all forest insects and researching the part they play in the functioning of the ecosystem. These studies have shown that natural forests on the one hand have a very different structure to those of managed forests, and on the other have a high level of biodiversity and contain many rare or threatened species. Among other recent fields of study in forest entomology, mention may be made of the influence of fire on the fauna, the effects of pollution, the effects of fragmentation of forest stands, and various aspects of the relationships between insects and trees, in particular in the case of beetles of the family Scolytidae, knowledge of which has made considerable progress.

It is all these aspects of forest entomology which we have sought to present in this book. A whole chapter deals with tropical forest insects which are of great interest: their biodiversity is exceptionally high and their biology is often very unusual or even unique.

I wish to thank all the entomologists who sent me reprints of their papers, and the Directors of American research stations where I could stay to study the insect fauna of the great forests of the West. I particularly thank Dr. Wade Sherbrooke at the Southwestern Research Station of the American Museum (Portal, Arizona), Dr. Allan Muth at Deep Canyon Research Centre (Palm Desert, California) and Dr. Daniel Dawson at Sierra Nevada Aquatic Research Laboratory (Mammoth Lakes, California). I also thank the persons who are responsible for National Parks and National Forests for their help. This book was written with the constant help of my wife Aline who has travelled and studied insects with me in the forests for so many years.

Sierra Nevada Aquatic Research Laboratory, California, September 1999

Paris, November 1999

1

Structure and evolution of temperate forests

Forest is an ecosystem in which trees predominate to such an extent that they modify conditions of life on the ground and create a special microclimate. Forest consists not only of large trees but also of shrubs and small trees, herbaceous plants and cryptogams. A special fauna develops in this environment which has a complex structure and, in particular, a characteristic vertical stratification. Numerous ties of dependency exist in this habitat due to the great wealth of species.

The earliest forests in the fossil record, and the earliest forest soils, date from the middle of the Devonian (figure 1.1). The progressive increase in the relative importance of woody plants in relation to herbaceous plants was accompanied by a diversification of animals, particularly insects, which demonstrates the importance of the forest environment in establishing animal diversity (Retallack, 1997). In the temperate regions of Europe, quaternary glaciations played a part as natural disruptions, to which were added those of man, and both profoundly modified and sometimes wiped out the primeval forest. These changes led to the total eradication of many large mammals and to the extinction of many invertebrates. The animals of today's temperate forests are, for the most part, no more than an impoverished sample of the primeval fauna, except in a few, relatively pristine stands of forest.

Studies on the fauna of Baltic amber (Upper Eocene, 40 million years) revealed insects which inhabited the forests of northern Europe. These mixed coniferous and broadleaf forests harboured species which have now disappeared from the region, such as the lucanid beetle *Palaeognathus succini*, belonging to a genus found today in South America. However the insect faunas had by that time already broadly established themselves, as is shown by the marked affinities of the amber species with those that inhabit North America today.

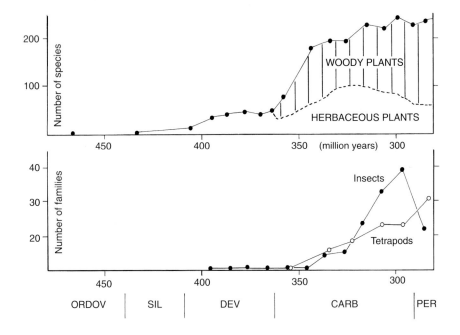

Figure 1.1.   *Above: diversity of plant species in the Paleozoic era expressed by the number of known species. Trees become numerous from the middle Devonian on. Below: diversity of insects and tetrapods expressed by number of families during the same period (Retallack, 1997).*

A remarkable peculiarity of insects is the morphological and genetic stability of species that have evolved little or not at all during the Quaternary. Baltic amber contains species, belonging to various orders, indistinguishable from present day species. This discovery enables reconstitution of paleo-environments when several species whose ecology is well known are spotted in a locality. It also shows how much the distribution ranges have varied in the course of time (Ponel, 1993). In a locality at Broomfield in England dated at 700,000 years, the presence of beetles such as the buprestids *Coroebus elatus* and *C. undatus* and the scarabaeids *Onthophagus similis* and *Ataenius horticola* suggest a hotter climate than the present one. This fauna was succeeded by communities of forest species such as the two beetles *Melasis buprestoides* and *Isorhipis melasoides* which are associated with decayed wood and point to the presence of a temperate forest similar to those of the present day. The deposit at Trafalgar Square, dated at 120,000 years, contains species of a much warmer environment than the present one, such as *Onthophagus massai* which today is endemic to Sicily, and the scolytid *Scolytus koenigi* which only survives in a few localities in southern Europe.

The discovery of Quaternary fossils shows that the beetle *Rhysodes sulcatus* which is heading for extinction in Europe (in France it only survives in three localities) was once more widespread.

1. FORESTS IN WESTERN EUROPE

Forests cover about 42 million square kilometres, i.e. 32% of dry land. Conifers account for one third of trees and broadleaf trees for two thirds. The distribution of forests varies greatly according to continent: only 4% of the world's forests are in Europe, 22% in the former Soviet Union, 16% in North America, 23% in South America, 17% in Africa, and 10% in Asia as a whole (but only 3% in China, where forests have suffered great destruction). In Europe forest spreads over 140 million hectares, and rates, in percentage of wooded land, vary greatly in different countries. France, with 14 million hectares, has 25%, whereas Great Britain with only 1.87 million hectares has 7.6%.

The balance between wooded and un-wooded areas changes frequently, often to the detriment of forest, sometimes to its advantage. Prior to the Neolithic revolution, which saw the intrusion of agriculture, Europe was covered with broadleaf forests (especially oak, elm, beech, lime and hazel). The development of agriculture coincided with elimination of part of the forest by fire, and the areas thus cleared were maintained by grazing cattle. According to Roman authors it took sixty days on foot to cross the Hercynian forest which covered present day Germany. The height of deforestation was probably reached in the early fourteenth century. During the following two centuries some forests regenerated as cultivated land was abandoned. At the end of the Middle Ages growing demand for wood led to a new regression of forest which has continued to our day, and contributed, for example, to the almost total disappearance of the huge forest of Scots pine which covered Scotland. In central Europe 75% of the territory of present day Hungary was covered with forests in the eleventh century, compared with only 17.3% in 1980. In Finland the great primary conifer forest was burnt at a very early date and the burnt parts used for temporary cultivation which would last only a few years; the deteriorated forest which regenerated consisted mainly of birch, and not of conifers (Brimblecombe & Pfister, 1990).

Many forests have been considerably altered, either in structure, or in species composition, by the introduction of alien trees. In Great Britain the area planted with conifers has increased fourfold since 1930, partly at the expense of old or semi-natural forests. The planting of taxa belonging to genera that no longer exist in the wild in Europe is one of the most obvious manifestations of this modification. Douglas-fir covered 200,000 hectares in France at the end of the 1970s, modifying the landscapes of whole regions such as the eastern and western borders of the Massif Central. By providing insects with new ecological niches these modifications allow native species to colonise the trees, and other species native to the same regions as the introduced trees to establish themselves and become potential pests. Plantations of eucalyptus in various regions of southern Europe have had the same result. Planting of native trees outside their natural range produces similar entomological problems. As early as the eighteenth century larch and spruce were used intensively in various parts of France and Europe.

2. DIFFERENT TYPES OF FOREST IN FRANCE

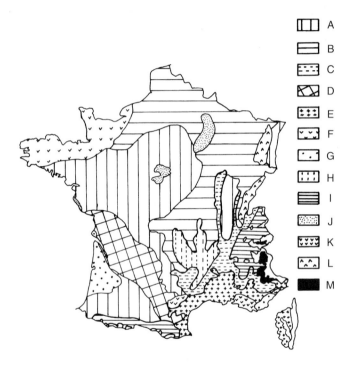

Figure 1.2. *Dominant trees in France. A: sessile and English oaks. B: beech and sessile oak with hornbeam. C: downy oak. D: Pyrenean or downy oaks. E: evergreen oak. F: beech and sessile oak without hornbeam. G: maritime pine. H: natural or planted conifers (fir, Scots pine, spruce). I: conifers (fir, spruce) and beech. J: conifers planted for reforestation. K: Aleppo pine. L: natural conifers (fir, Scots pine, spruce). M: larch.*

The percentage of wooded land varies according to region. It is only 3% in the Département of the Manche and 63% in the Landes. French forest is composed of about 25% conifers, 10% of tall broadleaf forest, 33% undergrowth under tall trees and 32% of non-productive areas (in other words badly managed forests or ones in poor condition). The main trees are the following: oak: 34%; beech: 15%; hornbeam: 8%; maritime pine: 12%; Scots pine: 7%; fir: 7%; spruce: 3%.

Lowland forests of the Atlantic zone are dominated by oak and beech, and in the Landes by maritime pine (Figure 1.2). Montane forests are mainly composed of conifers (fir, spruce, larch) mixed with beech. Mediterranean forests include evergreen oaks and various pines. Stands of beech with sessile oak and hornbeam occupy a large part of northern France, sessile oak often being more abundant than beech. Oak woods composed of sessile oak and English oak are found in the West, the Saône valley and Alsace. In the Garonne drainage valley oak woods are mostly composed of Pyrenean oak associated with downy oak. Forests in which downy oak are dominant surround the lower Rhône valley. Evergreen oak is characteristic of a large part of the Mediterranean region. In mountain areas conifers are often associated with beech; larch occupies those parts of the Alps having a very bright climate. Aleppo and maritime pine dominate primarily in the Maures and the Esterel in South East France and secondarily in the Landes in the South West. The distribution of forest tree species is essentially determined by temperature, rainfall and insolation. The requirements of a few trees of French forests are as follows:

Species	Average temperature (°C)	Rainfall (mm)	Insolation (hours per year)
Evergreen oak	13 to 14	400 to 700	2,800
Sessile oak	8.5 to 13	500 to 1,300	1,700
Beech	7 to 11.5	700 to 1,600	1,700
Fir	7 to 9.5	1,000 to 2,000	1,700
Mountain pine	5 to 8	500 to 1,100	2,700

Various factors such as soil and altitude also play their parts at the local level. The tree line is between 2,200 and 2,400 metres in the inner Alps where larch is found mixed with Arolla pine; in the Pyrénées it is mountain pine which survives at these altitudes.

3. THE MAIN FOREST TREE SPECIES

In Europe forests are composed of a limited number of woody species which make up almost the entire tree population; other species are scattered here and there as isolated individuals. This structure contrasts with that of equatorial forests which consist of a great many species of tree, each of which is represented by scattered individuals. In European temperate forest scarce and scattered species such as wild service tree (*Sorbus*), ash, elm, maple and lime are often considered (from a traditional point of view which regards forest only as a producer of wood) to play a negligible role.

3.1. Conifers

Conifers occur, but on the whole not abundantly, in central and western Europe. They are more abundant and dominant in the continental climate of eastern Europe, the boreal forest of Scandinavia, and in the Mediterranean region. In the latter, relict populations containing numerous species of *Pinus* and *Abies* survive here and there (Quézel, 1980).

Conifers include four native genera in Europe. The common silver fir (*Abies alba*) requires a moist climate and tolerates relatively low temperatures. This species lives in montane forests between 300 and 1,500 metres and in a few lowland localities in Normandy. Norway spruce (*Picea abies*) favours cold soils; all the lowlands and mountains of northern Europe suit it. In France it is spontaneous in the Vosges, the Jura and the Alps. Scots pine (*Pinus sylvestris*) which requires light, has been planted everywhere the soil is too dry to establish broadleaf trees. Maritime pine (*Pinus pinaster*) covers 900,000 hectares in the forest of the Landes. The mesogean pine of the Maures and the Esterel in South East France is either a race of maritime pine or a distinct species. Laricio pine (*P. laricio*) forms vast stands in Corsica; Salzmann's pine (*P. salzmanni*) survives as a relict species in the Cévennes and the Pyrénées. Mountain pine (*P. uncinata*) occupies the subalpine zone in the Alps and the Pyrénées. Aleppo pine (*P. halepensis*) and stone pine (*P. pinea*) are strictly Mediterranean. Arolla pine (*P. cembra*) grows between 1,200 and 1,500 metres in the Alps. European larch (*Larix decidua*), which is highly demanding of light, forms pure stands in the montane and alpine zones of the Alps.

3.2. Broadleaf trees

Broadleaf trees are represented by two main genera. Common beech (*Fagus sylvatica*) is the dominant tree and the one with the widest range of distribution in western and central Europe. It establishes itself everywhere summer rainfall is high enough. It reaches the southern slopes of the Pyrénées, southern England and Scandinavia, central Greece, and in the East is limited more or less by the isotherm corresponding to a mean January temperature of – 2 °C (Teissier du Cros, 1986). Oaks include deciduous and evergreen species. English or pedunculate oak (*Quercus robur*) and sessile oak (*Q. petraea*) predominate in the whole of temperate Europe. Downy oak (*Q. pubescens*) is more southern. Pyrenean oak (*Q. pyrenaica*) occupies south-western France. Kermes or holly oak (*Q. coccifera*) and cork oak

(*Q. suber*) are strictly Mediterranean; evergreen oak (*Q. ilex*) reaches Brittany in the North.

4. NORTH AMERICAN AND EUROPEAN FORESTS

Notwithstanding large-scale destruction, the forests of North America cover a substantially greater area than that of European forests. In some areas the proportion of wooded land is between 36% and 86%, and conifers may account for up to 91% of trees. These forests harbour a greater number of genera and species than European forests because they were not subjected to the same degree of catastrophic extinctions caused by Quaternary glaciations. Survivors there include six genera of Gymnosperms (*Chamaecyparis, Sequoia, Taxodium, Thuja, Torreya, Tsuga*) and fourteen genera of Angiosperms (*Liriodendron, Liquidambar, Magnolia, Diospyros, Persea, Robinia,* etc.), representing a total of 58 species which became extinct in Europe. Certain genera (*Quercus, Pinus*) which occur on both continents are richer in species in North America than in Europe. In North America the average number of species per genus is considerably higher among the Gymnosperms (6.06 versus 3.75 in Europe) and slightly higher among the Angiosperms (3.52 versus 3.28 in Europe). However Europe and North America share the same genera of dominant trees.

Three quarters of the 300 species of insect introduced to North America that live on trees are of European origin. These are mainly scale insects (45 species), sawflies (34 species), weevils (29 species), tortricid moths (25 species) and aphids (23 species). Conversely, only 44 species of American forest insect have colonised Europe; they include 17 Homoptera, 15 Hymenoptera, 6 Lepidoptera, 4 Coleoptera, 1 Thysanoptera and 1 Diptera. This may be partly explained by the greater volume of commercial exchanges from Europe to America, but also by the insects' biological characteristics. A general rule emerges: although North American phytophagous insects are on the whole richer in species than those of Europe (43,000 versus 32,000 species) the number of insect species per tree (or genus) is higher in Europe. 100 species of insect live on birch in Scandinavia, and only between 30 and 40 in Canada; between 8 and 11 species are specialised feeders on pine cones in France versus only between 5 and 8 on *Pinus resinosa* in North America. The number of sawflies living on trees in Great Britain is on average several times greater than that encountered on trees belonging to the same genera in the North-eastern United States.

An overall view of these characteristics allows us to understand why so many species of phytophagous insects have established themselves in North America. Many trees, given their low insect population and the great areas they still cover, act as "vacancies", whereas in Europe, the reduced area of forested land and the great number of native insect species prevent new arrivals from establishing themselves. A remarkable peculiarity of immigrant species established on trees in North America is the frequency of parthenogenic reproduction, which exceeds 50% in weevils, bark beetles and scale insects, and is 100% in sawflies and aphids (Niemelä & Mattson, 1996).

5. THE ROLES OF FOREST

Forest has long been regarded almost solely as a producer of wood, and has been managed in such a way as to produce a maximum output of trees of commercial value. Any biotic or abiotic factor which reduced wood production was regarded as "harmful" and had to be eliminated. Today this narrow view is gradually being discarded. Forest is – or ought to be – regarded as an ecosystem having multiple roles and which should be conserved or restored.

Apart from its role as a producer of firewood (almost totally obsolete, at least in our regions) and of timber, forest plays a protective role in the prevention of soil erosion and flooding, and a regulatory role in the water cycle. Forest reduces warming of the soil by solar rays, contributes to cloud formation, and moderates local climate. It acts in the regulation of levels of carbon dioxide in the atmosphere and limits the consequences of the greenhouse effect. Forest has also become a place for walking and recreation which provides ever greater numbers of city dwellers from enormous urban sprawls an opportunity to learn about nature. Thus the role of forests surrounding cities has increased, but excessive numbers of visitors create delicate problems in management.

Forest is a reservoir of biodiversity. Many more plant and animal species live in forests than in open environments. This is particularly evident in the tropics, but is true also of temperate regions. Conservation of this biodiversity has long been neglected; its importance is now recognized (Office National des Forêts, 1993). Insects, which are the richest in species of any group of animals, play very diverse and very important roles in forest, as plant eaters, agents of decomposition, pollinators, predators, parasites and vectors of pathogenic organisms.

6. NATURAL FORESTS AND MANAGED FORESTS

European forests have, almost without exception, been profoundly modified by methods of forestry, and contain only a few rare vestiges of the original "primary", or "natural" forest. Such forests are known as "Urwald" in German-speaking countries, "Urskog" in Sweden, "Prales" is Slovakia, "Pra-Gozd" in Slovenia and Croatia, and "virgin" or "old-growth" forests in North America.

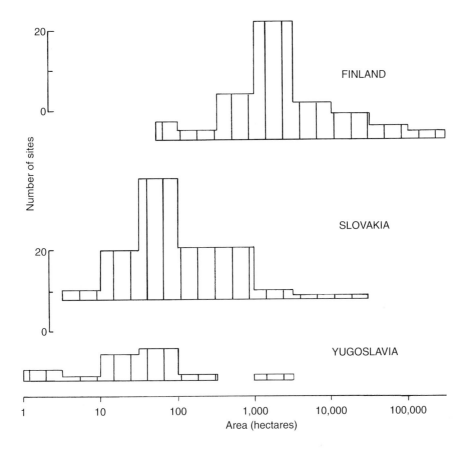

Figure 1.3. *Areas and numbers of primary forests remaining in three European countries in which they are still substantial. In the case of Finland localities covering less than 50 hectares are not shown. The situation in Yugoslavia corresponds to the year 1959, before recent destruction (Peterken, 1996).*

The largest vestiges of primary forests in Europe are mainly concentrated in Scandinavia (more than 100,000 hectares) and in central Europe (Czechoslovakia, Austria and Yugoslavia). Bialowieza Forest in Poland includes 1,250 square kilometres of exploited forest and a natural reserve of 2,800 hectares; in Slovenia 74 protected sites cover more than 26,000 hectares (figure 1.3). A fine, little modified, forest of evergreen oak covering 1,171 hectares survives near Dubrovnik. Such forests have disappeared from Britain and have become very rare in France, Spain and Italy. In France only reserves covering very small areas still survive in a few forests such as Fontainebleau near Paris, the beech forest of the Sainte Baume in Provence, the forest of la Massane in the Pyrénées-Orientales, the 300 hectares of the Ventron reserve in the Vosges and some alluvial forests in the Rhine valley in Alsace (Peterken, 1996). The forest of la Grésigne in the department of the Tarn was a very ancient forest of which little remains following recent forestry treatment (Torossian, 1977).

Managed forests are mostly spatially uniform woods consisting of trees of the same age which are harvested by clear felling before they reach their maximum age. By contrast, one of the characteristics of primary forest is a mosaic structure (Meyer & Neumann, 1981; Remmert, 1991). The effects of planting in the structure of forest is demonstrated by comparison of two parts of Bialowieza Forest: a managed and exploited area covering 530 square kilometres and an unexploited area of 47.5 square kilometres which is a National Park. Areas covered with different trees as percentages of the total area are as follows:

Tree	National Park	Exploited forest
Oak	20%	11%
Hornbeam	19%	2%
Lime and Maple	9%	0%
Pine	11%	26%
Spruce	16%	28%
Birch and Aspen	7%	13%
Alder	12%	17%
Ash	6%	3%

The most significant variations in numbers are those of hornbeam, lime, maple and pine (Jedrzejewska *et al.*, 1994). In Croatia the 38 hectare primary forest of Corkova-Uvala (which was destroyed very recently) consisted of a mosaic of plots measuring only between 0.1 and 0.5 hectares, some of which consisted of young regenerating trees, others of very old, dead or dying trees. This structure is due to the presence of natural disruptions (tornados, forest fires, insects) which create clearings at irregular intervals of time and where forest regenerates gradually according to a universal process encountered in all the world's forests, called sylvigenesis (Oldeman, 1990). Sylvigenesis consists of four main phases (or units). (a) The *youth stage*, or *regeneration unit* is characterised by seedlings and young trees. (b) The *maturity stage*, or *aggradation unit*, is a mixture of various species of growing trees which have not reached their maximum height, together with older specimens which are still increasing in diameter. (c) In the *senescent stage*, or *maturity unit*, all trees have reached their full height. This stage consists of a mixture of mature trees and of others which are aging, old, sick, hollow, or broken. (d) In the *death stage*, or *degradation unit*, most trees die, which again opens up the canopy (figure 1.4).

In these primary forests tree species diversity is greater than in managed forests, and the trees reach great heights and venerable ages: beech and fir live 350 years; in Finland Scots pine reaches 450 years (figure 1.5). The complexity of the plant mass thus formed allows a great diversification of ecological niches and the presence of a large number of birds, saproxylic insects (i.e. those associated directly or indirectly with dead wood), and other organisms such as mammals and lignicolous fungi. One hundred and seventy species of nesting bird live in Bialowieza (only 75 occur in Britain), including 8 species of woodpecker. This richness in cavernicolous birds is due to the abundance of hollow trees, which may number up to 70 per hectare.

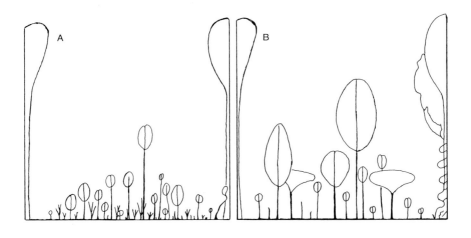

Figure 1.4. *Schematic illustration of the four stages of sylvigenesis.*
A: Youth stage in a clearing
B: Maturity stage and gradual canopy formation
C: Senescent stage
D: Death stage

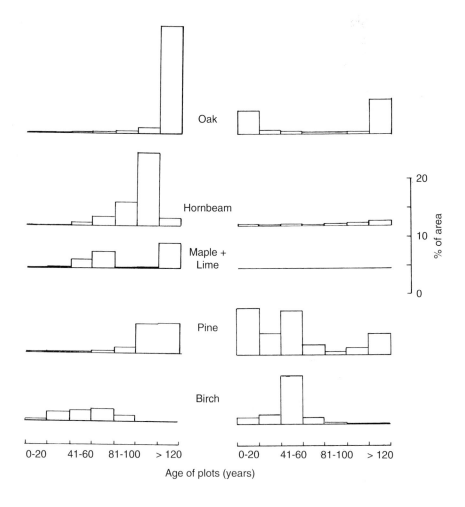

Figure 1.5.   *Comparison of age structure of 5 species of tree in an exploited part of Bialowieza Forest (right) and in unexploited primary forest (left). Very old trees are numerous in the primary forest (Jedrzejewska et al., 1994).*

Spatial diversity, the presence of trees of all ages and of ones of great size, and an abundance of dead wood are all characteristics seen in the 'old-growth forests' of North America (figure 1.6). On that continent this type of forest is more widespread than in Europe and protected stands, amounting to more than 10,000 hectares, are numerous.

Primary forests are of considerable scientific interest, as they provide the possibility of studying the structure, functioning and fauna of a type of forest which has maintained itself for thousands of years *without any human intervention* (Carbienier, 1996).

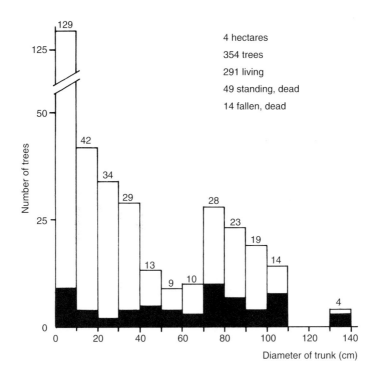

Figure 1.6. *Number of trees according to their diameter measured 1.3 m above ground in a 4 hectare plot of an unexploited population of Pinus jeffreyi in the Californian Sierra Nevada. Standing (in white) or fallen (in black) dead trees are numerous. Some trees reach great heights and ages. Two periods of regeneration appear; the smallest number of trees is of ones between 50-60 cm in diameter (Original).*

RÉFÉRENCES

BRIMBLECOMBE P., PFISTER C., (1990). *The silent countdown. Essays in European environmental history.* Springer, Berlin.

CARBIENIER D., (1996). Pour une gestion écologique des forêts européennes. *Courrier de l'environnement de l'INRA*, **29**: 19-38.

JEDRZEJEWSKA B., OKARMA H., JEDRZEJEWSKI W., MILKOWSKI L., (1994). Effects of exploitation and protection on forest structure, ungulate density and wolf predation in Bialowieza primeval forest, Poland. *J. Appl. Ecol.*, **31**: 664-676.

KOOP H., (1989). *Forest dynamics.* Springer, Berlin.

MEYER H., NEUMANN N., (1981). Struktureller und entwicklungsdynamischer Vergleich der Fichten-Tannen-Buchen Urwälder Rothwald / Niederösterreich und Corkova-Uvala / Kroatie. *Forstw. Central.*, **100**: 111-132.

NIEMELÄ P., MATTSON W. J., (1996). Invasion of North American forests by European phytophagous insects. *Bioscience*, **46**: 741-753.

OLDEMAN R. A. A., (1990). *Forests: Elements of sylvology*. Springer, Berlin.

OFFICE NATIONAL DES FORÊTS, (1993). *Prise en compte de la diversité biologique dans l'aménagement et la gestion forestière. I. Instructions*, 18 pages ; II. *Guide*, 37 pages.

PETERKEN G. F., (1996). *Natural woodlands*. Cambridge University Press.

PONEL P., (1993). Les Coléoptères du quaternaire: leur rôle dans la reconstitution des paléoclimats et des paléoécosystèmes. *Bull. Écol.*, **24**: 5-16.

QUÉZEL P., Biogéographie et écologie des Conifères sur le pourtour méditerranéen. *In*: P. PESSON, (1980). *Actualités d'écologie forestière. Sol, flore, faune*. Gauthier-Villars, Paris, 205-255.

REMMERT H., (1991). *The mosaic cycle concept of ecosystems*. Springer, Berlin.

RETALLACK G. J., (1997). Early forest soils and their role in Devonian global change. *Nature*, **276**: 583-585.

TEISSIER DU CROS E., (1986). *Le hêtre*. Éditions de l'INRA, Paris.

TOROSSIAN C., (1977). Étude préliminaire des conséquences entomologiques des coupes pratiquées en forêt de Grésigne. *Bull. Soc. H. N. Toulouse*, **113**: 366-373.

2

The forest environment: effects of abiotic factors

One of the aims of ecology is the study of the factors that come into play in determining abundance and distribution of species. These factors may be abiotic, i.e. independent of the intervention by living organisms, or biotic, i.e. caused by the intervention of living organisms. Abiotic ecological factors, which in a forest environment affect the flora and fauna, and consequently insects, may be grouped under three headings. The first involves the elements of climate, to which we will add a catastrophic factor: fire. The second involves various physical alterations brought about by reduction in area and the fragmentation of forest stands. The third involves changes brought about by various forms of pollution and by global change, and their effects on trees and insects (cf. chapter 6).

1. FOREST CLIMATE AND ITS INFLUENCE ON INSECTS

In forest, as in any other terrestrial environment, a distinction should be made between a general, or forest macroclimate, and a whole range of microclimates which correspond to often major modifications of the macro-climate brought about by the structure of the habitat.

The influence exerted on regional climate by a sufficiently large expanse of forest has been the subject of much research (Roussel, 1953; Kittredge, 1962, Pardé, 1974, etc.). The most important climatic factors are light, temperature, rainfall and relative humidity. Wind and several other climatic factors also influence forest fauna. Atmospheric pressure may also be considered to be a climatic factor, for it affects, for example, the response of bark beetles and moths to pheromones (Kennedy, 1974; Lanier & Burns, 1978). Forest provides insects with an environment which tempers climatic variations: it reduces wind speeds and variations in temperature and light, and increases relative humidity. In nurseries and during clearing in forests these climatic conditions change, and insects as well as trees are exposed to a more open environment, which affects their chances of survival favourably or unfavourably according to the species.

1.1. Light

Sunlight which penetrates the forest is modified by selective absorption by leaves. Under the cover of trees light is richer in infra-red and poorer in ultraviolet rays. Light intensity under the canopy varies widely according to the nature of the trees, and, in the case of deciduous broadleaf trees, with seasons, relative light being greater in winter when the leaves have fallen. Light intensity, expressed as percentages of light intensity in open situations is as follows:

Under 115 year-old oaks	winter	:	43 to 69%
	summer	:	3 to 35%
Under 60 year-old beeches	winter	:	26 to 36%
	summer	:	2 to 30%
Under Scots pine	all year	:	22 to 40%
Under a dense stand of spruce	all year	:	less than 1%

These figures show that beech casts more shade than oak. These alterations have considerable consequences on vegetation and on insects. The scolytid beetles *Dryocoetes hectographus* and *Pityogenes chalcographus*, which are only able to swarm when light intensity is sufficient, will only settle on branches exposed to direct sunlight. In oak woods egg laying density of the moths *Lymantria dispar* and *Tortrix viridana* depends on the amount of light which penetrates the canopy. Levels of oviposition are greater where light is more intense, in other words at forest edges and at the margins of clearings.

More than 60% of insects regarded as a threat in Sweden prefer to settle on trees exposed to light rather than on trees situated in dense forest, and 25% prefer dense forest to more open environments (Gärdenfors & Baranowski, 1992). Differences may also depend on the species of tree: 70% of species associated with beech prefer dense and shady forest, and 70% of those living on oak prefer open forest. The two cerambycids *Cerambyx cerdo* and *Xylotrechus antilope* live on oak and prefer trees exposed to direct sunlight; the elaterids *Denticollis rubens* and *Ischnodes sanguinicollis* and the melandryid *Melandrya barbata* live on beech and prefer shaded trees.

1.2. Temperature

Temperature and other climatic factors have multiple effects on the physiology and behaviour of insects and other animals. We will give only a few examples of the more typical cases.

In temperate forests maximum summer temperatures are lowered and minimum winter temperatures are raised. This is brought about by interception of solar rays by leaves, by reduction of rays from the ground, and by windbreak. High levels of transpiration by trees lowers temperature in summer. This lowering of temperature is high in the case of beech, average in spruce, and low in Scots pine. In one vegetative season the amount of water transpired by trees is equivalent to a layer of water 27.4 cm deep in the case of beech, 21 cm in the case of spruce, and 7.4 cm in Scots pine.

In the forest of Fontainebleau average annual temperature measured over a period of 50 years is 1.5 °C lower than that of the Paris basin; this increases the number of days of frost, and produces late spring and early autumn frosts. These alterations of thermal climate are seen in temperate forests, but not in Mediterranean forests, which harbour trees adapted to drought that do not transpire in summer, but only in autumn when rain triggers a renewal of vegetative activity. In a dense stand of evergreen oak typical of the Mediterranean *maquis*, temperature is higher than that of unwooded land in spring and summer; in autumn a sudden drop in temperature occurs, lowering it to about one degree below that of open land (Pavari, 1962).

A few geometrid moths such as *Biston*, *Operophtera* and *Erannis* are active at temperatures as low as 5 or 6 degrees, but most Lepidoptera, both nocturnal and diurnal, require temperatures above 15 degrees for flight. Hot, still and sunny days and close, stormy nights provide optimum conditions for movements of Lepidoptera and other insects. The weevil *Brachyderes incanus* causes great damage in plantations of Scots pine. The adults devour the needles and the larvae live underground where they feed on the roots. The latter only develop when the temperature is high, in other words when solar rays reach the ground directly; this occurs in populations of young trees. As soon as the crowns meet, the ground cools and development of the weevils is impeded, which reduces the risk of outbreaks of this species.

The range of the moth *Rhyacionia buoliana* is limited northward by minimum temperatures below − 28 °C, and that of *Lymantria dispar* by the − 32 °C isotherm. The ranges in latitude of three species of diprionid sawfly are a function of their tolerance of low temperatures during embryonic development and of the duration of the latter. Ranges in latitude and requirements of the three species are as follows:

Species	Embryonic tolerance	Range in latitude	Time of year of embryonic development	Duration of embryonic development at 18 °C
Neodiprion sertifer	12 to 28 °C	40 to 75° N	End of March	11 days
Gilpinia pallida	14 to 32 °C	?	April	15 days
Diprion pini	14 to 34 °C	35 to 65° N	End of April - May	18 days

Temperature is a factor limiting the activity of the bark beetle *Ips acumi-natus*. The number of insects caught in revolving nets set above trunks depends both on the number of hours during which the temperature exceeds the 18 °C threshold, which allows insects to take to the wing, and on the maximum air temperature during the day (figure 2.1).

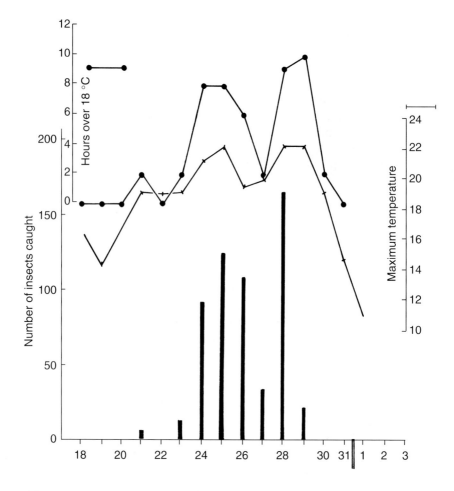

Figure 2.1.   *Effects of temperature on the activity of the scolytid* Ips acuminatus.
*Vertical columns represent the number of insects caught each day
by revolving nets set above trunks of* Pinus sylvestris *at a locality in
Norway. Temperature acts as a limiting factor. The quantity of
insects caught depends on two factors: the number of hours during
which temperature is above a threshold which allows flight (18 °C),
and maximum air temperature during the day (Bakke, 1968).*

Climatic characteristics at ground level, particularly temperature, explain why rhythms of activity of certain insects differ in forest and in herbaceous environments. In the herbaceous environment carabid beetles are essentially diurnal, and in forest many are nocturnal. Species of *Notiophilus* are mostly diurnal and numerous in the fauna of meadows, whereas *Nebria brevicollis* is mainly nocturnal and lives in forest. *Pterostichus madidus* is nocturnal in forest and diurnal in meadows (Wallwork, 1976).

The influence of temperature may manifest itself on a large scale and in spectacular ways. Parts of a birch (*Betula pubescens*) forest in North Sweden were entirely defoliated by caterpillars of *Oporinia autumnata*. Areas that were spared lie at the bottom of a valley where a dense stream of cold air flows permanently. By impeding the development of *O. autumnata*, the low temperatures prevented outbreaks of the species and enabled the birches to retain their leaves (Tenow, 1975). Those parts of Norway where outbreaks of *Operophtera brumata* and *Oporinia autumnata* occur have a typical maritime climate in which minimum winter temperatures do not fall below – 33 °C. This accords with the results of an experimental study which show that mortality at the egg stage increases considerably below – 33 °C in these two species (Tenow, 1972).

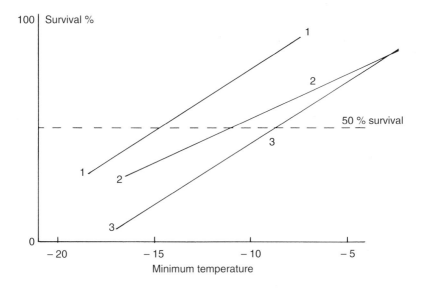

Figure 2.2. *Survival percentages in populations of the aphid* Elatobium abietinum *as a function of the minimum temperature to which the insects are exposed and of the temperature to which they had previously been acclimatized.*

Line 1: acclimatization temperature of 0 °C. Lethal temperature 50 (or LT50) at which 50% of insects are killed is −14.5 °C.
Line 2: acclimatization temperature 10 °C and LT50 is −11 °C.
Line 3: acclimatization temperature −15 °C and LT50 is − 8 °C (Carter, 1972, modified).

Some populations of the green spruce aphid *Elatobium abietinum* are controlled by low winter temperatures. Temperature that allows a survival rate of 50% lies between – 6.5 and – 14.5 °C according to the regime previously experienced by the insects (figure 2.2). The winter of 1970-71 was a mild one in Britain, and in the spring of 1971 parts of the country where outbreaks of *Elatobium* accompanied by severe defoliation occurred coincided with areas in which the minimum temperature had not fallen below – 8 °C between December 1970 and March 1971 (Carter, 1972).

Forest insects, many of which have vast distributions, must adapt to very different thermal climates. This entails changes in life cycles and population dynamics. The pine weevil *Hylobius abietis* takes one to three years to develop in Norway, and two to four years in Sweden, and populations are always greater in sites exposed to sunlight. The number of generations of bark beetles per year varies according to climate; most species in the French fauna have one or two generations. *Scolytus rugulosus* has two generations in the Paris region and three in the Mediterranean. *Hylurgops glabratus*, which lives on Arolla pine at high altitudes in the Alps, takes two years to grow; *Hypoborus ficus* has three generations per year in Montpellier and four in Algiers (Balachowsky, 1949). The effects of temperature on the annual number of generations of scolytids may be even greater. The American species *Ips calligraphus* which lives in *Pinus elliottii* has four generations per year in the mountains of California and at least six in the South-eastern United States; it reaches 9 in Florida and 12 in Mexico. Moreover variations in the thickness of phloem within a population of trees may modify the annual number of generations; growth is more rapid in trees with a thick layer of phloem (Haack, 1985).

Activity in scolytids is a function of temperature. Swarming, which enables them to colonise new trees, occurs at clearly determined temperatures (Chararas, 1962):

Lower lethal zone (death caused by cold)	between – 15 °C	and – 10 °C
Torpor caused by cold (hibernation)	between – 10 °C	and + 5 °C
First activity	between + 5 °C	and + 9 °C
Normal activity without swarming	between + 10 °C	and + 15 °C
Swarming	between + 16 °C	and + 18 °C
Optimal zone of activity	between + 18 °C	and + 29 °C
Hyperactivity zone	between + 30 °C	and + 40 °C
Torpor caused by heat	between + 40 °C	and + 49 °C
Upper lethal zone	between + 50 °C	and + 51 °C

Comparable data were provided by Miller (1931) for the North American scolytid *Dendroctonus brevicomis*. They may be summarized as follows:

Temperatures			Effects of temperature
		– 29 °C	Fatal to all stages
– 26°	to	– 22 °C	Survival of a small percentage of eggs and well adapted larvae
– 22°	to	– 15 °C	Critical temperature for larvae and pupae; mortality rate at –20 °C exceeds 50%
– 15°	to	– 12 °C	Critical temperature for adults; mortality rate of adults high at – 12 °C; mortality rate lower for larvae at – 15 °C
– 12°	to	– 9.5 °C	Larvae freeze but recover if warmed
– 9.5°	to	+ 4.5 °C	Beginning of dormancy in all stages
+ 4.5°	to	+ 7 ° C	Larvae move slowly
+ 7°	to	+ 12.5 °C	Larvae and adults active
+ 12.5°	to	+ 35 °C	Maximum activity
+ 35°	to	+ 38 °C	Larvae less active
+ 38°	to	+ 40 °C	Larvae move slowly
+ 40°	to	+ 43 °C	Larvae paralysed in 1 to 2 hours; recovery possible following return to lower temperature
+ 43°	to	+ 46 °C	Larvae paralysed following short exposure, but recovery possible at lower temperature
+ 46°	to	+ 48 °C	Larvae irrecoverably paralysed
		+ 49 °C	Fatal to all stages

Although it is distributed mainly in the Mediterranean region, the pine processionary caterpillar does not tolerate mean monthly maximum temperatures in excess of 25 °C (Demolin, 1969). The upper lethal temperature is 32 °C, and the lower lethal temperature – 6 °C for isolated caterpillars, and – 10 °C for caterpillars in colonies of 200 individuals gathered in a winter nest which acts as a solar radiator. Solar radiation acts on the nest by raising maximum daily temperature by 1.5 °C per hour of insolation. When the temperature is between 10° and 20 °C the caterpillars leave in procession, and only stop when they find a location where the temperature is in excess of 20 °C. Low temperatures during a season slows larval development, and the caterpillars go into an enforced diapause, which leads to a spread of the emergence of moths over two or three years.

The body of data gathered on the effects of climatic factors helps to explain the distribution of the processionary pine caterpillar in France. It occurs in the whole of the Mediterranean basin, and infestations are mainly severe South of a line stretching from Brest to Orléans. It is excluded from areas where annual insolation is less than 1,800 hours, in other words from North, East and central France, Brittany, and mountain regions where the mean minimum temperature in January is below – 4 °C. Where January temperatures are between 0 °C and – 4 °C, each degree below 0 °C must be compensated by 100 hours of annual insolation for the species to establish itself. This explains the presence of the caterpillar on the Mediterranean slopes of the Alps, the Massif Central and the Pyrénées.

Resistance to low temperature brings a phenomenon of gradual acclimatization into play. In aquatic insects, which live in an environment in which variations in temperature are slight, the lower lethal temperature in the winter is a little below 0 °C. Lethal temperature is lower (around – 10 °C) for the chafer *Popillia japonica* which hibernates underground, and very much lower (– 40 °C) for the beetle *Dendroides canadensis*, which lives and hibernates under bark where it is subjected to very low temperatures (Payne, 1926).

1.3. Rainfall and relative humidity

In forest, some of the precipitated rain is intercepted by foliage, some trickles down trunks, and the rest reaches the ground directly. Whether or not rainfall is higher in forest than elsewhere is highly controversial. Various authors record increases relative to surrounding areas of 17% for the forest of Fontainebleau, 12.5% for the forest of Haguenau in Alsace, and 30% for the wooded Haye plateau near Nancy. Other authors find no significant changes in rainfall in forest (Pardé, 1974).

Relative humidity is generally higher in forest than in open situations, especially in summer when transpiration from trees is at its height. Increases of relative humidity reach 9.35% in beech woods but only 3.87% in stands of Scots pine.

> In primary rainforest in Sulawesi the number of Homoptera and Heteroptera attracted to a light trap is proportional to the amount of rainfall, and the size of insects caught is greater when rainfall is high. This curious observation may be explained if bats are brought into the picture. Bats are predators of nocturnal insects, and use ultrasound to detect their prey. During heavy downpours raindrops of a diameter similar to the size of the larger insects reflect the ultrasounds and impede prey detection. The forays of larger insects only during heavy rainfall thus appears to be a protective reaction against a potential predator (Rees, 1983).

The activity rhythms of many forest insects are controlled by climatic factors such as relative humidity. Various xylophagous beetles such as the longhorn *Cerambyx cerdo* are nocturnal and wander at night on old tree trunks within which they remain concealed during the day. Park *et al.* (1931) have shown that activity of nocturnal insects such as the passalid beetle *Boletotherus cornutus* increases as relative humidity rises and temperature falls.

Rainfall also affects xylophagous insects indirectly through the changes it brings about in trees. The buprestid beetle *Agrilus viridis* only swarms on beech in Bavaria when less than 900 mm of rain falls per year. Dry years and permeable soils favour the establishment of this insect which, generally, only lays eggs in the bark of trees in which the water metabolism is perturbed (Bovey, 1971). The establishment of scolytids on trees suffering from physiological deficiencies resulting from lack of water is well known. In Canada *Choristoneura fumiferana* is more active and feeds more readily in a humid than in a dry atmosphere (figure 2.3).

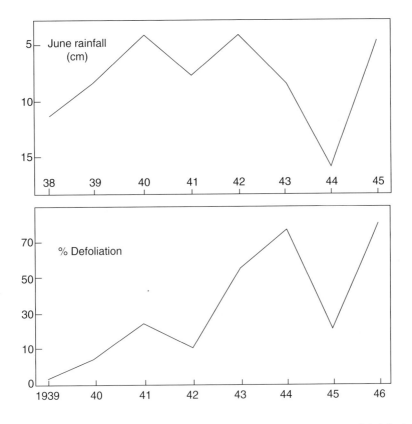

Figure 2.3. *Relationship between June rainfall and percentage of defoliation of spruce by* Choristoneura fumiferana *in Canada (Wellington et al., 1950).*

The simultaneous effects of temperature and relative humidity have been studied in many insects, and particularly in the two defoliators *Lymantria monacha* and *Panolis flammea* (figure 2.4). Areas in which plagues of *L. monacha* are to be feared are limited by the July 16 °C isotherm; outbreaks are greatest in areas with annual rainfall of 400 to 600 mm, followed by areas with rainfalls of 600 to 700 mm. No outbreaks occur where rainfall exceeds 1,000 mm per year (Breny, 1963).

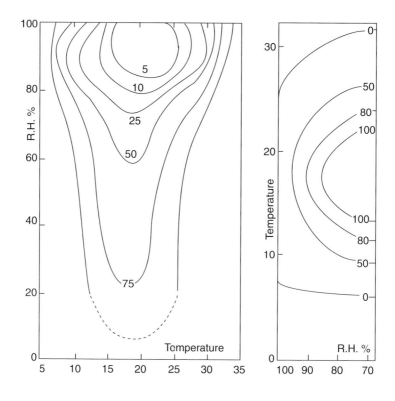

Figure 2.4.   *Effect of temperature and relative humidity. Left, percentage mortality for the first instar larvae of* Lymantria monacha; *right, number of eggs laid by female of* Panolis flammea *(Zwolfer, 1931, 1934).*

1.4. Wind

Forest acts as a windbreak. In a stand of oak wind speeds may fall to 11% of that measured on open land. High winds have an inhibiting effect on insect activity. Mosquitoes and many other insects stop flying when the wind speed exceeds a certain threshold. Wind plays a role in the dispersal of mobile larval stages of the felted beech scale insect. A study of a beech wood according to vertical gradient shows a slight increase in temperature from ground level to the upper part of the canopy (figure 2.5). Wind speed increases considerably above the canopy, and abundance of these scale insect larvae decreases progressively. Most individuals are thus dispersed to small heights and over small distances, but about 1% of insects are carried above the canopy and are dispersed over long distances (Wainhouse, 1980).

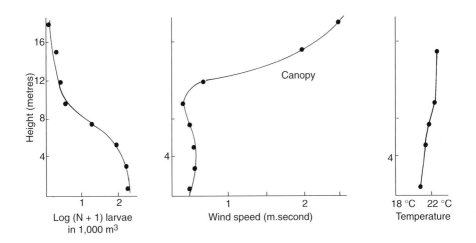

Figure 2.5. *Vertical variations in temperature and wind speed in an English beech wood and abundance of mobile larvae of* Cryptococcus fagisuga *(Wainhouse, 1980).*

Wind plays an important part in the dispersal of certain insects. In the Alps adults of the larch bud moth migrate over distances of more than 100 km (figure 2.6). These migrations, which explain the synchronism of infestations in various localities within the arc formed by the Alps, are made upwind (Baltensweiler & Fischlin, 1979); it is probable that the moths are guided by sex pheromones carried by the wind (Kennedy, 1974).

The distribution of the larch thrips *Taeniothrips laricivorus* in the Swiss Alps is a function of wind. Spruce trees situated at the edges of plantations and in the lee of the larch trees contain more thrips than spruce trees not in their lee. In spring the insects congregate on the fringe of larch plantations to which they are carried by the wind from neighbouring spruce trees where they hibernated (Maksymov, 1965). In the Canadian state of Alberta, masses of *Malacosoma disstria* caterpillars have been carried along a cold front over at least 500 km (Brown, 1965). Large movements of the scolytids *Dendroctonus ponderosae* and *D. rufipennis* carried by wind have been observed in North America. Young larvae of *Lymantria dispar* are also borne by wind; individuals whose morphology is that of the South-east Russian race of the species have been found in Finland, which they reached by means of a warm air front (Mikkola, 1971). Wind may also play an important part by blowing down trees which are the main attractants of xylophagous insects.

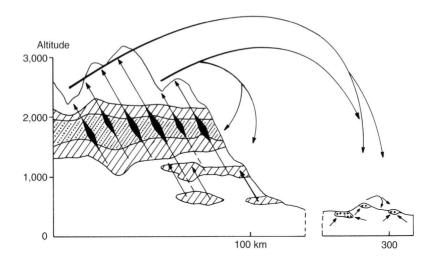

Figure 2.6. *Migrations of* Zeiraphera diniana *in the Alps as a function of altitude, distance and wind. Oblique hatching: range of spontaneous occurrence of larch; oblique hatching with dots: zones in which infestations occur; crosses: plantations of larch at low altitudes (Baltensweiler & Fischlin, 1979).*

1.5   Spatial distribution of insects in forest

Spatial distribution of insects in forest is not uniform; it is determined both by the behaviour and the requirements of insects in relation to the elements of microclimate. Knowledge of the types of distribution is essential for any research of a quantitative nature such as that relating to population dynamics. The following are a few examples.

1.5.1. The case of Rhyacionia buolina caterpillars

Distribution of caterpillars of this moth is of the aggregative type; individuals tend to live more or less in groups. It has been observed (Nef, 1959) that the distribution of 2,015 caterpillars on 2,074 trees shows a negative binomial distribution:

Number of caterpillars per tree	Number of trees	Calculated values of negative binomial distribution
0	1,063	1,062.6
1	511	509.8
2	264	251.3
3	123	125.0
4	56	62.4
5 or more	57	62.9

The tallest trees are most prone to attack, and the upper branches bear the greatest number of caterpillars (figure 2.7). These observations must be taken into account where one wishes to study the effects of treating trees with a fertilizer intended to reduce populations of *Rhyacionia buoliana*. Comparisons of treated and untreated plots must therefore be made using trees of the same height. The experiment shows that, on trees of equal height, caterpillars are more numerous on untreated plots (Nef, 1967), but that no significant differences result from the different fertilizers used.

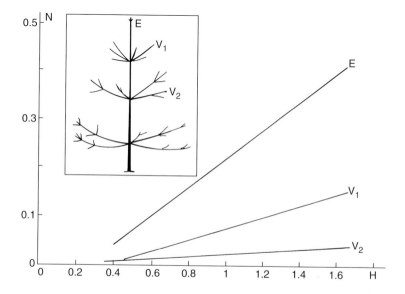

Figure 2.7.   *Distribution of* Rhyacionia buoliana *caterpillars on Scots pine. Average number N of caterpillars per branch extremity in relation to height of tree H. Lines representing N in relation to H were drawn for tree tip E, the tip of a one year old twig V_1, and the tip of a two year old twig V_2 (Nef, 1959).*

1.5.2. The case of Stilpnotia salicis

This moth, which lives on poplar, sometimes produces spectacular defoliation. Egg laying is concentrated on the largest trunks and on those parts most exposed to light. Maximum oviposition is generally on the middle of a trunk with two minimum ovipositions at the top and bottom of the tree. This aggregative type of concentration may be explained by the behaviour of the females, which seek maximum exposure to light to lay their eggs (Nef, 1975).

1.5.3. The case of Lymantria dispar

The spatial distribution of the gypsy moth and its parasites is determined by their behaviour and especially by their requirements of temperature, relative humidity and light (Weseloh, 1972, 1976). The adult moth and its parasite *Apanteles melanoscelis* are more frequently caught in the upper parts of trees. The parasite *Ooencyrtus kuvanae* predates the moth's eggs at various elevations, but is less frequent in clearings than in forest. Another parasite, *Brachymeria intermedia*, emerges mainly from pupae situated near the tops of trees and is absent from those situated less than five metres above ground. Knowledge of these data is important when it comes to sampling techniques and to effective methods of control: insecticides, if envisaged, should not be sprayed in the upper part of trees where the parasites are most active.

1.5.4. Vertical distribution of Diptera

Vertical distribution of insects in forest is often conditioned by the gradients of temperature and relative humidity which occur between ground level and the canopy. In Ugandan forest the mosquito *Aedes africanus* is localised at a height of about 18 metres near the canopy, whereas *Anopheles gambiae* lives mainly near ground level. In Monks Wood in England tipulids are most active between 20.00 and 23.00 hours; the density of all species is greatest near ground level, where mating takes place, and decreases rapidly until the insects disappear completely at about 3 metres above ground level. Tipulids are weak fliers which only move about when air turbulence is slight (Service, 1973).

In tropical forests, where trees are often of a great height, vertical stratification of temperature, relative humidity and light is much greater than in temperate forests, and the vertical distribution of various species is often very obvious. This distribution has been demonstrated in the case of scolytids and platypodids in a forest in the Ivory Coast. For each of the ten species studied the factor that best explains vertical distribution is evaporation, which is a function of relative humidity and temperature (Cachan, 1974). Evaporation being minimal near ground level, species requiring high levels of humidity such as *Platyscapulus auricomus* are concentrated in the lower parts of trees; species such as *Hypothenemus cavipennis* which tolerate lower levels of humidity occur in the upper parts of trees; *Platyscapulus camerunus*, which has average requirements as far as humidity is concerned, is distributed throughout the height of trees.

The behaviour of many parasitoid insects in forest is determined by humidity and temperature. Ichneumonids are more active at medium temperatures and in high humidity, whereas braconids and chalcids are more active at high temperatures and in low humidity. The distribution of the parasitoids of the sawfly *Neodiprion swainei* is governed by humidity (Price, 1971). The tachi-nid *Compsilura concinnata* is a polyphagous parasite which is more frequent in groups of trees than in isolated ones, because this fly prefers conditions provided by grouped trees, in other words lower temperature and higher humidity (Schwenke, 1958).

2. MICROCLIMATES OF FOLIAGE AND TREE TRUNKS

Forest incorporates many different microclimates. In winter, the tempera-ture of conifer needles exposed to sunlight may exceed that of the ambient air by more than 2 °C. In summer, leaf temperature is on average 8 °C higher than that of the air during the day and 3 °C lower at night.

The thermal microclimate of leaves has been studied in Canada (Wellington, 1950). When leaves of *Picea glauca* are exposed to direct sun-light their temperature rises by as much as 5.6 degrees above that of the ambient air; shaded leaves may be 2.2 degrees lower than that of the air. This affects the temperature in the galleries of mining insects, and consequently the insects' body temperature. Relative humidity is higher in mined galleries than in the air (table 2.1). The thermal microclimate of leaves may change when the tree undergoes stress. This was revealed by comparing *Picea engelmanni* trees infested by *Dendroctonus rufipennis* with others which had not been infested; during a day in August the temperature of needles of infested trees was significantly lower (between 0.1 and 2.4 degrees) than that of other trees between 08.50 h and 09.30 h and from 10.40 h to 11.30 h; it was higher (between 0.7 and 4.9 degrees) at other times of the day (Schmid, 1976).

Time	Air temperature	Temperature in galleries	Temperature of larvae	Relative humidity of air (%)	Relative humidity in galleries (%)
11 h 45	30.6	29.0	29.9	49.0	89.0
12 h 44	28.0	28.1	30.3	53.0	63.5
13 h 47	28.8	31.1	29.5	52.0	65.0
14 h 45	27.0	26.9	28.0	73.0	89.0
15 h 27	32.2	30.5	29.9	53.0	73.0

Table 2.1. *Leaf microclimate. Temperature and relative humidity of the air and in galleries, and body temperature of sixth stage larvae of* Choristoneura fumiferana *living on* Picea glauca *(Wellington, 1950).*

The microclimate of tree trunks is a very important type of forest micro-climate (Graham, 1925; Savely, 1939; Dajoz, 1967, etc.). Temperature under bark is a function of the exposure of the tree to sunlight, and the structure of the bark. When exposed to sunlight, daily variations in temperature are much greater than those recorded on sheltered trees, the difference sometimes reaching 20 °C. Under bark sheltered from sunlight (for instance on the underside of a felled trunk) swings in temperature are slighter than those of the ambient air; the same goes for temperature inside the trunk (figure 2.8).

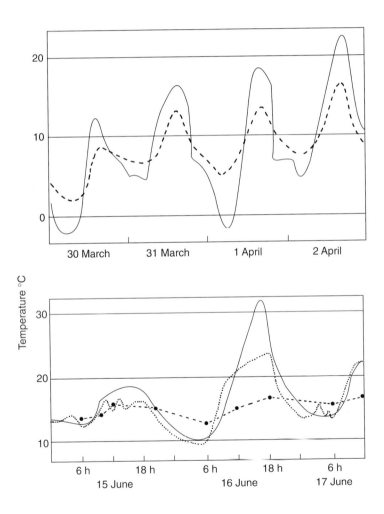

Figure 2.8.   *Thermal microclimate of tree trunks. Above, air temperature (solid line) and at a depth of 5 cm (dashes) in a pine tree trunk (Savely, 1939). Below, temperature under the bark of a felled beech tree in a clearing; upper side of trunk (solid line), underside of trunk (dashes), and air temperature (dotted line) (Dajoz, 1967).*

Temperature plays an important part in the localisation of insects that live in tree trunks. Heat-loving larvae of buprestid beetles of the genus *Chrysobothris* mostly settle in the upper part of trunks exposed to sunlight, whereas cerambycids of the genus *Rhagium* and species of the genus *Pyrochroa* (Pyrochroidae) seek shady areas on the sides or underside of fallen trunks.

When the two scolytids *Tomicus minor* and *Ips acuminatus* settle to bore their galleries, they show the same preference in regard to thickness of pine bark, but seek differently exposed faces of the trunk according to their different temperature requirements. *Tomicus minor* chooses temperatures between 14 and 26 °C and a mean temperature of 20.5 °C, whereas *I. acuminatus* chooses temperatures between 20 and 38 °C and a mean temperature of 29.9 °C. In the same pine trunk galleries of *T. minor* are generally found on the underside, and those of *I. acuminatus* on the lateral and upper sides. This choice of location for the galleries favours *I. acuminatus*, whose larval development requires higher temperatures (Bakke, 1968).

Relative humidity in galleries bored by xylophagous insects is often high, for there is a constant exchange of water vapour between the cellulose of the wood and atmospheric water. The tree acts as a sponge which reabsorbs water lost in dry weather during humid weather. For a water content in wood equal to 15%, the hygrometric state of the air corresponding to hydrostatic equilibrium varies no more than 75 to 80% between temperatures of 10 and 40 °C. Xylophagous insects therefore live constantly in an atmosphere close to saturation point, at least in the case of recently dead trees. Sometimes the water content of the wood may drop drastically, and the relative humidity of the air in the galleries falls to less than 50%. In this case the fauna is impoverished, and contains species associated with dry wood such as anobiids. Humidity affects the duration of larval growth; certain cerambycids live for 6 to 7 or more years in dry wood but complete their life cycles in only 2 to 3 years in moist wood.

Optimum growth rate in the cerambycid *Hylotrupes bajulus* occurs when the humidity of the wood is around 35%, and that of *Ergates faber* in a humidity of 60%. As for many other beetles of the family Anobiidae, the optimum growth rate in *Anobium punctatum* occurs in wood when humidity reaches 30% (Schwerdtfeger, 1963). Larval development of many xylophagous insects such as siricid wasps and cerambycid and scolytid beetles ceases when wood humidity drops below 20%. The larval life of the cerambycid *Monochamus scutellatus* lasts one year in trunks exposed to sunlight, and three years in shaded trunks (Graham, 1925).

A correlation exists between the characteristics of micro-habitats chosen by insects living in dead trees and their resistance to high temperatures and dehydration. The upper lethal temperature for these insects increases with relative humidity; it is higher (50 to 54 °C) in a species such as *Chrysobothris affinis*, which settles under sunlit bark, than in

Pyrochroa coccinea (44 to 48 °C), which seeks shady areas. Weight loss due to loss of water at high temperatures is greater in *Pyrochroa* than in *Chrysobothris* (figure 2.9).

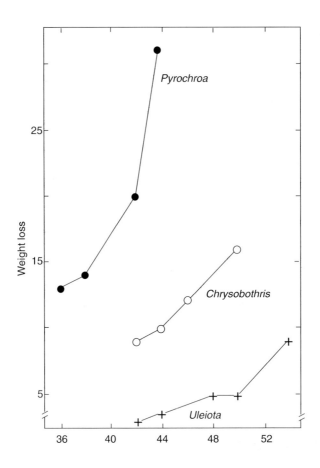

Figure 2.9.   *Effect of temperature on weight loss (expressed in % of initial weight) in the larvae of three species of beetle subjected to a relative humidity of 10% for three hours. Upper lethal temperature is 44 °C in* Pyrochroa coccinea, *50 °C in* Chrysobothris affinis, *and 54 °C in* Uleiota planata *(Dajoz, 1967).*

The fly *Andrenosoma bayardi*, belonging to the family Asilidae, lives in dead wood during its larval stage. The adults settle on dead tree trunks; at night they remain concealed on the lower surfaces of the trunks in dry areas, and on the western sides. They begin to stir when they are exposed to the first rays of the sun, and then move to sunny parts of the trunk; during the hottest hours they move gradually to shaded portions of the trunk. The location of insects on a trunk is explained by search for a temperature lying between 33 and 35 °C. The upper part of the trunk is deserted during the hottest part of the day when the temperature may reach 50 °C (Musso, 1972).

3. SEASONAL CLIMATIC VARIATIONS AND INSECT PHENOLOGY

Phenology is the study of the relationships between different stages of insect life cycles and the stages of tree development, seasons and climatic conditions.

3.1. Phenology of insects in a beech wood

Seasonal movements and the phenology of various insects of beech woods have been described in Denmark (Nielsen, 1974). Eight phenological stages may be recognized.

1. Between mid November and early March trees bear dormant buds, causing reduced activity in foliage arthropods which seek refuge on the branches, in bark crevices, in mosses and epyphitic lichens, or in the soil.

2. Between March and mid April the buds have still not opened, but in favourable climatic conditions a few species become active: the weevil *Rhynchaenus fagi*, the earwig *Chelidurella acanthopygia*, larvae of cantharid beetles, spiders and isopods. Cantharid larvae in particular are remarkable for their winter activity.

3. Between mid April and early May buds open and the leaves unfold. Insect activity increases and arboreal species such as the weevils *Rhynchaenus fagi*, *Phyllobius argentatus*, *P. mollis* and *P. undatus*, which have sought refuge in the soil during the winter are seen to move up into the trees. This is also the emergence period of the cecidomyiid gall midge *Mikiola fagi* which can form veritable swarms during sunny days in April.

4. May is characterised by full leaf development and maximum depredations by leaf-eating insects. It is the active period of the moth *Chimabacche fagella*, the flightless females of which climb up the trunks. Canopy fauna is dominated by the adults of *Rhynchaenus fagi* and *Phyllobius argentatus* and by the larvae of *Operophtera fagata*. These four species, together with the beetles *Phyllobius mollis*, *Strophosomus melanogrammus*, *S. capitatus*, *Coccinella septempunctata*, *C. 14-punctata* and the pentatomid bug *Troilus luridus*, form 80 to 90% of the total number of insects. At the end of this period the elaterid beetles *Athous subfuscus* and *A. vittatus* appear and infestations by the aphid *Phyllaphis fagi* begin.

5. In June foliage insects reach maximum density, but the activity of many defoliating species slows when sclerification of the leaves is completed. *Phyllobius argentatus* and *Operophtera fagata* end their activity; the former

descends to the soil to lay its eggs and the latter pupates. A new generation of *Rhynchaenus fagi* appears, and elaterids of the genus *Athous* and cantharids such as *Cantharis* spp. and *Malthodes* spp. abound in the foliage.

6. In July the main phytophagous insects are the larvae of tortricids and of the geometrid moth *Chimabacche fagella*, and of the sap feeders *Phyllaphis fagi* and *Typhlocyba cruenta*. The earwig *Chelidurella acanthopygia* becomes abundant.

7. In August and during the first half of September this same earwig is at the height of its activity; it feeds mainly on epiphytic plants.

8. The period between 15 September and the end of October corresponds to leaf fall. A few species associated with the upper part of the trees are still abundant, while those that hibernate underground return to the soil.

3.2. Phenology of beetles in an oak wood

In an oak and hazel wood in Belgium beetles are active throughout most of the year. In all there are two peaks of abundance, the first between 5 and 25 July, the second from 16 August to 5 September. At family level, carabids, chrysomelids, curculionids and leiodids are most abundant between 12 April and 23 May, scarabaeids and staphylinids between 27 June and 25 July, hydrophylids, cantharids and elaterids from 24 May to 13 June. At species level phenology is very variable. The dung beetle *Geotrupes stercorosus* is present throughout the year, with no clear period of maximum abundance; the leiodid *Catops tristis* is commonest between 1 and 21 March; the cantharid *Podabrus alpinus* is present only from 24 May to 4 July (Gaspar & Verstraeten, 1972).

4. FIRE: A LITTLE KNOWN ECOLOGICAL FACTOR

Fire is a quasi universal ecological factor that until recently remained little known, but has played an important role in the diversification and maintenance of plant communities. It has exerted its influence since the earliest appearance of terrestrial vegetation, long before the intervention of man. Lightning is the main cause of spontaneous fires from the tundra to the tropics. Volcanic eruptions may also cause large forest fires.

Forest fires were once common, especially in the boreal forests of Europe and North America. In the Sierra Nevada of California the "Valentine Camp" reserve is composed mainly of conifers including *Pinus jeffreyi*. The scars left by fires on trees that survived made it possible to establish the occurrence of forest fires (either spontaneous of started by Indians): in 1624, 1658, 1745, 1758, 1796, 1827 and 1844. The Paiute Indians would light fires in order to gather the caterpillars of the saturniid moth *Coloradia pandora*, a pest species on various species of pine, with major outbreaks at 20 to 30 year intervals, and one of the many insects used as food by man. In Sweden, before the nineteenth century when control of forest fires began, the conifer forest burned at average intervals of 80 years; thereafter the average interval between fires rose to 155 years. Spontaneous or man-made fire has shaped many Mediterranean landscapes. It is the action of fire combined with drought and poor soils that produced, at the expense of the original forest, extensive formations of the vegetation type known as *maquis, garrigue,* or *matorral,* as well as soil erosion and desertification. In the forest of Fontainebleau fires facilitated the spread of heather and birch at the expense of oak and beech.

Much of the diversity of the flora and the mosaic structure of boreal forest is attributable to recurrent fires. The natural cycle of forest fires in North America is on average 50 to 200 years, with a maximum of 500 years in very wet regions. By consuming a part of the organic matter of the soil and burning trees, fire alters soil temperature and humidity, as well as the cycles of minerals and productivity in the environment. One of the causes of forest fires is massive depredation by pest species, which are numerous in boreal forests, such as *Choristoneura fumiferana* in North America or *Dendrolimus sibiricus* and *Monochamus urussovi* in Siberia; the latter destroyed 3 million hectares of coniferous forest in western Siberia between 1954 and 1957. In a forest that has been ravaged and stripped of its canopy, sunlight and wind desiccate the wood, thereby increasing the chances of fire.

Forest fire is a catastrophic factor when it occurs at very close intervals and over large areas, as in the Mediterranean region. The same is true when fires in the Amazon destroy areas equal to half the forests of the United States. Fire is a disruptive element which prevents spontaneous growth of plants, but in moderation it is also an agent which, especially in forests, ensures the structural heterogeneity essential to the maintenance of biodiversity. In pine forests in the south-eastern United States where fire is recurrent, trees never become old except where they are protected from fire, in which case they reach old age and become enfeebled and vulnerable to insect depredations (Schowalter, 1985).

Forestry methods applied since the nineteenth century have profoundly altered natural forests; dead trees are removed; forest fires are controlled or eradicated, which explains the threats to or extinction of species associated with those environments. The suppression of forest fires in Sweden and Finland is the likely cause of decreasing populations of the bug *Aradus crenaticollis,* a rare species associated with conifers in burned forest, as well as of those of *Aradus signaticornis* and *Aradus anisotomus* (Heliövaara & Väisänen, 1983); these three species require newly burned trunks for egg laying. Absence of old and burned trees affects not only insects but also other organisms such as fungi, lichens and mosses as well as the animals associated with them. Burned wood is an environment which seems indispensable to the development of certain species. Burned areas favour the increase of other species which remain rare or virtually non-existent in areas that have not been burned (Lundberg, 1984).

Evans (1972) identifies two groups of "pyrophiles": some, such as cerambycids, are attracted to smoke; others, such as buprestids, are attracted to heat. The list of pyrophiles includes platypezid flies of the genus *Microsania*, and especially beetles: carabids of the genus *Sericoda*, cerambycids such as *Spondylis upiformis*, *Tetropium* spp., *Monochamus* spp., and buprestids of the genus *Melanophila*.

Some pyrophilous beetles are winged and have great powers of dispersal, which explains their vast geographic distribution. The genus *Sericoda* includes the only truly pyrophilous members of the family Carabidae. Four out of the seven known species have been observed arriving on the wing at sites of forest fires or already installed on sites containing scorched wood. Following a fire on May 22 in a Manitoba forest, *Sericoda quadripunctata* was the most abundant carabid in July, but only a single specimen was caught the following year. It may be that the insect reproduces in burned areas and then emigrates in the second year. In the forest of Fontainebleau in France this species was abundant after the fire of 1946. Its eggs are not laid on the ground as with most carabids, but on scorched twigs and leaves. Breeding populations of *Sericoda quadripunctata* have been observed in Sweden in most of fifteen localities consisting of unexploited and burned forest. *Sericoda* do not settle in exploited plots that have been burned after the felling of trees; these are colonised by carabids associated with open situations, which breed in spring as opposed to the typical autumn breeders of the forest environment. It is probable that these open environment species interfere with the settling of *Sericoda* (Wikars, 1995).

In the United States buprestids of the genus *Melanophila* are known as "smoke beetles" because of their attraction to fire and smoke. Their larvae can only develop in recently scorched trees; the females lay their eggs on wounds in the bark caused by fire. Hundreds of these insects have been observed wandering on scorched trunks and hot ashes during a fire. *Melanophila* possess an infra-red-sensitive sensory organ on the metathorax (this organ appears to function differently from the sensory organ of rattlesnakes, which also detect infra-red rays), which enables them to detect areas of fire at a distance of fifty kilometres (Evans, 1966, 1972; Schmitz *et al.*, 1997). In experimental conditions the temperatures sought by *Melanophila acuminata* vary between 33.5 °C and 49.5 °C with an optimum of 40.5 °C. In France *Melanophila cuspidata* is a "carbonicolous" species whose larvae live on burned *Pinus halepensis* and *Juniperus oxycedrus*. It prefers recently scorched standing trunks; more rarely it feeds on broadleaf trees damaged by fire. Like the carabid *Sericoda quadripunctata* it ranges over nearly the whole northern hemisphere.

The melandryid beetle *Phryganophilus ruficollis* is a very rare species which lives mainly in birch and spruce. It prefers scorched trees in which the wood has turned to a suitable consistency and is frequently attacked by fungi such as *Polyporus betulinus* or *Trametes versicolor* (Lundberg, 1993). Diptera of the genus *Microsania* are attracted in great numbers to smoke, which has earned them the name "smoke flies" (Komarek, 1969).

The beetles of burned forest have been studied in Sweden (Lundberg, 1993) and Finland (Muona & Rutanen, 1994); both authors provide long lists of species. Those attracted to scorched wood include xylophagous species such as buprestids of the genus *Melanophila* and the cerambycid *Acmaeops marginata*; corticolous species, which settle under the bark of damaged trees, such as *Cryptophagus corticinus*, *Pediacus fuscus* and *Laemophloeus muticus*; species associated with fungi which develop on scorched trees, and finally predators such as *Sericoda quadripunctata*. Heliövaara & Väisänen (1984) provide a list of insect species attracted by spontaneous forest fires in northern Europe (table 2.2). In a study of the successions of arthropods which settle in a burned pine forest Winter *et al.* (1980) observed the arrival of the pyrophilous species *Sericoda quadripunctata* and the great number of lathridiids that live in fungi following a fire.

Heteroptera	Elaterids
Anthocorids	*Denticollis borealis*
Scoloposcelis phryganophila	Buprestids
Aradids	*Melanophila acuminata*
Aradus anisotomus	*Melanophila cyanea*
Aradus signaticornis	Bostrychids
Aradus angularis	*Stephanopachys substriatus*
Aradus aterrimus	*Stephanopachys linearis*
Aradus lugubris	Cucujids
Aradus laeviusculus	*Laemophloeus muticus*
Coleoptera	Lathridiids
Carabids	*Corticaria planula*
Pterostichus quadrifoveolatus	Salpingids
Sericoda bogemanni	*Sphaeriestes stockmanni*
Sericoda quadripunctata	Anthribids
Staphylinids	*Platyrhinus resinosus*
Paranopleta inhabilis	

Table 2.2. *Insects attracted to forest fires in northern Europe.* Aradus *spp. (except* A. lugubris*) and the carabid* Sericoda bogemanni *have been in strong regression following control of forest fires (Heliövaara & Väisänen, 1984).*

Fire damage may be compounded by further damage to trees enfeebled by insects. Bark beetles kill trees which might otherwise have survived. In western Oregon and in Washington State forest fires provoke outbreaks of *Dendroctonus pseudotsugae* on Douglas-fir. Other beetles such as the cerambycids *Arhopalus productus* and *Tetropium velutinum* penetrate the wood of trees killed by fire and, in association with fungi, damage it, thus greatly reducing the time available to salvage wood (Kimney & Furniss, 1943).

Changes in populations of carabid ground beetles following a fire have been observed in Manitoba. Some species are significantly more abundant in burned areas. *Harpalus laticeps* and *H. egregius* arrive in the immediate wake of fires and remain abundant for several years. Other species become rarer, and some remain unaffected. Numbers of winged species and ones of small size are greater in burned areas. These characteristics are those of species with a demographic strategy type *r*, which are capable of flying great distances (Holliday, 1991).

5. EDGE EFFECT AND AREA EFFECT

Research on woodland edges and the characteristics of their populations began with the study of birds. Many studies have also been based on invertebrates: insects, spiders, and even gastropods. Loss of area, which turns extensively wooded areas into isolated patches of woodland, some of less than one hectare, is an increasingly widespread phenomenon. These fragments of woodland are surrounded usually by cultivated land, and may be separated from each other by distances that prevent any dispersal of fauna from one to another. Insects appear to be very sensitive to the fragmentation of large forests; four groups of insects are particularly affected: pollinators, seed-eaters, parasitoids and decomposers of organic matter.

5.1. Edge effect

The interface between forest and the adjacent community is a transitional zone, the forest edge, or ecotone. Forest rides and the borders of clear-felled areas may be regarded as forest edge as well as those which face cultivated land. When not interfered with by man, forest edge has a complex structure which, when well developed, is composed of two parts: the mantle and the skirt. The mantle consists of shrubs and of trees of a lesser height than those of the forest; the skirt is dominated by herbaceous, often perennial, plants. The flora of these two zones is rich in species, some of which are dependent on them. To the south of Paris *Ligustrum vulgare* (privet) and *Prunus mahaleb* (St. Lucie cherry) are two shrubs characteristic of the mantle on calcareous soils, while the skirt contains *Lithospermum purpureo-coeruleum* and *Vincetoxicum officinale*. As a corollary of this richness in plant species, the number of insects is high, particularly in phytophagous species. The forest edge has particular microclimatic characteristics; the values of temperature, relative humidity, wind speed and light are often intermediate between those that prevail inside the wood and on the surrounding cultivated land.

5.1.1 Favourable consequences of edge effect

Edge effect is often characterised by greater populations and greater species diversity at the forest edge than in the woods or on the cultivated land, due to the mixture of woodland species and open environment species, in addition to species not found elsewhere. The oribatid soil mites of two forest populations, a beech wood and a pine wood, as well as those of the transitional zone between them have been studied in Belgium (Lebrun, 1988). The ecotone has more species, a greater abundance of mites, and a higher diversity index, as is shown by the following figures:

	Beech wood	Pine wood	Ecotone
Number of species	30	42	52
Shannon diversity index	3.07	3.75	4.27
Abundance per square metre	120,000	100,000	160,000
Number of characteristic species	3	2	9

Edge effect has been demonstrated by the study of the carabid beetles of a wood isolated by cultivated land near Paris (Dajoz, 1993). The forest edge, 4 metres wide, composed of tall grasses and shrubs, harboured four species which were encountered neither in the cultivated land nor in the wood. The main characteristics of carabid populations, sampled over one year by means of flight interception traps, are given in table 2.3. Woodland species are of a greater average size than those of cultivated land, which is a general characteristic of carabid populations, at least in temperate regions. Forest edge is the environment richest in species, and in which populations are most abundant, and the diversity index highest.

Populations of edges of unexploited forests where clear felling has taken place have been studied in Finland (Helle & Muona, 1985). In all the taxa studied (Homoptera, Diptera, ants, beetles and gastropods) abundance is greater in the forest than in the clear-felled area. In many groups greatest abundance is found in the edges. One explanation for this richness is that trees situated at the edge are exposed to wind and sunburn which causes them to wilt and facilitates colonisation by many insects. This rich invertebrate fauna provides birds with an abundant food supply, which explains the greater density of nesting species often observed at forest edges.

Edge effect has been observed in other insects. In a transect consisting of an alder wood, the adjacent meadow and the ecotone separating them, each zone has a distinct mosquito fauna (Dabrowska-Prot *et al.*, 1973). In a stand of Austrian pine on the southern slopes of Mont Ventoux in South-East France 14 species of ant were counted. Forest is the poorest environment, with 6 species, while clearings contain 11 species and edges 10. Some species live exclusively in clearings, some are common to clearings and edges, one to edges and forest, and a few are ubiquitous and nest in all three environments. The structure of populations and abundance of nests vary throughout the year; a migratory species, *Leptothorax unifasciatus*, occupies clearings in spring and in summer moves to the forest in search of higher levels of humidity and lower temperature (Du Merle *et al.*, 1978).

	Cultivated land (5 traps)	Forest edge (2 traps)	Woodland (3 traps)
α diversity index	5.6	7.1	5.1
Average size of carabids	10.37 mm	10.79 mm	14.68 mm
Number of carabids per trap	311	376	94
Number of species	44	41	24
Theoretical number of species (based on rarefaction method)	31	35	24
Cultivated land species	42	28	3
Forest species	2	9	21

Table 2.3.   *Main characteristics of populations of carabid beetles in a wood, in adjacent cultivated land and at the wood edge in the Paris area.*

Fragmentation of habitats and the creation of forest edges may have beneficial effects on insects and on other animals such as birds. Many forest edge insects seek sunlight and avoid the shade cast by trees. Butterflies require open sunny places in which to mate; Heteroptera and curculionid and chrysomelid beetles mostly live on plants which require light (Greatorex-Davies *et al.*, 1993, 1994). Abundance of such plants and insects decreases when shade cast by trees in forest rides is too great. Conservation of all these species requires management of forest rides to ensure maximum diversity and penetration of light.

In natural populations of *Pinus radiata* in California insect communities have more species and individuals in the vicinity of the forest edges than in the centre (Ohmart & Voigt, 1981). This may be explained by differences in microclimate which alter physiology and phenology of trees as well as the survival rate of insects (Schowalter *et al.*, 1986). Väisänen and Heliövaara (1994) have shown that in the boreal forest in Finland the number of insect species is 1.2 to 1.3 times greater in the vicinity of the forest edge than at 250 metres from the edge inside the forest.

5.1.2. Unfavourable consequences of edge effect

Harmful consequences of edge effect may be grouped under four head-ings: (a) changes in microclimate: temperature, humidity and wind speed; (b) increased mortality in trees due to wind storms which may uproot many trees on the edge; (c) disruptions at ground level, particularly of the forest floor litter which is scattered by wind and can no longer provide sufficient cover for hibernating invertebrates; (d) changes in the flora and fauna, accompa-nied by extinction of inner forest species and invasions by open environment species.

All studies relating to edge effect lead to the same conclusion: conserva-tion of the inner fauna of forests is only possible where the spatial dimensions of the forest are great enough to prevent edge effect reaching the centre and causing extinction of its endemic species. This raises the question of minimum area of forest reserves. Where edge effect is felt to a depth of 200 metres, a reserve of a more or less circular shape will require an area greater than 300 m x 300 m x 3.14, i.e. at least 28 hectares if the core preserving true forest fauna is to have a diameter of 200 metres. Edge effect is felt over a distance of about 30 metres in the case of an orthopteran and to a depth of 200 to 500 metres in the case of species in Australian trop-ical rainforest. In the latter case a reserve must cover between 2,000 and 4,000 hectares in order that at least half its area remains unaffected by edge effect due to fragmentation.

5.2. Effects of area and of isolation

Insects, like other animals, are sensitive to the fragmentation of forests. It is often difficult to distinguish between the consequences of edge effect, those of spatial reduction due to fragmentation of forests, and those due to isolation. Species that inhabit a forest fall into three categories: (a) species of the inner forest which are sensitive to disruptions produced by edge effect – their abundance and diversity are a function of the core area of the forest fragment; (b) forest species which are not sensitive to disruptions caused by edge effect – the abundance and diversity of these species depend on the total area of woodland; (c) forest edge species, which include those of the forest edge itself and others from open environments which penetrate the forest to a greater or lesser depth – their abundance and diversity is a func-tion of the area subjected to edge effect. In addition to these three categories, there are other insects which depend on particular and more or less scarce structures such as old, dying trees. This explains why, contrary to the theory of island biogeography, species diversity of insects (or other taxa) is not always linked to the area of forest islands. This is the case of xylophagous beetles in which species diversity is not related to the area of the forest they inhabit. Diversity in these insects is especially a function of the abundance of decaying dead trees for them to colonise (Stevens, 1986).

Populations of carabid beetles inhabiting forests of areas between 0.1 and 1,500 hectares have been studied in a few localities in western Europe. A linear relationship does exist between the logarithm of species number, and area, but the low correlation coefficient shows that area is only one factor among others that determine richness in species. Forest structure, and in particular the kind of trees (broadleaf or conifer) matters: populations of conifers contain fewer species than oak woods, and the dominant species are different. Moreover there is a relationship between abundance of carabids and area of forest. In a forest such as Fontainebleau, which includes large areas of tall trees and many old trees, forest interior species are abundant. In isolated woods covering small areas and mostly consisting of young trees or coppice south of the forest, populations of carabids consist of species not sensitive to edge effect, and forest interior species are rare or absent. Fontainebleau harbours several species of *Carabus* and three species of *Abax*, including one ubiquitous species, *A. ater*, and two forest interior species, *A. ovalis* and *A. parallelus*. The outlying fragments of woodland are devoid of *Carabus*, and harbour only the ubiquitous species *A. ater*. The absence of two species of *Abax* in these fragments is compensated for by a greater abundance of *A. ater*, which may form up to 40% of the population, whereas this species only forms about 10% of the population in the forest of Fontainebleau (Dajoz, 1993).

In fragments of woodland, interaction with the surrounding environment may lead to colonisation by outside species. Paradoxically the number of species may increase as the area of woodland diminishes (Webb, 1989). This observation, based on arthropods, is sometimes also true of birds. A study conducted in North America has shown that the abundance and diversity of arthropods are significantly less in medium-sized woods than in small ones. However their abundance and diversity are greater in large, isolated woods. These results are difficult to interpret.

No relationship could be established between area of forest fragments and numbers of species or abundance of forest spiders in Finland (Pajunen *et al.*, 1995). However species composition of populations is different at the edges and the interior. Large sized species (gnaphosids, lycosids) prefer to colonise edges, while small species (linyphiids) are less numerous at edges, in open forests, and in plantations.

The effects of fragmentation on insects associated with beech have been studied in forest fragments and hedges in Scotland (Dennis & Watt, 1993). The species studied were leaf miners such as the weevil *Rhynchaenus fagi*, microlepidoptera such as the nepticulids *Stigmella tityrella* and *S. hemagyrella*, and the gracillariids *Phyllonorycter messaniella* and

P. maestingella, as well as gall insects such as the cecidomyiid *Hartigiola annulipes*. Large fragments harboured more species than small ones. Fragments of less than 5 hectares and situated more than 2 km from the nearest wood were devoid of mining microlepidoptera. This may be explained if one considers the following factors: (a) the poor dispersal powers of micro-moths which reduces the flux of colonisers; (b) unfavourable climatic factors which prevail in the foliage in small fragments; (c) sparseness of forest floor litter, which is dispersed by wind in winter in such fragments, thus depriving insects of cover during hibernation. Other research has revealed identical dispersal patterns in leaf miners and gall insects of birch and oak. Fragmentation of woodland isolates populations and prevents their dispersal. A clear-felled strip 80 metres wide in tropical forest is enough to prevent movement of many birds, mammals and insects. Forest roads often present insects with an impassable barrier and create, on either side, forest edges which are uninhabitable for certain species owing to changes in microclimate. Abundance of typical forest carabids declined on either side of a newly built road in Switzerland, and the resulting habitat loss was estimated at 7 hectares per kilometre of road inside the forest (Eyholzer, 1995). Spatial distribution and genetic structure of the moth *Operophtera brumata* are functions of the area of islands and their isolation. The greater the isolation, the lower the population density, and some fragments are not occupied at all. Genetic heterogeneity of isolated populations declines, and body weight of males is reduced. This is the consequence of local extinctions of small populations followed by recolonisation by a small number of immigrants, which leads to a "bottleneck" effect in populations (Van Dongen *et al.*, 1994).

Isolation also reduces the percentage of parasitism to which phytophagous insects are subjected. The percentage of parasitism in two weevils of the genus *Apion* which live on cultivated clover declines sharply as isolation from other clover fields increases, dropping from 75% in contiguous fields to about 30% in fields 500 metres apart (Kruess & Tscharntke, 1994). These results may certainly apply to phytophagous forest insects as has been shown by observations made by Faeth & Simberloff (1981), who studied isolated oak trees situated 165 metres from the edge of a wood composed of mixed oak and pine. Numbers of species of leaf-mining insects did not differ significantly on isolated trees and on those of the wood, but the percentage of parasitism of mining insects by various parasitoids was markedly higher in the wood than on the isolated trees. The authors interpret this result by suggesting that parasitoid insects generally restrict their search for host species to the vicinity of old trees inside the wood (figure 2.10).

In Canada vast areas were covered in aspen forests before man intervened. These forests are now fragmented into islands of different sizes, separated by cultivated areas. *Malacosoma disstria* is an aspen defoliator which occasionally occurs in massive outbreaks. This moth is attacked by various parasitoids, the main ones being tachinid and sarcophagid flies which are thought to play an important part in reducing the abundance of the moth following an outbreak. Four species of these parasitoids have been studied, showing that the fragmentation of forests affects their behaviour. Levels of predation by the three larger species drop in smaller forest islands whereas the rate of parasitism by the smaller species rises (Roland & Taylor, 1997). These results confirm theoretical arguments showing that changes in the structure of the landscape (in this case the structure of forest fragmented by man) may alter normal functions of an ecosystem by affecting major processes such as parasitism (Kareiva & Wennergren, 1995; Godfray & Hassel, 1997).

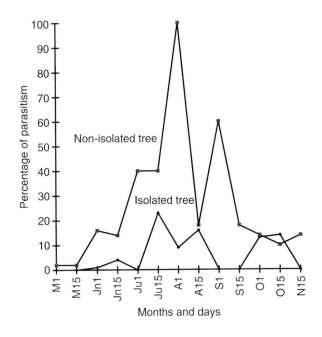

Figure 2.10. *Percentage of parasitised caterpillars mining the leaves of three species of oak from 1st May (M1) until 15th November (N15). Caterpillars living on isolated trees are less parasitised than those living on trees inside a wood (Faeth & Simberloff, 1981).*

Reduction of forest surface area and cuts made by roads and footpaths lead to increased public access and trampling which deeply disturb the fauna. Wood ants (of the *Formica rufa* group) may serve as indicators of degradation of montane forests (Torossian, 1977). These ants, which live in coniferous woods, build large nests above ground formed of conifer needles mixed with various debris. *Formica polyctena* mostly prefers deep forest and *Formica rufa* prefers forest edge. The nests of these ants are smaller and fewer as forests are more disturbed and degraded by activities such as winter sports and the presence of nearby buildings, which leads to overuse by visitors. In the small forest of Osseja (Pyrénées-Orientales) there is an average of five nests per hectare, but only 0.3 nests per hectare in the nearby but heavily degraded forest of Font-Romeu. Simultaneously the volume of the nests decreases from 0.284 m³ to 0.077 m³ at Font-Romeu.

Major decline or disappearance of a species of tree through total deforestation may lead to the loss of species of insect unable to adapt to another host species. In North America 7 species of microlepidoptera associated with American chestnut became extinct following the loss of their host tree through a fungal disease (Opler, 1978). Disappearance of several plant species in Hawaii led to the extinction of five species of Lepidoptera (Gagné & Howarth, 1985).

5.3. Practical consequences

It is important that edge effect be reduced to a minimum in forests turned into reserves. Wilson & Willis (1975) presented the hypothesis that the ideal shape of a reserve is circular. Narrow, elongated reserves are to be avoided as edge effect is felt over a larger area, preventing the maintenance of inner forest species and favouring invasion by open environment species. This notion arises from the discovery of peninsula effect in the course of a study of North American mammals. Narrow peninsulae such as Florida and Baja California often harbour fewer species, per area, than neighbouring parts of the continent, and the number of species decreases distally from the base (Simpson, 1964). This peninsula effect recurs in insects such as the butterflies of the Iberian peninsula (Martin & Gurrea, 1990), and the advantage of rounded fragments as opposed to narrow, elongated fragments has been confirmed, at least in the case of butterflies (Baz & Garcia-Boyero, 1995). Forest fragmentation leaves islets, none of which contains the original fauna in its entirety. Fifteen species of arthropod are associated with juniper in a group of 25 populations of this tree in southern England, but the number of species present in any one islet varies between 3 and 13 according to the fragment and its area.

5.4. Hedgerow, a faunal reserve and corridor

In a copse or small wood isolated populations are few and their extinction due to the effects of various risks may be predicted and evaluated. When populations are in contact by means of individuals which disperse from one site to another, the period of survival of the metapopulation thus formed is longer than that of isolated populations. This dispersal occurs in the case of many species by means of communication channels known as corridors. Hedgerows, which are common in certain traditional farm areas and link a number of fragments of forest, are structures of considerable importance in acting as corridors for woodland fauna.

A hedgerow may be considered to be a double forest edge containing at its centre a fragment of forest of varying width. Hedgerows act as refuges to many species by providing favourable climatic conditions and a source of food. The black-veined white butterfly *Aporia crataegi* lives on hawthorn and is disappearing together with the hedgerows which are the last refuge of its host plant; it has become extinct in Britain. Hedgerows are also corridors which enable the dispersal of certain woodland species from one island to another. Among carabid beetles, strictly woodland species such as *Abax parallelus* do not venture further than 100 metres from the forest into hedgerows. "Peninsular" species such as *Platysma nigrum* may travel up to 500 metres along hedgerows, and "corridor" species such as *Abax ater* may be found as far as 15 km from their original homes (Burel & Baudry, 1990). Evidence of the role of corridors in the dispersal of organisms has given rise to new concepts relating to the spatial structure of reserves. The ideal solution consists in linking different reserves by corridors so that their populations may function as metapopulations, which considerably increases their chances of survival.

REFERENCES

BAKKE A., (1968). Ecological studies on bark beetles (Col. Scolytidae) associated with Scots pine (*Pinus sylvestris* L.) in Norway with particular reference to the influence of temperature. *Medd. fra. det. Norsk. Skogfors*, **21**: 443-602.

BALACHOWSKY A.-S., (1949). *Coléoptères Scolytidae*. Faune de France, **50**. Lechevalier, Paris.

BALTENSWEILER W., FISCHLIN A., (1979). The role of migration for the population dynamics of the larch bud moth, *Zeiraphera diniana* Gn. (Lep. Tortricidae). *Bull. Soc. Ent. Suisse*, **52**: 259-271.

BAZ A., GARCIA-BOYERO A., (1995). The effects of forest fragmentation on butterfly communities in central Spain. *J. Biogeog.*, **22**: 129-140.

BLEUTEN W., (1989). Minimum spatial dimension of forests from point of view of wood production and nature preservation. *Ekologia* (CSSR), **8**: 375-386.

BOVEY P., (1971). L'impact de l'insecte défoliateur sur la forêt. La lutte biologique en forêt. *Ann. Zool. Écol. Anim.*, numéro spécial, 11-29.

BRENY R., (1963). Microclimats entomologiques. *Bull. Ann. Soc. R. Ent. Belgique*, **99**: 117-137.

BROWN C. E., (1965). Mass transport of forest tent caterpillar moths, *Malacosoma disstria* Hübner, by a cold front. *Canad. Ent.*, **97**: 1,073-1,075.

BUREL F., BAUDRY J., Hedgerow networks as habitats for forest species: implications for colonizing abandoned agricultural land. *In*: R. G. H. BUNCE, HOWARD D. C., (1990). *Species dispersal in agricultural habitats.* Belhaven Press, London, 238-255.

CACHAN P., Importance écologique des variations verticales micro-climatiques du sol à la canopée dans la forêt tropicale humide. *In*: P. PESSON (1974). *Écologie forestière.* Gauthier-Villars, Paris, 21-42.

CARTER C. I., (1972). Winter temperatures and survival of the green spruce aphid *Elatobium abietinum* (Walker). *Forestry Commission, Forest Record,* **84**. HMSO, London.

CHARARAS C., (1962). *Scolytides des conifères.* Lechevalier, Paris

DABROWSKA-PROT E., LUCZAK J., WOJCIK Z., (1973). Ecological analysis of two invertebrate groups in the wet alder wood and meadow ecotone. *Ekol. Polska,* **21**: 753-812.

DAJOZ R., (1967). Écologie et biologie des Coléoptères xylophages de la hêtraie. *Vie et Milieu,* **17**, sér. C: 523-763.

DAJOZ R., (1993). Les Coléoptères Carabidae d'une région cultivée à Mandres-les-Roses (Val de Marne). II. Comparaison de la faune des cultures, des lisières et d'un bosquet. *Cahiers des Naturalistes,* **48**: 67-78.

DEMOLIN G., (1969). Bioécologie de la processionnaire du pin *Thaumetopoea pityocampa* Schiff. Incidences des facteurs climatiques. *Bol. Serv. Plagas. Forest.,* **12**: 1-14.

DENNIS P., WATT A.-D., (1993). Effects of woodland fragmentation on insect abundance and diversity. *Institute of Terrestrial Ecology, Annual Report 1992-1993:* 18-20.

DU MERLE P., JOURDHEUIL P., MARRO J.-P., MAZET R., (1978). Évolution saisonnière de la myrmécofaune et de son activité prédatrice dans un milieu forestier: les interactions clairière-lisière-forêt. *Ann. Soc. Ent. Fr.,* **14**: 141-157.

ELENS A.-A., (1953). Étude écologique des Lophyres en Campine belge. I. Résistance à la dessication des éonymphes de *Diprion pini* L., *Diprion pallidum* Kl. et *Diprion sertifer* Geoffr. (Hymenoptera Symphyta). II. Incubation des œufs et adaptation à la température chez *Diprion pini* L., *Diprion pallidum* Kr. et *Diprion sertifer* Geoffr. *Agricultura* (Louvain), **1**: 3-32.

EVANS W. G., (1966). Morphology of the infrared sense organ of *Melanophila acuminata* (Buprestidae: Coleoptera). *Ann. Ent. Soc. Amer.,* **59**: 873-877.

EVANS W. G., (1972). The attraction of insects to forest fires. *Proc. Tall Timber Conference on Ecology. Animal control by habitat management,* **3**: 115-127.

EYHOLZER R., (1995). Auswirkungen der Erschliessung von Wäldern der montanen Stufe auf die Laufkäfer (Col., Carabidae). *Mittel Schweiz Entom. Ges.,* **68**: 83-102.

FAETH S. H., SIMBERLOFF D., (1981). Experimental isolation of host plants: effects on mortality, survivorship, and abundances of leaf-mining insects. *Ecology*, **62**: 625-635.

GAGNÉ W. C., HOWARTH F. G., (1985). Conservation status of endemic Hawaiian Lepidoptera. *Proc. 3rd Cong. European Lepidoptera*, Cambridge 1982: 74-84.

GÄRDENFORS U., BARANOWSKI R., (1992). Beetles living in open deciduous forests prefer different tree species than those living in dense forests. *Ent. Tidsk.*, **113**: 1-11.

GASPAR C., VERSTRAETEN Ch., (1972). Recherches sur l'écosystème forêt. Biocénoses des Coléoptères. *Bull. Soc. R. Sc. Liège*, **41**: 227-249.

GODFRAY H. C., HASSELL M. P., (1997). Hosts and parasitoids in space. *Nature*, **386**: 660-661.

GRAHAM A. S., (1925). The felled tree trunk as an ecological unit. *Ecology*, **6**: 397-411.

GREATOREX-DAVIES J. N., SPARKS T. H., HALL M. L., MARRS R.-H., (1993). The influence of shade on butterflies in rides of coniferised lowland woods in southern England and implications for conservation management. *Biol. Cons.*, **63**: 31-41.

GREATOREX-DAVIES J. N., SPARKS T. H., Hall M. L. (1994). The response of Heteroptera and Coleoptera species to shade and aspect in rides of coniferous lowland woods in southern England. *Biol. Cons.*, **67**: 255-273.

HAACK R.-A., (1985). Voltinism and diurnal emergence-flight patterns of *Ips calligraphus* (Coleoptera: Scolytidae) in Florida. *Florida Ent.*, **68**: 658-667.

HELIÖVAARA K., VÄISÄNEN R., (1983). Environmental changes and the flat bugs (Heteroptera, Aradidae and Aneuridae). Distribution, and abundance in Eastern Fennoscandia. *Ann. Ent. Fenn.*, **49**: 103-109.

HELIÖVAARA K., VÄISÄNEN R., (1984). Effects of modern forestry on north-western European forest invertebrates: a synthesis. *Acta Forestalia Fennica*, **189**:1-32.

HELLE P., MUONA J., (1985). Invertebrate numbers in edge between clear-fellings and mature forests in northern Finland. *Silva Fennica*, **19**: 281-294.

HOLLIDAY N. J., (1991). Species responses of carabid beetles (Coleoptera: Carabidae) during post-fire regeneration of boreal forest. *Canad. Ent.*, **123**: 1,369-1,389.

KAREIVA P., WENNERGREN U., (1995). Connecting landscape pattern to ecosystem and population processes. *Nature*, **373**: 299-302.

KENNEDY J. S., (1974). Pheromone-regulated anemotaxis in flying moths. *Science*, **184**: 999-1,001.

KIMNEY J. W., FURNISS R. L., (1943). Deterioration of fire-killed Douglas-fir. *US Dep. Agric., Tech. Bull.*, **851**, 61 pp.

KITTREDGE J.,. L'influence de la forêt sur le climat et les autres facteurs du milieu. *In: Influences exercées par la forêt sur son milieu*. FAO, Rome (1962), 95-156.

KOMAREK E.-V., (1969). Fire and animal behavior. *Proc. Annual Tall Timber Fire Ecology Conference no. 9*: 161-190.

KRUESS A., TSCHARNTKE T., (1994). Habitat fragmentation, species loss, and biological control. *Science*, **264**: 1,581-1,584.

LANIER G.-N., BURNS B. W., (1978). Barometric flux. Effects on the responsiveness of bark beetles to aggregation attractants. *J. Chem. Ecol.*, **4**: 139-147.

LEBRUN Ph., (1988). L'effet d'écotone. *Probio-Revue*, **11**: 23-42.

LUNDBERG S., (1984). The beetle fauna of burnt forests in Sweden. *Ent. Tidskr.*, **105**: 129-141.

LUNDBERG S., (1993). *Phryganophilus ruficollis* (Fabricus) (Coleoptera, Melandryidae) in north Fennoscandia. Habitat and developmental biology. *Ent. Tidskr.*, **114**: 13-18.

MAKSYMOV J.-K., (1965). Die Uberwinterung des Larchenblasenfusses *Taeniothrips laricivorus* Kratochvil und Frasky. *Mitt Schweiz. Anst forst Vers. Wes.*, **41**: 1-17.

MARTIN J., GURREA P., (1990). The peninsular effect of Iberian butterflies (Lepidoptera: Papilionidae and Hesperoidea). *Biogeogr.*, **17**: 115-128.

MIKKOLA K., (1971). The migratory habit of *Lymantria dispar* (Lep. Lymantriidae) adults of continental Eurasia in the light of a flight to Finland. *Acta. Ent. Fennica*, **28**: 107-120.

MILLER J. M., (1931). High and low lethal temperatures for the western pine beetle. *J. Agric. Res.*, **43**: 303-321.

MUONA J., RUTANEN I., (1994). The short-term impact of fire on the beetle fauna in boreal coniferous forest. *Ann. Zool. Fennici*, **31**: 109-121.

MUSSO J.-J., (1972). Étude des migrations journalières d'*Andrenosoma bayardi* Séguy (Diptère, Asilidae). *Bull. Soc. Zool. Fr.*, **97**: 45-53.

NEF L., (1959). Étude d'une population de larves de *Retinia buoliana* (Schiff.). *Zeit. ang. Ent.*, **44**: 167-186.

NEF L., (1967). Comparaison de populations de *Rhyacionia buoliana* Schiff. en réponse à une fumure minérale. *XIV IUFRO Kongress*, München, vol. V: 650-658.

NEF L., (1975). Étude écologique des pontes de *Stilpnotia* (= *Leucoma*) *salicis* L. *Ann. Soc. R. Zool. Belgique*, **105**: 129-146.

NIELSEN B.-O., (1974). The phenology of beech canopy insects in Denmark. *Vidensk. Medd. fra. Dansk Naturh. Forening*, **137**: 95-124.

OHMART C.-P., VOIGT W.-G., (1981). Arthropod communities in the crowns of the natural and planted stands of *Pinus radiata* (Monterey) in California. *Canad. Ent.*, **113**: 673-684.

OPLER P. A., (1978). Insects of American chesnut: possible importance and conservation concern. *Proc. Amer. Chesnut. Symp.*, Morgantown, WV: 83-85.

PAJUNEN T., HAILA Y., NIEMELA J., PUNTILA P., (1995). Ground-dwelling spiders (Arachnida, Araneae) in fragmented old forests and surrounding managed forests in southern Finland. *Ecography*, **18**: 62-72.

PARDÉ J., Le microclimat en forêt. *In*: P. Pesson (1974). *Écologie forestière.* Gauthier-Villars, Paris, 1-20.

PARK O., LOCKETT J. A., MYERS D. J., (1931). Studies in nocturnal ecology with special reference to climax forest. *Ecology*, **12**: 709-727.

PAVARI A., Introduction. *In*: *Influences exercées par la forêt sur son milieu.* FAO, Rome (1962), 3-33.

PAYNE N. M., (1926). The effect of environmental temperatures upon insect freezing points. *Ecology*, **7**: 99-106.

PRICE P. W., (1971). Niche breadth and dominance of parasitic insects sharing the same host species. *Ecology*, **52**: 587-596.

REES C. J., Microclimate and the flying Hemiptera fauna of a primary lowland rain forest in Sulawesi. *In*: S. L. SUTTON, T.C. WHITMORE, A. C. CHADWICK, (1983). *Tropical rain forest: ecology and management.* Blackwell, Oxford, 121-136.

ROLAND J., TAYLOR P. D., (1997). Insect parasitoid species respond to forest structure at different spatial scales. *Nature*, **386**: 710-713.

ROUSSEL L., (1953). Recherches théoriques et pratiques sur la répartition en quantité et en qualité de la lumière dans le milieu forestier. Influences sur la végétation. *Ann. École Nat. Eaux et Forêts*, **13**: 1-110.

SAVELY H. E., (1939). Ecological relations of certain animals in dead pines and oak logs. *Ecological Monographs*, **9**: 327-385.

SCHMID J.-M., (1976). Temperatures, growth and fall of needles of Engelmann spruce infested by spruce beetles. *USDA For. Serv., Note* **RM-331**: 1-4.

SCHMITZ H., BLECKMANN H., MÜRTZ M., (1997). Infrared detection in a beetle. *Nature*, **386**: 773-774.

SCHOWALTER T.D., Adaptations of insects to disturbance. *In*: S.-T.-A. PICKETT, P.-S. WHITE, (1985). *The ecology of natural disturbance and patch dynamics.* Academic Press, London, 235-252.

SCHOWALTER T.-D., HARGROVE W. W, CROSSLEY D. A., (1986). Herbivory in forested ecosystems. *Ann. Rev. Ent.*, **31**: 177-196.

SCHWENKE W., (1958). Local dependence of parasitic insects and its importance for biological control. *Proc. 10th Int. Congr. Entomol.*, **4**: 851-854.

SERVICE M. W., (1973). Spatial and temporal distribution of aerial populations of woodlands tipulids (Diptera). *J. anim. Ecol.*, **42**: 295-303.

SIMPSON G. G., (1964). Species diversity of North American recent mammals. *Syst. Zool.*, **13**: 57-73.

SOUZA O.-F.-F., BROWN V. K., (1994). Effects of habitat fragmentation on Amazonian termite communities. *J. Trop. Ecol.*, **10**: 297-206.

STEVENS G. C., (1986). Dissection of the species-area relationship among wood-boring insects and their host-plant. *Amer. Nat.*, **128**: 35-46.

TENOW O., (1972). The outbreaks of *Oporinia autumnata* Bkh. and *Operophtera* spp. (Lep. Geometridae) in the Scandinavian mountain chain and northern Finland. *Zool. Bidrag fran Uppsala*, suppl. **2**, 107 p.

TENOW O., (1975). Topographical dependence of an outbreak of *Oporinia autumnata* Bkh. (Lep. Geometridae) in a mountain birch forest in Northern Sweden. *Zoon*, **3**: 85-110.

TOROSSIAN C., (1977). Les fourmis rousses des bois (*Formica rufa*) indicateurs biologiques de dégradations des forêts de montagne des Pyrénées-Orientales. *Bull. Écol.*, **8**: 333-348.

VÄISÄNEN R., HELIÖVAARA K., (1994). Assessment of insect occurrence in boreal forests based on satellite imagery and field measurements. *Acta For. Fennica*, **243**: 1-39.

VAN DONGEN S., BACKELJAU T., MATTHYSEN E., DHONDT A.-A., (1994). Effects of forest fragmentation on the population structure of the winter moth *Operophtera brumata* L. (Lepidoptera, Geometridae). *Acta Œcologica*, **15**: 193-206.

VITÉ J.-P., (1956). Populationstudien am Larchenblasenfuss *Taeniothrips laricivorus* Krat. *Zeit. ang. Ent.*, **38**: 417-488.

WAINHOUSE D., (1980). Dispersal of first instar larvae of the felted beech scale, *Cryptococcus fagisuga*. *J. Anim. Ecol.*, **17**: 523-532.

WALLWORK J.-A., (1976). *The distribution and diversity of soil fauna.* Academic Press, London.

WARD L. K., LAKHANI K.-H., (1977). The conservation of juniper: the fauna of food-plant island sites in southern England. *J. Appl. Ecol.*, **14**: 121-135.

WELLINGTON W. G., (1950). Effects of radiation on the temperatures of insect habitats. *Sci. Agric.*, **30**: 209-234.

WELLINGTON W. G., FETTES J.-J., TURNER K. B., BELYEA R.-M., (1950). Physical and biological indicators of outbreaks of the spruce budworm. *Canad. J. For. Res.*, ser. D, **28**: 308-331.

WESELOH R. M., (1972). Spatial distribution of the gypsy moth (Lepidoptera: Lymantriidae) and some of its parasites within a forest environment. *Entomophaga*, **17**: 339-351.

WESELOH R. M., Behavior of forest insect parasitoids. *In*: J.-F. ANDERSON, H.-K. KAYA, (1976.). *Perspectives in forest entomology.* Academic Press, New York, 99-110.

WIKARS L. O., (1995). Clear-cutting before burning prevents establishment of the fire-adapted *Agonum quadripunctatum* (Coleoptera: Carabidae). *Ann. Zool. Fennici*, **32**: 375-384.

WILSON E. O., WILLIS E. O., Applied biogeography. *In*: M.-L. CODY, J.-M. DIAMOND, (1975). *Ecology and evolution of communities*. Harvard Univ. Press, 522-534.

WINTER K., SCHAUERMAN J., SCHAEFER M., (1980). Sukzession von Arthropoden in verbrannten Kiefernforsten. I. Methoden und allgemeiner Uberblick. *Forst Centr.*, **99**: 324-340.

ZWÖLFER H., (1931). Studien zur Okologie und Epidemiologie der Insekten. I. Die Kieferneule, *Panolis flammea* Schiff. *Zeit. ang. Ent.*, **17**: 475-562.

ZWÖLFER H., (1933). Studien zur Okologie, insbesondere zur Bevolkerungslehre der Nonne, *Lymantria monacha*. *Zeit. ang. Ent.*, **20**: 1-50.

◇

3

The role of biotic factors

The biotic factors that affect populations of insects in forest are the same as those in non-forest ecosystems, but their relative importance appears to differ. Prime consideration must be given to relationships between insects and trees. These relationships are particularly complex given the great number of species involved, and the peculiar biology of trees, which are very long-lived organisms. Other biotic factors such as competition, predation and parasitism also play a part in regulating the populations of forest insects.

1. INSECT / TREE RELATIONSHIPS

Until 1960 the generally accepted idea was that all parts of foliage are equally available and acceptable to phytophagous insects which are therefore able to consume the entire foliage. According to this hypothesis insect distribution and abundance are essentially controlled by competition, predation and parasitism; "secondary substances" are seen as by-products of metabolism which have no known roles. The development of chemical ecology (Fraenkel, 1959) and the concept of chemical defence in plants (Feeny, 1970) caused this view to change. The discovery of inducible resistance (Haukioja & Niemela, 1979) and the ability of plants attacked by herbivores to communicate (Baldwin & Schultze, 1983) added to this body of knowledge. The notion that the effects of phytophagy on plants are not merely negative, but may on the one hand stimulate plant growth and on the other act as a regulator of primary production in the ecosystem (Chew, 1974; Golley, 1977) appeared soon after, and in 1981 Schowalter proposed the hypothesis that phytophagy may be an element responsible for the evolution of ecosystems and for the phenomenon of succession.

1.1. Diet diversity

Every species of phytophagous insect is a specialist feeder on a particular part of one or more plant species. Monophagous species are mainly sap suckers, such as the scale insect *Cryptococcus fagisuga*, associated with beech, and many cicadellid and typhlocybid Homoptera. Most insects are oligophagous and confined to a few plant species belonging to a single genus or family. The tortricid moth *Tortrix viridana* feeds on many species of oak. Polyphagous species, which feed on many plant species, are relatively few. *Hyphantria cunea* feeds on some hundred different plants, many of them trees.

Monophagous species may change diet when their host plant disappears. The weevil *Rhynchaenus fagi*, for example, which is normally associated with beech, may feed on raspberry, and the oak processionary caterpillar may eat hazel when oaks are defoliated, but larval mortality then becomes high. Unexplained changes in host plant have occurred: in Austria and Holland *Coleophora laricella* switched from larch to Douglas-fir.

Insect specialization on a single host plant varies with geographical latitude in circumstances for which no general explanation can be given (Beaver, 1979). In bark beetles specificity is more pronounced in temperate regions than in the tropics, whereas specificity in Lepidoptera does not appear to vary (table 3.1).

Exclusive association	Tropics		Temperate regions	
	Malaysia	Fiji	France	California
To species	0	2	2	12
To genus	4	14	51	58
To family	35	16	38	24
Polyphagous	61	68	9	6
Number of species	275	56	132	173

Table 3.1.   *Specialization of scolytid beetles in two temperate and two tropical regions. Numbers indicate the percentage of the total number of scolytid species associated with a single plant species, genus, family, or which are polyphagous.*

Certain kinds of behaviour enable insects to recognize their host plant and reject others, and to recognize the organ of the plant to be eaten and avoid those containing inferior nutrients. Analysis of recognition and plant colonisation behaviour has been particularly developed for insects which feed on cultivated herbaceous plants (Robert, 1986), and, among species that live on trees, for scolytid beetles. Search for the host plant is supposed to depend on a choice generally due to olfactory stimuli. The chemical composition of a tree plays an essential part either in attracting or in repelling insects. Chemical composition may vary according to the age of a tree. This explains why pines planted in 1921 matured and suffered no damage until 1953 with the appearance of *Bupalus piniarius*, an insect that prefers trees about thirty years old.

Gluphisia septentrionalis is a notodontid moth which feeds on American trembling aspen, *Populus tremuloides*. The leaves of this tree contain sodium in a concentration from 2.9 ± 1.2 ppm (dry weight), which is considerably less than the average concentration in other trees. On the other hand the high concentration of potassium, 8,120 ± 990 ppm, is higher than average. *Gluphisia* thus suffers a deficiency in sodium and an excess of potassium in its diet. To compensate for this, males drink large quantities of water from puddles on the ground, which provides the necessary sodium. This water is eliminated in spurts immediately following ingestion by means of a particular structure of the moth's digestive tract. The amount of water expelled per minute may reach 12% of the insect's body weight. Measurements made confirm gains of sodium and partial loss of potassium. Females do not share this behaviour; they acquire sodium through the male spermatophore during copulation, and sodium is then passed on into the eggs (Smedley & Eisner, 1995).

Aphids detect nutrient-rich tissue and concentrate on it (Dixon, 1970). American holly (*Ilex opaca*) is well protected by its physical characteristics (tough spiny leaves) and chemical compounds (paucity of nitrogenous compounds, richness in saponins and phenol compounds). The agromyzid fly *Phytomyza ilicicola* is one of the few species that feeds on this holly. It avoids the plant's defences, in particular the tough leaves, by emerging very early and laying its eggs on tender young developing leaves. Its mining larvae live on the palisade parenchyma which is richer in proteins and in water than the other parts of the leaf, and is devoid of fibres that might act as a physical impediment to movement. This parenchyma is richer in saponins, to which the insect appears to be adapted (Kimmerer & Potter, 1987). The gregarious larvae of *Malacosoma americanum* prefer the young leaves of various trees. Individuals that have located a suitable food source deposit a guiding pheromone which leads other members of the colony to the source. Deposits of the pheromone are more concentrated on young leaves than on older ones (Peterson, 1987).

1.2. The concept of palatability

The concept of palatability describes the degree to which tree leaves are eaten or rejected by herbivorous animals such as insects. Experiments have shown (figure 3.1) that trees harbouring the greatest numbers of phytopha-

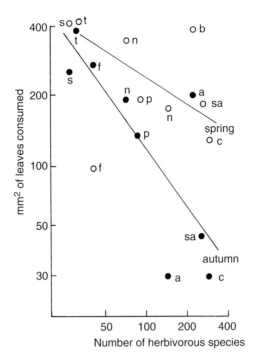

Figure 3.1. *Relationships between palatability of the leaves of 9 tree species and number of associated insect species in Britain. Palatability is evaluated according to the surface of leaves consumed by snails. The higher the number of phytophagous insect species is, the smaller the surface of eaten leaves is (Wratten et al., 1981). a: hawthorn ; b: birch, c: oak, f: ash, n: hazel, p: apple tree, s: rowan, sa: willow, t: lime.*

gous species are those whose leaves are least suitable as food for polyphagous herbivores such as snails (Wratten *et al.*, 1981). It is thought that natural selection has favoured the evolution of defences in plants that are eaten by large numbers of herbivorous species. In such cases predation by many phytophagous species does not necessarily imply greater damage to the leaves or to other organs. Deciduous tree leaves are more palatable than those of evergreens; those of temperate regions are more palatable than those of the tropics; trees that grow rapidly and have rapidly renewed leaf growth are more palatable than those that grow slowly and whose leaves are evergreen, the latter having a larger arsenal of chemical defences and probably containing higher levels of nitrogen compounds (Reader & Southwood, 1981; Southwood *et al.*, 1986; Basset & Burckhardt, 1992; Basset, 1994).

1.3.   Effects of levels of nitrogen compounds

Levels of absorbable nitrogen are a limiting factor in insect growth. Adult weight is a function of the nitrogen levels in foliage consumed by larvae, and the heaviest females lay most eggs. In insects that feed at the adult stage, fertility is related to nitrogen richness in the adult diet (McNeill & Southwood, 1978; Mattson, 1980). In the Australian chrysomelid beetle *Paropsis atomaria*, which feeds on *Eucalyptus blakelyi*, an increase of 1.5 to 4% in nitrogen levels in leaves leads to an increase of about four times the number of eggs laid per female (Ohmart *et al.*, 1985).

There is a relationship between nitrogen levels in the leaves of 14 species of host tree of the elongate hemlock scale insect *Fiorina externa* and 4 characteristics of the insect: its abundance, the number of eggs per female, the number of females producing eggs, and the survival rates of larvae (McClure, 1980b).

Most aphids live on elaborated sap which is drawn from phloem. This sap contains sugars in fairly high concentrations, but few amino acids or other nitrogen compounds. Levels of the latter change according to the stage of growth of leaves and terminal shoots; they are low in summer when leaves are fully grown, and high during the growth period in spring and during the senescent period in autumn. In oak and maple only glutamic and aspartic acids and alanin are present in summer; in autumn when leaves change colour the proteins they contain are activated and numbers of amino acids rise in the sap. In spruce, amino acid levels reach their maximum in May. The growth of populations of *Elatobium abietinum* on this tree is rapid in spring and limited only by temperature. At this time of the year the aphids are wingless. From May onwards amino acid levels in the needles drop; winged aphids appear and disperse to other trees, thus reducing the population, which maintains itself at a low level throughout the summer. Population levels may rise slightly in autumn together with levels of amino acids in the needles, but this rise in numbers of individuals is slowed by low temperatures which act as a limiting factor (McNeill & Southwood, 1978).

Another aphid, *Drepanosiphum platanoides*, which lives on *Acer pseudoplatanus*, shows variations in reproductive rate which are correlated with levels of soluble nitrogen in the leaves (figure 3.2). These levels are higher in spring when leaves are in active growth, and in autumn when dying leaves again contain a high level of nitrogen compounds (Dixon, 1970). In spring the petioles act as conduits for abundant nutrients while leaves are developing, and the aphids settle as close as possible to the petioles. In summer the petioles are not colonised, for they lose their value as a source of food. The aphids then settle on the leaf blades, as close as possible to the veins. A new migration occurs in autumn when the petiole once more becomes a rich source of food.

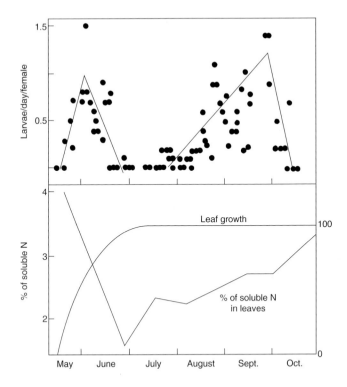

Figure 3.2.   *Variations in levels of amino acids (soluble nitrogen) in sycamore*
leaves and in the reproductive rate of the aphid Drepanosiphum
platanoides, *expressed by numbers of larvae produced per day*
and per female (Dixon, 1970).

Various ecological factors may cause tree stress which manifests itself in
particular by activation of nitrogen compounds in a soluble form. Broken
branches, fire, wounds caused by lightning, mineral deficiencies, pesticides
and atmospheric pollution are all stress agents (cf. chapter 6). Trees that are
stressed and enriched in soluble nitrogen are more often attacked by phytophagous insects than normal trees (White, 1984).

1.4.   The defensive role of secondary plant compounds

Resistance in plants may be defined as the combination of the features
that enable reduction of the possibilities of exploitation of a plant by an
insect (Beck, 1965). These features may be the nutritional value of the plant,
its anatomical or morphological characteristics, and the presence of toxic or
repellent secondary substances. A mixture of defensive characteristics
belonging to these three categories may occur in the same plant species. The
study of plant resistance to attacks by insects has been much elaborated with

a view to explaining certain peculiarities of insect populations. It has also been undertaken for a practical purpose, which is the search for resistant plants in order to avoid the use of insecticides in pest control.

Substances referred to as "secondary plant compounds" (or SPC) are considered to be chemical means of defence against insect attacks. At least 10,000 of these substances are known, of which 4,500 are alkaloids and 1,100 are terpenes. Some are compounds which act "qualitatively", and usually occur in tissues in low concentrations of less than 2% in dry weight. These are mostly alkaloids and certain nitrogen compounds found mainly in annual plants. Other compounds act "quantitatively", reducing the digestive powers of insects which ingest them, or act as repellents which prevent the insects from feeding. These are tannins (polyphenols), terpenes, and lignins which occur usually in high concentrations and are especially prevalent in woody plants. Tannins are classed in two categories according to their molecular weight and biological activity. Hydrolysable tannins of light molecular weight are biodegradable and effective as repellents to insects and other plant eaters. Condensed tannins are polymers whose molecular weight may reach 2,500 or more, and have the property of combining with proteins and inhibiting digestive enzymes.

The numbers of species of moth that exploit the leaves of the oak *Quercus robur* in England vary characteristically, with a maximum in May, a decrease in August, and a slight increase in September. These changes are accompanied by variations in the phenology of the various species and coincide with changes in the chemical composition of the leaves (figure 3.3).

Young leaves in spring are rich in water and proteins but contain few tannins; older leaves are tougher, poor in water and proteins, and richer in tannins (Feeny, 1970). The latter reduce the digestibility of the plants and especially that of their proteins (Bernays, 1983). Species whose caterpillars feed on older, tannin-rich leaves grow slowly despite the high temperatures of midsummer during which they develop. These are relatively rare species such as *Chimabacche fagella*, *Dasychira pudibunda* and *Phalera bucephala*. Nearly all the common species such as *Tortrix viridana* and *Operophtera brumata* eat the young leaves before they become charged with tannins; they grow quickly and develop in spring. A third category is represented by *Campaea margaritata* which grows slowly during its early stages in summer, hibernates as a young larva, and completes its growth rapidly on young leaves the following spring (figure 3.4).

The defensive role of SPCs is confirmed by observations made on trees other than oaks, and on other insects. The leaves of various clones of cultivated poplars contain varying amounts of polyphenols (figure 3.5). In clones containing high levels of polyphenols, mining larvae of *Phyllocnistis suffusella* grow more slowly and consume more vegetable matter, which explains why their galleries are longer than those of insects living on other cones less rich in polyphenols. In Scots pine, resistance to attacks on cones

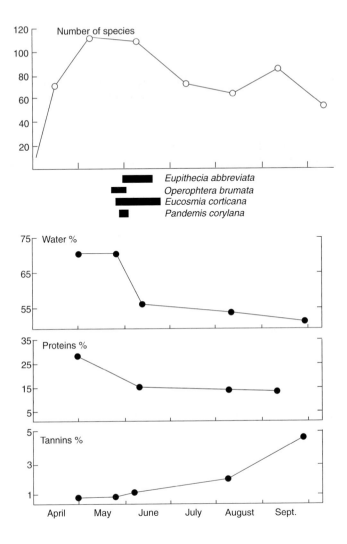

Figure 3.3   *Above: number of species of Lepidoptera exploiting oak leaves in a wood near Oxford. The period at which larvae go to ground to pupate is given for 4 species. Larvae feed during the period to the left of the black band and are absent in the period to the right of it. Below: changes in water, protein and tannin content in oak leaves (Feeny, 1970).*

by *Pissodes validirostris* is related to the composition of terpenes extracted from needles, particularly α-pinene which is the most abundant monoterpene. The infestations of cones decrease as the concentrations of α-pinene rise (Annila & Hiltunen, 1977).

The aphid *Elatobium abietinum* settles and prefers to feed on older needles of Sitka spruce, and shuns young needles; this appears to be due to

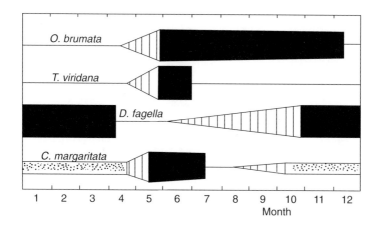

Figure 3.4. *Life cycles of four oak defoliating moths:* Tortrix viridana, Diurnea fagella, Operophtera brumata *and* Campaea margaritata. *Period of larval growth in vertical lines; pupal and imaginal period in black.* C. margaritata *has a long winter diapause as larva (stippling).*

the presence of volatile compounds (particularly monoterpenes such as myrcene) in the waxy layer covering the needles. These compounds, which are present in relatively high concentrations in young needles and are rare in old needles, probably act as repellents in high concentrations and as attractants in low concentrations (Jackson & Dickson, 1996). These differences in sensitivity to attacks by plant eaters are often genetic; trials have been made to use them for the selection of resistant clones.

The chemical composition of leaves may affect insects directly. Tannins in oak leaves reduce fertility in *Lymantria dispar* by 30%. It may also act indirectly; the activity of many parasitoids increases when their hosts' growth is slowed by poor quality nutrients (Rossiter *et al.*, 1988). On the other hand the pathogenic activity of viruses is reduced by leaf tannins (Keating & Yendol, 1987).

The theory according to which moths that feed on oak leaves feed in spring because the nutritional value of leaves decreases in summer due to increases in tannins and decreases in proteins has been contested (Bernays, 1981; Faeth, 1985). Observations on *Quercus emoryi* made in the United States have shown that variations in chemical composition do not follow the pattern described by Feeny. Levels of tannins and proteins are higher in damaged leaves than in intact ones. This suggests a reaction of the tree to attacks by plant eaters which has been demonstrated in the case of predation by *Lymantria dispar* (Schultze & Ricklefs, 1984).

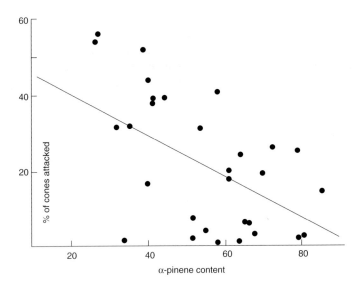

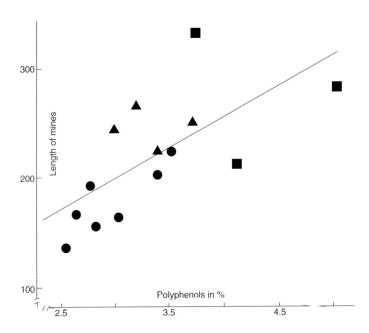

Figure 3.5.   *Resistance of trees to pests. Above: percentage of Scots pine cones attacked by* Pissodes validirostris *as a function of levels of α-pinene in the needles (Annila & Hiltunen, 1977).*
Below: length of galleries of Phyllocnistis suffusella *larvae as a function of levels of polyphenols in the leaves of different poplar clones. Squares:* Populus deltoides x nigra; *triangles:* Populus trichocarpa; *circles:* Populus trichocarpa x deltoides *(Nef, 1988).*

Diversity and abundance of moths are not dependent on the tannin levels in leaves or on their nutritional value (Karban & Ricklefs, 1984). In a study of mining insects on three species of oak in the United States Faeth *et al.* (1981) confirmed a decrease in the levels of nitrogen compounds in leaves during development, but also found an inverse relationship between the levels of nitrogen in leaves and the density of mining moths, which appears to contradict Feeny's concepts. Tannins in *Eucalyptus* leaves do not affect growth in the Australian chrysomelid beetle *Paropsis atomaria* (Fox & McCauley, 1977).

> Given what has been referred to as the "tannin imbroglio" what are the other possible explanations for Feeny's results which showed a greater abundance of defoliating moths in spring and, to a lesser extent, in autumn? Three hypotheses have been made. (a) Intense predation by birds in summer induces the insects to grow and feed in spring. (b) Larval growth and feeding in spring may be related to the thermal requirements of the caterpillars; excessively high temperatures in summer would inhibit development and / or lower the fertility of defoliating insects. (c) Most leaf eating insects concentrate larval growth and feeding in spring in order to avoid leaves previously damaged by other species and with less nutritional value. This hypothesis is confirmed by many observations. Larval growth of the geometrid moth *Oporinia autumnata* slows when they are fed with leaves that have previously been attacked by other insects. Female *Panolis flammea* are able to discriminate and lay fewer eggs on *Pinus contorta* trees which were defoliated the previous year than on unaffected trees. Caterpillar body weight, growth rate and survival rate are lower on trees that have been defoliated. These differences are due to variations in sap monoterpene profiles and to the abundance of tannins in the terminal shoots of pines (Leather *et al.*, 1987). Despite these various observations and critiques of Feeny's original theory, the defensive role of SPCs is a fact that is no longer disputed.

Does the intensity of attacks on leaves by insects differ in tropical and temperate forests? Various evaluations have shown that the average percentage of leaves consumed per year is equal to 10.9% of their surface in tropical forests, and only 7.5% in temperate forests. The difference is highly significant (Coley & Aide, 1991; Coley & Barone, 1996). It is due to more effective chemical defences which evolved as a result of more intense pressure of predation by herbivores in the tropics. Tropical plants also contain more polyphenols, more non absorbable fibres, less water and fewer nitrogen compounds. Leaf life-span is longer in the tropics, which favours predation by herbivorous insects, but levels of defensive substances such as tannins and fibres increase during the life of the leaf, and this limits the possibilities of predation by insects. The role of ants as protective agents of plants against herbivores is great in the tropics, where arboricolous ants are numerous, whereas it is virtually nonexistent in temperate regions. The true symbiosis which exists between several species of *Acacia* and certain ants is a classic example of this (Janzen, 1966); the defensive role of ants is confirmed by the absence of cyanogenic glucosides in the *Acacia* species that harbour ants (Rehr *et al.*, 1973).

Variations in the choice of host plants by phytophagous insects have been studied in Costa Rica. Hawk-moths (Sphingidae) tend to settle on plants that are poor in phenols but fairly rich in alkaloids as well as in other toxic molecules of low molecular weight. Saturniid moths appear to prefer plants that are rich in phenols, particularly certain tanins, and poor in toxic molecules such as alkaloids (Janzen & Waterman, 1984).

1.5. Interactions between endophytic fungi and mining Lepidoptera

Many micro-organisms live on the surfaces of leaves as well as within the leaf parenchyma. These are mainly endophytic fungi which have recently received attention owing to the effects they have on plant-insect relationships (Barbosa & Letourneau, 1988). Endophytic fungi have been found in all trees studied; they include ascomycetes and various micromycetes. Sixty four species have been found in beech; colonisation of leaves is by spores which may be dispersed by rain. Infestation by endophytes is heavier in leaves that have been mined by gracillariid moths of the genus *Cameraria* and that contain dead larvae than in mined leaves containing live larvae. The presence of larvae of mining Lepidoptera aids penetration of endophytic fungi by breaching the leaf surface and by allowing the fungal hyphae to develop. This type of interaction between plants, insects and micro-organisms was unknown until very recently, and studies have only recently begun (Faeth & Hammon, 1997).

1.6. Other defence strategies in trees

Certain morphological or anatomical characters serve as means of defence against phytophagous insects. Various characteristics of plant surfaces, such as wax, suberin or lignin impede access of insects to the nutritive elements. Lignified cell masses in the bark of Sitka spruce act by reducing the survival rates of larvae and density of egg laying of the bark beetle *Dendroctonus micans* (Wainhouse & Ashburner, 1996). Number, size and disposition of resin canals in pine needles play a part in resisting predation by *Rhyacionia buolina* and *Exoteleia pinifoliella*. High resin flux in resistant trees expels young larvae from their galleries and also acts as an irritant (Benett, 1954; Harris, 1960). The poor phenological coincidence between *Choristoneura fumiferana* and *Picea mariana*, the buds of which open too late, explains why this tree is seldom attacked, whereas *Picea glauca* and *Abies balsamea* are attacked (Blais, 1957).

Reactions of insects to SPCs vary. SPCs which were originally repellent or toxic may become attractants or even indispensable to insects specialised in the exploitation of a particular plant species. This phenomenon of coevolution between plants and insects is fairly widespread. Bark beetles of conifers use toxic monoterpenes in sap to locate host trees, and they convert the monoterpenes, by oxidation, to non-toxic compounds which they use as aggregation pheromones. In a further stage, these pheromones are used by the predators and parasitoids of bark beetles to locate their prey. In the case of *Dendroctonus micans* and its specific predator *Rhizophagus grandis* the pheromone emitted by the bark beetle stimulates egg-laying in *Rhizophagus* (Baisier *et al.*, 1988). Two volatile compounds, D-α-pinene and L-β-pinene act as egg-laying stimulants in *Choristoneura fumiferana* (Stadler, 1974).

1.7. Demes and physiological races

Genetic heterogeneity in trees favours the evolution of defence mechanisms which vary from one individual to another. Some insects react by forming small populations, or demes, each of which is adapted to a single tree. Deme formation is more frequent in species that live and feed inside the tissues of the host plant, for these species are subjected to greater selection pressure than those that live and feed outside. The presence of demes has been recorded in the scale insect *Nuculaspis californica* which lives on *Pinus ponderosa* (Edmunds & Alstad, 1981); specialisation is so narrow that the deme may become extinct with the death of the host tree, the insects being unable to colonise another tree. The tendency to reproduce descendants of insects on the same tree as the parent generation has also been recorded, no doubt with similar consequences, in *Matsucoccus feytaudi* which feeds on maritime pine in Provence (Riom & Fabre, 1977), and in *Oporinia autumnata* which lives on birch in Scandinavia (Ayres *et al.*, 1987). Race formation according to the phenology of the host tree is known in *Tortrix viridana* (Du Merle, 1981), and in the beech scale insect *Cryptococcus fagisuga* (Wainhouse & Howell, 1983). This evolutionary phenomenon which impedes insect dispersal is to be taken into account in sylviculture. The cloning of conifers, which has been considered with a view to obtain fast-growing trees, reduces genetic variability and possibilities of defence against insect infestations. Maritime pine shows variations in susceptibility to the scale insect *Matsucoccus* that appears to be linked to the geographic origin of trees. The bark of this pine has a structure which makes it more easily penetrated by the insects' styles in trees from various parts of the Mediterranean region than in trees from the Landes in South West France where the insect lives in small, benign populations (Schvester, 1988).

Deme formation leads to the evolution of races and even to sympatric species if the gene flux between populations is interrupted. Final stage larvae of *Zeiraphera diniana* belong to two distinct forms; those that live on larch are almost black, while those that live on Arolla pine are greyish yellow to orange. These two colour forms are independent of food, for caterpillars retain their characters if the host plant is changed. They represent two genetically distinct but mutually fertile forms, although natural hybrids are very rare, which indicates fairly strong sexual isolation. Various ecological characteristics separate the two forms: different hatching dates of larvae (due to adaptation to the different phenology of Arolla pine and larch), and de-synchronisation of all stages; different compositions of parasite complexes, and different sex pheromones. The pheromone of insects that live on Arolla pine is (E)-9-dodecenylacetate [or E9-12Ac], whereas that of individuals that develop on larch is (E)-11-tetradecenylacetate [or E11-14Ac]. The larch form has difficulty surviving on Arolla pine, whereas the pine one adapts well to larch. These are either two races which are distinct physiologically and genetically, and probably arose though selection induced by foliage characteristics of the two host trees, or even two distinct species which evolved through sympatric speciation (Emelianov *et al.*, 1995). Another example is that of *Choristoneura fumiferana*, which is known by two forms, one associated with fir and a few spruces, the other with various pines. The two forms constitute two sympatric populations which are sufficiently isolated to be raised to the rank of twin species separated by subtle morphological differences (Freeman, 1953).

Observations carried out in the Mont Ventoux region of the South of France have revealed two coexisting populations of *Tortrix viridana* which are genetically distinct but interfertile, each of which has adapted to two different species of oak with different phenologies. One population is adapted to downy oak, and its eggs hatch early; the other is adapted to evergreen oak, with late-hatching eggs. The phenological disjunction between the two populations remains throughout the life cycle; this ensures the necessary synchronism between hatching and unfolding of buds in the host plant. The two populations are probably further distinguished by the length of summer and winter diapause of the embryo (Du Merle, 1981).

The nature of SPCs in leaves plays an important part in determining individual and species richness in populations of phytophagous insects. Monophagy, which is common in Typhlocybinae, Homoptera that suck the mesophyll (the inner part of a leaf blade consisting of chlorophyll cells) proves this. Of 55 species which occur in Britain, 33 are associated with a single species of tree, 9 are associated with two species, and only 2 are highly polyphagous, feeding respectively on 16 and 17 tree species. The great majority of mining insects are also monophagous (Claridge & Wilson, 1982). In northern and central Europe about 70 species of chrysomelid beetle live on about thirty species of willow as well as on a few species of poplar, which, from a taxonomic point of view, are closely related to willows. These beetles do not feed indiscriminately on any species of willow. Four species of willow and four chrysomelids living on the trees were studied in Finland (Tahvanainen *et al.*, 1985). The nature, relative proportions and abundance of leaf phenylglycosides characteristic of these trees vary greatly from one species to another. The four chrysomelids *Phratora vitellinae*, *Plagiodera versicolor*, *Lochmaea capreae* and *Galerucella lineola* may feed on all four species of willow, but the populations they develop are very variable in abundance. The willow *Salix pentandra* is never much colonised;

Phratora vitullinae prefers *Salix nigricans* and *Plagiodera versicolor* prefers *Salix caprea*. As for *Lochmaea capreae* and *Galerucella lineola*, they also prefer *Salix caprea* but they may also feed on *Salix phylicifolia*, albeit less frequently.

1.8. Inducible defences

The term "inducible defences" designates preexisting but expanded defences, or new defences that arise in response to aggression produced by an outside agent termed an "elicitor". The elicitor may be an herbivorous animal or a pathogenic organism (Karban & Myers, 1989). Various agents are responsible for aggressions. They may be polysaccharides leaking from damaged cell membranes, or enzymes released by the plant or enzymes released by the herbivorous insect (Berryman, 1988; Lieutier & Berryman, 1988). The release of the polysaccharides or enzymes induces the plant to produce chemical defences that may act locally or may be carried by the sap and act at a distance, which explains general reactions observed in some plants (Edwards & Wratten, 1983).

Defence in conifers, and particularly in species of the genera *Pinus* and *Picea*, against bark beetles occurs in two stages: A preexistent defence system represented by the resin acts first, and an acquired defence system comes into operation following an attack by the scolytids. This system consists of a "hypersensitive" reaction which confines the attacker in a resin flux. It is efficient enough to turn many bark beetles into secondary pests which only exploit trees that are physiologically deficient or dead, but some species, particularly in the genera *Ips* and *Dendroctonus*, have become primary pests which infest trees on a massive scale and inoculate them with a pathogenic fungus of the genus *Ceratocystis* (= *Ophiostoma*) which weakens their defence system. The physiological mechanisms behind these defence reactions are little known. Reserves of glucosides such as starch are essential for the energy-costly production of terpenes in resin. Reductions in starch reserves in phloem following inoculation of spruce with *Ceratocystis* pathogens give a proof of this (Christiansen & Ericsson, 1986). Increased mortality in trees following infestations appears to be due to lack of glucoside reserves (Christiansen *et al.*, 1987; Langstrom *et al.*, 1992). A histological investigation has established the formation of resin in phloem and xylem cells of pines after an attack (Lieutier & Berryman, 1988).

During attacks by bark beetles the first relative increase is in monoterpenes which are normally rare and are the most toxic to herbivores (figure 3.6). In giant fir, *Abies grandis*, concentrations of monoterpenes rise from 2.4 mg/g before attacks by insects to 9.2 mg/g following attacks (Lerdau *et al.*, 1994), with the following percentages:

Monoterpene	% preceding attack	% following attack
Tricyclene	1.4	0.54
α-pinene	48.1	46.4
Camphene	1.2	0.0
β-pinene	42.1	30.5
Myrcene	4.5	14.7
Sabinene	2.2	0.0
δ-carene	0.3	2.8
Limonene	0.2	2.3
β-phellandrene	0.0	0.6
Terpinolene	0.2	1.0

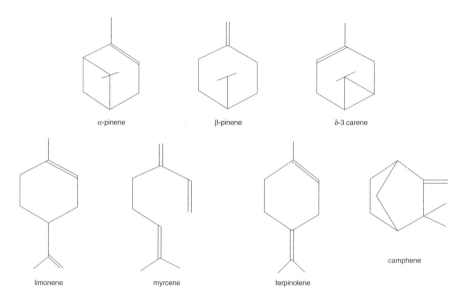

Figure 3.6. *Formulae of some monoterpenes*

1.8.1. *Hypersensitivity*

The term "hypersensitivity" was first used to describe the reaction of a plant to a pathogenic organism, and later reactions to other attacks. Hypersensitivity includes a range of morphological and histological changes which lead to necrosis of infected tissue as well as to inactivation or localisation of the infecting agent. Hypersensitivity is the first reaction during attacks by pathogenic fungi. It ends in the production of toxic substances called phytoalexins which impede growth of the pathogenic organism (Fernandes, 1990). Substances causing hypersensitivity are known in herbaceous plants; those in trees are probably the same. Chitosan, a constituent of cell membranes in many fungi, and a "protein inhibitor inducing factor", or

"PIIF", causes reactions very similar to those caused by pathogenic fungi such as *Ceratocystis*, which are carried by bark beetles, when they are inoculated into various conifers (Lieutier & Berryman, 1988). These reactions are particularly strong during infestations by bark beetles carrying pathogenic fungi; they also occur in some cases when trees are attacked by other phytophagous insects. Hypersensitive reactions occur in spruces attacked by the gall aphid *Sacchiphantes abietis* during early stages in gall formation. Affected cells plasmolyse and necrose and build up phenolic compounds. The necrosed area around the insect prevents the latter from reaching the soluble proteins it feeds on (Rohfritsch, 1988). Hypersensitivity has also been observed as a result of attacks by *Dreyfusia piceae* on firs (*Abies* spp.) and of attacks by *Sirex* on various conifers (cf. chapter 16).

Conifers, and pines in particular, have two defence mechanisms when they are attacked by bark beetles such as *Dendroctonus frontalis*. The first is a preexistent system which consists of a network of interconnected resin canals which are constantly filled with resin exuded under pressure when canal walls are breached. Resin is toxic or repellent to insects and microorganisms. The second is a system induced by the insect, consisting of hypersensitivity to the *Ceratocystis* fungus carried by bark beetles. The subcortical part of the trunk invaded by the fungus is filled with resin and surrounded by a reactive tissue formed by cambium. The reaction isolates the area invaded by the fungus from the rest of the tree. Resinous substances secreted inhibit the fungus, while a deficit in nutrient elements for the fungus occurs simultaneously (Raffa & Berryman, 1983). Injection of *Ceratocystis* into the trunk of *Pinus taeda* triggers a hypersensitive reaction which acts on the bark beetle *Dendroctonus frontalis* (Paine & Stephen, 1988); females of this species bore shorter galleries, lay fewer eggs and deposit fewer eggs per centimetre of gallery than in control trunks (table 3.2).

Tree		Gallery length (centimetres)	Number of eggs	Eggs per cm of gallery
Experiment 1	Control	26.50	63.44	2.44
	Induced reaction	10.94	10.94	0.91
Experiment 2	Control	9.59	24.50	2.53
	Induced reaction	1.73	0.82	0.22

Table 3.2 *Length of galleries bored by females of* Dendroctonus frontalis *on* Pinus taeda, *total numbers of eggs laid and number of eggs per centimetre of gallery in sub-cortical areas of normal (control) trees and trees inoculated with the fungus* Ceratocystis minor *(Paine & Stephen, 1988).*

1.8.2. Some examples of inducible defence

Larches defoliated by the larvae of *Zeiraphera diniana* during major out-breaks grow, in the following year, needles which are richer in fibres and with a lower content of nitrogen compounds. These leaves are less palatable and less nutritious; their composition produces a higher mortality rate in lar-vae and the surviving moths have low fertility (figure 3.7). This explains, at least in part, the collapse of populations following an outbreak. The acquired resistance of larch is long lasting; it disappears only after four or five years during which populations of the moth recover gradually (Baltensweiler *et al.*, 1977). The poor nutritional value of Austrian pine needles grown fol-lowing defoliation by the pine processionary caterpillar causes a change in sex ratio and lowered fertility, both of which contribute (together with other factors such as competition and scarcity of food) to the collapse of the population of this insect (Battisti, 1988).

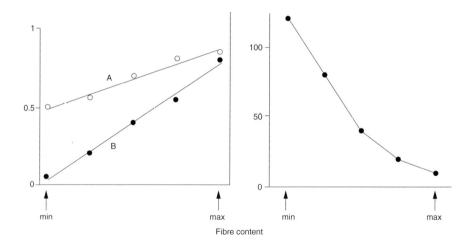

Figure 3.7.   *Effects of nutritional value of larch needles. Left: mortality rate of larvae of* Zeiraphera diniana *as a function of levels of non absorbable fibres in needles. A: larvae in stages L_1 and L_2. B: larvae in stages L_3 to L_5. Right: numbers of eggs laid by females as a function of fibre levels in needles eaten at the larval stage (Fischlin & Baltensweiler, 1979).*

A similar phenomenon has been described in birch during infestations by the larvae of *Epirrita autumnata*. The abundance of this insect fluctuates cyclically in parts of Scandinavia. In these parts birch shows a great intensi-fication of its chemical defence mechanisms for at least three years following infestation by the caterpillars. Phenol levels in leaves increase, while levels of absorbable nitrogen drop. A return to normal conditions occurs only in the fourth year following defoliation (Tuomi *et al.*, 1984). The fertility of moths which eat modified leaves at the larval stage may be reduced by more than 70% (Haukioja *et al.*, 1985).

Defence reaction of beech to the aphid *Phyllaphis fagi* occur in two phases. A change in C/N ratios due to a decrease in the level of amino acids and an increase in the level of monosaccharides occur in these leaves. Moreover procyanadins, which are compounds that inhibit digestion, appear and spread not only in the affected leaves but throughout the whole tree. In order to obtain sufficient quantities of amino acids the aphids must absorb greater amounts of sap, and reject excess sugars in the form of honeydew; this involves greater expense of energy and probably slows growth rate of the aphids (Gora *et al.*, 1994).

Infestation of terminal shoots of *Picea glauca* by the weevil *Pissodes strobi* provokes an inducible defence reaction. This response is triggered rapidly as soon as the insect has fed and laid eggs in the shoot. The response takes the form of a change in the cambium, which instead of normal parenchyma and tracheids, produces traumatic resin canals arranged in a circle in the growing xylem. These resin canals empty their contents into cavities formed by the weevils for egg laying, as well as in the larval cavities, killing both eggs and larvae. If the terminal shoot survives it reverts to producing normal xylem (Alfaro, 1995). Resistant trees have more resin canals than non resistant trees (Plank & Gerhold, 1965).

1.8.3. Leaf abscission: an acquired resistance mechanism?

Premature leaf abscission has been observed in several trees as a response to infestations of mining insects such as larvae of moths of the genus *Phyllonorycter* living on oak, buprestid beetle larvae, and *Pemphigus* gall aphids on poplar. Leaf abscission is considered to be an induced defence reaction which benefits the tree, as the fallen leaves desiccate, causing a high mortality rate in the larvae and a drop in numbers of the pest (Williams & Whitham, 1986). Moreover the fallen leaves enrich the soil with minerals useful to the tree (Owen, 1978). An opposite view holds that leaf abscission is a selective advantage to mining insects which thus evade their parasites (Kahn & Cornell, 1983). However Stiling & Simberloff (1989) point out that leaf abscission cannot be considered to be an acquired defence mechanism, for it usually occurs after emergence of the imaginal insects, or when the larvae have already succumbed due to interaction with other mortality factors. Premature leaf abscission would then be merely a consequence of lesions on leaves caused by the insects.

1.8.4. Effects of the phenology and age of trees

There must be a phenological coincidence between the period when the attacking stage of a phytophagous insect occurs and the period during which the plant can be exploited. Trees that have late developing buds are little damaged, for the young larvae are unable to feed and so starve to death. This phenomenon has been observed in North America in *Choristoneura fumiferana* which defoliates Douglas-fir and in *Operophtera brumata* which defoliates red oak, and in Europe in the spruce defoliator *Lymantria monacha*. The necessity for phenological coincidence between the development of larvae and that of oak trees is also known in *Tortrix viridana*.

There is great individual phenological variability in the oak *Quercus robur*, and each tree has its own characteristics which remain constant from year to year, particularly in the date at which buds unfold. These individual differences are accompanied by differences in their populations of phytophagous insects which are not apparent when the fauna of several trees is studied as a whole. Such differences affect the relative abundance of species and the specific composition of populations. The leaf-mining moth *Eriocrania subpurpurella* is more abundant in early-budding trees, while the aphid *Phylloxera quercus* and the gall wasps *Neuroterus quercusbaccarum* and *N. numismalis* are more abundant on late-budding trees. Individual variations in tree phenology are controlled by micro-climatic factors (trees that grow in colder environments develop later), but also by genetic factors (Crawley & Akhteruzzaman, 1988).

Changes in resistance to phytophagous insects occur as trees age. In the poplar *Populus angustifolia* resistance to the gall aphid *Pemphigus betae* diminishes with age; galls may be 70 times more abundant on old trees. The chrysomelid beetle *Chrysomela confluens* shows the inverse phenomenon, its density being 400 times greater on young trees than on old ones. These facts are important, as they enable us to explain certain variations in the structure of populations (Kearsley & Whitham, 1989).

Simulation of infestation by phytophagous insects by stripping a plant of its leaves often only approximates reality, for the effects of herbivorous insects on plants are not merely mechanical. During infestations by an herbivorous insect such as the moth *Spodoptera exigua*, maize emits a cocktail of volatile substances of the terpene group which attracts the parasitoid wasp *Cotesia marginiventris*. The signal which induces the plant to secrete the attractant is a molecule released by the feeding caterpillar. This molecule has been isolated and named volicitin; it is a derivative of glutamin with the function of a fatty acid provided by linoleic acid. Volicitin is a key element in a succession of chemical signals which regulate interactions between the three trophic levels, which are the plant, the herbivorous insect, and the herbivore's parasitoid. It is possible that such systems also exist in trees and their associated herbivorous insects (Farmer, 1997). The reaction of maize to the production of volicitin by the caterpillars is related to hypersensitivity.

1.9. Are there pheromone-type reactions between trees?

Communication between neighbouring trees by means of volatile compounds emitted by individuals attacked by insects is a recent hypothesis (Baldwin &

Schultz, 1983; Wratten & Edwards, 1984; Rhoades, 1985). Infestation of willow by larvae of *Malacosoma pluviale* leads to chemical changes, not only in the leaves eaten by the insects, but also in neighbouring intact trees. Caterpillars infest these trees less than they do trees further away. This can only be explained by supposing that affected trees emit a volatile messenger which acts at some distance as a hormone on uninfested trees. Ethylene, which is produced by damaged tissue in many plants and may affect synthesis of compounds such as tannins, is one of the possible chemical mediators. Another molecule that might allow communication between trees could be a volatile derivative of salicylic acid, which is already known as an agent secreted by plants infected by pathogenic micro-organisms (Shulaev *et al.*, 1997).

2. INSECT RELATIONSHIPS MODULATED BY PLANTS

Relationships between phytophagous insects on the one hand, and their predators and parasitoids on the other, may be indirect and modulated by characteristics of the plants. This type of interaction, which seems to be fairly widespread, is known as "interaction between three trophic levels", of which the following are a few examples. In North America the chermesid aphid *Dreyfusia piceae* feeds on the needles of the fir *Abies fraseri*; the chalcid wasp *Megastigmus specularis* destroys the cones of the same tree. Cones are smaller and the seeds lighter in trees infested by *Dreyfusia*; this facilitates exploitation of the cones by the chalcid whose females can reach the seeds more easily with their ovipositors. The level of infestation of the cones is on average 28% higher on trees infested by *Dreyfusia* than on healthy trees (Fedde, 1973).

Distribution of leaf mining caterpillars of the genus *Phyllonorycter* on oak is a function of the damage caused by defoliating caterpillars, especially those of *Tortrix viridana* and *Operophtera brumata*. The greater the damage caused by the two latter species, the lower the survival rate of *Phyllonorycter*. This competition between the two categories of moth is effected by chemical changes induced in oak leaves by the defoliating caterpillars (West, 1985). Hybrid poplars belonging to sensitive clones may harbour thousands of galls of the aphid *Pemphigus betae*, whereas resistant clones harbour very few or none. Trees devoid of galls harbour two or three times fewer arthropods, fungi and insectivorous birds than those with galls (Dickson & Whitham, 1996).

3. COMPETITION AND RESOURCE SHARING

The concept of inter- and intraspecific competition, long regarded as a phenomenon of little importance in determining the structures of herbivorous insect populations has been revived (Denno *et al.*, 1995; Stewart, 1996). As insects generally consume only a very limited part of available foliage, it was assumed that food could not be a limiting factor, that competition would not occur, and that population control must be regulated essentially by means of plant / insect relationships. Analysis of 104 plant / herbivorous insect systems has shown 193 pairs of potentially competing insects. In cases of large-scale or total defoliation, the quantity of available resources becomes a limiting factor, and intraspecific competition occurs. Such is the case in *Zeiraphera diniana* which destroys larch, in *Lymantria dispar* and *Operophtera brumata* which defoliate oak (Dempster, 1983), and in the pine processionary caterpillar (Geri, 1983). Competition may result in lowered fertility, changes in sex ratio, and dispersal of insects. In *Bupalus piniarius,* reared caterpillars produce pupae whose weight decreases with the density of caterpillars, and females that emerge from small pupae lay fewer eggs than large females (Klomp, 1966). Young larvae of *Lymantria dispar* disperse more readily by allowing themselves to be borne by wind when density is high. With aphids one sees development of winged individuals able to disperse to other plants when populations of wingless individuals become too abundant. Herbivorous insects may engage in competition to exploit a common limited resource such as elaborated sap. Competition may even make itself felt through the host plant, between aphids that live on the roots and gall aphids living on the leaves, without the species themselves coming into contact (Moran & Whitham, 1990).

The weevil *Balaninus elephas* and the moth *Laspeyresia splendana* both live in chestnut. There is an inverse relationship in abundance of the two species: the number of *Balaninus* decreases when that of *Laspeyresia* increases. The mechanism of this interaction is little known, but it may be a form of competition (Debouzie & Pallen, 1987).

Experiments (McClure & Price, 1975) have shown that competition leads to significant drops in numbers of offspring per female in various species of leafhopper belonging to the genus *Erythoneura* which live on the leaves of *Platanus orientalis* in North America. Competition occurs in population densities comparable to those found in nature. The limiting factor appears to be food quantity, and not the number of sites available for egg laying.

Insects which appear at different times of the year may compete if they exploit the same resource or trigger long delayed reactions in the plant. *Operophtera brumata* caterpillars which devour young oak leaves have a harmful effect on the species of leaf miner that appear later in the season (West, 1985). This is a case of asymmetrical competition in which only one of the species is harmful to the other.

Larval competition for food resources has been established in bark beetles. In *Ips typographus* spatial distribution of larval gallery systems is such that competition is reduced to a minimum (Light *et al.*, 1983; De Jong & Grijpma, 1986). Intraspecific competition lowers fertility in the females of *Ips typographus*. In *Tomicus piniperda* an increase in density of infestations results in a decrease in the number of galleries bored by females, in the average length of galleries, and in the number of eggs and larvae produced by each female. Decrease in individual fertility is however insufficient to prevent large increases in the total number of eggs per unit of surface area of bark, which leads to high mortality rates in larvae and a decrease in the size of imagines resulting from surviving larvae. This regulation of larval survival allows an almost constant number of insects emerging per unit of surface area whatever the density of the initial infestation (Sauvard, 1990). Infestation of trees by *Scolytus ventralis* is all the more successful when numbers of invading insects are highest, which shows that there is a measure of cooperation between the various individuals. Female fertility increases until a certain density of insects is reached. Competition then occurs and reduces fertility (Berryman, 1979). The result of these two processes is a characteristic productivity curve (figure 3.8).

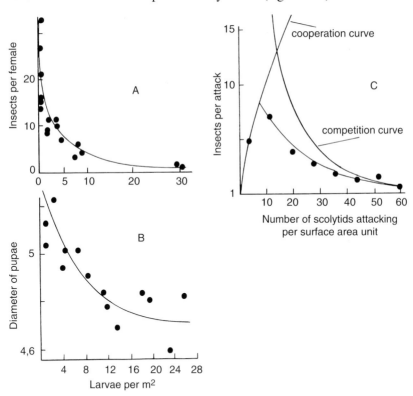

Figure 3.8 *Examples of competition in forest insects. A: number of offspring per female* Ips typographus *as a function of numbers of attacks per surface area unit. B: relationship between density of larvae h(individuals per m²) and size of pupae (diameter in mm) in* Bupalus piniarius *(Klomp, 1966). C: number of offspring of* Scolytus ventralis *per infestation as a function of surface area unit of bark. The productivity curve results from the interaction between the cooperation curve and the competition curve (Berryman, 1979).*

The relative importance of competition and predation varies with population density in species in which variations of abundance are great. In *Ips typographus* there is a latency phase during which spruce trees liable to infestation are few: predation, parasitism and intraspecific competition are low, but interspecific competition between larvae is intense and this maintains low population levels. When favourable circumstances (such as a storm which uproots many trees, or a drought) occur, an abundance of colonisable trees produces an expansion stage during which the population grows rapidly; intraspecific competition now takes the place of interspecific competition. The intensive phase corresponds to a massive colonisation of all available trees, including those that are in good physiological condition; intraspecific competition and predation now reach maximum intensity, which contributes to extinction of the outbreak (Thalenhorst, 1958).

When many species exploit the same resource, competition may be avoided by sharing resources and by different spatial distributions. Several species of *Neuroterus* cause lenticular galls on oak. *Neuroterus numismalis*, *N. lenticularis* and *N. laeviusculus* tend to settle on the same trees. Competition is reduced by the distribution pattern of galls. *N. numismalis* is more abundant near the top of the tree and on leaves situated on the periphery of the crown; *N. laeviusculus* settles mainly near the base and near the trunk, and *N. lenticularis* occupies intermediate situations. When the three species cohabit on the same leaf, the galls of *N. numismalis* are concentrated near the tip, those of *N. laeviusculus* around the base, and those of *N. lenticularis* in the centre (Askew, 1962). Beech leaves in Denmark are exploited by gall dwellers (three mites belonging to the genus *Aceria* and the three flies *Pegomyia fagicola*, *Mikiola fagi* and *Hartigiola annulipes*) and by leaf miners (the weevil *Rhynchaenus fagi* and the moths *Phyllonorycter maestingella*, *Stigmella hemargyrella* and *Stigmella tityrella*). Analysis of the position of galls and mines on leaves reveals a subtle but effective sharing of resources (figure 3.9). Among the acarian mites one species of *Aceria* forms

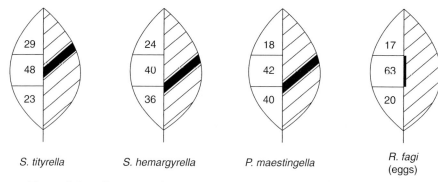

| S. tityrella | S. hemargyrella | P. maestingella | R. fagi (eggs) |

Figure 3.9.   *Resource sharing among three moths and a beetle that mine beech leaves. Location of mines (or eggs of Rhynchaenus fagi) is indicated by a black band and preferred areas are indicated on the left in % of total number of mines (Nielsen, 1978).*

galls preferably on the edge of the blade in its basal third; a second species settles along the basal half of the midrib, while a third species settles between two veins situated in the basal half of the leaf blade (Nielsen, 1978).

Distribution of various species of bark beetle on different parts of the same tree is one way of avoiding competition. Species which colonise a spruce tree spread to different levels which are functions of bark thickness and the body size of the insects. The largest species, *Dendroctonus micans* (5 to 9 mm), settles near the base, and the smallest, *Pityogenes chalcographus* (1.6 to 2.9 mm) occupies small branches. When only one species of scolytid exploits a spruce tree it occupies the whole of its *potential ecological niche*; when other species are present competition reduces available space; this defines the *real ecological niche* of the species. This is confirmed by the behaviour of *Ips typographus* and *Pityogenes chalcographus* installed on the same tree trunk. *I. typographus* may settle under bark as thin as 2.5 mm, but it is driven out of areas where the bark is less than 5 mm thick (figure 3.10).

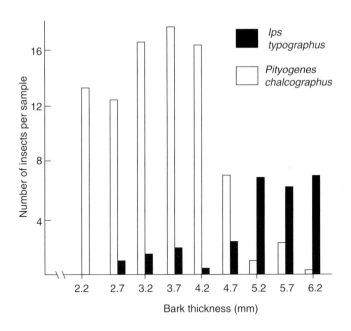

Figure 3.10. *Exploitation of spruce by scolytids according to bark thickness. Average number of* Ips typographus *(black columns) and* Pityogenes chalcographus *(white columns) emerging from spruce logs according to bark thickness (Grunwald, 1986).*

The scolytid *Dendroctonus frontalis* and the cerambycid *Monochamus titillator* compete, for galleries bored by the cerambycid disrupt those of the scolytid and cause high rates of larval mortality (Coulson *et al.*, 1976). Intensity of competition between *Ips typographus* and insects associated with it under bark has been estimated by means of exclusion experiments on certain species. The average mortality rate in *Ips typographus* is 85% in the presence of all associated species, and *Monochamus titillator* alone accounts for 51% of mortality (Miller, 1986).

Cases of facilitation have been recorded. The two aphids *Schizolachnus pineti* and *Eulachnus agilis* are closely associated in the exploitation of pine needles and buds. Thanks to this association *E. agilis* enjoys higher survival and growth rates, for it benefits from the increased nutritional value of pine needles attacked by *S. pineti* (Kidd *et al.*, 1985).

Congregation of aphids is a common occurrence. In the pine aphid *Schizolachnus pineti* generations of parthenogenetic females succeed each other in spring and in summer, and prefer to gather on the same needles:

Number of aphids per needle	Number of pine needles
0	38,400
1	25
2	11
3	7
4	6
5	7
6	13
7	8
8	23

The causes of congregation in aphids are little known: search for the most suitable feeding places, mutual attraction of individuals, or simply inertia of emergent insects which move only a little distance before settling down? Syrphid flies are major predators of these aphids. Analysis of predation shows that the number of aphids that avoid predation is proportional to the number of aphids that congregate in a single group. Group formation thus constitutes an effective protection against predators. The protective mechanism appears to be as follows: an aphid that is attacked struggles violently, colliding with others nearest to it in the group; this triggers withdrawal of the insects' styles from the plant and often the insects fall to the ground, which enables them to escape the predator (Kidd, 1982). Protection provided by group formation against attacks by predators has also been recorded in larvae of the sawfly *Neodiprion swainei* (Lyons, 1962).

4. THE ROLE OF PREDATORS AND PARASITES

Predators and parasites of insects in forest are very numerous (Blandin *et al.*, 1980; Dajoz, 1980). This chapter only presents a number of general data; various particular cases are given in later chapters.

4.1. Predators

The main predators of insects in forest are various vertebrates, and, among arthropods, insects, spiders and mites.

4.1.1. Spiders and mites

Spiders and mites (Acarina) are the main predators after insects, but their importance is still little known. A study of the impact of spiders in a tulip tree (*Liriodendron tulipifera*) wood showed that forest floor litter species consume 43.8% of the annual production of arthropods; small spiders (less than 1 mg) play a greater part than large ones (over 10 mg) (Moulder & Reichle, 1972).

The vertical distribution of spiders has been studied in Solling beech forest in Germany. Spiders are present at all levels, from forest floor litter to the canopy, and on the surfaces of tree trunks (Albert, 1976). The family Linyphidae is often dominant, both in the number of species and of individuals. The diet of some species is known. Diptera make up 80 to 88% of the prey of *Tetragnatha montana* (Dabrowska-Prot & Luczak, 1968). *Coelotes terrestris* has a more specialised diet and feeds mainly on beetles, and in particular on curculionids and elaterids, and on the defoliating moth *Chimabacche fagella* (Weidemann, 1976). It has been shown, by means of serological techniques, that in Canada many spiders eat the larvae of *Choristoneura fumiferana* (Morris, 1963), and that species belonging to the families Therididae and Salticidae are the most successful.

Predators among mites are mostly gamasids which feed mainly on organisms of the soil microfauna such as collembolans, proturans and pauropods.

4.1.2. Insects

Coleoptera are the most numerous predators in forest. The best studied are predators of scolytids and other insects that live under bark (cf. chapter 15). The family Carabidae is rich in species 5 most of which live on the ground, in forest floor litter, or in deep layers, at least in temperate regions. Tree-dwelling species are few; the most remarkable of these are *Calosoma*, with two common species, *C. sycophanta* and *C. inquisitor*, which are predators of defoliating moth caterpillars. Outbreaks of *Calosoma* occur quite often during outbreaks of processionary caterpillars. In the Paris region *C. inquisitor* is abundant in some years, and hunts the oak processionary caterpillar. In the forest of Porto Vecchio in Corsica *C. sycophanta* plays an essential role as predator in the natural extinction of populations of the gypsy moth which break out periodically every eight to ten years, in particular in 1953, 1962 and 1970, with gradations lasting three to four years. However in 1973 gradation spread to various parts of the island, and *Calosoma* became rare for unknown reasons. *Calosoma sycophanta* can fly several kilometres in search of its prey, which consists of about twenty species of Lepidoptera as well as cockchafer larvae. An adult with an active period of 50 days may devour between 235 and 336 gypsy moth caterpillars or pupae, and a third instar larva eats between 25 and 30 caterpillars. The offspring of one female *calosoma* may thus destroy the offspring of twenty female Gypsy moths in one year. Adults live up to four years, which is unusual in carabids. *Calosoma sycophanta* has been introduced to the United States to control the gypsy moth.

A few other tree-dwelling carabid predators may be mentioned. In France the genus *Lebia* includes several species. Adults of *Lebia scapularis* live mainly in the foliage of elm where they seek the eggs and larvae of *Galerucella luteola*. The first instar larva eats a *Galerucella* pupa and settles inside the pupal walls where it spins a cocoon. It then moults, and the second instar larva, which is immobile and bears reduced appendages, lives on its food reserves; pupation occurs thereafter. There are two generations per year. *Dromius* are found under the bark of various trees where they hibernate, and in the galleries of scolytids and other beetles. *Dromius quadrinotatus* hunts the larvae of the weevil *Pissodes notatus*, whereas *Dromius melanocephalus* seeks larvae of Diptera and Coleoptera in decayed wood, and *Dromius schneideri* hunts *Dendroctonus micans*. *Tachyta nana* is found under the bark of various deciduous and evergreen trees where it devours small living and dead arthropods as well as their droppings. The role of insect predators, and of carabids in particular, has been studied in Solling beech forest (cf. chapter 5).

The family Staphylinidae is richer in species than the Carabidae, but the biology of these insects is less well known. Terrestrial species are numerous in forest, particularly large species belonging to the genera *Ocypus*, *Quedius* and *Philonthus*. The best known species is certainly *Philonthus decorus* (Frank, 1967), a predator of the pupae of *Operophtera brumata* (cf. chapter 8.3.3). *Philonthus decorus* consumes various small arthropods in addition to *O. brumata* which is its main prey.

A certain number of arboreal staphylinids are predators of xylophagous insects. *Phloeonomus punctipennis* is found under the bark of various trees, and *Nudobius collaris* under the bark of conifers. *Baptolinus alternans* and *Placusa adscita* hunt scolytids in their galleries, in particular *Tomicus minor*.

The larvae of a number of beetles of the family Elateridae live in soil. These are mainly species of *Athous*, such as *Athous haemorrhoidalis*, which are common. In Solling beech forest the biomass of pterygote insects that emerged from soil in 1973 was 5 kg (dry weight) per hectare, and *A. haemorrhoidalis* accounted for 7% of the biomass (Thiede, 1977). Larvae of *A. haemorrhoidalis* feed on plant roots but they also attack the larvae of insects such as weevils, and practise cannibalism. Other elaterids live in dead wood. They are, at least partly, predators, their prey consisting mainly in beetle larvae. This category includes the genera *Ampedus*, *Melanotus*, *Lacon* and *Elater*.

The most remarkable member of the family Silphidae is *Xylodrepa quadri-punctata* which hunts in tree foliage for defoliating caterpillars as well as for sawfly larvae. The larvae of a few species of cantharid (particularly various *Cantharis* and *Rhagonycha* which live on the ground in woodland) are mainly abundant in winter and are active predators.

Many species of ladybird have an important role as predators of scale insects and aphids. Some are used to protect cultivated plants and in biological control. Their role in forest is less well known. Mention may be made of *Chilocorus renipustulatus* which is abundant in beech forests in the North of the Paris region where it preys on *Cryptococcus fagisuga*.

Predatory Diptera have a lesser role than Coleoptera. Robber flies (family Asilidae) include species with bristly bodies that capture their prey in flight with the aid of a rigid proboscis which is capable of perforating even such tough integuments as those of buprestids. Larvae of the subfamily Laphriinae have a mixed phytophagous and predatory diet consisting of the larvae of cerambycids and buprestids in the galleries they patrol (Melin, 1923). Most species of the family Cecidomyiidae are phytophagous, and often gall formers or leaf miners; a few are predators. This is the case in *Lestodiplosis* which preys on scale insects, and one species is a predator of the beech scale insect (cf. chapter 11). The most characteristic woodland representatives of the family Dolichopodidae belong to the genus *Medetera*, the larvae of which live at the expense of xylophagous insects, and particularly of beetles. A few species feed on dead insects or faeces, but the most interesting ones are predators, and might eventually be used in biological control. Adult *Medetera* are predators like their larvae, and are attracted to infested trees.

In the family Lonchaeidae a few species of *Lonchaea* prey on scolytids. The larvae of members of the family *Xylophagidae* are easily recognized by their conical, highly sclerotised and black-pigmented heads and bidentate anal plates. They reach a length of 2 cm and are carnivorous. Xylophagous species are common in the dead wood of various trees. The family Syrphidae includes important predators of aphids.

Forest Neuroptera belong to three main families: chrysopids, hemerobiids and coniopterigids. Chrysopids feed on aphids and play a part in regulating aphid populations that is on a par with coccinellids, syrphids and anthocorids. Chrysopid preferences for prey appear secondary; it is preference for habitat that seems to determine their abundance, as is shown by the fact that they do not follow aphids in their migrations from one plant to another. Some Neuroptera are associated with broadleaf woods and with shrubs; these are mostly hemerobiids such as *Hemerobius lutescens*, coniopterygids such as *Semiadalia aleyrodiformis*, and chrysopids such as *Chrysopa ventralis* or *Nineta vittata*. Other Neuroptera are associated with conifer woods; they are chrysopids such as *Chrysopa gracilis* or *Nineta pallida*, hemerobiids such as *Hemerobius pini*, and coniopterygids such as *Coniopteryx parthenia*.

Species such as *Chrysopa carnea* are found in all types of woodland. *Nineta pallida* is present wherever spruce grows, and *Chrysopa dorsalis* lives on Scots pine (Zeleny, 1978).

True bugs (Heteroptera) include predatory species belonging to various families. In the family Pentatomidae *Picromerus bidens* preys on tenthredinid sawfly larvae. Anthocorids include several species which may play a part in controlling pests. *Anthocoris nemoralis* is found on many trees, usually deciduous ones. This bug preys on various arthropods: psyllids, aphids, Thysanoptera, Lepidoptera and mites. On pine trees *Elatophilus nigricornis* is an active predator of scale insects; its role as enemy of *Matsucoccus feytaudi* has been studied (Biliotti & Riom, 1967). Also on pines, *Acompocoris pygmaeus* hunts aphids such as *Cinara pinicola* or *Schizolachnus pineti*. A sub-cortical species, *Xylocoris cursitans*, lives on dead or moribund deciduous trees as well as on pines. It preys on various arthropods such as scolytids, springtails and thrips. This species, which measures 2 to 2.5 mm, is dark brown to black as an adult, and its larvae are reddish, a common characteristic of anthocorids which makes them readily recognizable.

The most remarkable predatory Hymenoptera in forests are wood ants of the *Formica rufa* group; their use in biological control has been much studied (cf. chapter 6).

Data on abundance of predatory arthropods living in canopy are scarce. In the course of a study of *Choristoneura fumiferana* in Canada, Morris (1963) determined an average abundance of predatory arthropods in a stand of fir for the period 1950-1958 (these values are underestimated, as more agile predators such as wasps and dragonflies escaped). The following figures of abundance were obtained:

Arthropods	Numbers of individuals per square metre of foliage		
	June	July	August
Spiders	2.85	2.52	
Mites			2.73
Pentatomid bugs	0.01	0.03	
Other Heteroptera	0.04	0.04	
Neuroptera	0.09	0.04	
Syrphid flies	0.15	0.04	
Coccinellids	0.02	0.26	

Spiders are more abundant on spruce with a number of 5.18 individuals in June and 13.43 in July. Predation by arthropods during an outbreak of *Choristoneura* varied from 1 to 12% with a mean value below 10%. Predation on the moth's eggs is due entirely to mites, which are more numerous than spiders, and insects play a negligible part compared with spiders. Birds probably only eat pupae. In a one hectare stand of firs of average density in which foliage surface was estimated at 70,000 square metres, the population of arboreal spiders was around 187,500 individuals; to this figure must be added spiders that hunt on the ground or on herbaceous plants, which were not counted.

4.1.3. Vertebrates

A few small insectivorous mammals may play a considerable role in the functioning of the forest ecosystem. Predation of pupae of the gypsy moth by the American rodent *Peromyscus leucopus* prevents moth populations from rapidly reaching peaks of abundance. When the number of rodents falls, control of the moth is halted (Ostfeld *et al.*, 1996). The shrew *Sorex araneus* is an efficient predator of many insects. Its introduction to Newfoundland has made effective control of the sawfly *Pristiphora erichsoni* possible.

Foxes consume large quantities of insects, mainly beetles. The same is true of polecats and martens, and, to a lesser degree, of other mustelids. Hedgehogs and shrews also eat many beetles as well as worms. European forest bats feed exclusively on insects; details of their diets are poorly known. Various North American shrews such as *Blarina brevicauda* are major predators of gypsy moth caterpillars. In general, predation of insects by mammals is of pupae, which are on the ground, while birds eat the insects in their larval and adult stages, which are found in trees. The influence of these two groups of vertebrates on populations of the gypsy moth has been confirmed by exclusion experiments. The average number of ovipositions per surface aera unit was 14.6 in a control area, 96.5 in an area from which mammals were eliminated, and 128.8 in areas from which both mammals and birds were excluded (Campbell & Sloane, 1977).

Insectivorous forest birds are nearly all passerines. Predation by birds entails a degree of regulation of insect populations, but its actual extent is subject to discussion (Dicksen *et al.*, 1979). Birds make choices, and exert a degree of predation pressure on the commonest species. Certain large and highly visible species are spurned because they are unpalatable. When a new species of insect appears, it is at first little sought for, then it is caught in abundance, for birds adopt a "specific search image". The relationship between the density of an insect population and its importance in diet is not linear. In the case of the great tit, *Parus major*, which eats *Panolis flammea* and *Neodiprion sertifer*, the percentage of the former species in its diet stabilizes at about 30% when abundance increases, while percentage of the second species is maintained at about 10%. This may be explained by a certain "saturation" in the bird and its preference for a more varied diet (Tinbergen, 1960).

Birds can hardly prevent or stall outbreaks of insects, given that their fecundity is much lower than that of insects. However provision of artificial nesting boxes which increases populations of species such as tits shows that birds may play a significant role. Insectivorous birds of temperate forest consume mainly insect larvae (88% of diet on average), and particularly caterpillars (73% of diet). Next come sawflies and spiders, and far behind, ants and beetles. Thirty eight species of bird are recorded as feeding on gypsy moth caterpillars, despite the presence of urticant hairs on the caterpillar's body. The black headed tit is a major consumer, for it is very active, seeking its food from ground level to the canopy, and may exist in large populations (Smith & Lautenschlager, 1978).

Highly specialised birds such as woodpeckers have peculiar diets consisting of insects, such as ants, which are gathered on or near dead trees. In a forest in the Aude in France between 18th and 31st May a black woodpecker (*Dryocopus martius*) brought the following prey to its brood:

Undetermined ant larvae	968	Cerambycids, undetermined larvae	5
Undetermined ant pupae	1,438	*Arhopalus*, larvae	4
Camponotus ligniperda, adults	94	*Acanthocinus aedilis*, larvae	10
Formica rufa, adults	97	*Tomicus piniperda*, adults	12
Lasius niger, adults	1,526	*Pityogenes bidentatus*, adults	3
TOTAL HYMENOPTERA	4,123	Scolytid larvae	745
Undetermined molluscs	5	*Barypithes pellucidus*, adult	1
		Pterostichus larva	1
		Thanasimus formicarius (?), larvae	4
		Corticeus pini, adults	2
		Coleoptera larvae	4
		TOTAL COLEOPTERA	791

Birds reduce the abundance of spiders and insects that live in tree foliage. In an oak wood in Britain in 1950 and 1951 various species of tit destroyed between 0.9 and 3.2% of oak defoliating caterpillars and between 0.3 and 1.7% of those of *Operophtera brumata*. An attempt to quantify predation by two species of tit on spruce budworm in the Briançon area was carried out in sub-alpine and alpine zones (Le Louarn, 1980). Estimates of the number of caterpillars present and of caterpillars eaten show that percentages taken by the birds vary between 10 and 20% according to year and altitude. Tits are polyphagous predators which feed mainly on caterpillars, but also on other insects.

Predation by birds is selective: the average size of spiders living on branches protected by nets is consistently greater than that of spiders living on branches to which birds have access (Gunnarson, 1996). Birds have adapted to their prey by regulating the season during which they mate and feed their young according to the phenology of the insects they consume in greatest quantity. In the case of the tit *Parus major* a correlation exists between dates on which the first egg is laid and the period of maximum availability of *Operophtera brumata* caterpillars (Noordwijk *et al.*, 1995).

In North America the total number of insects collected on the oak *Quercus alba* protected from birds by cages is twice that of insects collected on unsheltered control trees. Protected trees lose about 25% of their foliage to insects, control trees lose 13%, and trees treated with insecticide 6%. These results show that birds can play a part in pest control, and they suggest that the decline in populations of insectivorous birds affecting North America might reduce productivity of forests, given the great abundance of defoliating insects whose numbers will be less and less controlled (Marquis & Whelan, 1994).

> Many birds consume caterpillars which often have cryptic colour patterns that enable them to go undetected by bird's eyesight. These caterpillars however leave characteristic traces of their attacks on plants. It has been shown experimentally, using tits maintained in captivity in large enclosures, that these birds quickly learn to recognize damage caused by caterpillars and to concentrate their search for food in the vicinity of damaged leaves. Caterpillars belonging to species preyed on by birds tend to minimize the extent of damage to the upper surfaces of leaves, and some go so far as to sever the petiole, so that the leaves fall to the ground, thus eliminating visible clues as to the location of caterpillars on the tree (Heinrich & Collins, 1983).

Selective predation by birds is the cause of considerable increase in the relative numbers of melanic forms in moths, particularly *Biston betularia*, in areas where tree trunks are blackened by atmospheric pollution, and of the virtual disappearance of pale forms which are highly visible on blackened trunks (Ford, 1972). More recent experiments (Howlett & Majerus, 1987) have also shown that birds find it more difficult to detect moths resting on the undersides of branches at their junction with the trunk than those resting on the trunk itself.

4.2. Parasites and parasitoids

True parasites of insects are various invertebrates such as nematode worms (whose roles are described in chapter 6). Parasitoids are insects mainly belonging to the orders Diptera and Hymenoptera.

The term parasitoid designates insects which are in some ways intermediate between predators and true parasites. Adult parasitoids live freely, while their larvae live as parasites dependent on a host which usually dies when the larva completes its development. Search for the host is left to the adult parasitoid, whereas predators seek their prey both at the larval and adult stage. Parasitoids live at the expense of a single host, and they often show more or less strict parasitic specificity. Predators attack and eat many prey, and are not usually confined to a single species of prey (exceptions do exist, but they are few).

Parasite complexes of insects that live on trees are usually richer in species than those of insects that live on herbaceous plants. A likely explanation is that trees and their phytophagous insects constitute a resource which is stable and in which prey are more easily detected and exploited by parasitoids than in ephemeral herbaceous environments in which they are less visible (Hawkins & Lawton, 1987). However a number of insects such as

Dreyfusia picea do not appear to be attacked by any parasitoids, whereas many defoliating species may have twenty or more parasitoids. This variation in number of parasitoids is partly linked to accessibility of prey, which explains why insects that live on roots are very little parasitised (Hawkins & Lawton, 1987). Parasitoids may be highly specific and only parasitise a single species (in only a very few cases), or they may be polyphagous and prey on a greater or lesser number of species (in most cases). In Europe *Operophtera brumata* has a suite of 26 parasitoids of which 16 are broadly polyphagous, 8 moderately so, and 2 only narrowly polyphagous.

4.2.1. Hymenoptera

The family Ichneumonidae is one of the richest in species of any family of animals, perhaps only surpassed by Curculionidae. It includes 5 to 8% of all known insects, and 9.6% of the British insect fauna; the French fauna includes over 3,000 species. All ichneumonids at the larval stage are parasites of other insects or, rarely, of spiders. The majority of hosts are caterpillars, sawfly larvae, and various beetles. Some ichneumonids are hyperparasites: *Gelis* species parasitise braconids and various ichneumonids. The host's appearance, habitat and behaviour are more important in the choice of species parasitised by ichneumonids than their taxonomic position. This explains why one species of ichneumonid may parasitise several species of insect.

The family Braconidae is close to Ichneumonidae, from which it is chiefly distinguished by wing venation. They are parasites of other insects. Several braconid larvae are often found in the body of a single host, either owing to multiple oviposition, or to polyembryony. The larvae of ichneumonids, on the other hand, are mostly solitary.

The superfamily Chalcidoidea is one of the largest in the order Hymenoptera, and it includes the smallest insects. Among the many families of Chalcidoidea, eurytomids, eupelmids and eulophids are particularly numerous.

4.2.2. Diptera

The Tachinidae constitute the main family of parasitic flies. Their larvae develop as endo- or ectoparasites of a great number of insects. Some tachinids are highly polyphagous: *Compsilura concinnata* parasitises at least 200 species belonging to three different orders of insect. Others may be confined to a single host or to a few closely related species. *Phryxe caudata* is associated with *Thaumetopœa pityocampa* and to the related species *T. wilkinsoni*. The most numerous hosts of tachinids are larvae of Lepidoptera and Coleoptera, followed by adult Coleoptera and Heteroptera. Some tachinids lay small eggs called microtype eggs (which vary in size

from 0.1 to 0.4 mm) which are deposited on soil or on plants and are ingested by the host. Others have larger eggs, called macrotypes (up to 1 mm in
size), which the females deposit on the host. The young larva which hatches
from it is a planidium which rapidly penetrates by breaking through the
host's body wall. Tachinids are solitary endoparasites, none of which is a
hyperparasite.

The multiplication potential of tachinids varies in inverse ratio to the
accessibility of hosts. Among those that are part of the parasite complex of
Lymantria dispar, the species *Blepharipa pratensis* lays its eggs on leaves,
and each female may lay up to 5,000 eggs. *Parasetigena silvestris* lays about
200 eggs on the bodies of caterpillars, while *Compsilura concinnata* deposits
90 to 110 larvae inside its host (Askew, 1971).

Bombyliids, commonly called bee-flies, are sun-lovers, and often have a
long proboscis and a rapid flight which makes them easily recognizable
when they hover around flowers. Their parasitic larvae pupate in the soil.
Villa brunnea is a bombyliid which parasitises the pine processionary caterpillar.

Flies are devoid of the strong ovipositor which exists in wasps. For this
reason they do not usually attack hosts which live in protected environments
such as galleries inside wood, leaf mines, or galls. Their hosts are more
varied than those of Hymenoptera, and besides insects, they also parasitise
gastropods, spiders, earthworms and myriapods.

5. SURVIVAL CURVES

Knowledge of abiotic and biotic mortality factors makes it possible to
draw survival curves. Analysis of available survival charts shows that there
are two main types of these (Price, 1975). The first type, which describes
low mortality rates (under 40%) in the early stages, is formed by a majority
of species which are protected by their way of life in galleries (in the case of
Scolytus scolytus and *S. ventralis*, which live under bark, and *Phyllonorycter
blancardella*, which is a leaf miner). The second type, characterised by
higher mortality rates (over 70%) in the early stages, includes species which
are unprotected, although some of them live inside buds. This group includes
defoliators such as *Bupalus piniarius*, *Operophtera brumata* and *Rhyacionia
buoliana* which lives in buds. The scale insect *Lepidosaphes ulmi* is unusual
in the high mortality rate of larvae in their mobile stage. Knowledge of survival curves may be useful in designing strategies for pest control, as it
enables detection of vulnerable stages (figure 3.11).

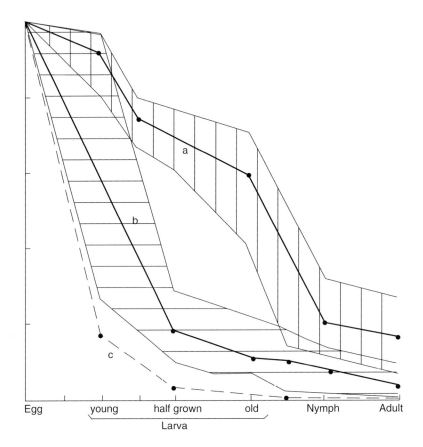

Figure 3.11. *General appearance of survival curves for 19 species of phytophagous insects. a: average curve for species which have low mortality rates in early stages. b: average curve for species which have high mortality rates in early stages. c: curve relating to the scale insect* Lepidosaphes ulmi. *Shaded areas indicate the zones to which curves in each group are limited (Price, 1975).*

REFERENCES

ALBERT R., (1976). Zusammensetzung und Vertikalverteilung der Spinnenfauna in Buchenwäldern des Solling. *Faun. - Okol. Mitt.*, **5**: 65-80.

ALFARO R., (1995). An induced defense reaction in white spruce to attack by the white pine weevil, *Pissodes strobi. Rev. Canad. Rech. For.*, **25**: 1,725-1,730.

ANNILA E., HILTUNEN R., (1977). Damage by *Pissodes validirostris* (Coleoptera, Curculionidae) studied in relation to monoterpene composition in Scots pine and lodgepole pine. *Ann. Ent. Fennici*, **43**: 87-92.

ASKEW R. R., (1962). The distribution of galls of *Neuroterus* (Hym: Cynipidae) on oak. *J. Anim. Ecol.*, **31**: 439-455.

ASKEW R. R., (1971). *Parasitic insects*. Heinemann, London.

AYRES M. P., SUOMELA J., MacLEAN S. F., (1987). Growth performance of *Epirrita autumnata* (Lepidoptera: Geometridae) on mountain birch: trees, broods and tree × brood interactions. *Œcologia*, **74**: 450-457.

BAISIER M., GRÉGOIRE J. C., DELINTE K., BONNARD O., The role of spruce monoterpene derivatives as oviposition stimuli for *Rhizophagus grandis*, a predator of the bark beetle, *Dendroctonus micans*. *In*: W.-J. MATTSON, J. LEVIEUX, C. BERNARD-DAGAN, (1988). *Mechanisms of woody plant defenses against insects*: *search for pattern*. Springer, Berlin, 359-368.

BALDWIN I. T., SCHULTZE J. C., (1983). Rapid changes in tree leaf chemistry induced by damage: evidence for communication between plants. *Science*, **221**: 277-279.

BALTENSWEILER W., BENTZ G., BOVEY P., DELUCCHI V., (1977). Dynamics of larch bud moth populations. *Ann. Rev. Ent.*, **22**: 79-100.

BARBOSA P., LETOURNEAU D. K., (1988). *Novel aspects of insect-plant interactions*. Wiley, New York.

BASSET Y., (1994). Palatability of tree foliage to chewing insects: a comparison between a temperate and tropical site. *Acta Œcologica*, **15**: 181-191.

BASSET Y., BURCKHARDT D., (1992). Abundance, species richness, host utilization and host specificity of insect folivores from a woodland site, with particular reference to host architecture. *Rev. suisse Zool.*, **99**: 771-791.

BATTISTI A., (1988). Host-plant relationships and population dynamics of the pine processionary caterpillar *Thaumetopœa pityocampa* (Denis et Schiffermüller). *J. Appl. Ent.*, **105**: 393-402.

BEAVER R. A., (1979). Host specificity of temperate and tropical animals. *Nature*, **281**: 139-141.

BECK S. D., (1965). Resistance of plants to insects. *Ann. Rev. Ent.*, **10**: 207-232.

BENETT W. H., (1954). The effect of needle structure upon the susceptibility of hosts of the pine needle miner (*Exoteleia pinifoliella* Chamb.). *Canad. Ent.*, **86**: 49-54.

BERENBAUM M., (1980). Adaptive significance of midgut pH in larval Lepidoptera. *Amer. Nat.*, **115**: 138-146.

BERNAYS E., (1981). Plant tannins and insect herbivores: an appraisal. *Ecol. Ent.*, **6**: 353-360.

BERNAYS E. A., Nitrogen in defence against insects. *In*: J.-A. LEE, S. MCNEILL, J.-H. RORISON, (1983). *Nitrogen as an ecological factor.* Blackwell, Oxford, 321-344.

BERRYMAN A. A., (1972). Resistance of conifers to invasion by bark beetle fungus associations. *Bioscience*, **22**: 589-602.

BERRYMAN A. A., (1979). Dynamics of bark beetle populations: analysis of dispersal and redistribution. *Mitt. Schweiz. Ent. Ges.*, **52**: 227-234.

BERRYMAN A. A., Towards a unified theory of plant defense. *In*: W.-J. MATTSON, J. LEVIEUX, C. BERNARD-DAGAN, (1988). *Mechanisms of woody plant defenses against insects. Search for pattern.* Springer, Berlin, 39-55.

BILIOTTI E., RIOM J., (1967). Faune corticole du pin maritime: *Elatophilus nigricornis* (Hem. Anthocoridae). *Ann. Soc. Ent. Fr.*, **3**: 1,103-1,108.

BLAIS J.-R., (1957). Some relationships of the spruce budworm *Choristoneura fumiferana* (Clem.) to black spruce *Picea mariana* (Münch) Voss. *Forest Chron.*, **33**: 364-372.

BLANDIN P., CHRISTOPHE T., GARAY I., GEOFFROY J.-J., Les arachnides et myriapodes prédateurs en forêt tempérée. *In*: P. PESSON (1980). *Actualités d'écologie forestière.* Gauthier-Villars, Paris, 477-506.

CHEW R. M., (1974). Consumers as regulators of ecosystems: an alternative to energetics. *Ohio J. Sci.*, **6**: 359-370.

CHRISTIANSEN E., ERICSSON A., (1986). Starch reserves in *Picea abies* in relation to defence reaction against a bark beetle transmitted blue-stain fungus, *Ceratocystis polonica. Canad. J. For. Res.*, **16**: 78-83.

CHRISTIANSEN E., WARING R. H., BERRYMAN A. A., (1987). Resistance of conifers to bark beetle attack: searching for general relationships. *Forest Ecology and Management*, **22**: 89-106.

CLARIDGE M. F., WILSON M. R., (1982). Insect herbivore guilds and species area relationships: leafminers on British trees. *Ecol. Ent.*, **7**: 19-30.

COLEY P. D., AIDE T. M., Comparison of herbivory and plant defenses in temperate and tropical broad-leaved forests. *In*: P. W. PRICE, T. M. LEWINSOHN, G. W. FERNANDES, W. B. BENSON, (1991). *Plant-animal interactions: evolutionary ecology in tropical and temperate regions.* John Wiley and Sons, New York, 25-49.

COLEY P. D., BARONE J. A., (1996). Herbivory and plant defenses in tropical forests. *Ann. Rev. Ecol. Syst.*, **27**: 305-335.

COULSON R. N., MAYYASI A. M., FOLTZ J. L., HAIN F. P., (1976). Interspecific competition between *Monochamus titillator* and *Dendroctonus frontalis. Environ. Ent.*, **5**: 235-247.

CRAWLEY M. J., AKHTERUZZAMAN M., (1988). Individual variation in the phenology of oak trees and its consequences for herbivorous insects. *Functional Ecol.*, **2**: 409-415.

DABROWSKA-PROT E., LUCZAK J., (1968). Studies on the incidence of mosquitoes in the food of *Tetragnatha montana* Simon and its food activity in the natural habitat. *Ekol. Pol.*, **16**: 843-853.

DAJOZ R., Les insectes prédateurs et leur rôle dans le milieu forestier. *In*: P. PESSON (1980). *Actualités d'écologie forestière*. Gauthier-Villars, Paris, 445-475.

DEBOUZIE D., PALLEN C., Spatial distribution of chesnut weevil *Balaninus* (= *Curculio*) *elephas* populations. *In*: V. LABEYRIE, G. FABRES, D. LACHAISE., (1987). *Insects-Plants*. W. Junk, Dordrecht, 77-83.

DE JONG M.-C.-M., GRIJPMA P., (1986). Competition between larvae of *Ips typographus*. *Ent. Exp. Appl.*, **41**: 121-133.

DEMPSTER J. P., (1983). The natural control of populations of butterflies and moths. *Biological Reviews*, **58**: 461-481.

DENNO R. F., MCCLURE M. S., OTT J. R., (1995). Interspecific interactions in phytophagous insects: competition re-examined and resurrected. *Ann. Rev. Ent.*, **40**: 297-331.

DICKSEN J. G., CONNER R.-N., FLEET R. R. *et al.*, (1979). *The role of insectivorous birds in forest ecosystems*. Academic Press, London.

DICKSON L. L., WHITHAM T. G., (1996). Genetically-based plant resistance traits affect arthropods, fungi and birds. *Œcologia*, **106**: 400-406.

DIXON A. F. G., Quality and availability of food for a sycamore aphid population. *In*: A. WATSON (1970). *Animal population in relation to their food resources*. Blackwell, Oxford, 271-287.

DU MERLE P., (1981). Variabilité génétique et adaptation à l'hôte chez *Tortrix viridana* L. (Lep. Tortricidae). *C. R. Ac. Sc.*, **292**: 519-522.

EDMUNDS G. F., ALSTAD D. N., Responses of black pineleaf scales to host plant variability. *In*: R. F. DENNO, H. DINGLE, (1981). *Insect life history patterns: habitat and geographic variation*. Springer, Berlin, 29-38.

EDWARDS P. J., WRATTEN D., (1983). Wound induced defenses in plants and their consequences for patterns of insect grazing. *Œcologia*, **59**: 88-93.

EMELIANOV I., MALLET J., BALTENSWEILER W., (1995). Genetic differentiation in *Zeiraphera diniana* (Lepidoptera: Tortricidae, the larch budmoth): polymorphism, host races or sibling species? *Heredity*, **75**: 416-424.

FAETH S. H., HAMMON K. E., (1997). Fungal endophytes in oak trees: long-term patterns of abundance and associations with leafminers. *Ecology*, **78**: 810-819.

FAETH S. H., MOPPER S., SIMBERLOFF D., (1981). Abundance and diversity of leaf-mining insects on three oak host species: effects of host-plant phenology and nitrogen content of leaves. *Oikos*, **37**: 238-251.

FARMER E. E., (1997). New fatty acid-based signals: a lesson from the plant world. *Science*, **276**: 912-913.

FEDDE G. F., (1973). Impact of the balsam wooly aphid (Homoptera: Phylloxeridae) on cones and seed production by infested fraser fir. *Canad. Ent.*, **105**: 673-680.

FEENY P., (1970). Seasonal changes in oak leaf tannins and nutrients as a cause of spring feeding by winter moth caterpillars. *Ecology*, **51**: 656-681.

FERNANDES G.-W., (1990). Hypersensitivity: a neglected plant resistance mechanism against herbivores. *Environ. Entomol.*, **19**: 1,173-1,182.

FISCHLIN A., BALTENSWEILER W., (1979). Systems analysis of the larch bud moth system. Part 1: the larch bud moth relationship. *Bull. Soc. Ent. Suisse*, **52**: 273-289.

FOX L. R., McCAULEY B. J., (1977). Insect grazing on *Eucalyptus* in response to variation in leaf tannins and nitrogen. *Œcologia*, **29**: 145-162.

FRAENKEL G.-S., (1959). The raison d'être of secondary plant substances. *Science*, **129**: 1,466-1,470.

FRANK J. H., (1967). The insect predators of the pupal stage of the winter moth, *Operophtera brumata* (L.) (Lepidoptera: Hydriomenidae). *J. Anim. Ecol.*, **36**: 375-389.

FREEMAN T. N., (1953). The spruce budworm, *Choristoneura fumiferana* (Clem.) and an allied new species on pine (Lepidoptera: Tortricidae). *Canad. Ent.*, **85**: 121-128.

GERI C., (1983). Répartition et évolution des populations de la processionnaire du pin, *Thaumetopœa pityocampa* Schiff. (Lep. Thaumetopœidae) dans les montagnes corses. I. Régimes d'apparition de l'insecte et dynamique des populations. *Acta Œcol. - Œcol. Appl.*, **4**: 247-268.

GOLLEY F. B., (1977). Insects as regulators of forest nutrient cycling. *Trop. Ecol.*, **18**: 116-123.

GORA V., KÖNIG J., LUNDERSTÄDT J., (1994). Physiological defence reactions of young beech trees (*Fagus sylvatica*) to attack by *Phyllaphis fagi*. *Forest Ecology and Management*, **70**: 245-254.

GRUNWALD M., (1986). Ecological segregation of bark beetles (Coleoptera, Scolytidae) of spruce. *J. Appl. Ent.*, **101**: 176-187.

GUNNARSON B., (1996). Bird predation and vegetation structure affecting spruce living-arthropods in a temperate forest. *J. Anim. Ecol.*, **65**: 389-397.

HANOVER J. W., (1975). Physiology of tree resistance to insects. *Ann. Rev. Ent.*, **20**: 75-95.

HARRIS P., (1960). Production of pine resin and its effect on survival of *Rhyacionia buoliana* (Schiff.). *Canad. J. Zool.*, **38**: 121-130.

HAUKIOJA E., NIEMELA P., (1979). Birch leaves as a resource for herbivores: seasonal occurrence of increased resistance in foliage after mechanical damage of adjacent leaves. *Œcologia*, **39**: 151-159.

HAUKIOJA E., SUOMELA J., NEUVONEN S., (1985). Long-term inducible resistance in birch foliage: triggering cues and efficacity of a defoliator. *Œcologia*, **65**: 363-369.

HAWKINS B. A., LAWTON J. H., (1987). Species richness for parasitoids of British phytophagous insects. *Nature*, **326**: 788-790.

HEINRICH B., COLLINS S. L, (1983). Caterpillar leaf damage, and the game of hide and seek with birds. *Ecology*, **64**: 592-602.

JACKSON D. L., DIXON A. F. G., (1996). Factors determining the distribution of the green spruce aphid, *Elatobium abietinum*, on young and mature needles of spruce. *Ecol. Ent.*, **21**: 358-364.

JANZEN D. H., (1966). Coevolution of mutualism between ants and acacias in Central America. *Evolution*, **20**: 249-275.

JANZEN D., WATERMAN P. G., (1984). A seasonal census of phenolics, fibre and alkaloids in foliage of forest trees in Costa Rica: some factors influencing their distribution and relation to host selection by Sphingidae and Saturniidae. *Biol. J. Linn. Soc.*, **21**: 439-454.

KAHN D. M., CORNELL H. V., (1983). Early leaf abscission and folivores: comments and considerations. *Am. Nat.*, **122**: 428-432.

KARBAN R., MYERS J. H., (1989). Induced plant responses to herbivory. *Ann. Rev. Ecol. Syst.*, **30**: 331-348.

KARBAN R., RICKLEFS R. E., (1984). Leaf traits and species richness and abundance of Lepidopteran larvae on deciduous trees in southern Ontario. *Oikos*, **43**: 165-170.

KEARSLEY M. J. C., WHITHAM T. G., (1989). Developmental changes in resistance to herbivory: implications for individuals and populations. *Ecology*, **70**: 422-434.

KEATING S. T., YENDOL W. G., (1987). Influence of selected host plants on gypsy moth (Lepidoptera: Lymantriidae) larval mortality caused by a baculovirus. *Envir. Ent.*, **16**: 459-462.

KIDD N. A. C., (1982). Predator avoidance as a result of aggregation in the grey pine aphid *Schizolachnus pineti*. *J. Anim. Ecol.*, **51**: 397-412.

KIDD N. A. C., LEWIS G. B., HOWELL C. A., (1985). An association between two species of pine aphid, *Schizolachnus pineti* and *Eulachnus agilis*. *Ecol. Ent.*, **10**: 427-432.

KIMMERER T. D., POTTER D. A., (1987). Nutritional quality of specific leaf tissues and selective feeding by a specialist leaf miner. *Œcologia*, **71**: 548-551.

KLOMP H., (1966). The dynamics of a field population of the pine looper, *Bupalus piniarius* L. (Lep. Geometridae). *Adv. Ecol. Res.*, **3**: 207-305.

LANGSTROM B., HELLQVIST C., ERICSSON A., GREF R., (1992). Induced defense reaction in Scots pine following stem attacks by *Tomicus piniperda*. *Ecography*, **15**: 318-327.

Leather S. R., Watt D., Forrest G. I., (1987). Insect-induced chemical changes in young lodgepole pine (*Pinus contorta*): the effect of previous defoliation on oviposition, growth and survival of the pine beauty moth, *Panolis flammea*. *Ecol. Ent.,* **12**: 275-281.

Le Louarn H., Les oiseaux prédateurs en forêt. *In*: P. Pesson (1980). *Actualités d'écologie forestière.* Gauthier-Villars, Paris, 389-406.

Lerdau M., Litvak M., Monson R., (1994). Plant chemical defense: monoterpenes and the growth-differentiation balance hypothesis. *TREE,* **9**: 58-62.

Lieutier F., Berryman A. A., Elicitation of defensive reactions in conifers. *In*: W.-J. Mattson, J. Levieux, C. Bernard-Dagan, (1988). *Mechanisms of woody plant defenses against insects: search for pattern.* Springer, Berlin, 313-319.

Lieutier F., Berryman A. A., (1988). Preliminary histological investigations of the defense reactions of three pines to *Ceratocystis clavigera* and two chemical elicitors. *Canad. J. For. Res.,* **18**: 1243-1247.

Light D. M., Birch M. C., Paine T. D., (1983). Laboratory study of intraspecific and interspecific competition within and between two sympatric bark beetle species, *Ips pini* and *I. paraconfusus. Zeit. ang. Ent.,* **96**: 233-241.

Lyons L. A., (1962). The effect of aggregation on egg and larval survival in *Neodiprion swainei* Meddl. (Hymenoptera: Diprionidae). *Canad. Ent.,* **94**: 49-58.

Marquis R.-J., Whelan C.-J., (1994). Insectivorous birds increase growth of white oak through consumption of leaf chewing insects. *Ecology,* **75**: 2,007-2,014.

Mattson W. J., (1980). Herbivory in relation to plant nitrogen content. *Ann. Rev. Ecol. Syst.,* **11**: 119-161.

McClure M. S., (1980a). Competition between exotic species: scale insects on hemlock. *Ecology,* **61**: 1,391-1,401.

McClure M. S., (1980b). Foliar nitrogen: a basis for host suitability for elongate hemlock scale, *Fiorina externa* (Homoptera: Diaspididae). *Ecology,* **61**: 72-79.

McClure M. S., Price P. W., (1975). Competition among sympatric *Erythroneura* leafhoppers (Homoptera: Cicadellidae) on American sycamore. *Ecology,* **56**: 1,388-1,397.

McNeill S., Southwood T. R. E., The role of nitrogen in the development of insect / plant relationships. *In*: J. B. Harborne, (1978). *Biochemical aspects of plant and animal coevolution. Phyto. Chem. Soc. Eur. Symp. Ser.,* **15**: 77-89.

Melin D., (1923). Contribution to the knowledge of the biology, metamorphosis and distribution of the Swedish Asilids. *Zool. Bidrag Uppsala,* **8**: 1-317.

MILLER M. C., (1986). Survival of within-tree *Ips calligraphus* (Col. Scolytidae): effect of insect associated. *Entomophaga*, **31**: 39-48.

MORAN N. A., WHITHAM T. G., (1990). Interspecific competition between root-feeding and leaf-galling aphids mediated by host-plant resistance. *Ecology*, **71**: 1 050-1 058.

MORRIS R. F., (1963). The dynamics of epidemic spruce budworm populations. *Mem. Ent. Soc. Canad.*, **31**:1-332.

MOULDER B. C., REICHLE D. E., (1972). Significance of spider predation in the energy dynamics of forest-floor arthropod communities. *Ecol. Monogr.*, **42**: 473-498.

NEF L., Interactions between the leaf miner *Phyllocnistis suffusella* and poplars. *In*: W.-J. MATTSON, J. LEVIEUX, C. BERNARD-DAGAN, (1988). *Mechanisms of woody plant defenses against insects: search for pattern.* Springer, Berlin, 239-251.

NIELSEN B.-O., (1978). Food resource partitioning in the beech leaf-feeding guild. *Ecol. Ent., 3*: 193-201.

NOORDWIJK A.-J.-VAN, McCleery R. H., PERRINS C. M., (1995). Selection for the timing of great tit breeding in relation to caterpillar growth and temperature. *J. Anim. Ecol.*, **64**: 451-458.

OHMART C. P., STEWART L. G., THOMAS J. R., (1985). Effects of nitrogen concentrations of *Eucalyptus blakelyi* foliage on the fecundity of *Paropsis atomaria* (Coleoptera: Chrysomelidae). *Œcologia*, **68**: 41-44.

OSTFELD R.-S., JONES C. G., WOLFF J. O., (1996). Of mice and mast. Ecological connections in eastern deciduous forests. *Bioscience*, **46**: 323-330.

OWEN D. F., (1978). The effect of a consumer, *Phytomyza ilicis*, on seasonal leaf-fall in the holly, *Ilex aquifolium*. *Oikos*, **31**: 268-271.

PAINE T. D., STEPHEN F. M., (1988). Induced defenses of loblolly pine, *Pinus taeda*: potential impact on *Dendroctonus frontalis* within-tree mortality. *Ent. Exp. Appl.*, **46**: 39-46.

PETERSON S. C., (1987). Communication of leaf suitability by gregarious eastern tent caterpillars (*Malacosoma americanum*). *Ecol. Ent.*, **54**: 156-159.

PLANK G.-H., GERHOLD H.-D., (1965). Evaluating host resistance to the white pine weevil, *Pissodes strobi*, using feeding preference tests. *Ann. Ent. Soc. Amer.*, **58**: 527-532.

PRICE P. W., (1975). *Insect ecology.* John Wiley and Sons, New York.

RAFFA K. F., BERRYMAN A. A., (1983). Physiological aspects of lodgepole pine wound responses of the southern pine beetle to a fungal symbiont of the mountain pine beetle *Dendroctonus ponderosae* (Coleoptera: Scolytidae). *Canad. Ent.*, **115**: 723-734.

READER P.-M., SOUTHWOOD T. R. E., (1981). The relationship between palatability to invertebrates and the successional status of a plant. *Œcologia*, **51**: 271-275.

REHR S. S., FEENY P.-P., JANZEN D.-H., (1973). Chemical defence in Central American non-ant-acacias. *J. Anim. Ecol.*, **42**: 405-416.

RHOADES D. F., (1985). Offensive defensive interactions between herbivores and plants: their relevance in herbivore population dynamics and ecological theory. *Amer. Nat.*, **125**: 205-238.

RIOM J., FABRE J.-F., (1977). Étude biologique et écologique de la Cochenille du pin maritime *Matsucoccus feytaudi* Ducasse 1942 dans le sud de la France. II. Régulation du cycle annuel, comportement des stades mobiles. *Ann. Zool. Écol. anim.*, **9**: 181-209.

ROBERT P. C., (1986). Les relations plantes-insectes phytophages chez les femelles pondeuses: le rôle des stimulus chimiques et physiques. Une mise au point bibliographique. *Agronomie*, **6**: 127-142.

ROHFRITSCH O., A resistance response of *Picea excelsa* to the aphid *Adelges abietis* (Homoptera: Aphidoidea). *In*: W.-J. MATTSON, J. LEVIEUX, C. BERNARD-DAGAN, (1988). *Mechanisms of woody plant defenses against insects: search for pattern*. Springer, Berlin, 253-266.

ROSSITER M. C., SCHULTZ J. C., BALDWIN I. T., (1988). Relationships among defoliation, red oak phenolics, and gypsy moth growth and reproduction. *Ecology*, **69**: 267-277.

SAUVARD D., (1990). Multiplication capacities of *Tomicus piniperda* L. (Col., Scolytidae). 1. Effects of attack density. *J. Appl. Ent.*, **108**: 164-181.

SCHOWALTER T.-D., (1981). Insect herbivore relationship to the state of the host plant: biotic regulation of ecosystem nutrient cycling through ecosystem succession. *Oikos*, **37**: 126-130.

SCHULTZ J. C., BALDWIN I. T., (1982). Oak leaf quality declines in response to defoliation by gypsy moth larvae. *Science*, **217**: 149-151.

SCHVESTER D., Variations in susceptibility of *Pinus pinaster* to *Matsucoccus feytaudi* (Homoptera: Margarodidae). *In*: W.-J. MATTSON, J. LEVIEUX, C. BERNARD-DAGAN, (1988). *Mechanisms of woody plant defenses against insects: search for pattern*. Springer, Berlin, 267-275.

SHULAEV V., SILVERMAN P., RASKIN I., (1997). Airborne signalling by methyl salicylate in plant pathogen resistance. *Nature*, **385**: 718-721.

SMEDLEY S. R., EISNER T., (1995). Sodium uptake by puddling in a moth. *Science*, **270**: 1,816-1,818.

SMITH H. R., LAUTENSCHLAGER R. A., (1978). *Predators of the gypsy moth*. USDA, Agricultural Handbook no. **534**: 1-74.

SOUTHWOOD T. R. E., BROWN V. K., READER P. M., (1986). Leaf palatability, life expectancy and herbivore damage. *Œcologia*, **70**: 544-548.

STADLER E., (1974). Host plant stimuli affecting oviposition behavior of the eastern spruce budworm. *Ent. Exp. Appl.*, **17**: 176-188.

STEWART A. J. A., (1996). Interspecific competition reinstated as an important force structuring insect herbivore communities. *TREE*, **11**: 233-234.

STILING P., SIMBERLOFF D., (1989). Leaf abscission : induced defense against pests or response to damage? *Oikos*, **55**: 43-49.

TAHVANAINEN J., JULKUNEN-TIITTO R., KETTUNEN J., (1985). Phenolic glycosides govern the food selection patterns of willow feeding leaf beetles. *Œcologia*, **67**: 52-56.

THALENHORST W., (1958). Grundzüge der populationsdynamik des grossen Fichtenborkenkäfer *Ips typographu*s L. *Schriftenreihe Forstl. Fak. Univ. Göttingen*, **21**, 126 p.

TINBERGEN L., (1960). The dynamics of insect and bird populations in pine woods. I. Factors influencing the intensity of production by songbirds. *Arch. Néerl. Zool.*, **13**: 265-343.

TUOMI J., NIEMELA E., SIREN S., NEUVONEN S., (1984). Nutrient stress: an explanation for plant anti-herbivore responses to defoliation. *Œcologia*, **61**: 208-210.

VARLEY G. C., The effects of grazing by animals on plant productivity. *In*: K. PETRUSEWICZ (1967). *Secondary productivity of terrestrial ecosystems*, 773-777.

WAINHOUSE D., ASHBURNER R., (1996). The influence of genetic and environmental factors on a quantitative defensive trait in spruce. *Funct. Ecol.*, **10**: 137-143.

WAINHOUSE D., HOWELL R. S., (1983). Intraspecific variation in beech scale populations and in susceptibility of their host *Fagus sylvatica. Ecol. Ent.*, **8**: 351-359.

WEIDEMANN G., (1976). Struktur der Zoozönose im Buchenwald Ökosystem des Solling. *Verh. Ges. Ökol. Göttingen*, 59-73.

WEST C., (1985). Factors underlying the late seasonal appearance of the lepidopterous leaf mining guild on oak. *Ecol. Ent.*, **10**: 111-120.

WHITE T. C. R., (1984). The abundance of invertebrate herbivores in relation to the availability of nitrogen in stressed food plants. *Œcologia*, **63**: 90-105.

WILLIAMS A. G., WHITHAM T. G., (1986). Premature leaf abscission : an induced plant defense against gall aphid. *Ecology*, **67**: 1,619-1,627.

WRATTEN S. D., EDWARDS P. J., (1984). Wound induced defences against insect grazing. *Antenna*, **8**: 26-29.

WRATTEN S. D., GODDARD P., EDWARDS P. J., (1981). British trees and insects: the role of palatability. *Am. Nat.*, **118**: 916-919.

ZELENY J., (1978). Les fluctuations spatiotemporelles des populations de Névroptères aphidiphages (Planipennia) comme élément indicateur de leur spécificité. *Ann. Zool. Écol. Anim.*, **10**: 359-366.

4

Diversity and abundance of insects in forest

Forest environments differ from open environments with herbaceous vegetation in their great structural heterogeneity in both the vertical and horizontal planes. Heterogeneity in the horizontal plane is especially marked in primary forest, where clearings alternate with plots in which trees have different ages and structures. Heterogeneity in the vertical plane takes the form of stratification which may be very marked when trees reach a great height. This stratification, creates microclimates and varied habitats which allow occupation by species which are often highly specialised.

1.   ECOLOGICAL DIVERSITY

Diversity of habitats and ways of life enables the distinction of groups of species, or guilds, which cohabit and exploit the same resources. The main guilds of forest insects are as follows:

– Canopy species live amidst the foliage trees high above the ground. From a biological standpoint this group is heterogeneous. It includes mainly leaf eaters dominated by caterpillars. To these may be added other insects such as Psocoptera, and other arthropods such as spiders, which may be considered true canopy animals, for they seek food other than leaves in the foliage.

– Meristem insects live either in buds (various microlepidoptera, Diptera, cecidomyiids) or in cambium. In the latter case, attacks often affect neighbouring tissues (phloem and xylem), and it is usual to class the insects in question as corticoles or subcorticoles.

WHAT IS BIODIVERSITY ?

The concept of biodiversity, or biological diversity, may apply to different levels of organisation of living organisms.

Species diversity relates to the number of species present in an area or in a particular biotope. In the case of some organisms such as insects, it is difficult or impossible to determine numbers; the number of insect species that exist even in a country such as France is still unknown, not to mention the tropics. Species diversity may also be evaluated by calculating a diversity index, which has the advantage of taking into account not only numbers of species but also abundance of different species in a population. Biodiversity is often evaluated by considering a taxonomic group that is easily sampled and whose species are easily identified. Whittaker (1977) has shown that diversity may be evaluated at the level of a site of limited area (α diversity), between neighbouring sites (β diversity) or on a regional level (γ diversity).

Genetic diversity, or intraspecific diversity, relates to genetic differences between individuals belonging to the same species. Most species have a high level of genetic polymorphism, which enables then to adapt to a changing environment. This is in contrast to cultivated plants and domestic animals which are often genetically impoverished.

Structural diversity, or diversity of ecosystems, is high in natural ecosystems such as unmanaged "primary" forests, and low in ecosystems which have been changed or created by man, such as agrosystems and managed forests. Structural diversity ensures stability in ecosystems and is a major element in the maintenance of species diversity at a local or regional level.

Biodiversity is now being considerably reduced on a global scale. The study of biodiversity and research into ways of conserving maximum biodiversity have become high priority subjects in the field of ecology.

– Feeders on roots are few; they mainly consist of beetle larvae.

– Sap suckers include Homoptera (scale insects, aphids), thrips and Heteroptera.

– Gall insects are to be found mainly among cynipid wasps, cecidomyiid flies and Homoptera, as well as a few beetles.

– Fruit and seed eaters (as well as eaters of flowers) are mainly weevils, microlepidoptera and a few Diptera.

– Strictly cortical species belong mainly to the families Scolytidae and Curculionidae and to the often rich fauna of commensals, predators and parasitoids associated with them. These two families include some species which are partly cortical and partly xylophagous, such as *Pissodes*. Strictly cortical species feed on phloem.

– Strictly xylophagous insects which feed on wood (or xylem) include many Coleoptera, a few Hymenoptera, Lepidoptera and Diptera, and, especially in the tropics, termites.

– Mycetophagous insects eat fungi.

– Rot holes in trees harbour a particular fauna which is sometimes treated as cavernicolous. These cavities may be dry or filled with water.

– Soil and litter fauna is composed mainly of Apterygota (Collembola), but also includes many Coleoptera and Diptera which are characteristic of the forest environment.

To these categories, which are not always well defined, should be added phytophagous insects of the herbaceous stratum. These various habitats are of course not isolated from each other. Relationships develop, particularly by means of migrations of species which hibernate in litter and return to the foliage in summer.

2. SPECIES DIVERSITY

Diversity of species and of structure in plants as well as species diversity in insects has been evaluated (by calculating Williams' α diversity index) in different stages of an ecological succession leading from herbaceous vegetation (pioneer stage) to forest (climax stage). Insect and plant diversity are low at the pioneer stage; they go through a maximum during an intermediate stage in the succession, and diminish slightly at the climax stage (Southwood *et al.*, 1979). Insect diversity remains greater than that of the plants, because the structural diversity of trees at the climax stage is greater than that of herbaceous plants at the pioneer stage (figure 4.1). Diversity in the larger fungi increases like that of insects during the succession. The structure of insect populations changes and their equitability increases.

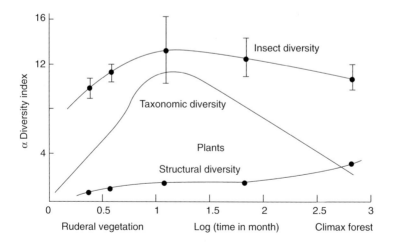

Figure 4.1. *Evolution of species diversity in plants and insects (Coleoptera and Heteroptera) evaluated by means of Williams' diversity index, and evolution of structural diversity of vegetation in the course of a succession which leads from the pioneer stage (ruderal vegetation) to the climax stage (forest) (Southwood* et al., *1979).*

A study conducted at medium altitude in the Swiss Alps relates to plants and butterflies in a succession leading from natural meadows which have been more or less modified to the climax spruce forest. The number of plants decreases fairly regularly throughout the succession; the number of species of butterfly increases slightly, then decreases to a minimum at the green alder stage, before increasing again at the climax stage (Ehrardt & Thomas, 1991).

Insect fauna is usually somewhat less rich in species at the climax stage than in intermediate stages of the succession. However, the climax stage includes species that are unique to this environment, therefore from the standpoint of biodiversity management conservation of tracts of climax forest is a necessity. The existence of this special fauna of old forests was established in the course of various research projects. A study by Niemala *et al.* (1993) was devoted to carabid beetles which live in plots of forest in the course of regeneration following clear felling, and in plots of non-exploited primary forest in Canada and Finland. Three categories of carabid beetles were found: (a) generalist species able to colonise plots of all ages – in Finland the two commonest species in this category are *Calathus micropterus* and *Pterostichus oblongopunctatus*; (b) species of open environments which mainly colonise stages of regeneration under 20 years old, and whose greatest abundance is in plots ten years old – these species belong mainly to the genera *Amara, Bembidion* and *Harpalus*; (c) rare species characteristic of primary forest such as *Cychrus caraboides* or *Agonum mannerheimi*; the latter, which is always scarce, is totally absent from plots less than 60 years old, and it is confined mainly to humid areas in non-exploited primary forest.

2.1. Effects of structural diversity of vegetation

When the structural diversity of vegetation increases, the number of insect species it harbours also increases. Comparisons of plants with similar distribution ranges show that monocotyledons, whose structures are simple, include fewer species than shrubby plants, and the latter have fewer representatives than trees (Strong & Levin, 1979). On cacti of the genus *Opuntia*, some species of which reach a height of five metres and have a complex structure, richness in insect species increases with structural complexity (Moran, 1980). Exceptions to this rule are few. Each species of oak in North America bears a number of cynipids as a function of the area occupied by the tree, but cynipids are just as numerous on bushy oaks with relatively simple structures as on large arborescent trees (Cornell & Washburn, 1979).

2.2. Effects of distributional range of trees

The relationships between forest area and richness in species have been interpreted using the model of island biogeography, which considers forests as islands (cf. chapter 2.5). Other hypotheses may be made. Two possible alternatives are the hypothesis of habitat heterogeneity (Strong *et al.*, 1984) and the hypothesis of passive colonisation of areas of variable extent (Connor & McCoy, 1979). A relationship between the distributional range of trees and the number of species they harbour has been established for all the trees and phytophagous insects of Britain, where the relationship between distribution of tree A and the number of phytophagous insects S takes the form $S = cA^z$ where c and z are constants (figure 4.2). At the level of groups of insects there is a relationship of the same kind between the number of insect species belonging to the leaf miner guild (which includes Lepidoptera, Hymenoptera and Coleoptera) and the extent of distributional range of their host trees (Claridge & Wilson, 1982). In North America there is an analogous relationship between the area occupied by the many species of pine and the number of species of bark beetle which each pine species harbours. (Sturgeon & Mitton, 1982)

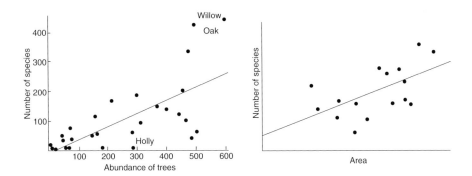

Figure 4.2.   Left: *relationship between the number of insect species associated with various British trees and estimated abundance of trees. Abundance is evaluated by counting the number of 10 km squares in which each tree is present. Note the particular position of yew and holly which are poorer in species than average, and the higher than average species richness of willow and oak (Kennedy & Southwood, 1984). Right: number of species of Scolytidae present on various pines in the western United States as a function of the distribution of trees (Sturgeon & Mitton, 1982).*

Equation $S = cA^z$ can be explained if one supposes that a tree species with a vast distributional range has more chances of receiving colonising insects than a tree having a more limited range, and that the tree may harbour larger insect populations which persist longer and have less chance of disappearing due to factors such as competition, natural enemies or the vagaries of climate.

The effect of surface area occupied by trees on species-richness of insects is shown by comparison of faunas associated with various trees in Britain and in Russia (table 4.1). In Russia where the dominant trees are conifers, these are two to three times richer in species than in Britain. Oak, which is much commoner in Britain than in Russia, harbours almost twice as many insects in Britain (Southwood, 1961). In Scandinavia, lime, hornbeam, elm, alder and hazel only occur as scarce isolated trees, and these species support a bark beetle fauna that is poor in species and often comprises very few individuals. Pine and spruce form vast populations and have a rich scolytid fauna. Birch is an exception: it has a poor fauna despite forming extensive populations. The numbers of scolytid species found on these trees are as follows: pine: 42; spruce: 34; lime: 8; elm: 8; hazel: 5; birch: 4.

Distribution of insects associated with various trees shows great variations: Lepidoptera is the family richest in species, followed by Coleoptera, at least in the British fauna, which is the best known (table 4.2).

Tree	Number of phytophagous insect species	
	Great Britain	Russia
Deciduous		
Oak (*Quercus*)	284	150
Willow (*Salix*)	266	147
Birch (*Betula*)	229	101
Hawthorn (*Crataegus*)	149	59
Poplar (*Populus*)	97	122
Apple (*Malus*)	93	77
Alder (*Alnus*)	90	63
Elm (*Ulmus*)	82	81
Hazel (*Corylus*)	73	26
Beech (*Fagus*)	94	79
Ash (*Fraxinus*)	41	41
Lime (*Tilia*)	31	37
Hornbeam (*Carpinus*)	28	53
Holly (*Ilex*)	7	8
Conifers		
Pine (*Pinus*)	91	190
Spruce (*Picea*)	37	117
Larch (*Larix*)	17	77
Fir (*Abies*)	16	42

Table 4.1.   *Number of phytophagous insect species living on different deciduous and coniferous trees in Great Britain and in Russia (Southwood, 1961).*

2.3. Effects of the age of a population

The length of time that a tree has been established may be estimated from discoveries of Quaternary fossils, or, in the case of planted species, from knowledge of the dates of their introduction. Faunal richness is a function of the length of time a tree has been established in the region. This has been shown for British trees and also for those of the Hawaiian islands. The longer the coexistence between trees and insects, the more coevolution occurs, allowing the establishment of greater numbers of species (Strong, 1974). Evergreen oak, introduced in England in 1580, and represented there by a small number of individuals, is only attacked by two species, whereas in the Mediterranean region its associated fauna is very rich.

Tree	Total	Heteroptera	Homoptera	Lepidoptera	Coleoptera
Quercus spp..	284	37	10	106 + 81	50
Betula spp.	229	12	4	94 + 84	35
Corylus avellana	73	16	2	18 + 28	9
Salix spp.	266	22	20	100 + 73	51
Alnus glutinosa	90	14	8	28 + 27	13
Crataegus spp.	149	17	1	64 + 53	14
Fraxinus excelsior	41	10	2	16 + 9	4
Pinus sylvestris	91	15	3	10 + 28	35
Ilex aquifolium	7	0	0	2 + 2	3
Taxus baccata	1	0	0	1 + 0	0
Prunus spinosa	109	4	2	48 + 43	12
Populus spp.	97	8	11	33 + 26	19
Ulmus spp.	84	11	4	33 + 26	10
Fagus sylvatica	64	4	3	24 + 16	17
Acer campestre	26	2	2	8 + 12	2
Carpinus betulus	28	1	0	7 + 16	4
Juniperus communis	20	6	0	4 + 8	2
Tilia spp.	31	7	2	15 + 5	2
Malus spp.	93	18	3	21 + 42	9
Sorbus aucuparia	28	0	1	2 + 17	8

Table 4.2.   *Distribution in different orders of insects living on various indigenous trees in Britain. The heading Homoptera includes only the sub-order Auchenorrhyncha and the superfamily Psylloidea. In the Lepidoptera column the first figure indicates the number of macrolepidoptera and the second figure the number of microlepidoptera (after Southwood, 1961).*

Some trees are exceptions to the rules stated above, and have faunas that are richer or poorer than the average value. Ash and rowan (*Sorbus aucuparia*) are very little attacked. The same is true of yew which contains substances analogous to the moulting hormone, phytoecdysones, which may play a repellent role. In holly the hardness of the wood appears to account for the low levels of attacks by insects. Oak and willow have larger than average faunas; this is certainly due, at least in willow, to the presence of substances such as derivatives of salicylic acid which are attractive to some insects, including various chrysomelid beetles. Few insects colonise both broadleaf and coniferous trees. This specificity is evident in moths and in sawflies as well as in scolytid bark beetles (table 4.3). Conifers harbour more species of scolytid than broadleaf trees, and some genera such as *Ips* are restricted to conifers. A very few species which live on broadleaf trees, such as *Anisandrus dispar* and *Xyleborus monographus* have occasionally been recorded on spruce. *Scolytus* species in Europe are restricted to broadleaf trees. Fifty three species of scolytid live on pines in France, while oaks and beech have only 19 species each.

Diet type	Oak	Pine
Defoliating Lepidoptera	*Tortrix viridana* *Operophtera brumata* *Lymantria dispar* *Thaumetopoea processionea*	*Rhyacionia buoliana* *Bupalus piniarius* *Lymantria monacha* *Thaumetopoea pityocampa*
Defoliating Coleoptera	*Melolontha vulgaris* *Attelabus nitens* *Orchestes quercus*	*Luperus pinicola*
Defoliating Hymenoptera	*Arge rustica*	*Diprion pini*
Sap feeders	*Lachnus roboris* *Stomaphis quercus*	*Cinara* spp.
Gall insects	*Biorrhiza* sp. *Cynips* sp. *Neuroterus* sp.	*Pineus pini* *Cecidomyia pini* *Petrova resinella*
Fruit and seed eaters	*Balaninus* sp. *Laspeyresia* sp.	*Pissodes validirostris*
Xylophagous scolytids	*Scolytus intricatus* *Platypus cylindrus* *Trypodendron lineatum*	*Ips typographus* *Ips sexdentatus* *Tomicus piniperda*
Xylophagous buprestids	*Agrilus viridis* *Coroebus* sp.	*Phaenops cyanea*
Xylophagous cerambycids	*Cerambyx cerdo* *Rhagium sycophanta*	*Ergates faber* *Rhagium inquisitor*
Xylophagous lucanids	*Lucanus cervus*	
Xylophagous Hymenoptera	–	*Sirex* sp.
Root eaters	Scarabaeidae Elateridae	Scarabaeideae Curculionidae Elateridae

Table 4.3.   *Comparison of the insect fauna of a broadleaf tree (oak), and a conifer (pine). Main associated species met with in the French fauna.*

Some insect guilds are particularly well represented in forests. Oaks in France harbour more than 200 species of gall insects. The moth *Tortrix viri-dana* is parasitised by 60 ichneumonids, 32 braconids, 23 chalcidians and 14 tachinids.

2.4. Speed of colonisation

Patterns of colonisation of introduced plants by native insects have been described by Strong (1974) and by Connor *et al.* (1980). Colonisation is rapid; the initial period of establishment of insects varies from 100 to 300 years where trees closely related to the introduced species are present in the area of introduction. *Eucalyptus*, which are remote from European trees from a botanical point of view, have a rich fauna of associated insects in their native Australia, but are almost devoid of insects in southern Europe, where they are introduced. *Nothofagus* are beeches from New Zealand or South America; they have been introduced to Britain, where they acquired a relatively rich fauna of defoliating moths and sap-feeding Homoptera of the sub-family Typhlocybinae associated with oak which is taxonomically close to *Nothofagus* (Welch, 1981).

Beetles associated with Scots pine in the forest of Fontainebleau form a community which has developed gradually since the tree was introduced in dense and extensive stands in 1786 on areas formerly occupied by heath and rocks (Méquignon, 1936). Progression of numbers of species was as follows:

Year	Total number of species	Number of new species in ten years
Before 1792	5	0.75
1802	8	6.19
1833	21	5.38
1846	28	3.33
1876	38	12.86
1890	56	11.0
1910	78	6.32
1929	90	25.00
1933	100	–

Speed of colonisation seems to have accelerated with time; this may be explained by the appearance of new habitats, such as old trees with thick bark, or dead trees. A certain time elapses before the appearance of predators which can only establish themselves when their hosts are already in place.

2.5. Effects of geographical latitude

The great richness of tropical faunas was noted by the first naturalist travelers as early as the nineteenth century. In an oft-quoted text dating from

1863, Bates notes that 700 species of butterfly are to be found "within an hour's walk" around the town of Manaus in Brazil, whereas Britain has only 66 species and Europe 321. This richness has been measured mainly using data relating to beetles, which are the most easily studied insects, and often the richest in species. The diversity gradient is also marked in termites (figure 4.3). On the American continent numbers of species of scolytid beetles in a 1,000 square kilometre area increase with lower latitudes in the following scale:

Country	Number of species	Species per 1,000 km^2
Alaska + Canada	179	0.016
United States	477	0.051
Mexico	605	0.307
Honduras	123	1.098
Guatemala	189	1.75
Panama	210	2.80
Costa Rica	393	7.70

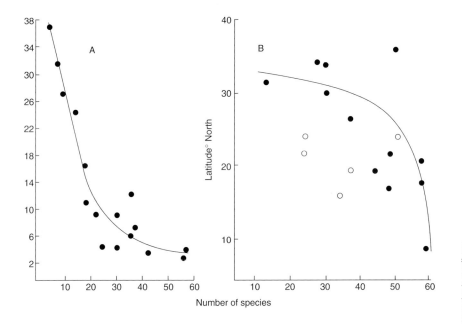

Figure 4.3.   *Biodiversity gradient as a function of latitude. A. The case of termites (Collins, 1983). B. The case of predators and commensals associated with the scolytids* Tomicus piniperda *and with various species of* Ips *living on conifers (black dots) and with scolytids living on broadleaf trees (circles). The fauna of broadleaf trees is poorer than that of conifers. The regression curve is drawn freehand, and only for the species on conifers (Dajoz, 1993).*

Added to this increase in biodiversity is a biological evolution characterised by an increase in mycetophagous species known as ambrosia beetles and of species in breeding, that is, reproduction between sibling individuals from a common family system (Kirkendall, 1993). In the tropics living organisms often have biological peculiarities that are rare or absent in temperate regions.

Numbers of families and species of beetle collected in a locality, and the corresponding diversity indices, are shown in the following table; localities are arranged in order of increasing latitude (the low values for France and Bohemia are abnormal, and correspond to very degraded or altered forests):

Region	Number of species	Number of families	α Diversity index
Sulawesi	1319	105	128
Borneo	859	?	359
Amazon	1080	57+	235
Australia	302	?	155
United States (Arizona)	329	40	73.3
Finland	195	33	41.1
England	117	?	23.6
Denmark	160	30+	21.9
France	98	30	18
Bohemia	68	21	15.5

These figures show that the average number of species of Coleoptera per locality is 171 in temperate forests and 890 in tropical forests, an average ratio of 5.2. Hammond (1992) also, by comparing a tropical forest in Sulawesi with a temperate forest in southern England, shows that there are approximately five times more species of Coleoptera in the former. In both these regions the ratio of numbers of species of Hemiptera is about the same as that for Coleoptera, while Lepidoptera and Hymenoptera are between two and four times as rich in species in the tropical forest. There is a ratio of between 7 and 8 in the number of species in boreal and in tropical forests (Hanski & Hammond, 1995). Given that temperature varies more or less with latitude, these results may be compared to those of MacArthur (1975), who showed that in animals as diverse as mammals, birds and gastropods there exists a relationship between species richness S and temperature T in the form:

$$S = A \ln (1 + B/T)$$

A and B being the constants.

Biodiversity, evaluated by means of α biodiversity index, in the case of nocturnal moths caught in light traps in a tropical forest in Malaysia and in a locality in England were compared. The differences are spectacular. Average diversity index for all families taken together is 30.9 in England and 301.3 in

Malaysia. Great differences are apparent at family level. Geometrids and noctuids have high diversity in England as well as in Malaysia. Pyralids are rare in England and numerous in Malaysia. The number of rare species is greater in Malaysia. Numbers of insects caught in a trap give indications of the levels of abundance and biomass of various species. Numbers of captures per trap per year are 29,000 in Malaysia and 2,600 in England, which suggests that the biomass of Lepidoptera is ten times greater in Malaysia (Barlow & Woiwod, 1989).

There are exceptions to the rule of increasing richness in species with lower latitudes. The most remarkable is that of the family Ichneumonidae, in which the peak of diversity in North America lies between latitudes 38° and 42° N (Janzen, 1981). Among other insect groups that are less rich in species in the tropics than in temperate regions are aphids, and probably, tachinids. The lower number of ichneumonids in the tropics is due to the lesser specificity of these insects, in which each species parasitises a greater numbers of insect species than in temperate regions (cf. chapter 21). In scolytid beetles, specialisation in a host tree is also less marked in tropical regions. There are on average three to five species of scolytid per species of tree in California and in France, and fewer than one species in Malaysia. The much greater number of tree species accounts for the greater richness in insect species in the tropics (Beaver, 1979).

3. BIODIVERSITY, ABUNDANCE, AND BODY SIZE

Relationships in a given habitat between numbers of species, body size, numerical abundance and geographical distribution have been researched in various groups of animals, and in particular in forest insects in various parts of the world (Morse *et al.*, 1988; Basset & Kitching, 1991; Gaston *et al.*, 1993).

3.1. Relationships between abundance and numbers of species

In all the types of forest that have been studied there are many rare species and a small number of abundant or very abundant species. The percentage of rare species, represented by single individuals in the samples is higher in tropical than in temperate forests. For Coleoptera, this percentage is 57% in Borneo, 35 to 45% in an Australian tropical forest and 65% in Costa Rica, whereas it is only 11 to 20% in temperate forests. In tropical forests, there are few very abundant species but many more in temperate forests. Among the 1,048 species of Lepidoptera represented by 9,461 individuals attracted to light in Malaysia, 416 species, that is to say almost 40%, are represented by single individuals and only 8 species by more than 127 individuals (Barlow and Woiwod, 1989). In a temperate forest of *Pinus radiata* in California, a single species is represented by 42% of collected individuals; in a boreal forest in Finland, the ten most abundant species represent 47% of all the individuals collected and a single species of Staphilinidae 20% (Sippola *et al.*, 1995). In a beech stand in Denmark ten species represent 85% to 90% of the beetles collected and two curculionid beetles represent respectively 57 and 21% of all the Coleoptera collected (Nielsen, 1975). In the Parisian region, in a small stand where 156 species of Coleoptera have been recorded, there are only 5 rare species and 24% of all the beetles collected are represented by the five most important species. In all cases when species are classified according to decreasing abundance the relation between the rank of species and their abundance N can be represented by the equation $N = a R^{-b}$. In the case of a Bornean forest $N = 470 R^{-0.9}$ and in an Austrian forest $N = 992.1 R^{-1.374}$.

3.2. Relationships between size and number of species

A general rule emerges: many species are small in size, and the average size of insects in the tropics appears to be smaller than that in temperate areas. In a Borneo forest, fogging samples provided about 3,800 arthropod species of which 859 were beetles. The latter included 384 species of leaf eaters, 200 species of predator, 120 detritivorous species and 141 myceto-phagous species. Species of small size were predominant (the most frequent size class was around 2 mm), as were rare species, about 500 species being represented by single individuals (Morse *et al.*, 1988).

A comparison of arboreal arthropod communities in England and in South Africa has shown that South African species are more numerous, especially among phytophagous insects such as curculionids, melolonthids and tenebrionids (Southwood *et al.*, 1982). In the Amazon the average size of canopy Coleoptera varies between 2.2 and 3.2 mm, and the size class richest in species is that between 3 and 4 mm. Seventy eight to eighty three percent of species are below 4 mm in length (Erwin, 1983). In Australia the most frequent size class in arboreal arthropods is that between 1.8 to 2.2 mm. In Indonesia the average size of canopy arthropods is 2.41 mm (Stork & Blackburn, 1993). In a forest in Finland the average size of beetles is 7.6 mm (Sippola *et al.*, 1995), and in three areas of Arizona (south-western United States) it varies between 10.9 and 13.2 mm.

Most species are of small size and are not abundant. Species of small size appear to be more widely distributed than species of large size (Gaston & Lawton, 1988).

3.3. Relationships between abundance and distribution range

A study relating to birds has shown that widely distributed species are, in general, locally abundant (Brown & Maurer, 1987). This rule also seems to apply, with perhaps a few exceptions, to insects.

4. SPECIES-RICHNESS OF FOREST INSECT FAUNA

No complete inventory exists of forest insect fauna, nor of invertebrates in general. In central European beech woods there are some 3,900 plant species and at least 6,800 animal species. 5,830 species of the latter are arthropods, including 560 arachnids, 60 myriapods and 5,210 insects, of which 100 are Apterygota, 820 Hemiptera, 1,550 Coleoptera, at least 700 Hymenoptera, 1,500 Lepidoptera and 1,000 Diptera. An inventory of the ichneumonids of a Polish forest revealed 680 species (Sawoniewicz, 1979).

Many temperate forests which have been highly modified by man now show only an impoverished reflection of the original fauna. Some primary forests still have a rich fauna. In the forest of Fontainebleau, which covers 20,000 hectares, records so far total 5,600 plant species (including many fungi) and 7,600 animal species, of which at least 6,000 species are insects. Only the beetles (3,200 species) and small orders such as Orthoptera may be considered to have been studied thoroughly. In the southern French beech forest of la Massane (Pyrénées-Orientales) there are 1,200 species of beetle in an area of 300 hectares. The many research projects that have been carried out in the 1,250 square kilometres of Bialowiesza forest in Poland have revealed 990 flowering plants, 254 bryophytes, 1,000 fungi and 11,000 species of animal, including 226 birds (of which 169 breed in the forest), 206 spiders and 8,500 insects. The beetles include 135 carabids, 250 staphylinids, 78 scolytids and 91 cerambycids; other insects include 234 species of Heteroptera, 107 butterflies, and 22 ants. The 150 hectares of protected forest at Monks Wood in England harbour 503 species of plant and 2,842 species of animal, including 149 vertebrates and 2,693 invertebrates.

A few Scandinavian forests still have rich insect faunas. Research carried out in Oulanka National Park in Finland has already resulted in a list of 820 species of Coleoptera, and the true number of species occurring there is certainly higher (Muona & Viramo, 1986). Saproxylic Coleoptera (those associated with dead wood) represent about 35% of beetles in the Finnish forest, between 20 and 25% in southern England, 30 to 40% in Arizona and the south-western United States, and 33% in a tropical forest in Sulawesi.

Thirty species of sawfly coexist in stands of larch in the Briançonnais (French Alps). In Belgium 615 species of Lepidoptera out of a total of 1,015 that have been recorded are forest insects. Entire groups, at genus or even subfamily level, are characteristic of forest. Notodontid, thaumetopoeid, lymantriid, nymphalid, lasiocampid, geometrid and noctuid moths are all composed mainly of forest species (Verstraeten, 1967). Oaks in Europe are colonised by 5,000 to 6,000 species of insect, and pines and spruce by 1,500 to 2,000 species.

An undoubtedly still incomplete inventory of the cork oak in Mamora forest in Morocco drawn up over several years revealed the existence of 350 species, of which 70 are leaf eaters, about 100 live in the crowns, another 10 live in the cork or in crevices of the bark, and about 50 are xylo-phagous (Villemant & Fraval, 1991). In a German forest 91 species of moth were found on oak, geometrids (18 species), noctuids (16 species) and tortri-cids (14 species) being the best represented families (Mœller & Lotz, 1968). In England 38 insects and mites are dependent on juniper, although this tree is fairly scarce, and more than 70 predators and parasitoids live on these plant eaters (Ward, 1977). A list of species associated with evergreen oak in Provence includes 101 Coleoptera, 38 Lepidoptera, 11 Hymenoptera, 11 Homoptera, 8 orthopteroids, 4 Thysanoptera and 4 Diptera (Favard, 1962). This list is very incomplete, since several of the insect orders are poorly known. Other inventories, all of them incomplete and often limited to pests or to very common species, have been made for other trees. In the Dolomites (northern Italy) alder harbours about a hundred species, most of them leaf eaters (Colpi & Masutti, 1984). A list of the main destroyers of Austrian black pine was compiled in northern Italy (Masutti, 1959). Beetles living on fir have been studied in southern Czechoslovakia (Pfeffer & Zumr, 1983), as well as those living on mountain pine in the Pyrénées (Dajoz, 1990). Saproxylic beetles in Sweden have been the subject of research by Palm (1951, 1959); those of mountain pine forest in Scotland were listed by Hunter (1977), and those of spruce in Finland by Saalas (1917-1923). A list of insect pests of pine was drawn up by Joly (1975).

5. NUMERICAL ABUNDANCE AND BIOMASS

Due to their small size, insects often go unnoticed in woodland except in cases of massive outbreaks. Their numerical abundance, however, fully com-pensates for their small size, and their biomass is often greater than that of vertebrates. In western European forests the average biomass of insects is estimated at 5 kg per hectare, and that of mammals and birds at 1.3 kg per hectare (Duvigneaud, 1974).

Equipment used in making quantitative samples aimed at indicating insect abundance is very diverse. It includes emergence traps (also known as "photoeclectors") consisting of four-sided boxes placed on the ground which collect the fauna that emerges from the soil. A frequently used type has an area of one square metre and a height of 0.5 m. The frame is topped by a pyramid of black cloth or metal sheeting with a container at its apex which collects insects drawn by the light. A container (known as a Barber trap) is buried in the soil at the centre of this emergence cage to trap flightless insects. The fauna of the aerial stratum is sampled by jarring the trunks of trees under which sheets have been spread. 'Fogging' (sending clouds of insecticide into the crowns of trees, which causes the insects to drop onto sheets laid on the ground) is another technique which is increasingly used. Insects which crawl on tree trunks are collected by means of funnels opening downwards with a flask acting as trap at their tops. These funnels are best used in groups of four so as to encircle the tree trunk. Tanks filled with water (which may be coloured), Malaise traps and revolving net traps are other methods used in sampling fauna on the wing.

A few, always partial, inventories produced by these methods give some idea of the abundance of forest insects (these data are enlarged on in later chapters). Emergence traps set on the ground have shown that in one hectare of the Ferage oak wood in Belgium abundance of Diptera varies in different plots from 5,687,000 to 6,620,000 individuals, having biomasses (dry weight) of 5,850 grams to 8,160 grams. The dominant families are Sciaridae, Empididae, Cecidomyiidae and Mycetophilidae (Krizelj, 1972). In this same oak wood in the spring of 1967 caterpillars formed a population of 1,100,000 individuals per hectare, having a biomass of 120 kg (fresh weight) or 37 kg (dry weight). This volume of caterpillars was produced in one and a half months on 390 kg of leaves (Verstraeten, 1972). In the same year the number of flying insects that circulated in this wood was estimated at 17 million per hectare, of which 94% were Diptera, 4% Hymenoptera and 1% Coleoptera, with a total biomass of 5 to 6 kg. The insect fauna on the wing at 20 metres above ground is remarkable for its density, which at certain seasons reaches nine times that measured in the first nine metres above ground level. Abundance varies greatly from one year to another: more than a million caterpillars per hectare were counted in 1963, but only 180,000 in 1967; three million sciarids per hectare were found in 1967, but only 300,000 in 1968.

In Ispina forest in Poland, the maximum biomass of caterpillars, mainly *Tortrix viridana*, in outbreak years amounts to 250 times that of birds, and their productivity is one hundred times higher. In the German oak and beech wood at Solling, abundance of winged insects was estimated at 4,123 individuals per square metre, with a biomass (dry weight) of 14.321 kg per hectare (table 4.4). The most abundant leaf-eating weevils in this forest are *Rhynchaenus fagi*, *Phyllobius argentatus* and *Strophosomus* sp. Their abundance is estimated at 296 individuals per square metre and their biomass at 279 mg (dry weight). The biomass of xylophagous Coleoptera in a non-exploited beech wood rich in dead wood is 7.8 kg per hectare. These beetles constitute at least 90 to 95% of the invertebrate biomass living in dead wood.

The richness of the tropical forest has been demonstrated by many research projects including those conducted on Seram in Indonesia. In one hectare of this forest, Stork (1988) estimates the number of arthropods to be 42.3 million, of which 23.7 million live in the soil, 6.0 million in litter, 0.1 million in the herbaceous stratum, 0.5 million on tree trunks, and 12.0 million in the canopy. Soil is by far the richest in number of individuals owing to the abundance of Collembola and mites which also dominate litter but are rare on trunks and especially in the canopy. Ants are numerous in the canopy and constitute a considerable part of the fauna of other habitats; they are more numerous than all other insects together (with the exception of

Collembola). Quantitative data obtained show the sparseness of tree trunk fauna, an astonishing absence of termites, and a fairly equal balance of most other insect orders.

Taxonomic group	Abundance (individuals per m^{-2})	Biomass (kg. hectare $^{-1}$ dry weight)
Diptera	3,028	3.742
Nematocera	2,961	3.424
Brachycera	67	0.318
Hymenoptera	82	0.571
Apocrites	79	0.162
Symphytes	3	0.409
Coleoptera	933	9.915
Lepidoptera	5	0.052
Thysanoptera	67	
Psocoptera	3	0.041
Homoptera	1	
Planipennia	4	
Total	4,123	14.321

Table 4.4. *Abundance and biomass of winged insects in the oak and beech wood of Solling, Germany (Thiele, 1977).*

REFERENCES

Barlow H. S., Woiwod I.-P.; (1989). Moth diversity of a tropical forest in peninsular Malaysia. *J. Trop. Ecol.*, **5**: 37-50.

Basset Y., Kitching R. L., (1991). Species number, species abundance and body length of arboreal arthropods associated with an Australian rainforest tree. *Ecol. Ent.*, **16**: 391-402.

Bates H. W., (1863). *The naturalist on the river Amazon.* Murray, London, 2nd edition.

Beaver R. A., (1979). Host specificity of temperate and tropical animals. *Nature*, **281**: 139-141.

Claridge M. F., Wilson M. R., (1982). Insect herbivore guilds and species-area relationships: leafminers on British trees. *Ecol. Ent.*, **7**: 19-30.

Collins N. M., Termite populations and their role in litter removal in Malaysian rain forests. *In*: S. L. Sutton, T. C.Whitmore, A. C. Chadwick (1983). *Tropical rain forest: ecology and management.* Blackwell, Oxford, 311-325.

COLPI C., MASUTTI L., (1984). Reperti sull'entomofauna epigea di popolamenti di *Alnus viridis* (Chaix) D.C. nel parco naturale di Paneveggio-Pale di S. Martino (Dolomiti trentine). *Studi Trentini Sci Nat.*, **61**: 197-237.

CONNOR E. F., FAETH S. H , SIMBERLOFF D., OPLER P. A., (1980). Taxonomic isolation and the accumulation of herbivorous insects: a comparison of introduced and native trees. *Ecol. Ent.*, **5**: 205-211.

CONNOR E. F, McCOY, (1979). The statistics and biology of the species-area relationship. *Amer. Nat.*, **113**: 791-833.

CORNELL H. V, WASHBURN J. O., (1979). Evolution of the richness area correlation for cynipid gall wasps on oak trees: a comparison of two geographic areas. *Evolution*, **33**: 257-274.

DAJOZ R., (1990). Coléoptères et Diptères du pin à crochets dans les Pyrénées-Orientales. Étude biogéographique et écologique. *L'Entomologiste*, **46**: 253-270.

DAJOZ R., (1993). Écologie et biogéographie des peuplements de Coléoptères associés à douze espèces de Scolytidae dans le sud de la Californie. *Bull. Soc. Ent. Fr.*, **98**: 409-424.

EHRARDT J., THOMAS J. A., Lepidoptera as indicators of change in the seminatural grasslands of lowland and upland Europe. *In*: N. M. COLLINS, J. A. THOMAS (1991). *The conservation of insects and their habitats*. Academic Press, London, 214-236.

FAVARD P., (1962). Contribution à l'étude de la faune entomologique du chêne vert en Provence. Thesis, Université Aix-Marseille.

FRANKLIN J.-F., Structural and functional diversity in temperate forests. *In*: E. O WILSON, F. M PETERS (1988). *Biodiversity*. National Academy Press, Washington, 166-175.

FREI-SULZER P., (1941). Erste Ergebnisse einer biocœnologischen Untersuchungen schweitzerischer Buchenwälder. *Ber. Schweiz. Bot. Gesells.*, **51**: 479-484.

HAMMOND P. M., Species diversity. *In*: B. GROOMBRIDGE, (1992). *Global biodiversity. Status of earth's living resources*. Chapman & Hall, London, 17-39.

HAMMOND P. M., (1994). Practical approaches to the estimation of the extent of biodiversity in species groups. *Phil. Trans. R. Soc. London*, B, **345**: 119-136.

HANSKI I., HAMMOND P. M., (1995). Biodiversity in boreal forests. *TREE*, **10**: 5-6.

HUNTER F. A., Ecology of pine wood beetles. In: R. G. H. BUNCE, J. N. R. JEFFERS (1977). *Native pinewoods of Scotland*. Institute of Terrestrial Ecology, Cambridge, 42-55.

JANZEN D. H., (1981). The peak in North American Ichneumonid species richness lies between 38° and 42° N. *Ecology*, **62**: 532-537.

Joly R., (1975). *Les insectes ennemis des pins.* Vol. 1, 222 pp. ; vol. 2, 40 pl. Nancy, ENGREF.

Kirkendall L. R., Ecology and evolution of biased sex ratios in bark and ambrosia beetles. *In*: D.-L. Wrensch, M.-A. Ebbert (1993). *Evolution and diversity of sex ratio in insects and mites.* Chapman & Hall, New York, 235-345.

Magurran A. E., (1988). *Ecological diversity and its measurement.* Princeton University Press.

Masutti L., (1959). Reperti sull'entomofaune del *Pinus nigra* Arn. var *austriaca* Hœss nelle Prealpi Giulie. *Accad. Ital. Sci. For.*, **8**: 263-308.

Méquignon A., (1936). Une biocénose en formation: les Coléoptères attachés au pin en forêt de Fontainebleau. *Trav. Nat. Vallée Loing*, **8** : 5-89.

Mœller J., Lotz G., (1968). Vergleichende Untersuchungen an der Kronenfauna der Eichen in Latenz - und Gradationsgebieten des Eichenwicklers. *Zeit. ang. Ent.*, **61**: 282-297.

Moran V. C., (1980). Interactions between phytophagous insects and their *Opuntia* hosts. *Ecol. Ent.*, **5**: 153-164.

Morse D. R., Stork N. E., Lawton J. H., (1988). Species number, species abundance and body length relationships of arboreal beetles in Bornean lowland rain forest trees. *Ecol. Ent.*, **13**: 25-37.

Muona J., Viramo J., (1986). The Coleoptera of the Koillismaa area (Ks), North-East Finland. *Oulanka Rep.*, **6**: 1-51.

Niemela J., Spence J. R., Langor *et al.*, Logging and boreal ground-beetle assemblages on two continents: implications for conservation. *In*: K. J. H. Gaston, T. R. New, M. J Samways (1993). *Perspectives on insect conservation.* Intercept Ltd., Andover, 29-50.

Noss R. F., (1990). Indicators for monitoring biodiversity: a hierarchical approach. *Cons. Biol.*, **4**: 355-364.

Palm T., (1951). Die Holz und Rindenkäfer der nordschwedischen Laubbäume. *Medd. Fran. Stat. Skogforkninginstitut*, **40**: 1-242.

Palm T., (1959). Die Holz und Rindenkäfer der sud und mittelschwedischen Laubbäume. *Opus. Ent.*, suppl. **16**: 1-374.

Pfeffer A., Zumr V., (1983). Communities of Coleoptera on the silver fir (*Abies alba*). *Acta. Ent. Bohem.*, **80**: 401-412.

Saalas U., (1917-1923). Die Fichtenkäfer Finlands. *Ann. Acad. Sc. Helsinki*, ser. A, **8**: 546 pp. and **23**: 746 pp.

Shreeve T. G, Mason C. F., (1980). The number of butterfly species in woodlands. *Œcologia*, **45**: 414-418.

Silbaugh J. M., Betters D. R., (1995). Quantitative biodiversity measures applied to forest management. *Environ. Rev.*, **3**: 277-285.

SOUTHWOOD T. R. E., (1961). The number of species of insect associated with various trees. *J. Anim. Ecol.*, **30**: 1-8.

SOUTHWOOD T. R. E., BROWN V. K., READER P. M., (1979). The relation-ships of plant and insect diversities in succession. *Biol. J. Lin. Soc.*, **12**: 327-348.

SOUTHWOOD T. R. E., MORAN V. C., KENNEDY C. E. J., (1982). The richness, abundance and biomass of the arthropod communities on trees. *J. Anim. Ecol.*, **51**: 635-649.

STORK N. E., (1988). Insect diversity: facts, fiction and speculation. *Biol. J. Lin. Soc.*, **35**: 321-337.

STRONG D. R., (1974). The insects of British trees: community equilibrium in ecological time. *Ann. Missouri bot. Garden*, **61**: 692-701.

STRONG D. R, LAWTON J. H, SOUTHWOOD T. R. E., (1984). *Insects on plants. Community, patterns and mechanisms.* Blackwell, Oxford.

STRONG D. R, LEVIN D. A., (1979). Species richness of plant parasites and growth form of their hosts. *Amer. Nat.*, **114**: 1-22.

STURGEON K. B, MITTON J. B., Evolution of bark beetle communities. *In*: B. B MITTON, K. B STURGEON (1982). *Bark beetles in North American conifers.* University of Texas Press, Austin, 350-384.

VERSTRAETEN C., (1967). Introduction à l'étude des Lépidoptères dans l'écosystème forêt en Belgique. *Bull. Recherches Agron. Gembloux*, **2**: 181-197.

VILLEMANT C., FRAVAL A., (1991). *La faune du chêne liège.* Actes Éditions, Rabat.

WARD L. K., (1977). The conservation of juniper: The associated fauna with special reference to southern England. *J.Appl. Ecol.*, **14**: 121-135.

WELCH R. C., (1981). The insect fauna of *Nothofagus. Institute of Terrestrial Ecology, Annual Report* 1980: 50-53.

WHITTAKER R. H., (1977). Evolution of species diversity in land communities. *Evol. Biol.*, **10**: 1-67.

◇ **5** ◇

Insects and the forest ecosystem

Because of their abundance and diversity, insects play a major role in the functioning of the forest ecosystem. In order to understand this role it is necessary to determine the abundance, biomass and productivity of species as a whole, the quantity of plant tissue consumed, as well as the situation of the various species in trophic networks. These studies have only been undertaken recently, and for a small number of forest ecosystems.

Insects play many roles in the forest ecosystem:

(a) Pest species, during outbreaks, may kill trees, or slow their growth (cf. chapter 6).

(b) By consuming a proportion of flowers, fruits and seeds insects affect population dynamics and may jeopardise the regeneration of trees. Fruit- and seed-eaters are more numerous in tropical than in temperate forests. Among the main insects in this category are beetles belonging to the families Bruchidae, Curculionidae and Scolytidae, bugs of the families Lygaeidae and Pyrrhocoridae, and various moth caterpillars. These insects may almost entirely destroy seeds fallen under a tree that have not been dispersed by mammals and birds. This predation is less intense on seeds that have been carried a distance, and this explains, at least in part, the high dispersal rate of various trees as well as the great richness in tree species of tropical forests (Janzen, 1970). This effect of insects on the dispersal and species-richness of trees appears to be generally insignificant or non-existent in temperate forests, in which populations are composed of only one or a few species of tree, and seed-eating insects are less numerous. Nevertheless, some seed-eating insects may interfere with tree regeneration.

(c) Insects may accelerate or retard ecological successions and the evolution of ecosystems by modifying the intensity of competition between plant species or by eliminating a species (Schowalter, 1981). In forests of the eastern United States predation by *Malacosoma americanum* reduces numbers of *Prunus pennsylvanica*. In the Alps *Zeiraphera diniana* attacks young Arolla pines growing under cover of larch trees. Most of the these pines succumb to predation by other insects, such as *Pissodes notatus* or various scolytids, which are exploiters of enfeebled trees. *Zeiraphera* thus contributes to retarding the establishment of Arolla pine, which is the normal climax stage species in the region (Baltensweiler, 1975). In Canada a mixed population of *Picea mariana* and *Larix laricina* was turned into a monospecific *Picea mariana* wood due to a high mortality rate of *Larix* and lack of recruitment caused by depredations by the larch sawfly *Pristophora erichsoni* (Jardon & Filion, 1994). Anatomical studies on larch attacked by *Pristophora* showed changes in structure of the wood, including decrease in the size of tracheids and an increase in summer wood (Cloutier & Filion, 1991).

Defoliating insects reduce the capacity for interspecific competition in certain trees and shrubs of the Chilean matorral, and lead to their exclusion from plant communities. Species heavily predated by insects are few, natural selection favouring those species which are little predated (Fuentes *et al.*, 1981).

(d) The role of predatory insects in forest is little known, but their impact may be intense (cf. chapter 3.4). In a German beech wood about 70% of the weevils *Phyllobius argentatus* and *Polydrusus undatus* which pupate in the soil are killed by various predators when the adults return to the tree foliage in spring (Ellenberg *et al.*, 1986). Population control of the winter moth (*Cheimatobia*) is due to well studied predators and parasitoids (cf. chapter 8). Carabid beetles have a considerable impact on populations of the insects they consume (Weidemann, 1972). The role of wood ants of the *Formica rufa* group has been demonstrated in Europe as well as in Canada (cf. chapter 6).

1. THE ROLE OF DEFOLIATING INSECTS

Quantitative data on the role of insects in the functioning of forest ecosystems are as yet few. Leaf-eating insects are those that have been most studied; a few example will illustrate their importance.

Determination of consumption and production of leaf-eaters can only be done by simultaneous use of suitable sampling methods and laboratory measurements. In order to estimate quantities of vegetable matter consumed and production one must know the species' life cycles and the duration of their various stages, their biomass, and respiratory losses and assimilation. One method consists in establishing an energetics table for an individual at different growth stages. A table can then be established for the whole of the population if the numbers of individuals at different stages is known.

Such energetics tables have been established for two moths, *Operophtera fagata* and *O. brumata* which cohabit in a broadleaf wood in Sweden

(Axelsson *et al.*, 1975). Measurements made on caterpillars fed on oak leaves provide an individual energetics table. An individual *O. brumata* consumes 115.5 mg (dry weight) of leaves in the course of its growth. The various elements of the energetics table are as follows (in calories):

	Leaves consumed (mg dry weight)	Energy equivalent (calories)
Consumption C	115.5	540.4
Non-utilised NU	79.3	313.4
Production P	19.8	114.2
Respiration R	–	227.0
Assimilation A	36.1	141.7

Except during major outbreaks, the number of females of these two species is 0.0054 per square metre, that of young larvae 0.77, and that of mature larvae 0.04. The total energetics balance for the whole population of the two species represents a consumption of 22.0 cal per square metre. Consumption amounts to only 0.004% of leaf production, which contrasts sharply with years of major outbreaks, when defoliation is almost total.

An energetics table has been drawn at individual and population level for the whole of root- and leaf-eating curculionids in a beech wood (cf. chapter 10).

1.1. Quantity of foliage consumed

During outbreaks, defoliating insects may consume the entire available foliage. In normal times they do not generally consume more than 10% of annual leaf production, and possibly less than 8%, although precise estimates are difficult to obtain (Bray, 1961; Carlisle *et al.*, 1966; Rafes, 1971; Reichle *et al.*, 1973; Larsson & Tenow, 1980; Larsson & Wiren, 1982; Ohmart *et al.*, 1983; Wiegert & Petersen, 1983; Schowalter *et al.*, 1986 etc.). An average value of 3.5% has been proposed by Wiegert & Petersen (1983). Consumption of Scots pine foliage in a Swedish forest by the whole leaf-eating population (12 species of moth, 11 sawflies and a few beetles) only amounts to 2.5% of annual production, or 0.7% of all available foliage (figure 5.1) for a density of defoliating insects estimated at 56,600 moths and 19,200 sawflies (Larsson & Tenow, 1980). Consumption by leaf-eaters on maple, ash and beech in southern Canada amounts to 5.0 to 7.7% of annual leaf production, and consumption on *Liriodendron tulipifera* in the eastern United States amounts to 5.6%. These figures show that quantity of available food is seldom a limiting factor in abundance of insect populations

The quantity of food ingested each day by many tree leaf-eaters ranges between 50 and 150% of their body weight. Insects that feed on sap have higher rates of consumption, ranging between 100 and 1,000 times their body weight (Mattson, 1980). An arboreal cricket, *Oecanthus*, in a population of *Liriodendron tulipifera* consumes an average of 7.35 mg per day (dry

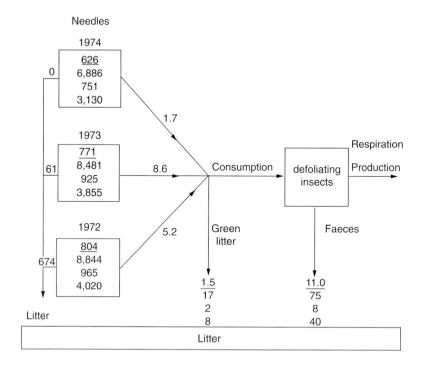

Figure 5.1   *Biomass of needles formed in 1974, 1973, 1972 and before in a population of* Pinus sylvestris *in Sweden; transfer of dry matter, nitrogen, phosphorus and potassium during needle fall; consumption by insects. "Green litter" corresponds to fragments of needles gathered but not ingested by insects and which fall to the ground. Consumption is only 1.7 + 8.6 + 5.2, or 15.5 kg per hectare, and the quantity of available needles is 626 + 771 + 804, or 2,201 kg per hectare. Insects consume only 0.7 of available foliage. Underlined figures represent kg per hectare of dry weight and non-underlined figures represent grams of nitrogen, phosphorus and potassium per hectare (Larsson & Tenow, 1980).*

weight) of foliage, or 81% of its own mass, and geometrid moths about 56%. These two categories of insect account for about 60% of total consumption by insects (Reichle & Crossley, 1967).

1.2. Leaf-eating insects as regulators of primary production

Phytophagous insects, and leaf-eaters in particular, act as regulators of primary production and of the cycle of mineral elements (Mattson & Addy, 1975). Consumption of tree foliage affects the ecosystem by (a) increasing the light flux which penetrates to the undergrowth and by increasing the trees' rate of photosynthesis (Collins, 1961); (b) reducing competition between plants and favouring the growth of young trees which had been dominated by older trees (Collins, 1961; Berryman, 1973); (c) accelerating the circulation of mineral elements; (d) stimulating the activity of sapropha-gous organisms which enjoy a more abundant litter provided by insect droppings. Numerous works on the role of caterpillar frass lend support to this theory.

Moderate or severe consumption of foliage may increase the nitrogen, phosphorus and potassium content of litter by 20 to 200%. Insect faeces and corpses as well as the various plant debris they cause to fall enrich litter and favour an increase of soil organisms, in particular bacteria. Enrichment provided by faeces has been established in the case of the sawfly *Gilpinia hercyniae* and the gypsy moth, *Lymantria dispar* (table 5.1).

	Elements in percentages				
	N	P	K	Ca	Mg
Leaves	1.38 to 2.65	0.05 to 0.14	1.04 to 1.79	1.15 to 2.12	0.13 to 0.38
Faeces	3.26	0.20	3.21	1.87	0.76
Corpses	9.62	1.53	2.98	0.39	0.27

Table 5.1. *Chemical composition of leaves of five species of tree, droppings and corpses of* Lymantria dispar *caterpillars. Content in percentage of the 5 main elements (Rafes, 1971).*

In the oak-hornbeam forest of Ferage (Belgium), litter fall in 1968 (figure 5.2) amounted to 7,592 kg per hectare, 330 kg of which consisted of caterpillar faeces (Duvigneaud *et al.*, 1969). The cork oak forest of La Mamora in Morocco has an average of 250 trees per hectare. Leaf biomass is 15 tons and wood biomass 100 tons. *Lymantria dispar* caterpillars have a mean biomass of 170 kg and their leaf consumption is 3.3 tons, that is 20% of the available tree foliage. The dead leaves, which fall to the ground and turn into litter, weigh 6.8 tons and the amount of caterpillar faecal matter in the litter is estimated at 1.4 tons per year (Villemant & Fraval, 1991).

In the Ispina forest in Poland, caterpillar faeces amounted to 531 kg per hectare in 1968 during an outbreak and only to 123 kg per hectare in 1969. Caterpillar faeces contain no tannins, which are bactericides, but contain twice more biologically active elements (nitrogen, phosphorus, potassium) than leaves. Therefore these faeces favour the decomposition of litter by micro-organisms. The rate of decomposition significantly increases the years when there are outbreaks of defoliators. In 1968 during an outbreak 78.4% of litter was broken down whereas in 1969, after the outbreak, this amount was only 42.8% (Bandola-Ciolczyk, 1974). This more rapid recycling is also due to changes in the microclimate at ground level, in a forest whose trees have shed their leaves. It enables oaks to recover faster the losses caused by severe defoliation.

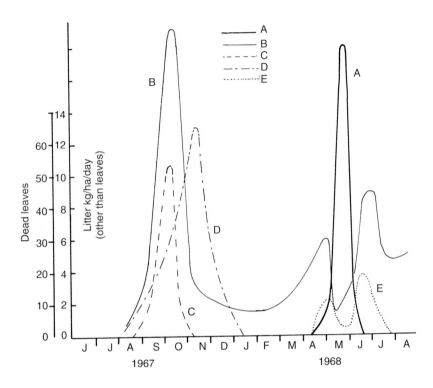

Figure 5.2.   *Fall of the various components of the litter from 1967 to 1968 in an oak forest in Belgium. Note the importance of caterpillar faeces in 1968 during an outbreak. A: caterpillar faeces; B: dead twigs; C: fruits (acorns); D: dead leaves; E: tree flowers (Duvigneaud et al., 1969).*

The role played by *Tortrix viridana* and by other less important Lepidoptera, as accelerators of biogeochemical cycles has been established in oak forests in the South of Russia (Zlotin & Khodashova, 1980). The average amount of faecal matter on the ground is 30 grams per square metre (dry weight). These faeces are rich in nitrogen (the ratio C/N is 20 whereas it is 42 in litter); they are poor in tannins whose bactericidal and toxic effects on the micro-organisms of soil are well known. They accelerate microbiological processes in soil and it has been proved that saprophagous soil invertebrates prefer caterpillar faeces to non-broken down litter. In forests where there are defoliating caterpillars, the amount of mineral elements important for the vegetation is 1.5 times higher than in forests not attacked by caterpillars.

During outbreaks, the reduction in tree growth is counterbalanced by an increase in the productivity of shrubs and herbaceous layer of the underground, and consequently the overall productivity remains constant.

During outbreaks, the American spring cankerworm *Paleacrita vernata* (Lepidoptera, Geometridae) ingests up to 33% of the foliage of a broadleaf forest. These alterations to the biogeochemical cycles increase the amount of nitrogen washed out of the forest stand watershed by five times (Swank *et al.*, 1981). This is not always the case since this negative effect was not observed when studies were made in the Hubbard Brook forest watershed (Bormann & Likens, 1979). The other consequences of defoliation by the spring cankerworm are a decrease in wood production, an increase in leaf production and caterpillar faeces, an increase in the metabolism of litter organisms and in the quantity of available nutrients. The alteration of the cycle of mineral elements by defoliating arthropods was also established by Schowalter *et al.* (1981). For several years, in two watersheds in the north east of the United States studies were conducted which showed that nitrogen compounds introduced to the soil by insect faeces are not washed out by surface runoff and are not found in the water of the streams that flow across the forest. The mecanisms that cause nitrogen to stay in soil have been recently elucidated (Lovett & Ruesink, 1995).

The percentage of leaves consumed by insects in eucalyptus forests in Australia is higher than in other parts of the world (table 5.2). According to Springett (1978) this is an adaptation which accelerates the cycle of mineral elements on a continent which has poor soils and a semi-arid climate. The commonest leaf-eaters in eucalyptus forests are the moth *Perthida glyphoba*, which may reduce tree growth by 70%, the sawfly *Perga affinis*, and stick insects such as *Didymuria violescens*; a few scarabaeid and chrysomelid beetles may also occur in outbreaks.

Forest type	Percentage of leaves consumed
Quercus forest	10.6 %
Acer and *Fagus* forest	6.6 %
Pinus and *Quercus* forest	3.4 to 10 %
Liriodendron forest	7.7 %
Tropical forest	3 to 8.5 %
Eastern Australian *Eucalyptus* forest	10 to 30 %
Western Australian *Eucalyptus* forest	15 to 30 %

Table 5.2. *Percentage of leaves consumed by leaf-eating insects in various types of forest in the world. Consumption is clearly higher in Australia (Springett, 1978).*

Increases in levels of nitrogen and of its microbial activity in the soil are common following depredations by phytophagous insects, but this is not a general rule. In a population of *Alnus rubra* in the western United States, aphids feeding on the trees release honeydew, which results in significant decreases in nitrogen content of the soil, reduced primary production and wood and bark growth, but no reduction in leaf production (Grier & Vogt, 1990).

The scolytid beetle *Dendroctonus ponderosae* may also play a part in regulating primary production in the ecosystem by selectively killing dominant trees and enabling a redistribution of minerals released among other trees (Romme *et al.*, 1986). Following an outbreak of this beetle in Yellowstone National Park, surviving trees did indeed grow faster, and annual wood production per hectare returned to what it had been before the *Dendroctonus* outbreak, or even surpassed it. One may think that though depredations by phytophagous insects during outbreaks lead to a decrease in primary production, this decrease is short lived.

2. TROPHIC NETWORKS AND ENERGY FLUXES

Forest ecosystems are characterised by a large plant biomass of which the two main constituents are leaves and tree trunks. This living biomass is consumed by leaf- and seed-eating insects or by other primary pests. The dead biomass is consumed by secondary pests which are mostly xylophagous or saproxylophagous species. Trophic networks and energy fluxes are divided accordingly into two main groups: consumers of living matter and consumers of dead matter.

2.1. The role of phytophagous organisms in trophic networks

A few general principles apply to all ecosystems:

– Only a small proportion of the energy in light received is used by plants. It lies between 0.001 and 3.5%, with an average value of 1%.

– The biomass of leaves is estimated at 2.2% of the total biomass, tree trunks and limbs forming almost 98% of that biomass.

– Consumption of living tissues by phytophagous organisms is small. It lies between 0.1% and 2.5% of net primary production with a probable average below 0.2%.

– The greater part of primary production passes in the form of dead litter by way of saprophagous chains of the soil.

– The part played by predators is generally small.

These principles apply in the German beech wood at Solling in which numerous studies have been made of biomass, production and energy flux, in which arthropods play a part (Grimm, 1976). It is obvious that saprophagous arthropods, which exploit dead organic matter in the litter predominate, and that phytophagous insects and predators play a smaller part (table 5.3). Leaf-eaters (weevils and moths), sap-feeders and root-eaters consume about 2% of net primary production. Litter fall accounts for about half of primary production of leaves; the greater part of energy flux is conveyed by saprophagous organisms (figure 5.3). Comparable results were obtained in a *Liriodendron* wood in North America.

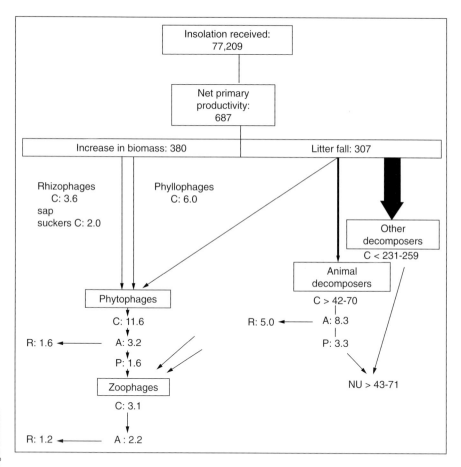

Figure 5.3. *Energy flux conveyed by arthropods in Solling beech wood. All values are in 10⁵ kcal ha⁻¹ year⁻¹. Note the predominance of saprophagous over phytophagous organisms, and, in the latter, of leaf-eaters. A: assimilation; R: respiration; C: consumption; NU: non-utilised; P: production (Grimm, 1976).*

	P	R	A	C	A/C %
Phytophagous organisms					
Leaf-eaters (weevils, moths)	75	75	150	600	25
Sap-feeders	40	40	80	200	40
Rhizophagous organisms	45	45	90	360	25
Saprophagous organisms					
Collembola	188	134	322	805	40
Oribatids	4	16	20	100	20
Diptera (sciarids and sciophilids)	105	263	368	2,450 to 5,260	7 to 15
Elaterids (*Athous subfuscus*)	35	87	122	814	15
Predators					
Gamasids	44	32	76	-	
Spiders	6	14	20	-	-
Carabids	6	14	20	30	66
Staphylinids (*Othius* spp.)	20	14	34	-	-
Opiliones and pseudoscorpions	4	10	14	-	-
Diptera (Empididae and Muscidae)	15	18	33	36 to 70	47 to 92
Chilopods	5.5	17	22.5	-	-
Total arthropods	592.5	779	1,371.5	5,643 to 8,453	

Table 5.3.   *Energetics in 10³; kcal ha⁻¹ year⁻¹ of arthropods in Solling beech wood (Germany). P: production; R: respiration; A: assimilation; C: consumption. Three trophic levels appear: phytophages and rhizophages, saprophages, and predators. Figures for saprophages are minimal values (Grimm, 1976)*

2.2. Trophic networks and productivity in an English oak wood

A simplified trophic network, including only the main species, was established for Wytham Wood, an oak wood in England (Varley, 1970). This trophic network is centred on the two main oak leaf-eating moths *Operophtera brumata* and *Tortrix viridana*. These two primary consumers serve as food to secondary consumers, the main ones being three ground-living beetles (two carabids belonging to the genera *Abax* and *Pterostichus* and a staphylinid of the genus *Philonthus*), tits (*Parus* sp.), the parasitoid fly *Cyzenis albicans*, and spiders. Tertiary consumers in the form of super-predators include small mammals such as moles (*Talpa*), shrews (*Sorex*) and weasels (*Mustela*), owls (*Strix*), and insect parasites of spiders and of *Cyzenis albicans*. The main characteristics of trophic networks are given in schematic form (figure 5.4). The two main phytophagous insects consume only a small part of available foliage (about 10%), and the greater part of this passes through soil detritivorous chains in the form of litter.

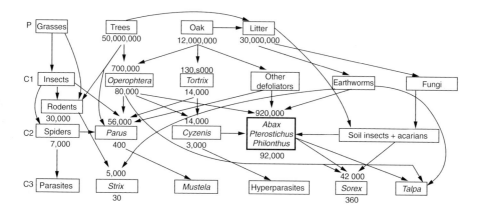

Figure 5.4. *Simplified trophic network centred on two main oak leaf eaters, Operophtera brumata and Tortrix viridana, in Wytham Wood, England (Varley, 1970). All figures are in kcal ha⁻¹ year⁻¹. Those above rectangles correspond to energy produced. Trophic levels of polyphagous species such as small mammals and birds are determined arbitrarily. In winter, tits consume mosses and seeds as well as insects and spiders and thus behave both as primary and secondary consumers.*

In this ecosystem:

– photosynthesis efficiency is equal to
(50,000,000 + 12,000,000) : 2,500,000,000, or 2.5%

– consumption is low, herbivores consuming only a small part of available food. For the whole of the two species *Tortrix viridana* and *Operophtera brumata* exploitation production is:
(700,000 + 130,000) : (50,000,000 + 12,000,000) = 1.2%

– growth production is around 10% for herbivorous caterpillars (respectively 80,000 : 700,000, or 11.4% and 14,000 : 130,000, or 10.7%)

– growth production is barely 1% in homeotherms. In the case of the tit *Parus* it is 400 : 56,000, or 0.71%. It is higher in carnivorous poikilotherms; in the predatory carabid *Abax* it is 92,000 : 920,000, or 10%, and in the parasitic tachinid *Cyzenis* it is 3,000 : 14,000, or 22%

– fallen organic material in the form of litter represents about one third of primary productivity, being:
30,000,000 : (12,000,000 + 50,000,000 + unquantified herbaceous plants), or about 30%.

Similar research has been carried out in Mirwart beech wood in Belgium (Josens & Pasteels, 1977). Insect production was 12 kg per hectare per year requiring consumption of 674 kg of vegetable matter, or 5.4% of primary production. This corresponds to a production of 11.6 million insects per hectare per annum. Saprophagous (mainly tipulid larvae) and mycetophagous insects dominate with an annual production of 2.3 kg per hectare, and predators (mainly Coleoptera and parasitoids) have a production of 2.1 kg per hectare.

2.3   Energetics of hazel defoliators

A study of insects which feed on hazel leaves in central Sweden has shown that the most abundant species is the chrysomelid beetle *Phytodecta pallidus*; the weevil *Phyllobius argentatus* is also abundant; the main moths are *Chelaria huebnerella* and *Chimabacche phryganella*; the aphid *Mysocallis coryli* is also present. Energetics of the *Phytodecta pallidus* population are as follows, assimilation being determined according to two theoretically possible methods:

At individual level (values in calories):

Consumption (C)	: 127.7	Respiration (R)	: 16.7
Non-utilised (NU)	: 58.4	Assimilation (A = C – NU)	: 69.3
Net production (P)	: 36.9	Assimilation (A = P + R)	: 53.6

At population level (values in kcal m^{-2}):

Consumption (C)	: 0.702	Respiration (R)	: 0.082
Non-utilised (NU)	: 0.331	Assimilation (A = C – NU)	: 0.271
Net production (P)	: 0.209	Assimilation (A = P + R)	: 0.298

The calorific equivalent of available foliage being estimated at 167 kcal m^{-2}, it will be seen that *Phytodecta* consumes only 0.42%. As the biomass of *Phytodecta* is about equal to that of all other leaf-eating insects taken together, total consumption by these insects amounts to only around 1% of primary production (Axelsson *et al.*, 1975).

In the same Swedish forest, caterpillars of *Operophtera brumata* and *O. fagata*, which feed on oak leaves, may almost entirely defoliate trees during periods of major outbreaks, whereas during periods in which they are scarce they consume only 0.004 to 0.006% of primary production.

2.4.   A forest food-chain

Outbreaks in populations of leaf-eating Lepidoptera, especially *Tortrix viridana*, occur periodically in Ispina Forest in Poland. Energetics of the larval development of *T. viridana* have been determined in the wild and in reared populations. The quantity of food assimilated A may be evaluated in two different ways using two theoretically equivalent formulae:

$$\text{In laboratory conditions} : \quad A = P + R$$
$$\text{In the wild} \qquad\qquad : \quad A = C - NU$$

P being production of living matter by caterpillars, R respiration, C what is ingested, and NU what is not assimilated (faeces). The value of A obtained from the first equation is 42.2% below that obtained from the second equation. This appears to be due to the behaviour of caterpillars which build a protective sheath of leaves from which only the head is extruded while feeding. In laboratory conditions, daily weighing of the larvae forces them to constantly rebuild their shelters, hence the extra expense of energy, and lower value obtained for A. In order to obtain a realistic result it is therefore necessary to use data gathered in the wild. In these conditions the energetics during the development of an individual *T. viridana* correspond to assimilation of a quantity of food equal to 228.7 calories.

The importance of *Tortrix viridana* in this forest is shown by the fact that this insect consumes about ten times more of primary production than do rodents. A food chain composed of oak leaves, *Tortrix* caterpillars and insectivorous birds has been studied (Medwecka-Kornas *et al.*, 1974). Biomasses and energy fluxes involved are shown in figure 5.5. It is interesting to calculate the efficiency of the main energy transfers from one trophic level to another in this chain:

– Ecological efficiency. Efficiency PN_2 / PN_1 between leaves and caterpillars equals 0.49 / 14.2, or 3.5%. Efficiency PN_3 / PN_2 between birds and caterpillars equals 0.045 / 0.49, or 0.9%;

– Working efficiency. Efficiency I_1 / N_1 equals 1.4 / 14.2, or 10%, and efficiency I_2 / PN_2 equals 0.117 / 0.49, or 24%.

These efficiency values agree with the body of other studies undertaken outside forest environments which show that working efficiency is usually between 1 and 10% for herbivores and between 10 and 80% for carnivores.

2.5. The role of xylophagous insects

Tree trunks and branches represent a far larger biomass than that of leaves. However these elements are nearly always overlooked in studies of trophic networks and energy flux, probably due to the difficulty in studying them. One of the few studies of energy flux through populations of xylophagous insects was carried out in Niepolomice Forest (Wittkowski & Borusiewicz, 1984). Insects that feed on the subcortical area and on wood consist of the following twelve species of Coleoptera in a part of the forest composed of associated oak and hornbeam:

Species	Numbers of larvae per hectare
Cerambycidae	
Leiopus nebulosus	65
Oplosia fennica	3
Plagionotus arcuatus	108
Rhagium mordax	85
Rhagium sycophanta	14
Saperda scalaris	37
Stenostola dubia	14
Buprestidae	
Agrilus angustulus	26
Agrilus biguttatus	1
Scolytidae	
Scolytus intricatus	90
Xyleborus dispar	6
Xyleborus saxaseni	2

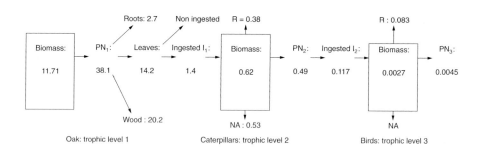

Figure 5.5. *Energy transfer in a food chain consisting of oak leaves (trophic level 1), larvae of* Tortrix viridana *(trophic level 2) and insectivorous birds (trophic level 3). Biomasses are expressed in kcal ha^{-1} year^{-1}. Net productivity of the three trophic levels is represented by PN$_1$, PN$_2$ and PN$_3$.*

The energy equivalent of wood consumed is $5.6.10^6$ kJ ha^{-1}. Annual consumption by cerambycids and buprestids is $6.7.10^3$ kJ ha^{-1}. This allows growth of an annual biomass of larvae equivalent to $0.18.10^3$ kJ ha^{-1} for the whole of Cerambycidae and Buprestidae, and to $0.17.10^3$ kJ ha^{-1} for Scolytidae. Xylophagous insects only consume 0.25% of the biomass of the trees they feed on, which corresponds to less than 0.001% of the biomass of all trees in the forest.

Other data on the role of insects and various invertebrates in the decomposition of dead wood are given in chapter 17.

REFERENCES

AXELSSON B., LOHM U., NILSSON A. *et al.*, (1975). Energetics of a larval population of *Operophtera* spp. (Lep. Geometridae) in central Sweden during a fluctuation low. *Zoon*, **3**: 71-84.

BALTENSWEILER W., (1975). Zur Bedeutung des grauen Lärchenwicklers (*Zeiraphera diniana*) für die Lebensgemeinschaft des Lärchen Arvenwaldes. *Bull. Soc. Ent. Suisse*, **48**: 5-12.

BANDOLA-CIOLCZYK E., (1974). Production of tree leaves and energy flow through the litter in *Tilio-Carpinetum* association (International Biological Program area). *Polska Akad. Nauk*, Studia Nat., ser. A, **9**: 29-91.

BERRYMAN A. A., (1973). Population dynamics of the fir engraver, *Scolytus ventralis* (Coleoptera). I. Analysis of population behaviour and survival from 1964-1971. *Canad. Ent.*, **105**: 1,465-1,470.

BORMANN F. H., LIKENS G. E., (1979). *Pattern and process in a forested ecosystem*. Springer, Berlin.

BRAY J. R., (1961). Measurement of leaf utilization as an index of minimum level of primary consumption. *Oikos*, **12**: 70-74.

CARLISLE A., BROWN A. H. F., WHITE E. J., (1966). Litter fall, leaf production and the effects of defoliation by *Tortrix viridana* in a sessile oak woodland. *J. Ecol.*, **54**: 65-85.

CLOUTIER C., FILION L., (1991). Recent outbreaks of the larch sawfly *Pristiphora erichsoni* (Hartig) in subarctic Québec. *Canad. Ent.*, **123**: 611-619

COLLINS S., (1961). Benefits to understorey from canopy defoliation by gypsy moth larvae. *Ecology*, **42**: 836-838.

DUVIGNEAUD P., DENAEYER DE SMETS S., MARBAISE J. L., (1969). Litière totale annuelle et restitution au sol des polyéléments biogènes. *Bull. Soc. R. Bot. Belgique*, **102**: 339-354.

ELLENBERG H., MAYER R., SCHAUERMANN J., (1986). *Okosystemforschung - Ergebnisse des Sollingprojektes 1966-1986*. Ulmer, Stuttgart.

FUENTES E.-R., ETCHEGARAY J., ALJARO M.-E., MONTENEGRO G., (1981). Shrub defoliation by matorral insects. *Ecosystems of the world 11. Mediterranean-type shrublands*, Elsevier, Amsterdam, 345-359.

GRIER C. C., VOGT D.-J., (1990). Effects of aphid honeydew on soil nitrogen availability and net primary production in an *Alnus rubra* plantation in western Washington. *Oikos*, **57**: 114-118.

GRIMM R., (1976). Der Energieumsatz der Arthropoden populationen in Okosystem Buchenwald. *Ver. Ges. Okol. Gottingen*, 125-131.

HORSTMANN K., (1977). Waldameisen (*Formica polyctena*) als Abundanz Faktoren für den Massenwechsel des Eichenwicklers *Tortrix viridana* L. *Zeit. ang. Ent.*, **82**: 421-435.

JANZEN D. H., (1970). Herbivores and the number of tree species in tropical forests. *Amer. Nat.*, **104**: 501-528.

JARDON Y., FILION L., (1994). Long-term impact of insect defoliation on growth and mortality of eastern larch in boreal Québec. *Ecoscience*, **3**: 231-238.

LARSSON S., TENOW O., (1980). Needle-eating insects and grazing dynamics in a mature Scots pine forest in Central Sweden. *Ann. Ent. Fenn.*, **48**: 119-125.

LARSSON S., WIREN A., (1982). Leaf-eating insects in a forest stand of *Salix viminalis* L. in Central Sweden. *Ann. Ent. Fennici*, **48**: 119-125.

LOVETT G., RUESINK A., (1995). Carbon and nitrogen mineralization from decomposing gypsy moth frass. *Œcologia*, **104**: 133-138.

MATTSON W. J., ADDY N. D., (1975). Phytophagous insects as regulators of forest primary production. *Science*, **190**: 515-522.

MEDWECKA-KORNAS A., LOMNICKI A., BANDOLA-CIOLCZYK E., (1974). Energy flow in the oak-hornbeam forest (IBP Project « Ispina »). *Bull. Acad. Polon. Sci.*, sér. sci. biol., Cl II, **22**: 553-566.

OHMART C. P., STEWART L. G., THOMAS J. R., (1983). Leaf consumption by insects in three *Eucalyptus* forest types in Southwestern Australia and their role in short-term nutrient cycling. *Œcologia*, **59**: 322-330.

RAFES P. M., Pests and the damage which they cause to forests. *In*: P. DUVIGNEAUD (1971). *Productivité des écosystèmes forestiers*. Unesco, Paris, 357-367.

REICHLE D. E., CROSSLEY D. A. Jr., Investigations on heterotrophic productivity in forest insect communities. *In*: K. PETRUSEWICZ, (1967). *Secondary productivity of terrestrial ecosystems*, 563-587.

REICHLE D. E., GOLDSTEIN R. A., VAN HOOK R. J., DODSON G. I., (1973). Analysis of insect consumption in a forest canopy. *Ecology*, **54** : 1,076-1,084.

ROMME W. H., KNIGHT D. H., YAVITT J. B., (1986). Mountain pine beetle outbreaks in the Rocky Mountains: regulators of primary productivity? *Am. Nat.*, **127**: 484-494.

SCHOWALTER T. D., (1981). Insect herbivore relationship to the state of the host plant : biotic regulation of ecosystem nutrient cycling through ecological succession. *Oikos*, **37**: 126-130.

SCHOWALTER T. D., WEBB J. W., CROSSLEY D. A., (1981). Community structure and nutrient content of canopy arthropods in clearcut and uncut forest ecosystems. *Ecology*, **62**: 1,010-1,019.

SPRINGETT B. P., (1978). On the ecological role of insects in Australian eucalyptus forests. *Australian J. Ecol.*, **3**: 129-139.

SWANK W. T., WAIDE J. B., CROSSLEY D. A., TODD R. L., (1981). Insect defoliation enhances nitrate export from forest ecosystems. *Œcologia*, **51**: 297-299.

VILLEMANT C., FRAVAL A., (1991). *La faune du chêne-liège*. Éditions Actes, Rabat.

WIEGERT R.-G., PETERSEN C.-E., (1983). Energy transfer in insects. *Ann. Rev. Ent.*, **28**: 455-486.

WITTKOWSKI Z., BORUSIEWICZ K., Ecology, energetics and the significance of phytophagous insects in deciduous and coniferous forests. *In:* W. GRODZINSKI, J. WEINER, P.-F. MAYCOCK, (1984). *Forest ecosystems in industrial regions*. Springer, Berlin, 103-113.

ZLOTKIN R.-I., KHODASHOVA K.-S., (1980). *The role of animals in biological cycling of forest-steppe ecosystems*. Dowden, Hutchinson et Ross, Stroudsburg, Pennsylvania.

6

Forest insect pests

Truly harmful insects amount to less than one percent of total species present on crops as well as in forest. In all types of forest very few, usually between one and half a dozen species, act as primary pests. By way of illustration, the following lists can be made for four regions in France (Decourt *et al.*, 1981):

– Pine forest of the Landes (SW France). Main pest: pine processionary caterpillar; potential pest: *Matsucoccus feytaudi*; lesser primary pest: *Dioryctia sylvestrella*; secondary pests: various scolytids.

– Austrian pine in the southern Massif Central. Main pests: pine processionary caterpillar and *Neodiprion sertifer*; lesser primary pest on young trees: *Rhyacionia buoliana*; secondary pests: scolytids.

– Lowland beech woods. Main pest: *Cryptococcus fagisuga*.

– Atlantic oak woods. Main pest: *Tortrix viridana*; occasional pest: *Operophtera brumata*.

1. TYPES OF DAMAGE

Damage caused by insects in woodland may be classed under four headings:

(a) The effects of some species lead to *deterioration* in the useful quality of wood after trees have been exploited. Examples are attacks by the weevil *Pissodes strobi*, the cerambycid *Saperda carcharias*, "black spots" on oak and chestnut wood caused by a few species of beetle including *Platypus cylindrus* and various scolytids such as *Xyleborus domesticus* and *Xyleborus monographus*, as well as changes in structure caused by the perforations made by agromyzids in poplar wood. In this category one may also include insects which attack building timber and furniture, such as various species of termite, the cerambycid *Hylotrupes bajulus*, known as the "house longhorn", and some anobiids known as "furniture beetles".

(b) Other species are responsible for *loss of production* due to reduced growth of trees as a result of repeated defoliation. Species which cause this type of damage are mainly moths such as *Thaumetopoea pityocampa*, *Lymantria dispar* and *Tortrix viridana*. Conifers are more vulnerable than broadleaf trees to attacks by defoliating insects, since, with the exception of larch, foliage is regenerated slowly, and they do not survive total defoliation. Destruction of their needles predisposes conifers to attacks by xylophagous insects which, following major outbreaks of defoliating insects such as *Lymantria monacha*, *Panolis flammea* or *Bupalus piniarius* in Europe, and *Choristoneura fumiferana* in North America, have wiped out vast populations. In North America defoliation of Douglas-fir by *Orgyia antiqua* leads to increased attacks by *Dendroctonus pseudotsugae* and *Scolytus ventralis*. Defoliation lowers a tree's resistance, and reduces photosynthesis and development of reproductive organs which may impair regeneration and allow colonisation by scolytids in low density. Attack by *Tortrix viridana* reduces acorn production in oaks. Control of *Tortrix viridana* in populations of evergreen oak in Spain increased acorn production from 150 to 600 kg per hectare per year.

Perforations by sap-sucking insects such as aphids may have the same effect as defoliating insects in reducing growth. Destruction of buds and terminal shoots by species such as the scolytid *Tomicus piniperda* or *Rhyacionia buoliana* also results in loss of growth. Damage caused to trees sometimes makes woodland inhospitable to man and even to animals. The irritant hairs of processionary caterpillars repel birds and other animals. In Mamora Forest in Morocco gypsy moth caterpillars defoliate cork oak. Production of cork and acorns, which are the forest's main produce, is greatly reduced during outbreaks of the insect; the appearance of defoliated trees and invasion of homes by caterpillars are much resented by nearby residents.

Phytophagous insects may alter the architecture, or crown structure, of trees. In destroying the terminal bud of young pines, *Rhyacionia buoliana* causes a characteristic "bayonet" deformation. Changes caused by *Dioryctria albovitella* in the pinyon pine *Pinus edulis* have been described by Whitham & Mopper (1985). Intense attack destroys terminal buds and gives trees a low branching aspect. Because female cones develop on dominant apical shoots, only male cones, which grow on lateral shoots, survive. The tree loses its female functions and becomes entirely male. Absence or scarcity of female cones bearing seeds causes the bird *Gymnorhinus cyanocephalus*, an efficient seed disperser, to leave.

Damage by a phytophagous insect may make a tree more vulnerable to attacks by other species. When the gall aphid *Dreyfusia piceae* attacks the fir

Abies fraseri, 30% of the trees' seeds become infested by *Megastigmus specularis* larvae, as against 3.1% in trees unaffected by the aphid (Fedde, 1973).

(c) Some insects may *kill trees* in cases of massive infestation. In such cases the wood, which is still exploited, is damaged and loses much of its value. This type of damage is caused by many xylophagous insects such as scolytids, the beech scale insect *Cryptococcus fagisuga* or the maritime pine scale insect *Matsucoccus feytaudi.* Attacks by these insects contribute to the spread of forest fires, where dead trees are left *in situ* and dry out.

(d) Other insects are *vectors of viruses or pathogenic fungi.* In North America various nitidulid beetles, which live on sap exudations or on sub-cortical mycelium on oak, propagate spores of the pathogenic fungus *Ceratocystis fagacearum* which causes oak wilt (Dorsey & Leach, 1956). The buprestid *Agrilus bilineatus* is also a vector of the same fungus. Elm phloem necrosis, which is a viral infection, is transmitted by the bites of the leaf hopper *Scaphoideus luteolus.* Dutch elm disease is caused by the fungus *Ceratocystis ulmi* which is spread by the bark beetles *Scolytus multistriatus* and *Scolytus scolytus.* The scale insect *Cryptococcus fagisuga* is conducive to infection of beech by the parasitic fungus *Nectria coccinea.* The mycelium and spores of the lignivorous fungus *Ungulina annosa* may be spread by the weevil *Hylobius abietis,* experiments having shown that the fungus is not killed as it passes through the weevil's digestive tract. Propagation may also be due to the scolytid *Tomicus piniperda* or to its predators such as *Rhizophagus ferrugineus.* Many other scolytids are also vectors of pathogenic fungi.

Various xylophagous insects spread the pathogenic nematode *Bursaphelenchus xylophilus* which causes pine wilt disease. The main culprits are cerambycids of the genus *Monochamus* as well as *Spondylis buprestoides, Arhopalus rusticus, Asemum striatum* (the last three occurring in Europe), and curculionids and buprestids (Linit, 1988).

2. MAIN PESTS OF EUROPEAN FORESTS

A few species of insect, either native or introduced, may cause severe damage. The main forest destroyers in Europe are as follows.

2.1. Indigenous species

Devastations caused by scolytid beetles have been known for many years. An outbreak of *Ips typographus* occurred in Bohemia following the hurricane of 7th December 1868 and the cyclone of 26th October 1870, destroying 100,000 hectares of spruce; this outbreak was only halted in

1875. In Switzerland the same species caused the loss of 663,000 cubic metres of wood in the years 1947-1948. Outbreaks of this scolytid in France occurred in the Vosges and the Jura. *Ips sexdentatus* spread in the Landes following the forest fires of 1944-1945. In the Massif Central *Dendroctonus micans* devestates spruce plantations. In North America scolytids are responsible for 85% of damage caused by insects, the genus *Dendroctonus* alone causing losses estimated at several hundred million dollars between 1860 and 1910. In Belgium the weevil *Hylobius abietis* is one of the main forest pests, ravaging 5,000 hectares of conifers which are replanted each year.

The defoliating moth *Bupalus piniarius* devastated Scots pine in Haguenau forest in Alsace between 1924 and 1926. In 1925 to 1927 *Dasychira pudibunda* occurred in great numbers in the beech woods of northern France, and outbreaks were recorded in 1848 in Lorraine and in 1926 at Fontainebleau. Between 1925 and 1933 *Choristoneura murinana* devastated plantations of fir on the Alsatian slopes of the Vosges. This moth is now widely distributed in the Alps and the Massif Central (Lemperière *et al.*, 1984), where numerous centres of outbreaks have been discovered, and it also occurs in Normandy and southern Brittany, where it followed plantations of fir. It also feeds on Atlas cedar, and constitutes a new pest in plantations of this tree in south-eastern France (Du Merle *et al.*, 1983). The pine processionary caterpillar occurs in various localities in southern France, and it reached northern France in stages during warm years since 1945. Between 1946 and 1949 *Lymantria dispar* appeared in Anjou and Brittany. *Lymantria monacha* destroyed about 580,000 hectares of spruce forest in Czechoslovakia where 14,800,000 cubic metres of wood had to be felled. Another defoliator, the sawfly *Diprion pini*, occurs in outbreaks at regular intervals in the Campine of Belgium and in the Fontainebleau Forest.

Among sap feeders, the chermesid *Dreyfusia nordmannianae* has spread in France in the fir plantations of the North East and the Cévennes, where it has killed many young trees. The beech scale insect has devastated many beech woods in Normandy and North East France, and the maritime pine scale insect has almost wiped out populations of maritime pine in Provence.

The forests of northern Europe, which represent about one third of all European forests (excluding Russia), are mainly composed of broadleaf trees (oak, beech and birch) in Denmark and in southern Sweden, and of conifers (pine and spruce) further North. The weevil *Hylobius abietis* is a major pest in these forests, where its spread has been favoured by clear felling (Bakke & Lekander, 1965). *Oporinia autumnata* is a pest of birch which may cause local large-scale defoliation. The moths *Operophtera brumata, O. fagata, Erannis defoliaria* and *E. aurantiaria* have invaded broadleaf forests and caused great losses, as has *Tortrix viridana*. Of the sawflies, *Neodiprion sertifer* and *Diprion pallipes* attack pine, and in Finland *Pristophora abietina* has caused some damage (Christiansen, 1970).

In Spain the main defoliators are *Tortrix viridana* which infests millions of hectares of oak; *Lymantria dispar* which also feeds on oak and occasionally on pine; *Lymantria monacha* which lives on *Pinus sylvestris*; *Malacosoma neustria*, a pest of oak; the pine processionary caterpillar; the sawflies *Diprion pini* and *Neodiprion sertifer*. The main xylophagous species live on conifers; they are *Pissodes notatus, Hylobius abietis, Rhyacionia buoliana, Dioryctria splendidella* and various scolytids (Romanyk, 1966). In Portugal, where *Pinus pinaster* forms 40% of forests, its two main enemies are the processionay caterpillar and *Tomicus piniperda*. Cork oak is attacked by *Tortrix viridana, Lymantria dispar* and the buprestid *Coroebus undatus*. In Yugoslavia *Lymantria dispar* is the main pest of oaks which suffer an average loss of growth of three cubic metres per hectare in years of total defoliation (Androic, 1966). This species is also the main destroyer of cork oak in Morocco.

Pests of saplings in nurseries are often polyphagous species which feed on various different trees. Insects of the soil fauna such as larvae of the cockchafer *Melolontha melolontha* eat the roots. *Agrotis segetum* caterpillars attack germinating plants. The weevil *Otiorrhynchus singularis* devours leaves. The springtail *Bourletiella hortensis* feeds on germinating plants. The most serious depredations on young trees appear to be those of aphids such as *Elatobium abietinum* or *Phyllaphis fagi*. The weevil *Hylobius abietis* is a common pest in young plantations in forest. *Rhyacionia buoliana* caterpillars deform young trees by attacking the terminal bud. Several defoliating insects, including *Panolis flammea*, the processionary caterpillar and *Neodiprion sertifer* appear on young trees. Old trees in forests are subject to attacks either by these same species or by others (table 6.1).

2.2. Introduced species

Introduced species in Europe, although few in number, are often pests. The moth *Hyphantria cunea*, originally from America, has spread in Europe since 1940, and its highly polyphagous caterpillar predates about a hundred different fruit-, ornamental- and forest trees. The advent of the scale insect *Matsucoccus feytaudi* in Provence leads to severe damage to maritime pine. The longhorn beetle *Phoracantha semipunctata*, which appears to be an innocuous species in its native Australia, was introduced accidentally to the Mediterranean basin where it is a serious threat to eucalyptus plantations. The Atlas cedar, which is native to North Africa, was introduced in large populations in Provence nearly a century ago. In the Atlas Mountains of Morocco this cedar has an associated fauna living in the foliage estimated at 107 species, many of which are specific to this tree. In France, 63 species are already known to live on cedar, but all of them come from other plants (Mouna, 1985). Added to this associated fauna are pests specific to cedar in North Africa which have arrived in Provence accidentally. These are the

Defoliators	Xylophagous insects	Sap feeders	Flower, cone and seed eaters	Phloem feeders
Lepidoptera	**Coleoptera**	**Homoptera**	**Coleoptera**	**Scolytidae**
Bupalus piniarius	*Phoracantha semipunctata*	*Dreyfusia piceae*	*Pissodes validirostris*	*Dendroctonus micans*
Coleophora laricella	*Tetropium gabrieli*	*Sacchiphantes abietis*	**Hymenoptera**	*Ips cembrae*
Lymantria dispar	**Hymenoptera**	*Elatobium abietinum*	*Megastigmus spermotrophus*	*Ips typographus*
Panolis flammea	*Sirex noctilio*	*Cryptococcus fagisuga*	*Andricus quercuscalicis*	*Ips sexdentatus*
Thaumetopoea pityocampa	**Lepidoptera**	*Matsucoccus feytaudi*	**Lepidoptera**	*Scolytus scolytus*
Thaumetopoea processionea	*Zeuzera pyrina*		*Cydia strobilella*	*Tomicus piniperda*
Tortrix viridana	*Cossus cossus*		*Tortrix viridana*	*Tomicus minor*
Zeiraphera diniana				*Trypodendron lineatum*
Oporinia autumnata				
Hymenoptera				
Diprion pini				
Gilpinia hercyniae				
Neodiprion sertifer				
Pristiphora erichsonii				

Table 6.1. *Main forest pests of old trees in western Europe.*

aphid *Cedrobium laportei* which has been found in the Luberon where it is responsible for defoliation of the trees, and the tortricid moth *Epinotia cedricola* which infests on a massive scale and totally defoliates trees in the South of France. In the Middle Atlas this species occurs in small populations, as it is controlled by numerous enemies (Fabre, 1976; Leclant *et al.*, 1977).

Sitka spruce (*Picea sitchensis*), a North American conifer introduced to Britain, has already been colonised by 90 species of insect, whereas it harbours only 59 in its homeland (Evans, 1987). Lodgepole pine (*Pinus contorta*), originally from the western United States, has been recently introduced, particularly in Normandy, in central France and in the Landes in SW France (Delplanque *et al.*, 1987). In 1987 it was already attacked by the following species which have all come from native pines:

Species	Usual host pines in France
Rhyacionia buoliana	*P. sylvestris, P. nigra*
Rhyacionia duplana	*P. pinaster*
Petrova resinella	*P. sylvestris*
Exoteleia dodecella	*P. sylvestris, P. pinaster*
Thaumetopœa pityocampa	*P. nigra, P. pinaster*
Hyloicus pinastri	*P. sylvestris, P. pinaster*
Dendrolimus pini	*Pinus* spp.
Bupalus piniarius	*P. sylvestris*
Diprion pini	*P. sylvestris*
Neodiprion sertifer	*P. nigra*
Acantholyda hieroglyphica	*P. nigra*
Dioryctria splendidella	*P. pinaster*
Pissodes pini	*P. sylvestris, P. nigra*
Pissodes validirostris	*P. halepensis, P. nigra, P. sylvestris*

While American species introduced to Europe are still few and of minor importance from an economic point of view, European species which have reached North America are formidable pests. Among the worst are the chermesid *Dreyfusia piceae*, known as the balsam woolly aphid, which attacks various firs; the scale insect *Cryptococcus fagisuga* which feeds on beech; the moth *Coleophora laricella*, known as the larch casebearer, which feeds on various larches; the sawflies *Diprion similis* (known as the "introduced sawfly"), *Neodiprion sertifer* ("European pine sawfly"), *Fenusa pusilla* and *Pristiphora erichsoni* ("larch sawfly"), the latter perhaps being indigenous rather than introduced; the moths *Lymantria dispar* (gypsy moth) and *Rhyacionia buoliana* ("European pine shoot moth"); and the bark beetle *Scolytus multistriatus* ("smaller European elm bark beetle") which is the vector of Dutch elm disease (Davidson & Prentice, 1967; Furniss & Carolin, 1977).

In North America as in Europe a few indigenous species account for most damage to forests. The main ones are *Choristoneura fumiferana*, the spruce budworm, which feeds on *Picea glauca* and defoliated 5,089,400 hectares of forest in Canada in 1994; various species of lasiocampid moths of the genus

Malacosoma; and scolytids of the genus *Dendroctonus*. In the southern United States the dominance of pines explains why the major pests are three scolytids of the genus *Dendroctonus,* accompanied by a few defoliating species such as *Orgyia pseudotsugata, Choristoneura occidentalis, Neophasia menapia* and *Coloradia pandora*. In the north-eastern United States damage is mainly due to moths such as *Lymantria dispar* and several species of *Malacosoma*. In Canada the weevil *Pissodes strobi* is a fearsome enemy of Sitka spruce, while *Choristoneura fumiferana, Neodiprion abietis* and *Pristophora erichsoni* are other major pests.

Losses caused outside Europe by some forest insects are often high. Damage caused by *Lymantria dispar* in the United States was estimated at 2.5 million hectares in 1910 and 10 million hectares in 1930. A few decades following its introduction to Australia *Sirex* wiped out several million standing Monterey pines (*Pinus radiata*). In 1958 damage caused by *Dreyfusia piceae*, introduced to North America at the beginning of the century, amounted to 160,000 hectares of trees that were destroyed or seriously weakened. Annual loss of wood in *Agathis* forests in Australia caused by the thrips *Oxythrips agathidis* and by the scale insect *Conniferococcus agathidis* was estimated at 10,000 cubic metres.

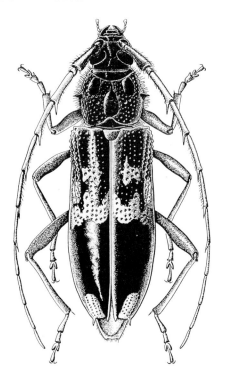

Figure 6.1. *The cerambycid* Phoracantha semipunctata, *native to Australia, has become a pest of eucalyptus in the Mediterranean region and has now come to North America.*

3. EFFECTS OF INSECTS ON TREE GROWTH

When they do not kill a tree, leaf-eating or sap-sucking insects may slow its growth and attempts have been made to measure this phenomenon (Rafes, 1970, 1972; Grison, 1972; Joly, 1970; Laurent-Hervouet, 1986; Rubtsov, 1996). Annual growth of wood in oak may be reduced to one third or one quarter during infestations by caterpillars. In a Belgian oak wood leaf-eating insects ingest up to four tons (dry weight) of leaves per hectare during major outbreaks of the insects. Defoliating moths in Russian oak woods (mainly *Operophtera brumata* and *Lymantria dispar*) reduce tree growth when the amount of leaves consumed exceeds 75% of total foliage. Following severe defoliation, production of annual tree rings is reduced, particularly at the level of wood formed during the summer months (figure 6.2). Insects which feed and grow in summer such as *Dasychira pudibunda* are more harmful than those which grow in spring. This is because thickening of the trunk occurs mainly early in the growing season using the tree's reserves, spring photosynthesis playing an insignificant part in growth. Summer growth depends more on photosynthesis, and leaf-eating insects therefore limit growth to a greater extent. Soil quality also plays a part. Growth loss due to *Lymantria dispar* in an oak wood in the Voronèje region was greater in plots situated on rich soils than in those on poor soils.

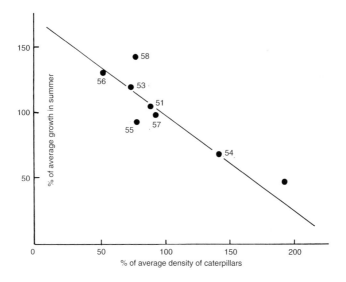

Figure 6.2.   *Effects of density of defoliating larvae of* Operophtera brumata, Erannis defoliaria *and* Tortrix viridana *on the thickness of tree rings of summer wood in oak in Wytham Wood in England in the years 1952 to 1958. An outbreak of* Tortrix viridana *occurred in 1952 (dot on the right of the graph) (Varley, 1967).*

Defoliation of oak by the common cockchafer results in a marked reduction in final wood and has no effect on initial wood. This is because the latter is formed by the tree's reserves before budding, whereas final wood is associated with development of photosynthesis by the year's growth of leaves (Huber, 1982).

In broadleaf trees the presence of starch reserves, which is easily detected, is a good indicator of a tree's vitality, for it reflects the degree of photosynthetic activity during the growth season and responses to various forms of stress. Low levels of starch reserves have been recorded following drought and following defoliation of maple (Wargo *et al.*, 1972) and in oaks defoliated by *Lymantria dispar* (Anonymous, 1979). Mortality rates in oak are higher in trees with little or no reserves than in those with reserves. Low levels of starch in roots have also been recorded in oaks attacked by the buprestid *Agrilus bilineatus* (Haak & Benjamin, 1982). This observation may be interpreted in two ways: production of defensive substances by oaks is energy costly and this is provided by starch reserves; or starch favours more rapid growth of phloem and of the cambial area, which crushes and destroys *Agrilus* larvae settled in this area (Dunn *et al.*, 1987). Infestation of oak by the pathogenic fungus *Endothia parasitica* does not kill the tree, but causes malformations in the bark and wood. The fungus also disrupts the tree's physiology by some little understood mechanism making it more vulnerable to attacks by *Agrilus bilineatus*.

Partial or total removal of foliage from maritime pines enables an approximate simulation of damage inflicted by processionary caterpillars and comparison of the growth of these trees with that of control trees. Reduction in growth is marked in trees having undergone total defoliation, and loss of growth increases in succeeding years, as trees continue to lack vigour. The experiments justify the practice of felling for immediate use spruces that have been entirely defoliated by *Lymantria dispar*. A study on Austrian pine devastated by processionary caterpillars on Mont Ventoux in France showed a zero growth ring in 1968 following heavy infestations in the years 1966-1967. Growth of trees was more affected by feeding by caterpillars than by climate. Defoliation occurring in autumn of year $n - 1$ and in spring of year n affects growth in year $n + 1$. In the most heavily affected sector global loss was 35%. Parallelism between abundance of caterpillars and growth loss is remarkable (figure 6.3). The impact of *Bupalus piniarius* on the growth of Scots pine is felt over one to two years following peak abundance of the moth. Reduction in growth occurs even when the scale of defoliation is small. Volume and height of trees are reduced respectively by 5.6% and 1.5% (Straw, 1996).

Depredations by *Neodiprion sertifer* on Scots pine can reduce growth in volume by 33% when the rate of defoliation reaches 55% during two successive years, and growth in height may be reduced by 50%. Defoliation also reduces root growth (Britton, 1988).

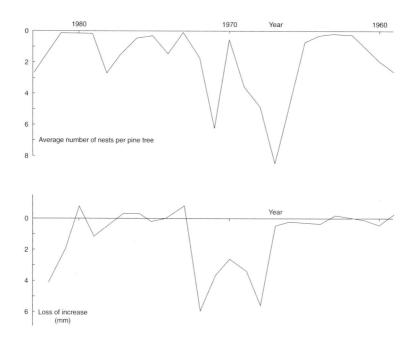

Figure 6.3. *Relationship between loss of annual radial growth of Austrian pine in the Mont Ventoux region and average number of nests of processionary caterpillars per tree (Laurent-Hervouet, 1986).*

In western Canada, defoliation by *Choristoneura fumiferana* led to a loss in volume of 134,000 cubic metres per year through reduction of growth rate and mortality. *Choristoneura pinus* caused a loss of 62,000 cubic metres through reduction of growth and 2,444,000 cubic metres through mortality. *Choristoneura conflictana* caused a loss of 24,000 cubic metres through reduced growth rate, and *Operophtera brumata* a loss of 163,000 cubic metres per year; these two species did not kill any trees. In comparison with these losses, those due to scolytids are insignificant: less than 1,000 cubic metres for *Dendroctonus ponderosa* and 2,000 cubic metres for *Dendroctonus rufipennis* (Brandt, 1995). The consequences of outbreaks of *Choristoneura fumiferana* on the growth and mortality of the fir *Abies balsamea* have been the subject of a theoretical model (figure 6.4).

Periods of intense defoliation of spruce by *Epinotia nanana* occurred in two consecutive years in Scandinavia (Austara, 1984). They were accompanied by a slowing of growth which went on for several years (figure 6.5). In Sweden, depredations by *Tomicus piniperda* on Scots pine are responsible for an annual reduction in growth varying between 9 and 27%. Attacks by this scolytid affect both needles and buds. Heavy and prolonged loss of growth is attributed to reduction in assimilation of nitrogen by the tree (Fagerström *et al.*, 1978). In Canada, the larch *Larix laricina* attacked by

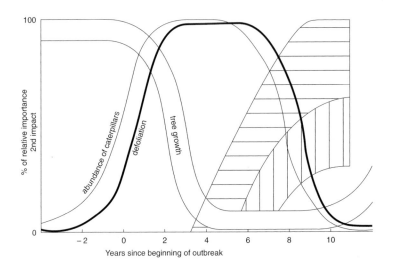

Figure 6.4.   *Periodic outbreaks of* Choristoneura fumiferana *on fir in Canada and their effects on levels of defoliation (thick line) and on tree growth. Tree mortality is represented by horizontal hatched lines for old trees and vertical hatching for young trees.*
Abscissa: time in years; ordinate: relative levels of abundance of the moth and its impact on defoliation and growth of trees (MacLean, 1981).

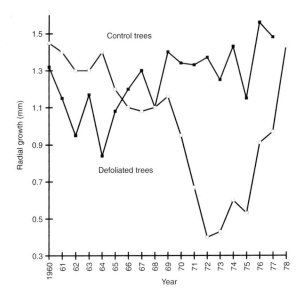

Figure 6.5.   *Consequences of defoliation of spruce by* Epinotia nanana. *Maximum defoliation occurred in 1970 and 1971, and slowing of radial growth continued for several years (Austara, 1984).*

Pristophora erichsoni shows changes in the anatomical structure of its wood: a decrease in the size of tracheids and an increase in relative quantity of summer wood (Filion & Cournoyer, 1995).

With the exception of the spruce feeder *Elatobium abietinum*, aphids do not usually cause defoliation or apparent damage. However their consumption of elaborated sap may affect growth. The impact of *Elatobium abietinum* on Sitka spruce results in both vertical and radial growth reduction. Following an infestation by this aphid which caused total defoliation of five to six year old trees, the growth of terminal shoots was slowed for at least two years. A study made on two year old trees grown in pots showed a real but slight (about 10%) reduction in growth of roots and terminal shoot, a reduction of up to 50% in mass (dry weight) of needles and buds, and a decrease of over 50% of total mass (Carter, 1977).

When the aphid *Drepanosiphum platanoides* which lives on maple (*Acer platanoides*) is abundant it induces formation of leaves with a surface reduced by 40% and the tree's wood production is reduced by 62% (Dixon, 1971). A maple tree 20 m high and bearing 116,000 leaves may harbour 2.25 million *Drepanosiphum*. A 16 year old spruce massively infested by the aphid *Cinara piceicola* showed decreased photosynthetic activity and consequent growth reduction estimated at 38% compared with non-infested trees (Eckloff, 1972).

4. PRIMARY AND SECONDARY PESTS

Primary pests are insects able to attack and live on healthy trees in good physiological condition (Escherich, 1914; Bovey, 1971). These are mainly sap-feeders and leaf-eaters. Secondary pests are those whose ability to develop is limited by a more or less deficient physiological state of their host. When environmental conditions favour them, secondary pests may become abundant and harmful. Secondary pests include nearly all xylophagous species such as scolytid, cerambycid and buprestid beetles and siricid wasps. The distinction between primary and secondary pests is hard to make. Some species are difficult to categorise, and secondary pests such as the scolytids *Ips sexdentatus* and *Ips typographus* may, during major outbreaks, attack healthy trees and thus become primary pests. The scheme proposed by Kangas (1950) acknowledges the difficulty in assigning various species to one or other category. Between the absolute primary nature of insects which only ever attack healthy trees in good physiological condition and the absolute secondary nature of insects associated with dead wood there are intermediate stages corresponding to populations of fallen, diseased or moribund trees.

5. POPULATION DYNAMICS OF FOREST INSECTS

In the wild no species maintains a constant population. There are always variations in abundance throughout the year and from one year to the next. Gradation corresponds to the duration of a fluctuation of abundance between two periods of minimal abundance. This may last several years and include periods of progradation, culmination and retrogradation. Periods of minimal abundance between two gradations are known as latency (figure 6.6).

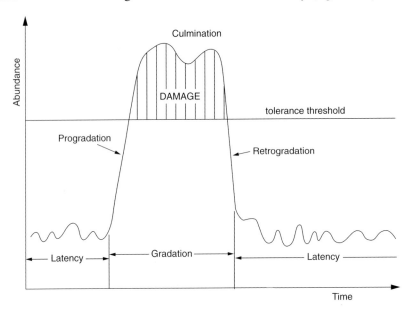

Figure 6.6. *Schematic representation of variations in abundance of an insect during a gradation phase. When numbers exceed the tolerance threshold damage appears (area in hatched lines).*

There are four types of fluctuation of abundance. The *latent* type is characterised by small fluctuations which always remain below the tolerance threshold. This type occurs in innocuous insects which do not usually draw attention but represent the great majority of species. The *permanent* type designates species whose populations do not vary much but are always above the tolerance level. These species are rarer in forests than on cultivated land (where they are known as permanent pests). Forest species include, for instance, *Tortrix viridana*, *Coleophora laricella* and *Dreyfusia nordmannianae*. The *temporary* type refers to occasional pests, in other words all secondary destroyers and some periodic primary pests. These insects, which are present permanently in dead or dying trees, occur in large numbers when conditions become favourable. One example of this type is provided by the scolytid *Ips typographus* which in southern Norway in 1978 was confined to a number of restricted localities and which, following favourable circumstances, spread over a vast area surrounding the original localities (Worrell, 1983). The *periodic* type is characterised by more or less regular cyclical gradations which vary in extent from one species to another and from region to region. An example of the periodic type of fluctuation is provided by four species of moth (figure 6.7). *Panolis griseovariegata* showed peaks of abundance in 1883, 1888, 1893 and 1897, suggesting a period of about five years. However oscillations stopped between 1899 and 1910, to be followed by more peaks of abundance later. *Hyloicus pinastri* showed similar variations in abundance. *Dendrolimus pini* is a usually rare species which only occurred in two gradation phases. *Bupalus piniarius* showed constant fluctuations and no stationary phases as in the other three species. In larch forests in the Alps *Zeiraphera diniana* has long occurred in remarkably regular gradations lasting eight or nine years with periods of damage lasting from one to three years.

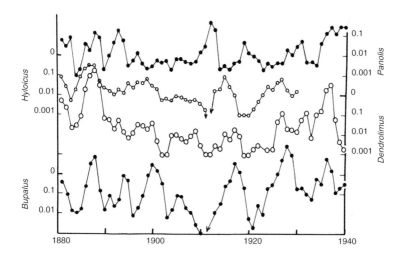

Figure 6.7. *Results of counts of pupae (or hibernating larvae in the case of* Dendrolimus*) over 60 successive winters in a German forest. Abundance is indicated by the logarithm of numbers of insects (Varley, 1949).*

Many forest Lepidoptera show large variations in abundance which are termed cyclical because they are often, although not always, remarkably regular. The best known case is that of *Zeiraphera diniana*, but there are others, such as those of *Lymantria dispar, Epinotia tedella, Oporinia autumnata, Operophtera brumata* and *Bupalus piniarius* in Europe, or *Acleris variana, Malacosoma californicum pluviale* and *Orgyia pseudotsugata* in North America. Despite several decades of research, the causes of cyclical fluctuations of animal populations (insects and also microtid rodents and various birds) are still not clearly understood. In the case of Lepidoptera, climatic factors, biological factors such as epizootic viral diseases, decrease in the nutritional quality of leaves and acquired resistance in plants, intra-specific competition and lack of resources, and the roles of predators and parasitoids have all been taken into account. Decrease in nutritional quality of leaves can only be of consequence if it is affected by defoliation occurring in the previous year and if the new foliage changes survival rates and fertility in the insect that consumes it, or leads to a change in genotype in genetically polymorphic species. The effects of pathogenic organisms have been con-sidered owing to the occurrence of spectacular epizootic diseases which kill a high percentage of larvae, but this theory is increasingly disregarded for want of convincing observations (Bowers *et al.*, 1993). Although climate may explain synchronism of outbreaks in areas at great distances from each other, variations in this factor do not appear to be regular enough to account for the periodically very regular outbreaks of some species. Effects of cli-matic factors may however explain more or less regular outbreaks such as those of *Tortrix viridana*. Leaf-eating Lepidoptera are predated by very many parasitoids, both Diptera and Hymenoptera. *Zeiraphera diniana* is host to 95 species of parasitoid. A number of predators also exist. An old theory that now appears increasingly convincing is that interaction with parasites is the essential factor in cyclical fluctuations in populations of forest Lepidoptera (Berryman, 1996).

6.   WHY DO SOME INSECTS BECOME PESTS?

The factors which cause an insect to become harmful are very diverse. Many are climatic, and act either by inducing a state of stress which makes trees less resistant to attacks, or by increasing the reproductive potential of insects (increased fertility, increased survival rate). Ill-considered interference by man, such as inappropriate forestry methods or the introduction of alien species, may have similar effects.

Stress is a reaction to any factor which affects the normal functioning of a plant and induces potentially harmful physical or metabolic changes. Stress as a factor may act for a relatively brief period or in a permanent way. Low fertility, too high or low pH levels in the soil, drought, floods, atmospheric pollution, and abnormally high or low temperatures are all factors of stress. The phenomenon known as fluctuating asymmetry designates variations in the bilateral symmetry of leaves resulting in the plant's inability to control symmetry when exposed to stress. In wych elm (*Ulmus glabra*) subjected to stress induced by excessive doses of fertiliser, leaf asymmetry is more fre-quent than in untreated trees, and asymmetrical leaves are more often attacked by mining larvae of the weevil *Rhynchaenus rufus* than are sym-metrical leaves (Møller, 1995).

6.1. Drought

The hypothesis that plant stress caused by a climatic factor such as drought leads to an outbreak of an insect was put forward as early as 1969 by White to explain outbreaks of the psyllid *Cardiaspina densitexta* which lives on *Eucalyptus fasciculosa* in southern Australia. According to this author, populations of many phytophagous insects are controlled by a deficiency in nitrogen compounds (amino acids) in the host plant. In response to stress caused by drought the plant produces more nitrogen compounds, which allows an increase in abundance of the insect. This theory was later applied to the moth *Selidosoma suavis* which feeds on *Pinus radiata* in New Zealand (White, 1974).

The notion that stress due to drought facilitates predation by phytophagous insects is supported by numerous observations (table 6.2). Outbreaks of the pine sawfly *Neodiprion sertifer* were preceded by periods of hot dry weather which weakened trees and raised growth and survival rates of the insects (Larsson, 1989). Outbreaks due to the same causes have been observed in *Dreyfusia nuesslini* in the Vosges, and in various parts of France spectacular proliferations of scolytids followed dry years such as those of the period 1945-1950 (Joly, 1977).

Orders	Families	Genera	Trees attacked
Coleoptera	Buprestidae Cerambycidae Scolytidae	*Agrilus* *Tetropium* *Corthylus* *Ips* *Dendroctonus* *Scolytus*	*Betula, Populus, Quercus* *Abies* *Acer* *Picea, Pinus* *Picea, Pinus* *Abies, Carya*
Homoptera	Psyllidae	*Cardiaspina*	*Eucalyptus*
Hymenoptera	Diprionidae	*Neodiprion*	*Pinus*
Lepidoptera	Geometridae Tortricidae Lymantriidae	*Bupalus* *Lambdina* *Selidosema* *Choristoneura* *Lymantria*	*Pinus* *Abies, Picea* *Pinus* *Abies, Picea, Pinus* *Picea*

Table 6.2. *Forest insect genera which have occurred in outbreaks following periods of drought.*

Following the exceptionally severe drought of the summer of 1976, conifer forests in France were invaded by various species of scolytid and "death spots" were discovered in many areas (figure 6.8). Attacks on spruce were the work of *Ips typographus,* often associated with *Pityogenes chalcographus* and *Dendroctonus micans*; those on fir were due to *Pityokteines*

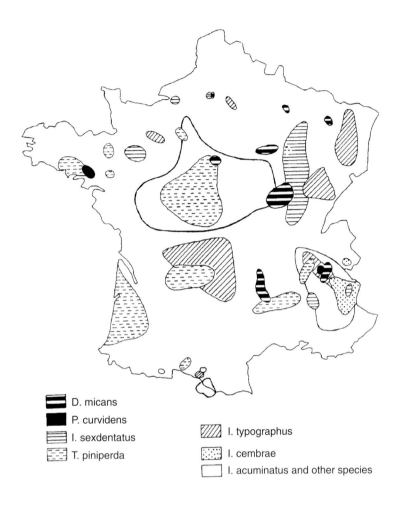

Figure 6.8. *Distribution of death spots of conifers caused by scolytids reported in France between 1974 and 1978. Spots are due (at least in part) to the great drought of the summer of 1976 and to depredations by scolytids on trees subjected to hydric stress (Anonymous, 1978).*

curvidens, often followed by *Cryphalus abietis*; those on pine were due to *Ips sexdentatus, Tomicus piniperda* and *Ips acuminatus,* and those on larch were due to *Ips cembrae.* Other scolytids as well as curculionids and buprestids settled in the areas, but only on dying or dead trees (Anonymous, 1978).

   Abnormal temperatures and rainfall may also enfeeble trees, which then become more vulnerable to pests. In the Sierra Nevada in California, annual mortality of Colorado white fir (*Abies concolor*) due to depredations by the scolytid *Scolytus ventralis* and the cerambycid *Tetropium abietis* is in inverse proportion to peaks of annual rainfall and to mean rainfall and temperature in spring (figure 6.9). In this region attack by scolytids on conifers is often favoured and preceded by invasions of pathogenic organisms responsible for disease in the roots, by colonisation by parasitic plants related to mistletoe belonging to the genera *Arceuthobium* and *Phoradendron*, and by parasitic fungi such as *Heterobasidion annosum, Armillaria* sp., *Leptographium* sp. or by the rust *Cronartium* sp.

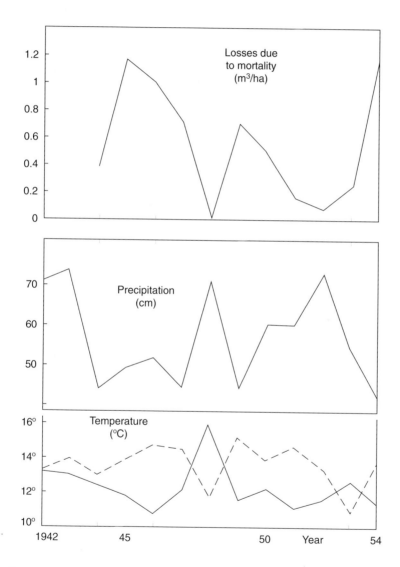

Figure 6.9. *Effects of temperature and rainfall on mortality in Colorado white fir* (Abies concolor) *in the Sierra Nevada. A: annual mortality expressed in numbers of cubic metres lost per hectare.*
B: annual variations in rainfall. C: annual variations in mean temperature in spring (dashes) and rainfall (solid line) (Ferrell & Hall, 1975).

Drought affects trees in different ways, the main ones being as follows: growth reduction (and particularly reduction in size of leaves, buds and annual growth rings); changes in chlorophyll synthesis causing yellowing of leaves; rise in temperature due to reduced transpiration through occluded stomata; increased levels of soluble nitrogen compounds in leaves; breaks in the liquid column of sap in xylem, which produces sounds audible to insects; variations in osmotic pressure; emission of volatile compounds, and, in coni-fers, changes in resin terpenes and decreased ability to exude oleoresins.

For the insects these changes result in improved nutritional quality of the plant, and in greater attraction to the plant, particularly due to the yellow colour of leaves, and perhaps also to the nature of sounds produced by trees when the sap column is broken. *Choristoneura fumiferana* reacts to sucrose levels ranging between 0.01 and 0.05 M. In a normal *Abies balsamea* tree sucrose concentration in leaves ranges between 0.004 and 0.011 M. Increase in this concentration in trees stressed by drought leads to greater consump-tion of leaves by caterpillars. Decreases in terpene levels in conifers suffering from drought make these substances attractive to scolytids. Concentrations of α-pinene, which is only slightly toxic, increase, while levels of mycene and limonene, which are highly toxic, decrease (Cates & Alexander, 1982). Female *Neodiprion sertifer* lay twice as many eggs on pines which only receive normal rainfall as they do on irrigated trees (Saikkonen *et al.*, 1995).

Drought may also benefit micro-organisms that live symbiotically with phytophagous insects. Stressed conifers have higher levels of glucose and fructose in the cortical area than do normal trees. This favours growth of the fungus *Ceratocystis minor*, an external symbiont of the American scolytid *Dendroctonus frontalis*. Rapid growth of the pathogenic fungus facilitates colonisation of the tree by the beetle by reducing resin production. Drought may also trigger changes in the genetic structure of insect populations and favour individuals whose genotype makes them more aggressive. In some insects such as the larch budworm the dominant genotype is not the same during gradation periods as in latency periods.

Abnormally low or high temperatures may produce stress and be fol-lowed by outbreaks of insect pests. Oaks enfeebled by severe cold in winter undergo changes in their bark, and may then be attacked by xylophagous insects such as buprestids of the genus *Agrilus* or pathogenic fungi such as *Armillaria* sp. or *Hypoxylon* sp. (Wargo, 1996).

6.2. Atmospheric pollution

Atmospheric pollution has become a widespread phenomenon; its effects on forest and on insects has been the subject of much research (Charles & Villemant, 1977; Villemant, 1980, 1981; Alstad *et al.*, 1982; Heliövaara & Väisänen, 1986; Nevalainen & Liukkonen, 1988; Riemert & Whittaker, 1989;

Nef, 1994, etc.). In central Europe pollution is thought to be responsible for the decline of forests known as "Waldsterben".

At the individual level pollutants may change insects' food preferences, growth rates, duration of life cycles, pupal body size, fertility, and survival rates. At population level pollution may lead to changes in genetic structure, spatial distribution, sex ratio, and age structure. The most commonly observed effect however is variation of abundance (Riemer & Whittaker, 1989). The main pollutants whose effects have been studied are sulphur dioxide, nitrogen oxides, ozone and some heavy metals.

The effects of pollutants may be classed under three headings:

(a) Pollutants may have a direct effect on insects. The little experimental proof that is available consists of feeding insects in artificial environments subjected to effects of pollutants. A direct effect has been established in some cases (Whittaker & Warrington, 1990).

(b) Pollutants may have a direct effect on host plants. Changes in chemical composition of plants subjected to pollutants are known. Tissues and sap are enriched in soluble nitrogen compounds, particularly in amino acids, as well as in soluble sugars, which increases their nutritional value and stimulates insect reproduction. The chemical composition of leaves may also be changed by pollutants such as heavy metals. Percentages of beech leaves attacked by the aphid *Phyllaphis fagi* correlate significantly with cumulative levels of calcium, copper, cadmium and zinc in the leaves (Nef, 1994). Furthermore pollutants may reduce a tree's ability to secrete defensive substances such as resin terpene in conifers (Larsson *et al.*, 1986; Rhoades, 1979).

(c) Pollutants may change the equilibrium between phytophagous insects and their predators or parasitoids; the latter are more sensitive to pollutants than phytophagous insects. Populations of parasitoid wasps are often much more strongly affected than their prey, and the resulting loss of equilibrium increases outbreaks of pest species.

Insects may be classed in four groups: (a) insects such as *Rhyacionia buoliana, Exoteleia dodecella* or *Pristophora abietina* that benefit from pollution, and which occur in outbreaks near centres of pollution; (b) those such as *Phaenops cyanea, Ips sexdentatus* and *Aradus cinnamomeus* which occur at high density in moderately polluted areas and at low density in unpolluted or heavily polluted areas; (c) those such as *Pityokteines curvidens* that are common in unpolluted areas and scarce or absent in polluted areas; (d) those whose population density does not alter in polluted areas. Aphids mostly belong to category a or b, but there are exceptions. *Cinara pini* and *Protolachnus agilis* are more abundant in polluted areas, while *Pineus pini* and *Schizolachnus tomentosus* show the opposite tendency (Villemant, 1981). Scolytids appear to benefit from pollution. Abundance of mining or boring species may increase or decrease in polluted areas (table 6.3).

Aphids	**Lepidoptera**
Cinara pini	*Zeiraphera diniana*
Eulachnus agilis	*Exoteleia dodecella*
Schizolachnus pineti	*Rhyacionia buoliana*
Sacchiphantes (=Adelges) abietis	*Petrova resinella*
Dreyfusia nordmannianae	*Coleophora fuscedinella*
Scolytids	*Dioryctria mutatella*
Tomicus piniperda	*Blastesthia turionella*
Tomicus minor	*Thaumetopœa pityocampa*
Ips typographus	**Curculionids**
Ips acuminatus	*Strophosomus capitatum*
Trypodendron lineatum	*Pissodes piniphilus*
Dendroctonus micans	**Buprestids**
Heteroptera	*Phaenops cyanea*
Aradus cinnamomeus	**Cerambycids**
Thysanoptera	*Acanthocinus aedilis*
Taeniothrips laricivorus	**Hymenoptera**
Diptera	*Sirex* spp.
Thecodiplosis brachyntera	Some ichneumonids, chalcidids, braconids

Table 6.3.   *Insects that occur frequently in woodland affected by atmospheric pollution (Nevalainen & Liukkonen, 1988).*

The responses of insects to pollutants depend largely on their diet. The following grouping of insects according to increasing resistance to pollutants has been proposed:

cambium feeders > sap feeders > miners > chewers > gall insects.

This scheme shows that atmospheric pollution has a positive effect on sap-sucking insects and a negative or no effect on chewing insects. Effects of pollutants may be compared to those of pesticides, as low doses of the latter cause metabolic changes in the plant to which they are applied and make it more vulnerable to attacks by insect pests.

Pollution involving fluorine and sulphur extends the sensitivity period of Scots pine cones and facilitates their predation by insects (Roques *et al.*, 1980). The bug *Aradus cinnamomeus* is a sucking insect harmful to pine. A transect extending from a polluting factory to an unpolluted area showed variations in abundance of this bug. *Aradus* density is very low in the vicinity of the source of pollution, where the latter has a lethal effect. Density increases rapidly and reaches a maximum (up to 100 times the density in the unpolluted area) at a distance of one to two kilometres from the source of pollution. Density then decreases gradually until it reaches a normal value (Heliövaara & Väisänen, 1986). Eggs of *Aradus cinnamomeus* are parasitised by the scelionid wasp *Telenomus aradi*. Percentages of parasitised eggs are considerably lower in the vicinity of the polluting factory, varying between about 18% at 500 metres to 30% at 2,500 metres from the factory. Avoidance of polluted areas by the parasite is one of the causes of population increase of *Aradus*.

Industrial atmospheric pollution favours many pests of conifers (Charles & Villemant, 1977; Villemant, 1980, 1981; Heliövaara, 1986). Among those species whose numbers are rising are microlepidoptera which feed on buds and shoots of Scots pine, such as *Rhyacionia buoliana* and *Petrova resinella*; scolytid beetles such as *Tomicus piniperda* and the buprestid *Phaenops cyanea*; sawflies such as *Pristophora abietina* which is a pest of spruce; *Sirex* on Scots pine; and chermesids on pine and spruce. In the Normandy forest of Roumare which is polluted by effluents of sulphur and fluorine there has been a significant increase in a few species associated with Scots pine such as *Rhyacionia buoliana, Schizolachnus tomentosus, Protolachnus agilis* and *Cinara pini*. Percentages of depredations by *Exoteleia dodecella* and *Rhyacionia buoliana* rise with the degree of pollution by sulphur dioxide and fluorine (table 6.4). The complex of entomophagous parasites of *Rhyacionia buoliana* is changed by pollution. Sensitive ectoparasitic species have regressed, while endoparasitic species have been favoured by or remain indifferent to pollution, according to different cases. It seems that obligate parasites of *Rhyacionia buoliana* have suffered decreases in population at the expense of occasional polyphagous parasites such as the ichneumonid *Macrocentrus thoracicus*. In Austria the sawfly *Pristophora abietina*, which lives on spruce, has become a serious pest since 1960. Atmospheric pollution by sulphur dioxide and nitrogen oxides increases the nutritional value of needles and plays a key role in population outbreaks. A correlation has been established between levels of compounds containing the radical SH in spruce needles in October and the number of sawflies the following year (Schafellner *et al.*, 1996).

	Unpolluted	Low pollution	Moderate pollution	polluted
Fluorine in needles (in ppm)	13.37	22.81	22.87	28.78
Sulphur in needles (in SO_2 ‰ dry weight)	1.35	1.72	1.85	1.87
% of attacks by *Exoteleia dodecella* in 1978	0.44	0.95	2.27	10.96
% of attacks by *Rhyacionia buoliana* in 1977	0.5	0.8	6.6	27.4

Table 6.4.   *Percentage of attacks on Scots pine by two moths in various more or less polluted environments in Roumare forest in Normandy (Villemant, 1980).*

Niepolomice pine forest in Poland is intensely polluted by local indus-
tries. Enfeebled trees have reduced osmotic pressure and the composition of
their resin terpenes is changed. This allows massive depredations by xylophagous secondary destroyers which cause a high mortality rate in the trees.
The group of pest species thus benefiting from pollution includes scolytids
(*Tomicus minor* and *T. piniperda, Pityogenes bidentatus, Pityophtrorus
pityographus*), a curculionid (*Pissodes pini*), and cerambycids (*Monochamus
galloprovincialis, Acanthocinus aedilis, Arhopalus rusticus, Rhagium inquisitor, Callidium aenum*). Concentrations of pollutants in plant tissues are high.
These pollutants are found in the insects' bodies and faeces, and return to the
soil when the insects die. They may thus be transmitted through food chains
and be present in insectivorous birds such as woodpeckers. Concentrations
of various heavy metals in insects' bodies and faeces may be higher than in
plant tissues such as phloem, particularly in the case of iron. Manganese and
copper, on the other hand, are eliminated, and lead does not seem to be
assimilated at all, as the following figures show (nd = "non-detectable"):

	Concentrations of elements (in ppm)						
	Zn	Cd	Ni	Fe	Mn	Cu	Pb
Phloem	69	3.6	5.3	137	152	3.4	4.8
Insects bodies	375	9.3	62.5	281	46	87.5	nd
Faeces	312	2.6	nd	2,589	72	89.3	nd

In the same Polish forest an increase in abundance of cynipid gall wasps
was recorded, particularly *Cynips lignicola* which swarms on young oaks,
where it damages buds; growth is thus inhibited and trees are reduced to
bushes. *Tortrix viridana* has also become much commoner than in the past
(Borusewicz, 1984). Niepolomice forest has for many years been subjected
to uncontrolled exploitation which has had a disastrous effect on the health
of the woods. Outbreaks of *Panolis flammea, Dendrolimus pini, Bupalus
piniarius, Neodiprion pini* and *Acantholyda nemoralis* have all occurred,
beginning in the last century.

Variations in population density of the chrysomelid beetle *Melasoma
lapponica*, which lives on willow, with distance from a source of pollution
have been recorded in northern Russia. In moderately polluted localities,
density is ten to twenty times higher than in highly polluted or scarcely polluted areas (table 6.5). When pollutant levels in the air reach 400 to 1,000 µg
per cubic metre the insect concentrates on *Salix borealis*, whereas in less
polluted areas it also feeds on other willows. *Salix borealis* is the only
species on which the insect can survive in highly polluted areas (Zvereva *et
al.*, 1995; Zvereva & Kozlov, 1996). Insects that live on ash bark vary in
abundance according to intensity of pollution. Algae and ephiphytic lichens
such as *Pleurococcus* and *Lecanora conizaeoides* become less abundant, as

do insects such as Psocoptera which feed on them. Abundance may decline from 400 to 2 individuals from an unpolluted to a polluted area (Gilbert, 1971). Several species of moth which rest on tree trunks tend to develop melanic populations, the pale forms being eliminated in industrial regions where pollutants cover tree trunks with a layer of dark grime. This industrial melanism is seen in several species of geometrid moth, including *Biston betularia* which lives on birch, oak, elm and beech, and *Gonodontis bidentata* which lives on oak (Ford, 1972). It is also known in the psocid *Mesopsocus unipunctatus* (Popescu *et al.*, 1978).

Distance from polluting factory	SO$_2$ µg/m^3	Metals in leaves of *Salix borealis* (mg/kg)		*Melasoma* found in ten minutes
		Ni	Cu	
1 kilometre	167	273.3	95.4	5.0
5 kilometres	162	145.7	58.8	15.0
14 kilometres	76	130.5	71.1	341.0
29 kilometres	36	27.7	16.9	7.3

Table 6.5. *Effects of atmospheric pollution on abundance of the chrysomelid* Melasoma lapponica *(Zvereva & Kozlov, 1996).*

Motor vehicle exhaust fumes fill the air with pollutants such as nitrogen oxides and lead. Levels of nitrogen compounds in plants increase in the vicinity of roads and are the probable cause of outbreaks of insects seen in such places. Lead passes from plants to herbivores and increases in concentration along the food chains. Outbreaks of *Phalera bucephala* on beech and of *Euproctis similis* on hawthorn have been recorded by the sides of a motorway in Britain (Port & Thompson, 1980). These observations confirm the hypothesis according to which abundance in populations of herbivores is limited by food quality and in particular by levels of soluble nitrogenous compounds (Mattson, 1980).

Defoliation of a tree by caterpillars is a stress factor which favours colonisation by destroyers which exploit weakness, and seek trees suffering from physiological deficiency. In this way, following severe defoliation of Douglas-fir by *Orgyia pseudotsuga* larvae, two scolytids, *Scolytus ventralis* and *Dendroctonus pseudotsugae*, infested trees that had been stripped of more than 90% of their foliage (Wright *et al.*, 1984).

6.3. Consequences of global change

Various activities by man are responsible for changes called "global" because they affect the whole biosphere. The two main changes are increases in carbon dioxide and ozone in the atmosphere. Concentration of carbon dioxide in the atmosphere is rising by an average of 0.5% per year. This will lead to a rise in average global temperature, changes in patterns of rainfall and wind speeds and frequency of exceptional climatic events, and it will have profound repercussions on the abundance and distribution of living organisms. Insects will certainly react very quickly due to their short life cycles and their mobility (Watt *et al.*, 1996). The possibility of some of them becoming pests of agriculture and forests has been envisaged. This is likely to be true of the aphid *Elatobium abietinum* which, in North West Europe, is the main enemy of European spruce and Sitka spruce, both of which are planted on vast areas (530,000 hectares in Britain). This aphid is at present controlled in Britain by cold periods which reduce hibernating populations. If climate becomes milder, winter mortality will decrease and the aphids will be able to multiply more intensely in spring and autumn. Earlier budding of spruce will favour the multiplication of the insect even further. Since spruce trees enfeebled by aphids are more vulnerable to attacks by the scolytid *Dendroctonus micans*, increased ravages by the latter are feared (Straw, 1995).

Increases in levels of carbon dioxide affect a tree's physiology and productivity (Eamus & Jarvis, 1989). More specifically, relative levels of carbon and nitrogen increase following accumulation of starch in leaves. It is thought that these changes will affect the biology of phytophagous insects. Gypsy moth larvae fed on young aspen and maple grown in a carbon dioxide enriched atmosphere show changes in growth rate, final body size and weight, and in the quantity of foliage they consume. The magnitude of these changes varies according to tree, which makes it difficult to make general predictions relating to the effects of increased atmospheric carbon dioxide (Kinney *et al.*, 1997).

Trees weakened by ozone are more vulnerable to attacks by scolytids and various pathogenic agents. This has been established in the San Bernadino Mountains in southern California, where 25% of ponderosa pines died between 1973 and 1992 (Stark *et al.*, 1968). Exposure of trees to ozone generally favours attacks by insects and multiplication of aphids such as *Phyllaphis fagi* on beech or *Schizolachnus pineti* on Scots pine. These reactions are associated with disruptions of the tree's metabolism, and more particularly with increases in relative ratios of amino acids to sugars in phloem (Braun & Flückiger, 1989).

Climatic change may alter the geographic distributions of trees and phytophagous insects. This possibility has been studied in the cases of the defoliating moth *Choristoneura occidentalis* (Williams & Liebhold, 1995) and of the chermesid *Dreyfusia piceae*, which has become a dangerous pest of *Abies lasiocarpa* in North America. *Dreyfusia piceae* is at present limited by its thermal requirements to low and medium altitudes, but will be able to colonise high altitude localities as temperatures rise, and thus extend the range of its damage (Dale *et al.*, 1991). Many specialists believe that the enhanced greenhouse effect will increase numbers of pest species and the extent of their depredations, and that this will be compounded by stress on trees which will be in a state of lowered resistance (Mattson & Haack, 1987).

Predictions of the reactions of phytophagous insects to global change are difficult to make because not all species react in the same way. A recent experiment (Lindroth *et al.*, 1993) was carried out on larvae of *Lymantria dispar* and *Malacosoma disstria*; these were reared on one year old saplings of *Populus tremuloides, Quercus rubra* and *Acer saccharum,* some of which were cultivated during sixty days in a normal atmosphere of 385 ppm of carbon dioxide and others in air enriched to 642 ppm of carbon dioxide. Both species increased consumption of leaves on poplar but their growth rates slowed. Responses differed on oak and maple for reasons which remain unknown. *Lymantria* grows more rapidly on oak but shows no reaction on maple; *Malacosoma* grows less well on maple and does not react on oak. It should be noted that these experiments only concern increase in carbon dioxide and not other climatic factors altered by global change. Watt *et al.* (1996) have studied *Panolis flammea*, which lives on many species of tree, in the light of temperature rise. If the latter is gradual, the moth may progressively adapt to variations in the date at which the host trees' buds open, and phenological coincidence between insect and tree will be maintained. If temperature variations are greater, synchronization may disappear and lead to a decline of the species.

7.   FOREST PEST CONTROL

Protection of forests against pests must take special features of the forest ecosystem into account. Woodland, even where artificial, is a permanent environment in which networks of complex interactions are created and in which numerous factors regulating populations are at play. Cultivated crops are threatened by permanent pests; forests are most often threatened by occasional or cyclical pests. Populations of primary destroyers such as defoliating insects maintain themselves for more or less long periods of latency and only occur in the form of outbreaks at fairly long intervals.

Methods of pest control must be low cost given the relatively low commercial value of forest products compared with cultivated ones. Implementation requires a thorough knowledge of insect biology and of the functioning of the forest ecosystem.

Forests are in a more precarious situation today than in the past for various reasons. International traffic and exchanges of goods have increased chances of introducing harmful organisms from other regions. Joly (1975) recorded the presence near Rouen in Normandy of several species of North American bark beetles over stocks of logs, which had not been stripped of bark, coming from Canada. The establishment of uniformly planted woods increases the risk of massive outbreaks of pests. Morris (1963) has shown how monoculture of the fir *Abies balsamea* in Canada provides optimum conditions for the spread of *Choristoneura fumiferana*. Similarly, intensively managed plantations of pines of the same age which have been established in the south-eastern United States are vulnerable to outbreaks of pests like *Dendroctonus frontalis* (Schowalter *et al.*, 1981). Most tits, which are major consumers of insects, cannot nest and settle in this type of woodland for lack of hollow trees. The nature of woodland and type of exploitation are the origins of differences in insect population (Voute, 1964). Primary forest, and traditional forest, which still resembles the former in its diversity and hetero-geneity, are better protected from outbreaks of pests than "cultivated" woodland, composed of a single species (often a conifer) and of trees of the same age, of the kind that tends to be planted everywhere today (Lattin & Oman, 1983).

Barbey (1942) has spelled out that "on the pretext of intensifying forestry production man has unwittingly caused the proliferation of insect pests. There is an undeniable correlation between man, his activities and his ap-petites and the propagation of insects; the study of forest entomology demonstrates the evidence of this".

Plantation of a tree species outside its natural range of distribution may favour outbreaks of insects when the latter find environmental conditions favourable, all the more since in most cases these pests are introduced without their natural enemies. This is what happened in the case of larch which, during the last 150 to 200 years, has been planted in lowlands. Inappropriate soils, unfavourable climatic conditions and bad forestry methods have allowed fungal diseases such as larch canker (caused by *Trichoscyphella laricella*) and continuous attacks by insects like the needle borer *Coleophora laricella*, various sawflies belonging to the genus *Pristiphora* such as *P. erichsoni* and *P. laricis*, the cerambycid *Tetropium gabrieli*, the scolytid *Ips cembrae*, and the chermesids *Adelges laricis* and *Sacchiphantes viridis* (Postner, 1963).

To local conditions which may be unfavourable to trees and favourable to insects one must add the effects of atmospheric pollution, which is spread-ing, and the possible consequences of global change.

7.1. Chemical control

Disruption brought about by ill-considered treatments with insecticides may act on forest zoocenoses with far greater intensity than on cultivated land. Destruction of salmon in several Canadian rivers and high mortality rates of birds in the Swiss Valais are just two consequences of aerial spraying. Chemical control has been used on a much greater scale in North America than in Europe, where it is often seen as an emergency measure that is only taken for want of anything better (Hobart, 1977; Way & Bevan, 1977).

Chemical control programmes should only be undertaken when normal pest control factors have proved ineffective. Applications of insecticide should also be as localised as possible, for instance by spraying only the tops of young pines in order to protect them from *Rhyacionia buoliana*, or by painting rings of insecticide and gum on pine trunks to catch adult *Dendrolimus pini* moths which climb the trunks when they emerge from hibernation.

One of the drawbacks of repeated treatment with insecticides is the appearance of resistance to DDT and other insecticides. Proliferation of *Tortrix viridana* in Spain resulted from the consequent loss of natural equilibrium following repeated spraying of insecticide intended to control *Lymantria dispar* (Dafauce, 1970). More than four million hectares of forest were treated with insecticides in that country between 1953 and 1968. Control of the lime aphid *Eucallipterus tiliae* with an aphicide (a selective insecticide aimed at aphids) favours selection of resistant individuals of the mite *Eotranychus tiliarium* which is just as destructive as the aphid. Tetranychid mites are not killed by treatment with one insecticide, benomyl, but the fertility of their predators is reduced, provoking outbreaks of the mite (Carter, 1975). This phenomenon, called "pest resurgence", is well known in agriculture. Conversely, sub-lethal doses of an aphicide called primicarbe increase the efficiency of the parasitoid wasp *Encarsia formosa* (Irving & Wyatt, 1973).

Some of the harmful and unexpected effects of insecticides may be shown by a single example. DDT was sprayed in Canada to control *Choristoneura fumiferana*. The insecticide caused a large drop in numbers of the sawfly *Gilpinia hercyniae*. This drop was accompanied by a virtual eradication of parasitoids and of the virus which had been an effective controller of the sawfly. When treatments of DDT ceased the sawfly rapidly reconstituted its population and became a pest, but the viral disease and the parasitoids did not recover their former populations (Neilson *et al.*, 1971).

One frequently overlooked drawback of chemical control is its high cost. In the United States treatment using the insecticide BHC (benzene hexachloride) to protect pines against the scolytid *Dendroctonus frontalis* costs from one to ten dollars per tree (Thatcher *et al.*, 1981).

Chemical control of the weevil *Hylobius abietis* is done by powdering with lindane or by dipping conifer saplings in permethrin, the latter method being cheaper and ecologically much more acceptable (Nef & Minet, 1992). Insecticides are frequently used against aphids, particularly *Elatobium abietinum*, which attacks Sitka spruce in nurseries and Christmas tree plantations. Some of the insecticides used seem to have low phytotoxicity, but others cause growth problems in young trees (Straw *et al.*, 1996).

Protection of "seed orchards" against destroyers of conifer cones using insecticides was carried out for some twenty years in North America. This method of control has only created new problems: toxicity of substances used on trees, insect resistance to insecticides and outbreaks of secondary destroyers which had previously been innocuous. Every year insects from untreated populations invaded the orchard again. Protection of these orchards ought to rely on mechanisms which control insect populations, and on a systematic harvesting of all cones every year even if they are few (Roques, 1988).

BHC used to control *Dendroctonus frontalis*, which attacks pines in Texas, proved ineffective, partly because the pest's natural enemies – predators, parasitoids and competitors – were killed by the insecticide. In six months the scolytid increased its population sixfold on treated trees and only twofold on untreated trees. In the words of Vité (1971), "The insecticide does probably more harm than good as it is more efficient against predators and competitors than against the pest".

Diflubenzuron is an insecticide which acts by interfering with moulting in insects. It is only active on larvae, and has been used to control *Lymantria dispar* and other defoliating insects.

7.2 Biological control

"To manage forest insect populations by sound ecological means in order to prevent rather than suppress outbreaks is the continuing dream of forest entomologists and, generally, forest resource managers" (Waters, 1971). The aim of biological control is not eradication of an undesirable species (we know such eradication is almost impossible) but the containment of its numbers below an economic tolerance level, below which damage is sufficiently reduced as to be negligible. Determination of this level is difficult and, for a given species, may vary according to the nature of cultivated plants. In the case of depredations by aphids the tolerance threshold is lower in young trees grown in nurseries and destined for reafforestation, or to be sold as Christmas trees, than it is for old trees in woodland.

7.2.1. Use of entomophagous insects

Recourse to predatory or parasitoid entomophagous insects is an old technique. It was two American entomologists, Howard and Fiske, who in 1910 first had the idea of importing the carabid beetle *Calosoma sycophanta* from Europe to control *Lymantria dispar* which had carelessly been introduced to the United States and had become a real plague. Natural enemies used in biological control are mostly insects and other insect-eating arthropods which may be predators or parasitoids, and pathogenic agents. A great amount of research has been devoted to biological control in forest (Franz, 1976; Grison, 1970; Pschorn-Walcher, 1977; Greathead, 1976; Guilbot, 1991; Fraval, 1994, etc.). Pest control using entomophagous insects involves at least forty species of forest insect in Europe. The most frequently used parasitoids are ichneumonid, braconid and chalcid wasps and tachinid flies.

SOME DEFINITIONS

Biological control. Use by man of natural enemies such as predators, parasitoids, or pathogenic agents to control populations of pest species and keep them below an *economic harm level.* Where the enemy used is an animal (nearly always an insect), it is controlled by an entomophagous insect, which may be a predator or a parasitoid. When the enemy used is a micro-organism, it is known as microbiological control.

Integrated control. Application of all appropriate techniques tending to increase specificity and decrease the incidence of interventions aimed at keeping damage by phytophagous pests below the tolerance level that cultivation of the host plant can bear without economic harm. Integrated control associates methods of sylviculture (fertilisers, selection of resistant varieties, changes in forest structure) and methods of changing insect behaviour (pheromones) with classic biological control. Recourse to selective insecticides is also possible in the framework of integrated control.

Pest management. This term means management of agrosystems which takes pests into account. It has been defined as reduction of problems caused by pests after understanding of their biology has been acquired and the ecological and economic consequences of any intervention have been predicted, by interventions chosen with maximum precision regarding the greatest interests of man's activity. This concept is implicit in integrated control.

OLD METHODS
OF PEST CONTROL

Outbreaks of caterpillars of some species in the past turned them into real plagues. In the nineteenth century, damage caused by the brown tailed moth in northern and central France, as well as damage by other Lepidoptera, gave rise to laws on "decaterpillaring", that is, gathering by hand and destruction of eggs and larvae. The law of 26 Ventose, year IV of the Republic, is phrased as follows:

"Article 1. Within ten days following publication of the present law, all landowners, farmers, tenants or others working their own land or that of others shall be personally responsible for decaterpillaring trees or have trees on the said land decaterpillared, under pain of a fine which shall not be less than three days' labour and not exceed ten days.

Article 2. They shall be bound, under pain of the same penalties, to burn all nests and webs removed from trees, hedges or shrubs at once, this to be done in such place where there be no danger of fire spreading either to woods, trees or heath, or to houses and buildings.

Article 3. Département administrations shall, within the same period, order trees on untenanted National estates to be decaterpillared."

While destruction of caterpillars by hand may have proved relatively effective, other methods used in attempts at control were surely less so. Towards the beginning of the sixteenth century caterpillars were so abundant in Dauphiné Province that the public prosecutor felt he had to start proceedings to "order them to decamp and vacate the premises". In 1543 a member of Grenoble municipality, seeing that caterpillars were causing great damage, requested that "the ecclesiastical judge be asked to excommunicate the said beasts, and proceed against them by way of censure, to obviate the damage they did daily and would do in future". The town council accordingly issued an order in response to this request.

Predatory insects, which are less often used, are beetles, a few bugs (Heteroptera), and ants. Earliest research into predators and parasitoids was almost certainly directed at controlling scolytids. As early as the first half of the nineteenth century the German entomologist J.-T. Ratzeburg showed that some species of beetle belonging to the family Cleridae are voracious predators of scolytids. Use of these auxiliaries however proved difficult. Control of *Dendroctonus micans* by a specific predator, *Rhizophagus grandis*, has recently been perfected.

The Japanese wasp *Ooencyrtus kuvanae* has been introduced to Europe, Morocco and the United States to control the gypsy moth, but with varying degrees of success, for it has proven unable to control outbreaks of its host. Among the pests of pine, the sawfly *Diprion pini* in Spain is controlled by a native species, *Dahlbominus fuscipennis* (Hymenoptera, Eulophidae). The sawfly *Acantholyda nemoralis* is controlled by *Trichogramma cacaeciae* in Poland. Among predatory beetles, ladybirds appear to be the most effective in controlling aphids. A recent success is control of the scolytid *Dendroctonus micans* using its specific predator *Rhizophagus grandis*. The beetle *Laricobius erichsoni* is a specific predator of *Dreyfusia piceae* in Europe. It has been introduced to North America where it has become the most effective predator of that species which is a major destroyer of pine forests (Franz, 1958).

Ants of the *Formica rufa* group are known as wood ants. The group includes several species in Europe, including *Formica rufa, F. aquilonia, F. lugubris* and *F. polyctena*. These ants gather a great number of diverse species as food, and their role as predators in controlling forest pests has been known for a long time. According to Pavan (1959) there are 135,000 hectares of forests in the Italian Alps in which there are some 250,000 wood ant nests. Assuming that each nest contains 300,000 workers, this represents 300 billion insects with a biomass of 2,400 tons which consume 24,000 tons of food each year, including 2,400 tons of insects. Such figures justify experiments in transplanting ants to woods where they did not exist, and also classing them as protected species, as has been done in some countries. Wood ants are effective against more than ten major pests, including processionary caterpillars, the sawfly *Pristiphora erichsonii*, and *Tortrix viridana* (Adlung, 1966). An average sized colony of *Formica polyctena* destroyed over a million caterpillars during an outbreak of *Tortrix viridana* (Horstmann, 1977). The mortality rate of *Pristiphora erichsonii* is 71.8% on larch trees on which ants can forage and only 24.7% on trees whose trunks were girdled by rings of glue to prevent ants climbing them (Pschorn-Walcher & Zinnert, 1971). *Formica lugubris* has been introduced to Quebec where it proved effective in controlling *Pristiphora erichsonii* (Finnegan, 1975). Some authors however are less optimistic. According to Wellenstein (1965), ants reduce population density of a few macro moths and sawflies by more than 50% but only in a radius of 5 to 35 metres around the nest. Their effect is less on beetles and micro moths. Lachnids, which are honeydew-producing aphids, are more abundant in the vicinity of ants, as are scale insects of the genus *Lecanium*. Forest soil undergoes undesirable change caused by formic acid which ants deposit near the nest, creating a "green belt" due to richness in nitrogen, and to the well aerated and friable structure of the soil.

The cedar aphid *Cedrobium laportei* became a serious pest of cedars when it arrived in Europe. The sap it draws and the honeydew it deposits cause needle fall, *Fumago* infections and sometimes the death of the tree. *Pauesia cedrobi* is an aphidiid wasp whose larva develops in the aphid's body and mummifies it. When the wasp was introduced to the Luberon hills where the aphid was rife, this parasite adapted to the area and has probably managed to eliminate it (Fabre & Rabasse, 1987). The moth *Operophtera brumata* was introduced to Canada, and was later controlled there by two of its European parasites, the fly *Cyzenis albicans* and the wasp *Agrypon flaveolatum*. The first of these parasites is effective when population density of the moth is high, and the second is effective when density is low (Embrée, 1966; Mesnil, 1967). *Agrypon* settled later than *Cyzenis*, and its rate of parasitism only rose after numbers of the moth had severely declined following attacks by *Cyzenis* (figure 6.10).

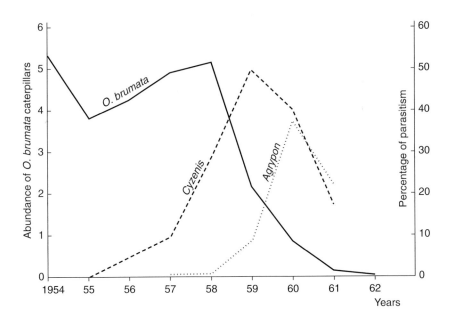

Figure 6.10. *Variations in abundance of* Operophtera brumata *in Canada and percentage of parasitism by* Cyzenis albicans *and* Agrypon flaveolatum *(Embrée, 1966).*

Some attempts at biological control using entomophagous insects have failed following promising beginnings for reasons that remain unknown. Massive releases of enemies of the gypsy moth begun in 1879 in the United States appeared to be successful, but this was belied later, and *Lymantria* went on spreading and occurring in periodic outbreaks. An unusual approach at biological control by manipulation of the pest's population was undertaken against the larch budworm (*Zeiraphera diniana*) (Auer *et al.*, 1981). This consisted in artificially maintaining populations of the insect at relatively high levels in the period between two outbreaks in order to favour its predators and parasitoids. In 1974, during a rarity phase in the insect's population, 35,000 pupae distributed over 650 hectares of sparse larch plantations produced at least 6,000 moths which provided an expected gradation from which trees did not suffer. For five years predators and parasitoids were more numerous than usual, but the essential cause of observed protection was thought to be deterioration of the nutritional quality of leaves as a result of early damage. This type of experiment has not been repeated.

In many cases in which biological control has succeeded (such as control of *Operophtera brumata* in Canada) a single species of entomophagous insect, or a small number of non-competing species were used. Competition between parasitoids has been observed in Canada during trials at controlling *Rhyacionia buoliana* which is parasitised by the ichneumonid *Temelucha interruptor* and by the braconid *Orgilus obscurator. T. interruptor* however shows a predilection for hosts already parasitised by *O. obscurator* which is *R. buoliana*'s most widespread and effective parasite. Unfortunately thousands of *T. interruptor* had already been released before this species' behaviour was discovered (Syme, 1974).

The richness of a species' complex of parasites varies from one region to another. In Europe, Britain has the poorest parasite complexes owing to the country's insularity and to the fact that several pest species, as well as the trees they feed on, were introduced recently. The sawfly *Pristiphora erichsonii* is only parasitised in Britain by three species, whereas it has eleven parasites in the Alps. *Gilpinia hercyniae* does not appear to have any parasite in Wales whereas it has at least twelve on the continent. This paucity of parasite complexes in areas where trees and their pests are not indigenous is also found on the continent. In the Alps, where larch is endemic, the larch budworm has many more parasites than in northern Germany or in Sweden where larch was introduced. Moreover genetic variability of parasites is lower in areas of introduction ("founder effect"), which gives less scope for adaptation. This is why parasites should be sought, whenever possible, in the pest's homeland. Entomophagous insects used in biological control in North America were first obtained in Britain, which may explain some of the failures recorded (Greathead, 1976).

Attempts have been made at controlling pests in North America using European parasitoids. These were sought in European species close to the American ones to be controlled. Enemies of *Orgyia antiqua* are used to combat *Orgyia pseudotsuga*, and enemies of *Choristoneura murinana* against *Choristoneura fumiferana* (Mills, 1983; Mills & Schönberg, 1985).

7.2.2. Use of pathogenic micro-organisms

The most widely used microbiological control preparation is made from the bacterium *Bacillus thuringiensis* (or Bt), which is the basis of many commercial products that can be stored and easily used as insecticides. The advantage of this technique is the specificity of the product, which only affects chewing insects, thus essentially the defoliators (moths and sawflies), and spares useful auxiliaries as well as vertebrates. The pine processionary caterpillar is effectively controlled with Bt. Application is by aerial spray at doses of one to three litres of product per hectare. Other moths such as *Lymantria dispar* and *Zeiraphera diniana* as well as sawflies can also be controlled with Bt.

Bacillus thuringiensis is an aerobic bacterium which is able to sporulate and can easily reproduce in an artificial medium. The toxic part is a proteic crystal the amino acid composition of which has been determined. When this crystal is ingested by some insects, mostly caterpillars, whose intestinal pH exceeds 9, enzymes in the digestive tract free the toxic parts of the proteins of which the crystal is made. This insoluble toxin paralyses the caterpillar's digestive tract which causes feeding to stop, followed by microbial infection which causes death by septicaemia. Preparations such as "Bactospeine" are available commercially in many countries.

The use of viruses in biological control is more tricky as they are difficult and costly to produce. Insect viruses are responsible for spectacular epizootic diseases. They trigger diseases called viroses. More than 300 species are known, most of which affect Lepidoptera. Some viruses, having what is known as "inclusion bodies", are large enough to be visible with an optical microscope. These inclusion bodies, which are proteic in nature, protect the virus against various stresses, including climatic ones. Other, "free", viruses are visible only with an electron scanning microscope. Insect viruses cannot be bred in artificial media, which makes production more difficult. Attempts to use the cytoplasmic specific virus *Smithiavirus pityocampae* against the processionary pine caterpillar on the slopes of Mont Ventoux produced spectacular results, but the epidemic did not maintain itself, and as it was not possible to renew treatment, the moths reconstituted their population (Grison *et al.*, 1959). Success was recorded in North America where populations of *Pinus banksiana* were periodically destroyed by *Neodiprion swainei*. A nuclear polyhedron virus, whose virulence was increased by selection, proved very effective against this sawfly, populations of which are now contained at low levels. The epidemic maintains itself, as dissemination of the virus is ensured by the cocoons of infected larvae (Smirnoff & Juneau, 1973; Smirnoff, 1975).

Some species of entomopathogenic nematodes are used in the control of agricultural pests and some have been tested as controllers of insects harmful to trees. In Canada the nematode *Steinernema carpocapsae* is used in the form of a suspension in plain water or with the addition of various products (oil, glycerine, *Bacillus thuringiensis*), which is sprayed on trees to control the larch sawfly, the gypsy moth, and the budworm *Zeiraphera canadensis*. Treatment of pine saplings by watering with a suspension of *S. carpocapsae* considerably decreases mortality due to infestations of the weevil *Hylobius*, and enables immediate replanting after felling. *Steinernema* acts in an unusual way. At the infectious juvenile stage the nematodes harbour a symbiotic bacterium of the genus *Xenorhabdus* in their gut. They penetrate the insect through natural orifices or by breaking through the integument. Once in the haemocoel they release their bacteria which, as they reproduce, cause septicaemia and kill the insect. The nematodes spend at least one generation in the insect's corpse and end by leaving it to return in the infectious juvenile form (Poinar, 1972; Eidt *et al.*, 1992). Control of *Sirex noctilio*, introduced to New Zealand and Australia where it ravages plantations of *Pinus radiata*, is achieved by using the nematode *Deladenus siricidicola*. Research has also been done on parasitic nematodes of scolytids, but apparently without practical results.

7.2.3. Use of ecomones to modify behaviour

Discovery, isolation and synthesis of sex pheromones of several species of Lepidoptera has led to the use of these substances for practical purposes. Three quarters of the pheromones that have been isolated, synthesised and commercialised belong to moths. Catching males of many defoliating moths by means of traps using sex pheromones produced by females is too costly to be used to reduce populations, but these traps are useful for monitoring variations in abundance and in predicting outbreaks, which makes it possible to intervene at the right moment. In Canada (where the traps are known as "Multipher" traps) more than twenty species of forest moth, including *Choristoneura fumiferana*, *Actebia fennica* and *Lambdina fiscellaria*, are monitored in this way (Sanders, 1994).

Sex pheromones are used in the technique known as "confusion" to lure males and prevent them from finding the females in some moths such as the apple worm, *Carpocapsa*. In woodland this method helps to control *Rhyacionia buoliana*, *Panolis flammea* and *Lymantria dispar*. For the latter species a synthetic substance, disparlure, has been manufactured. Location of females by males is prevented in *Choristoneura fumiferana* by using a synthetic attractant, *trans*-11-tetradecenal, which constitutes the essentials of the moth's pheromone system (Sanders, 1975). In *Zeiraphera diniana*, *trans*-11-tetradecenyl acetate is highly attractive to males of the form which lives on larch, while the *cis* isomer has no effect (Benz & Von Salis, 1973). *Trans*-9-dodenyl acetate attracts males of the form that lives on Arolla pine

(Baltensweiler *et al.*, 1978). In seed orchards in Canada attempts were made to either trap on a massive scale or to sexually confuse pests of conifers, particularly the moth *Dioryctria abietivorella*, which attacks Douglas-fir cones, and the scolytid *Conophthorus coniperda*.

ECOMONES

The term *ecomone*, or semiochemical substance, designates any chemical substance which animals use to perform vital processes essential to their survival. The best known (and until now the most important in controlling pest insects) are *pheromones*, compounds produced by an organism and released into the environment, which affect the behaviour of other members of the same species. *Sex pheromones*, released by individuals of one sex, signal the individuals' whereabouts and indicate that they are ready to mate. Social insects such as ants emit *trail pheromones* on their trails. *Aggregation pheromones*, characteristic of scolytid bark beetles, incite individuals to congregate, which facilitates mating and increases chances of overcoming the host tree's resistance. *Allelochemical substances* induce a reaction in individuals of another species. The term *allomone* is given to allelochemical substances which are beneficial to the emitter. *Kairomones* are ecomones that are beneficial to the receiver to the detriment of the emitter, and *synomones* are ecomones which are beneficial both to the emitter and receiver. Chemical substances given out by a host plant which a parasitoid uses to locate its prey are synomones, whereas those given out by the prey itself and exploited by the parasitoid are kairomones. Some chemical substances may belong to several categories: a substance which is a pheromone for members of the same species may also act as a kairomone for predators or parasites of that species (Nordland & Lewis, 1976).

Use of ecomones to modify insect behaviour and control pest species began in the 1950s and 60s. Many ecomones are used in integrated anti-parasite control.

Experiments are made in the use of allelochemical substances that regulate relationships between plants and insects. Conifer monoterpenes often act as kairomones in the choice of host tree and laying sites by *Dioryctria* in North America and by *Panolis flammea* in Europe (Fatzinger & Merkel, 1985; Leather, 1987). These kairomones may be used in glue traps that catch females, or as bait that induces females to lay in egg traps. Location of host insect and parasitoid oviposition are often controlled by kairomones emitted by the host (Weseloh, 1981). In the moth *Choristoneura fumiferana* kairomones are present in scales that cover the laid eggs and affect the behaviour of the egg parasite *Trichogramma minutum* which is used in biological control of this insect. Identification and use of this kairomone might increase the effectiveness of *Trichogramma* (Zaborski *et al.*, 1987; Smith *et al.*, 1990). The tachinid fly *Cyzenis albicans,* which is a parasite of *Operophtera brumata*, locates its host by means of synomones produced by leaves of the food plant. Synomones in the needles of silver firs stimulate activity in the search for eggs by *Choristoneura fumiferana* and *Trichogramma minutum*.

Scolytids possess anti-aggregative pheromones such as 3-methyl-2-cyclo-hexene-1-one, (or MCH), which is attractive in low concentrations and repellent in high concentrations. Verbenone is the most common anti-aggregative pheromone. Attempts to use verbenone to protect pines against *Dendroctonus* have produced varying results, from complete success to total failure. In many cases significant decreases in infestations were recorded; failures may be due to inappropriate techniques in the use of verbenone (Amman *et al.*, 1989; Shea *et al.*, 1992; Kostyk *et al.*, 1993).

During an outbreak of *Dendroctonus pseudotsugae* in a Douglas-fir wood in Oregon the density of this scolytid was estimated at 12,000 individuals per hectare. Traps baited with the insect's aggregation pheromone caught 277,921 individuals in one year. These traps thus make it possible to reduce tree mortality caused by scolytids. One drawback should however be mentioned: the traps also caught 43,527 individuals of the clerid beetle *Thanasimus undulatus*, which is the main predator of *D. pseudotsugae* (Ross & Daterman, 1997).

7.3. Integrated control

A few examples of integrated control in woodland environments are known. In the case of aphids integrated control consists in combining the effects of carefully chosen selective pesticides with measures designed to improve the physiological condition and resistance of the tree (Carter, 1975). Integrated control of *Ips typographus* is conducted using trap trees (trees that have been felled and left on the ground, or weakened trees that are left standing) on which aggregation pheromone diffusers are stapled in order to increase the trees' powers of attraction. These trap trees are sprayed with insecticides of the pyrethrenoid group, destined to kill the *Ips* that attempt to settle on them (Drumont *et al.*, 1991).

The scolytid *Ips typographus* is a major pest of spruce in Europe. Outbreaks were recorded in the eighteenth century. This beetle caused the loss of four million cubic metres of wood in Germany during the years 1857-1862, and thirty million cubic metres in 1976-1979. In Sweden, losses amounted to two million cubic metres in 1976-1979; in Norway, losses were five million cubic metres during the period 1970-1982. The recent outbreak in Norway and Sweden was triggered mainly by drought in 1974-1976 and storms in the 1970s which uprooted many trees (Bakke, 1983). Outbreaks were facilitated by the structure of spruce populations consisting of old trees which are very prone to infestation. "Forest hygiene" measures were taken on extensive areas in order to stem infestations: infested trees were removed, a network of attractant traps designed to catch *Ips* on a massive scale was installed, and very old trees were eliminated. In Norway 600,000 pheromone traps dispersed over an area of 140,000 square kilometres caught 2.9 thousand million scolytids in 1979 and 4.5 thousand million in 1980. A detailed study of the invasion enabled calculation of a "risk index", and the identification of the parts of the forest most threatened by future outbreaks (Worrell, 1983).

7.4. Use of fertilisers

Application of nitrogenous fertilisers affects the physiological state of trees, increases growth rate and the volume of foliage. In conifers it increases resin quantity and alters its chemical composition, making it more toxic. Effects on phytophagous insects are varied (Kyto *et al.*, 1996). Numbers of conifer pests such as scolytids and sawflies are reduced owing to the toxic properties of resin. Abundance of pests of broadleaf trees is often greater following application of fertiliser because food enriched with nitrogen compounds increases their growth and survival rates (Stark, 1965; Björkman *et al*, 1991). A low level in populations of *Neodiprion sertifer* in a *Pinus sylvestris* wood was attributed to fertilisation (Larsson & Tenow, 1984). Numbers of *Diprion pini* were reduced by 30 to 50% following spreading of various fertilisers in woodland. Numbers of the moths *Lymantria monacha* and *Lymantria dispar* were reduced by 50 to 80% following spreading of chalk and ammonium nitrate. Nef (1959) has shown that fertilisers applied to pine reduce the abundance of *Rhyacionia buoliana* and that reduction of abundance is greater following application of nitrogen alone than it is following a complete treatment including nitrogen, potassium and calcium.

These results are difficult to generalise. In the case of the bug *Aradus cinnamomeus*, fertilisation of pines with nitrogen compounds increases the trees' vigour, which makes them grow taller, but also results in an increase of *Aradus* (Heliövaara *et al.*, 1983), as can be seen in the following table which gives results of treatment with various insecticides (lindane and dimethoate) or with fertilisers:

Treatment	Annual growth in height of trees (cm)	Number of *Aradus* per dm^2
Nitrogen alone	49.5	15.1
Nitrogen + phosphorus + potassium	48.5	17.8
Lindane	37.4	5.0
Dimethoate	39.7	7.3
Control	38.6	11.5

Fertilisation of false acacia (*Robinia pseudoacacia*) with a mixture of nitrogen, potassium and phosphorus provokes a two-stage response from insects. In the first stage, increase in the nutritional value of leaves causes increased attack; then a late defence reaction by the tree reduces levels of attack even when the nutritional value of leaves remains high (Hargrove *et al.*, 1984). Additions of nitrogen to the fir *Abies grandis* are favourable or unfavourable to *Dreyfusia piceae* according to the form in which the nitrogen is used. With urea, populations of *Dreyfusia* increase threefold in five generations, while with ammonium nitrate their numbers decrease. Spreading urea on *Pinus banksiana* at a concentration of 400 kg per hectare reduces leaf consumption by the sawfly *Neodiprion swainei* and increases the insect's mortality rate by 50%. On the other hand urea favours infestation by the scale insect *Toumeyella numismaticum*, increasing the population nine-fold. Fertilisation with potassium at a concentration of 100 kg per hectare greatly lowers levels of infestation by the scale insect (Smirnoff & Valero, 1975).

Fertilisation of spruce induces changes in the chemical composition of the bark, particularly in mineral elements. Knowledge of average quantities of a few elements such as zinc and phosphorus enables predictions of average numbers of attacks by *Ips typographus* to be made. Zinc and phosphorus act by facilitating production of substances such as terpenes, the role of which in attracting or repelling scolytids is known. Significant increases in levels of potassium and phosphorus, and a decrease in levels of aluminium, occur in the bark following attacks. These results should help in the detection of trees resistant to infestations by *Ips typographus* (Snauwert *et al.*, 1992).

7.5. The search for resistant varieties

Selection of trees resistant to insect predation is a method of control that appears promising, but has not yet given many concrete results. A remarkable example is the production of hybrids of European elm and a Japanese elm which are resistant to Dutch elm disease. Selection of resistant varieties assumes that the characters which give resistance are controlled genetically and are inherited. Attempts at selection are often hampered by lack of understanding of resistance mechanisms, either in isolated individuals, or in the species as a whole. Few trees successfully resist infestation by aphids. Those that do include *Platanus orientalis, Castanea sativa* and *Tsuga heterophylla*, but their resistance mechanisms remain virtually unknown. Inter-tree variability in resistance to infestations by *Ips typographus* has been shown to occur in spruce, which suggests that genetic selection of resistant trees may be possible (Nef, 1990). Genetic differences have been found between beech trees heavily infested by the scale insect *Cryptococcus fagisuga* and trees that remained little infested. Study of eleven different genes showed that the most sensitive trees are those having the highest levels of heterozygosity (Gora *et al.*, 1994). Vulnerability to infestation by the scale insect is also a function of bark structure, which seems to be controlled genetically (Lonsdale, 1982). Resistant clones have a greater quantity of lignified cells, which hamper the scale insect's feeding.

Resistance in pines to *Rhyacionia buoliana* is conditioned by some of the tree's physiological characteristics such as water content of needles and their hardness, the amount of resin in buds at the time of penetration by larvae, and the number and size of buds. Experiments have shown differences in sensitivity of different species of pine, and differences within a single species according to their geographical origins. These observations may provide the grounds needed for selection of those pines most resistant to *Rhyacionia buoliana* in a given region (Charles, 1976).

Resistance of various species of pine to infestations by the mining larvae of *Exoteleia pinifoliella* is a function of the size and number of resin canals present in the needles. Species such as *Pinus sylvestris* and *Pinus resinosa*, which are little attacked, have numerous and broad resin canals (seven visible canals in the cross section of a needle). A species such as *Pinus contorta*, which is heavily attacked, has only two resin canals, situated at both ends of the needle.

The search for resistant varieties is made all the more difficult by the fact that resistance to one pest may be associated with high susceptibility to another pest. It has been shown, by using clones of the poplar *Populus angustifolia*, that when resistance to the gall aphid *Pemphigus betae* increases, resistance to grazing by red deer (*Cervus elaphus*) diminishes. Clone number 1,008 is the most resistant to the aphid; its branches situated under two metres from the ground are destroyed by deer. Clone number 996 on the other hand shows great resistance to both predators, whereas clone number 1,000 is very vulnerable to both the aphid and deer (Witham *et al.*, 1991). Rousi *et al.* (1997) have found that in two species of birch in Finland (*Betula pendula* and *Betula pubescens*) there are clones that are resistant and clones that are sensitive to predation by weevils of the genera *Phyllobius* and *Polydrusus*, and to grazing by blue hare (*Lepus timidus*) and by short-tailed vole (*Microtus agrestis*). Some clones resist predation by both the insects and mammals, while others only resist one of the two groups of animals.

7.6. Forestry methods

Understanding the factors that regulate insect populations makes it possible to research forestry methods that strengthen the resistance of trees and reduce populations of pests. These techniques are referred to by German entomologists by the term *Waldhygiene*. They include, for instance, mixing age groups, careful choice of tree species and mixing broadleaf trees and conifers in order to avoid monoculture, and cultivation in the understorey of nectar-producing plants in order to increase the longevity and fertility of entomophagous parasites. As Decourt *et al.* (1981) point out: "most of the woodland management measures mentioned make sense... But given the socio-economic pressures some of these measures amount to no more than a pious wish... In the particular case of infestation of young plantations of all conifers in the northern half of France by the weevil *Hylobius abietis*, the ecologically ideal solution would be not to replant for three years following felling; however the cost of clearing plots after that time appears to be greater than that of preventive chemical control of *Hylobius* during the early years of plantation".

Outbreaks of pests are triggered by events which lower the resistance of trees or reduce activity of the pest's natural enemies (figure 6.11). Measures that can maintain stability of the forest ecosystem have been listed by Berryman (1988): planting trees in conditions which do not jeopardise their defence mechanisms, that is, appropriate soils in regions where climatic conditions are favourable; avoiding overly dense plantation; maintaining or increasing populations of insectivorous organisms (insects, birds etc.); monitoring and preventing the introduction of alien species which might become harmful; and conservation of genetically resistant trees during selection operations – selection of fast-growing clones may reduce resistance to pests, since energy spent on growth cannot be invested in the production of defensive substances.

The advantage of mixed populations has often been reiterated (Auclair, 1983). The reasons for better resistance in mixed forest are various. Parasites of several very harmful insects require changes of hosts which often live on other species of plant. Radical weeding reduces numbers of useful species. Uniformly planted woods, turned into monocultures, increase the risk of massive outbreaks. Extensive planting of conifers in Europe since the beginning of the nineteenth century has resulted in outbreaks of insects, particularly moths and scolytids. In every case the introduction of a mixture of broadleaf trees has considerably reduced damage (Baltensweiler, 1976). In the Palatinate in Germany, invasions of *Bupalus piniarius, Dendrolimus pini, Lymantria dispar* and *Diprion pini* have been surveyed since 1810. The frequency and intensity of these invasions have recently declined. The only explanation for this regression is the increase of areas devoted to broadleaf trees at the expense of those planted with conifers (Klimitzek, 1972). In Lower Saxony, spruce planted on poor soils produced shallowly-rooted trees which are easily uprooted by wind; those that remain standing are affected by sunburn which predisposes them to infestation by *Ips typographus*. Where mixed with beech, spruce trees root better and are less vulnerable to infestation by scolytids (Otto, 1985).

It should be noted that there are a few rare exceptions. When two species of tree that play host to an insect are combined this favours propagation of the insect. This is the case of the chermesid *Gilletteella cooleyi* (cf. chapter 12.2).

The last few decades have shown the extraordinary abundance of pest insects and the instability of spruce populations in the mountains of central Europe. This has led to consideration of a series of measures such as soil restoration, search for the best adapted tree species, and conversion of pure populations into populations of mixed broadleaf and coniferous trees, or even of broadleaf trees alone (Führer, 1990, 1996).

The possibility of re-establishing populations of naturally resistant trees is illustrated by a case cited by Bovey (1974). A German forest with an area of 3,000 hectares originally consisting of broadleaf trees was turned into a pure

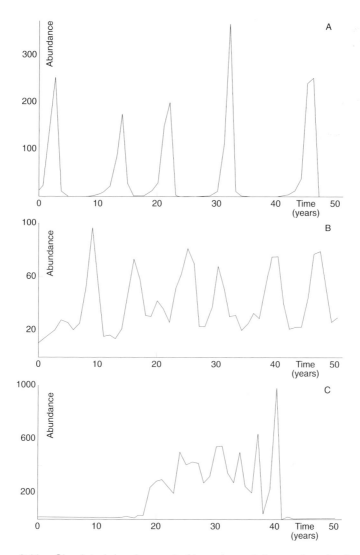

Figure 6.11. *Simulated development of insect populations using simple models that show the effects of environmental conditions.*
A. Cyclical outbreaks of Orgyia pseudotsugata *on Douglas-fir in localities favourable to the insects and unfavourable to their natural enemies and to the host tree (for example warm and arid localities). Outbreaks occur at intervals of 8 to 12 years and peaks of abundance vary from one outbreak to another.*
B. More stable populations of O. pseudotsugata *in localities favourable to natural enemies and to Douglas-fir (for example humid and mild localities). Outbreaks occur at intervals of 5 to 8 years and peaks of abundance are lower.*
C. Outbreak of Choristoneura fumiferana *triggered by an immigration of moths from a neighbouring infested locality, which overwhelmed control factors such as birds or predatory insects (Berryman, 1988).*

pine plantation two hundred years ago, and has ever since become more and more threatened by defoliating insects. The danger was averted by introducing a lower stage of broadleaf trees and putting a stop to harvesting litter in order to improve soil fertility. Whereas for 100 years outbreaks had been recorded every four years, none has been seen since 1949. In Denmark, plantations of Sitka spruce are more subject to infestations by *Dendroctonus micans* in dry locations than in humid ones. *Rhyacionia buoliana* prefers sparsely planted stands of young pines, where it causes greater damage than in dense stands. It is therefore unwise to make clearings too large before trees are old and vigorous enough. Following felling it is important not to leave debris of exploited wood and stumps which may harbour xylophagous insects such as scolytids.

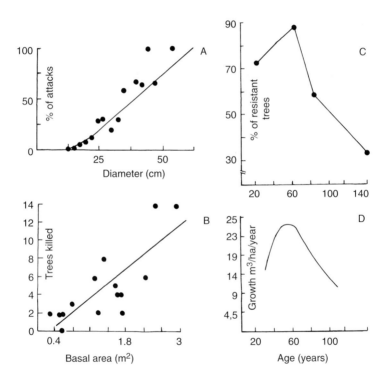

Figure 6.12. *Effects of age and trunk diameter on susceptibility of* Pinus contorta *to infestation by* Dendroctonus ponderosae. *A. Relationship between percentage of trees infested and trunk diameter. B. Relationship between numbers of trees infested and basal area. C. Relationship between percentage of resistant trees and tree diameter. D. Relationship between average annual growth and age of tree (Safranyik et al., 1974).*

Resistance of *Pinus contorta* to infestations by *Dendroctonus ponderosae* is increased when clearings are made to open the forest canopy and allow penetration of sunlight, which increases the trees' photosynthetic activity (Waring & Pitman, 1985). Trees over eighty years old are more prone to attack. Risk is therefore reduced if trees are exploited as soon as possible (figure 6.12). The age at which resistance is greatest is also that at which the tree's annual increase in volume is greatest (Safranyik *et al.*, 1974).

REFERENCES

ALSTAD D. N., EDMUNDS G. F., WEINSTEIN L.-H., (1982). Effects of air pollutant on insect populations. *Ann. Rev. Ent.*, **27**: 369-384.

AMMAN G. D., THIER R. W., MCGREGOR M. D., SCHMITZ R. F., (1989). Efficacity of verbenone in reducing lodgepole pine infestation by mountain pine in Idaho. *Canad. J. For. Res.*, **19**: 60-62.

ANDROIC M., (1966). Los mas importantes problemas de entomologia forestal en Yugoslavia. *Bol. Ser. Plagas Forest.*, **9**: 43-53.

ANONYMOUS, (1978). Les problèmes posés par la recrudescence des attaques de Scolytides dans les forêts résineuses françaises. *Rev. For. Fr.*, **30**: 37-41.

ANONYMOUS, (1979). Gypsy moth. Program accomplishment report. USDA, agriculture information bulletin no. **421**, 20 pp.

AUCLAIR D., Natural mixed forests and artificial monospecific forests. *In* : H.-A. MOONEY et M. GODRON, (1983). *Disturbance and ecosystems.* Springer, Berlin, 71-82.

AUER C., ROQUES A., GOUSSARD F., CHARLES P.-J., (1981). Effets de l'accroissement provoqué du niveau des populations de la Tordeuse du mélèze *Zeiraphera diniana* Guénée (Lep. Tortricidae) au cours de la phase de régression dans un massif forestier du Briançonnais. *Zeit. angew. Ent.*, **92**: 286-303.

AUSTARA Ø., (1984). Diameter growth and tree mortality of Norway spruce following mass attacks by *Epinotia nanana*. *Research papers from Norwegian forest research institute*, **10 / 84**, 9 pp.

BAKKE A., LEKANDER B., (1965). Studies on *Hylobius abietis*. II. The influence of exposure on the development and production of *Hylobus abietis*, illustrated through one Norwegian and one Swedish experiment. *Medd. Fra det Norske Skorgsf.*, **20**: 117-135.

BALTENSWEILER W., (1976). Natural control factors operating in some European forest insect populations. *Proc. XV Int. Congr. Ent.*, 617-621.

BALTENSWEILER W., PRIESNER E., ARN H., DELUCCHI V., (1978). Unterschiedliche Sexuallockstoffe bei Lärchen und Arvenform des Grauen Lärchenwicklers (*Zeiraphera diniana* Gn., Lep. Tortricidae). *Bull. Soc. Ent. Suisse*, **51**: 133-142.

BARBEY A., (1942). *Traité d'entomologie forestière.* Berger-Levrault, Paris.

BENZ G., VON SALIS G., (1973). Use of synthetic sex attractant of larch bud moth *Zeiraphera diniana* (Gn.) in monitoring traps under different conditions; and antagonistic action of *cis* isomere. *Experientia,* **29**: 729-730.

BERRYMAN A. A., (1988). Pests and stability of forest ecosystems. *Northwest Environ. J.,* **4**: 351-355.

BERRYMAN A. A., (1996). What causes population cycles of forest Lepidoptera ? *TREE,* **11**: 28-32.

BJÖRKMAN C., LARSSON S., GREF R., (1991). Effects of nitrogen fertilization on pine needle chemistry and sawfly performance. *Œcologia,* **86**: 202-209.

BORUSEWICZ K., Effects of industrial pollution on the insect fauna. *In*: W. GRODZINSKI, J. WEINER, P.-F. MAYCOCK, (1984). *Forest ecosystems in industrial regions.* Springer, Berlin, 112-113.

BOVEY P., (1971). L'impact de l'insecte déprédateur sur la forêt. *Annales de zoologie-écologie animale,* numéro hors série, 11-29.

BOVEY P., (1974). Incidences des traitements pesticides sur les zoocénoses forestières. *Bull. Soc. Zool. Fr.,* **99**: 9-18.

BOWERS R. G., BEGON M., HODGKINSON D. E., (1993). Host-pathogen population cycles in forest insects? Lessons from simple models reconsidered. *Oikos,* **67**: 529-538.

BRANDT J.-P., (1995). Forest insect- and disease-caused impact to timber resources of west-central Canada: 1988-1992. *Information Report* NOR-X-341. Canadian Forest Service, Edmonton.

BRAUN S., FLÜCKIGER W., (1989). Effect of ambient ozone and acid mist on aphid development. *Environ. Pollut.,* **56**: 177-187.

BRITTON R. J., (1988). Physiological effects of natural and artificial defoliation on the growth of young crops of lodgepole pine. *Forestry,* **61**: 165-175.

CARTER C.-I., (1975). Towards integrated control of tree aphids. *Forestry Commission,* Forest Record, **104**, 17 pp.

CARTER C.-I., (1977). Impact of green spruce aphid on growth. *Forestry Commission, Research and Development Paper,* **116**. HMSO, London.

CATES R. G., ALEXANDER H., Host resistance and susceptibility. *In*: J. B. MITTON, K. B. STURGEON, (1982). *Bark beetles of North American Conifers.* University of Texas Press, Austin, 212-263.

CHARLES P.-J., (1976). Incidence des attaques de la tordeuse des pousses *Rhyacionia (Evetria) buoliana* Schiff. sur le choix des espèces de pins susceptibles d'être plantées en région méditerranéenne. *C. R. IVᵉ journées Phytiatrie - Phytopharmacie,* Montpellier, 468-471.

CHARLES J.-P., VILLEMANT C., (1977). Modifications des niveaux de population d'insectes dans les jeunes plantations de pins sylvestres de la forêt de

Roumare (Seine-Maritime) soumises à la pollution atmosphérique. *C. R. Ac. Agr. Fr.*, **63**: 502-510.

CHRISTIANSEN E., (1970). Insect pests in forest of the Nordic countries. *Norsk. Ent. Tidssk.*, **17**: 163-168.

CONNOR E. F., FAETH S. H., SIMBERLOFF D., OPLER P. A., (1980). Taxonomic isolation and the accumulation of herbivorous insects: a comparison of introduced and native trees. *Ecol. Ent.*, **5**: 205-211.

DAFAUCE C., (1970). Action et organisation du « servicio de plagas forestales ». *Rev. For. Fr.*, no. spécial: *La lutte biologique en forêt*, 297-308.

DALE V. H., GARDNER R. H., DE ANGELIS D. L. *et al.*, (1991). Elevation mediated effects of the balsam wooly adelgid on the southern Appalachian spruce-fir forests. *Canad. J. Forest. Res.*, **21**: 1,634

DAVIDSON A. G., PRENTICE R.-M., (1967). *Important forest insects and diseases of mutual concern to Canada, the United-States and Mexico.* Department of forestry and rural development, Canada, 248 pp.

DECOURT N., BECKER M., CHARLES P.-J., Bases écologiques du développement des ressources sylvicoles. *In: Écologie et développement*, (1981). CNRS, Paris, 193-217.

DELPLANQUE A., GIOVANNI J.-P., ROQUES A., AGUSTIN S., (1987). Primeros problemas entomologicos relacionados con la introduccion de *Pinus contorta* Dougl. en Francia. *Boletim de la Estacion Central de Ecologia*, **16**: 123-132.

DIXON A. F. G., (1971). The role of aphid in wood formation. *J. Appl. Ecol.*, **8**: 165-179 and 393-399.

DORSEY C. K., LEACH J. G., (1956). The bionomics of certain insects associated with oak wilt with particular reference to Nitidulidae. *J. Econ. Ent.*, **49**: 219-230.

DRUMONT A., GONZALEZ R., DE WINDT N. *et al.*, (1991). Contrôle intégré d'*Ips typographus* (Col., Scolytide) en Belgique. *Parasitica*, **47**: 113-128.

DU MERLE P., BRUNET S., CHAMBON J.-P. *et al.*, (1982). Colonisation d'un végétal introduit (*Cedrus atlantica*) et de nouveaux milieux bioclimatiques par un insecte phytophage: *Choristoneura murinana* (Lep. Tortricidae). *Acta Œcologia, Œcol. Appl.*, **10**: 289-301.

DUNN J. P., KIMMERER T. W., POTTER D. A., (1987). Winter starch reserves of white oak as a predictor of attack by the twolined chesnut borer, *Agrilus bilineatus* (Weber) (Coloptera : Buprestidae). *Œcologia*, **74**: 352-355.

EAMUS D., JARVIS P. G., (1989). The direct effects of increase in the global atmospheric CO_2 concentration on natural and commercial temperate trees and forests. *Adv. Ecol. Res.*, **19**: 1-55.

ECKLOFF W., (1972). Beitrag zur Okologie und forstlichen Bedeutung bienenwirtschaftlich wichtiger Rindenläuse. *Zeit. Ang. Ent.*, **70**: 134-157.

EIDT D.-C., THURSTON G.-S., BOIVIN G., (1992). *Lutte contre les insectes nuisibles dans les régions de climat froid au moyen de nématodes entomopathogènes*. Service Canadien des Forêts, rapport d'information M-X-187F, 72 pp.

EMBRÉE D.-G., (1966). The role of introduced parasites in the control of the winter moth in Nova Scotia. *Canad. Ent.*, **98**: 1,159-1,168.

ESCHERICH K., (1914). *Die Forstinsekten Mitteleuropas*. Band 1. Paul Parey, Berlin.

EVANS H. F., (1987). Sitka spruce insects: past, present and future. *Proceedings Royal Society Edinburgh*, B: 157-167.

FABRE J.-P., (1976). Extension du cèdre et risques d'attaques d'insectes. *Rev. For. Fr.*, **28**: 261-269.

FABRE J.-P., RABASSE J.-M., (1987). Introduction dans le sud-est de la France d'un parasite *Pauesia cedrobii* (Hym. Aphidiidae) du cèdre de l'Atlas: *Cedrus atlantica. Entomophaga*, **32**: 127-141.

FAGERSTRÖM T., LARSSON S., LOHM U., TENOW O., (1978). Growth in Scots pine (*Pinus silvestris* L.): A hypothesis on response to *Blastophagus piniperda* L. (Col. Scolytidae) attacks. *Forest Ecology and Management*, **1**: 273-281.

FATZINGER C.-W., MERKEL E.-P., (1985). Oviposition and feeding preference of southern pine coneworm (Lepidoptera: Pyralidae) for different host-plant materials and observations on monoterpenes as an oviposition stimulant. *J. Chem. Ecol.*, **11**: 689-699.

FEDDE G. F., (1973). Impact of balsam wooly aphid (Homoptera: Phylloxeridae) on cone and seed produced by infested Fraser fir. *Canad. Ent.*, **105**: 673-680.

FERRELL G. T., HALL R. C., (1975). Weather and tree growth associated with white fir mortality caused by fir engraver and roundheaded fir borer. *Research Paper PSW-109*. U S Forest Service.

FILION L., COURNOYER L., (1995). Variation in wood structure of eastern larch defoliated by the larch sawfly in subarctic Québec, Canada. *Canad. J. For. Res.*, **25**: 1,263-1,268.

FRAVAL A., (1994). La lutte biologique contre les ravageurs des forêts. *Bull. Tech. ONF*, **26**: 1-7.

FÜHRER E., (1990). Forest decline in central Europe: additional aspects of its cause. *Forest Ecology and Management*, **37**: 249-257.

FÜHRER E., (1996). Entomologische Aspekte der Umwandlung montaner Fichtenforste in Mitteleuropa. *Entomol. Gen.*, **21**: 1-15.

FURNISS R. L., CAROLIN V. M., (1977). *Western forest insects*. USDA, Miscellaneous publication no. 1,339, 654 pp.

GORA V. *et al.*, (1994). Influence of genetic structures and silvicultural treatments in a beech stand (*Fagus sylvatica*) on the population dynamics of beech scale (*Cryptococcus fagisuga*). *Forest Genetics*, **1**: 157-164.

GREATHEAD D. J., (1976). *A review of biological control in western and southern Europe.* Commonwealth Agricultural Bureaux, Slough, England, 182 pp.

GRISON P., (1970). La lutte biologique en forêt. *Rev. For. Fr.*, **22**: 256-271.

GRISON P., Observations sur l'impact des insectes défoliateurs sur la productivité primaire. *In: Productivité des écosystèmes forestiers* (1972). Editions de l'UNESCO, Paris, 369-375.

GUILBOT R., (1991). La lutte biologique en milieux urbain et péri-urbain. *Courr. Cell. Env. INRA*, **16**: 9-26.

HAAK R.-A., BENJAMIN D.-M., (1982). The biology and ecology of the two-lined chestnut borer, *Agrilus bilineatus* (Coleoptera : Buprestidae), on oaks, *Quercus* spp. in Wisconsin. *Canad. Ent.*, **114**: 385-396.

HARGROVE W. W., CROSSLEY D. A., SEASTEDT T. R., (1984). Shifts in herbivory in the canopy of black locust, *Robinia pseudoacacia* L. following fertilization. *Oikos*, **43**: 322-328.

HELIÖVAARA K., (1986). Occurrence of *Petrova resinella* (Lepidoptera, Tortricidae) in a gradient of industrial air pollutants. *Silva Fennica*, **20**: 83-87.

HELIÖVAARA K., TERHO E., ANNILA E., (1983). Effects of nitrogen fertilization and insecticides on the population density of the pine bark bug, *Aradus cinnamomeus* (Heteroptera, Aradidae). *Silva Fennica*, **17**: 351-357.

HELIÖVAARA K., TERHO E., KOPONEN M., (1982). Parasitism in the eggs of the pine bark-bug, *Aradus cinnamomeus* (Heteroptera, Aradidae). *Ann. Ent. Fenn.*, **48**: 31-32.

HELIÖVAARA K., VÄISÄNEN R., (1986). Industrial air pollution and the pine bug *Aradus cinnamomeus* Panz. (Het. Aradidae). *Zeit. ang. Ent.*, **101**: 469-478.

HOBART J., Pesticides in forestry: an introduction. *In:* F.-H. PERRING, K. MELLANBY (1977). *Ecological effects of pesticides.* Academic Press, London, 61-88.

HORSTMANN K., (1977). Waldameisen (*Formica polyctena*) als Abundanzfaktoren für den Massenwechsel des Eichenwicklers *Tortrix viridana* L. *Zeit. ang. Ent.*, **82**: 421-435.

HUBER F., (1982). Effet de défoliaisons des chênes par les hannetons sur la structure du bois. *Rev. For. Fr.*, **34**: 185-190.

IRVING S. N., WYATT I. J., (1973). Effects of sublethal doses of pesticides on the oviposition behaviour of *Encarsia formosa. Ann. Appl. Biol.*, **75**: 57-62.

JOLY R., (1970). Action des déprédations dues aux insectes défoliateurs sur le pin maritime. *Rev. For. Fr.*, **22**, numéro spécial: 205-220.

JOLY R., (1975). Les insectes ennemis des pins. Editions de l'ENGREF, Nancy. Vol. 1, 222 pp.; vol 2, 40 pl.

JOLY R., (1977). Sécheresse et danger de prolifération des insectes corticoles et xylophages. *Rev. For. Fr.*, **29**: 5-14.

KANGAS E., (1950). Die Primarität und Sekundarität als eigenschaften der Schadlinge. *VIIIth Int. Congr. Ent.*: 792-798.

KINNEY K. K., LINDROTH R. L., JUNG S. M., NORDHEIM E. V., (1997). Effects of CO_2 and NO_3 availability on deciduous trees phytochemistry and insect performance. *Ecology*, **78**: 215-230.

KLIMETZEK D., (1972). Die Zeitfloge von Ubervermehrungen Nadel-fressender Kiefernraupen in der Pfalz seit 1810 und die Ursachen ihres Rückganges in neueres Zeit. *Zeit Ang. Ent.*, **71**: 414-428.

KOSTYK B. C., BORDEN J. H., GRIES G., (1993). Photoisomerism of the antiaggregation pheromone verbenone: biological and practical implications with respect to the mountain pine beetle, *Dendroctonus ponderosae* Hopkins (Coleoptera: Scolytidae). *J. Chem. Ecol.*, **19**: 1,749-1,759.

KYTO M., NIEMELA P., LARSSON S., (1996). Insects on trees: population and individual response to fertilization. *Oikos*, **75**: 148-159.

LARSSON O., (1989). Stressful times for the plant stress-insect performances hypothesis. *Oikos*, **56**: 277-283.

LARSSON O., TENOW O., (1984). Areal distribution of a *Neodiprion sertifer* Geoffr. (Hym, Diprionidae) outbreak on Scots pine as related to stand condition. *Holarct. Ecol.*, **7**: 81-90.

LARSSON O., BJORKMAN C., GREF R., (1986). Responses of *Neodiprion sertifer* Geoffr. (Hym. Diprionidae) larvae to variation in needle resin acid concentration in Scots pine. *Œcologia*, **70**: 77-84.

LATTIN J. D., OMAN P., Where are the exotic insect threats? *In:* C. WILSON & C. GRAHAM (1983). *Exotic plant pests and north American agriculture.* Academic Press, New York, 93-137.

LAURENT-HERVOUET N., (1986). Mesure des pertes de croissance radiale sur quelques espèces de Pinus dues à deux défoliateurs forestiers. I. Cas de la processionnaire du pin en région méditerranéenne. *Ann. Sci. For.*, **43**: 239-262.

LEATHER S., (1987). Pine monoterpenes stimulate oviposition in the pine beauty moth, *Panolis flammea. Ent. Exp. Appl.*, **43**: 295-303.

LECLANT F., EL IDRISSI A., MIERMONT Y., (1977). Les Aphides et les Tortricides du cèdre au Maroc et en France et leurs parasites. *C. R. Ac. Sc.*, ser. D, **284**: 647-649.

LEMPERIÈRE G., GIVORS A., SABATIER R., (1984). Une infestation de *Choristoneura murinana* Hbn. tordeuse du sapin dans le nord de l'Ardèche. *Rev. For. Fr.*, **36**: 206-210.

LINDROTH R. L., KINNEY K. K., PLATZ C. L., (1993). Responses of deciduous trees to elevated atmospheric CO_2: productivity, phytochemistry and insect performance. *Ecology*, **74**: 763-777.

LINIT M. J., (1988). Nematode-vector relationships in the pine wilt disease system. *J. Nematol.*, **20**: 227-235.

LONSDALE D., (1982). Wood and bark anatomy of young beech in relation to *Cryptococcus* attack. *Proc. IUFRO beech bark disease*, 43-49. USDA General technical Report WO-37.

MacLean D. A., (1981). Impact of defoliation by spruce budworm populations on radial and volume growth of balsam fir: a review of present knowledge. *Mitteilungen der Forstlichen Bundesversuchanstalt Wien*, **142**: 293-306.

Mattson W. J., (1980). Herbivory in relation to plant nitrogen content. *Ann. Rev. Ecol. Syst.*, **11**: 119-161.

Mattson W. J., Haack R.-A., (1987). The role of drought in outbreaks of plant-eating insects. *Bioscience*, **37**: 110-117.

Mesnil L. P., (1967). History of a success in biological control: the winter moth project for Canada. *Tech. Bull. Commonwealth Institute of Biological Control*, **8**: 1-6.

Mills N. J., (1983). Possibilities for the biological control of *Choristoneura fumiferana* (Clemens) using natural enemies from Europe. *Commonwealth Institute of Biological Control*, **4**: 104-125.

Mills N. J., Schönberg F., (1985). Possibilities for the biological control of the Douglas-fir tussock moth *Orgyia pseudotsugata* (Lymantriidae), in Canada, using natural enemies from Europe. *Commonwealth Institute of Biological Control*, **6**: 7-18.

Møller A.-P., (1995). Leaf-mining insects and fluctuating asymmetry in elm *Ulmus glabra* leaves. *J. Anim. Ecol.*, **64**: 697-707.

Morris R. F., (1963). The dynamics of epidemic spruce budworm populations. *Mem. Ent. Soc. Canada*, **31**: 1-332.

Mouna M., (1985). Comparaison des communautés frondicoles du cèdre (*Cedrus atlantica* Lanetti) en France (Provence) et au Maroc (Moyen-Atlas). *Vie et Milieu*, **35**: 99-106.

Nef L., (1959). Étude d'une population de larves de *Rhyacionia buoliana* (Schiff.). *Zeit. angew. Ent.*, **44**: 167-186.

Nef L., (1990). Variabilité inter et intra arbres des attaques d'*Ips typographus* L. (Col. Scolytidae) sur *Picea abies* Karst. *J. Appl. Ent.*, **110**: 516-523.

Nef L., (1994). Relations entre les populations de *Phyllaphis fagi* L. et la composition minérale des feuilles de *Fagus sylvatica* L. en pépinières forestières. *Parasitica*, **50**: 41-45.

Nef L., Minet G., (1992). Évaluation des risques de dégâts d'*Hylobius abietis* L. dans les jeunes plantations de conifères. *Silva Belgica*, **99**: 15-20.

Neilson M. M., Martineau R., Rose A. M., (1971). *Diprion hercyniae* (Hartig) European spruce sawfly (Hymenoptera: Diprionidae). *Commonwealth Institute Biological Control*, technical communication, **4**: 136-143.

Nevalainen S., Liukkonen M.-H., (1988). The effects of air pollution on biotic forest diseases and pests. A literature review. *Folia Forestalia* no. 716: 1-25.

Nordland D. A., Lewis W. J., (1976). Terminology of chemical releasing stimuli in intraspecific and interspecific interactions. *J. Chem. Ecol.*, **2**: 211 220.

Otto H.-J., (1985). Sylviculture according to site conditions as a method of forest protection. *Zeit. ang. Ent.,* **99**: 190-198.

Pavan M., (1959). Attivita italiana per la lotta biologica con formiche del gruppo *Formica rufa* contro gli insetti dannosi alle foreste. *Collana Verde,* **4**: 1-80.

Poinar G.-O., (1972). Nematodes as facultative parasites of insects. *Ann. Rev. Ent.,* **17**: 103-122.

Port G. R., Thompson J. R., (1980). Outbreak of insect herbivores on plants along motorways in the United States. *J. Appl. Biol.,* **17**: 649-656.

Postner M., (1963). Insektenschaden an der Lärche ausserhalb ihres natürlichen Verbreitungsgebietes. *Forstw. Cb.,* **82**: 27-33.

Pschorn-Walcher H., Zinnert K.-D., (1971). Investigations on the ecology and natural control of the larch sawfly (*Pristiphora erichsoni* Htg., Hym. Tenthredinide) in Central Europe. Part II. Natural enemies: their biology and ecology, and their role as mortality factors in *P. erichsoni. Tech. Bull. no. 14, Commonwealth Institute of Biological Control,* 1-50.

Rafes P. M., Estimation of the effects of phytophagous insects on forest production. *In:* D.-E. Reichle (1970). *Analysis of temperate forest ecosystems.* Springer, Berlin, 100-106.

Rafes P. M., Pests and the damage which they cause to forest. *In: Productivité des écosystèmes forestiers* (1972). Editions de l'UNESCO, Paris, 357-367.

Rhoades D. F., Evolution of plant chemical defense against herbivores. *In:* G. A. Rosenthal, D. H. Janzen (1979). *Herbivores: their interaction with secondary plant metabolites.* Academic Press, London, 3-54.

Riemert J., Whittaker J. B., Air pollution and insect herbivores: observed interactions and possible mechanisms. *In:* E.-A. Bernays (1989). *Insect-plant interactions.* Volume I. CRC Press, Boca Raton, Florida, 73-105.

Romanyk N., (1966). Plagas forestales mas importantes de Espana. *Bol. Serv. Plagas For.,* **9**: 83-96.

Roques A., (1988). *La spécificité des relations entre cônes de conifères et insectes inféodés en Europe occidentale. Un exemple d'étude des interactions plantes-insectes.* Thesis, Université de Pau, 242 pp.

Roques A., Kerjean M., Auclair D., (1980). Effets de la pollution atmosphérique par le fluor et lc dioxyde de soufre sur l'appareil reproducteur femelle de *Pinus sylvestris* en forêt de Roumare (Seine-Maritime, France). *Env. Poll.* (ser A), **21**: 191-201.

Ross D. W., Daterman G.-E., (1997). Using pheromone-baited traps to control the amount and distribution of tree mortality during outbreaks of the Douglas-fir beetle. *Forest Science,* **43**: 65-70.

Rousi M., Tahvanainen J. et al., (1997). Clonal variation in susceptibility of white birches (*Betula* spp.) to mammalian and insect herbivores. *Forest Science,* **43**: 396-402.

RUBTSOV V. V., (1996). Influence of repeated defoliations by insects on wood increment in common oak (*Quercus robur* L.). *Ann. Sci. For.*, **53**: 407-412.

SAFRANYIK L. D., SHRIMPTON D. M., WHITNEY H. S., (1974). *Management of lodgepole pine to reduce losses from the mountain pine beetle*. For. Tech. Rep, Pac. For. Res. Cent., Canad. For. Serv., **1**, 24 pp.

SAIKKONEN K., NEUVONEN S., KAINULAINEN P., (1995). Oviposition and larval performance of European pine sawfly in relation to irrigation, simulated acid rain and resin acid concentration in Scots pine. *Oikos*, **74**: 273-282.

SANDERS C. J., (1994). *Recherche, développement et commercialisation des écomones dans la lutte contre les insectes nuisibles au Canada*. Resources Naturelles Canada, Service des forêts, 102 pp.

SCHAFELLNER C., BERGER R., MATTANOVICH J., FÜHRER E., Variations in spruce needle chemistry and implications for the little spruce sawfly, *Pristiphora abietina*. *In:* W.-J. MATTSON, P. NIEMELÄ, M. ROUSSI (1996). *Dynamics of forest herbivory: quest for pattern and principle*. USDA For. Serv., Gen. Tech. Rep. NC-183, 249-256.

SCHOWALTER T. D., COULSON R. N., CROSSLEY D. A., (1981). Role of southern pine beetle and fire in maintenance of structure and function of the southeastern coniferous forest. *Environ. Entomol.*, **10**: 821-825.

SHEA P. J., McGREGOR M. D., DATERMN G. E., (1992). Aerial application of verbenone reduces attack of lodgepole by mountain pine beetle. *Canad. J. For. Res.*, **22**: 436-441.

SKINNER G. J., (1980). The feeding habits of the wood-ant, *Formica rufa* (Hymenoptera: Formicidae), in limestone woodland in north-west England. *J. Anim. Ecol.*, **49**: 417-433.

SKINNER G. J., WHITTAKER J. B., (1981). An experimental investigation of interrelationships between the wood-ant (*Formica rufa*) and some tree-canopy herbivores. *J. Anim. Ecol.*, **50**: 313-326.

SMIRNOFF W.-A., (1975). Les perspectives d'utilisation des prédateurs, des parasites et des entomopathogènes dans la répression des insectes ravageurs des forêts. *Ann. Soc. Ent. Québec*, **20**: 86-97.

SMIRNOFF W., JUNEAU A., (1973). Quinze années de recherches sur les micro-organismes de la province de Québec (1957-1972). *Ann. Soc. Ent. Québec*, **18**: 147-181.

SMIRNOFF W., VALERO J., (1975). Effets à moyen terme de la fertilisation par urée ou par potassium sur *Pinus banksiana* L. et le comportement de ses insectes dévastateurs: tel que *Neodiprion swainei* (Hymenoptera, Tenthridinidae) et *Toumeyella numismaticum* (Homoptera, Coccidae). *Canad. J. For. Res.*, **5**: 236-244.

SMITH S. M., WALLACE D. R., HOWSE G., MEATING J., (1990). Suppression of spruce budworm populations by *Trichogramma minutum* Riley, 1982-1986. *Mem. Ent. Soc. Canada*, **153**: 56-80.

SNAUWERT M., VANDE MAELE R., NEF L., (1992). Composition minérale de l'écorce de *Picea abies* et attaques d'*Ips typographus*. *Conférence IUFRO, Zakopane 1991*, 67-72.

STARK R. W., (1965). Recent trends in forest entomology. *Ann. Rev. Ent.*, **10**: 303-324.

STARK R. W., MILLER P. R., COBB F. W., WOOD D. L., PARMENTER J. R., (1968). Photochemical oxydant injury and bark beetle (Coleoptera : Scolytidae) infestation of ponderosa pine. I. Incidence of beetle infestation in injured trees. *Hilgardia*, **39**: 121-126.

STRAW N. A., (1995). Climate change and the impact of green spruce aphid, *Elatobium abietinum* (Walker) in the UK. *Scottish Forestry*, **49**: 134-145.

STRAW N. A., (1996). The impact of pine looper moth, *Bupalus piniarius* L. (Lepidoptera: Geometridae) on the growth of Scots pine in Tentsmuir Forest, Scotland. *Forest Ecology and Management*, **3**: 1-24.

STRAW N. A., FIELDING N. J., WATERS A., (1996). Phytotoxicity of insecticides used to control aphids on Sitka spruce, *Picea sitchensis* (Bong.) Carr. *Crop Protection*, **15**: 451-459.

STRONG D. R., (1974). Non asymptotic species richness models and the insects of British trees. *Proc. Nat. Acad. Science USA*, **71**: 2,766-2,769.

SYME P. D., (1974). Interaction de trois parasites de la tordeuse européenne des pousses du pin. *Rev. Bimest. Recherches*, **30**: 20.

THATCHER R. C., SEARCY J. L., COSTER J. E., HERTEL G. D., (1981). *The southern pine beetle*. USDA, Forest Service, technical bulletin, **1,631**, 266 pp.

VARLEY G. C., (1949). Population changes in German forest pests. *J. Anim. Ecol.*, **18**: 117-122.

VARLEY G. C., The effects of grazing by animals on plant productivity. *In:* K. PETRUSEWICZ (1967). *Secondary productivity of terrestrial ecosystems*, 773-777.

VILLEMANT C., (1980). Modifications du complexe entomophage de la tordeuse des pousses du pin *Rhyacionia buoliana* Schiff. (Lépidoptère Tortricidae) en liaison avec la pollution atmosphérique en forêt de Roumare (Seine-Maritime). *Acta Œcologia, Œcol. Applic.*, **1**: 139-160.

VILLEMANT C., (1980). Influence de la pollution atmosphérique sur les microlépidoptères du pin en forêt de Roumare (Seine-Maritime). *Acta Œcologia, Œcol. Applic.*, **1**: 291-306.

VILLEMANT C., (1981). Influence de la pollution atmosphérique sur les populations d'Aphides du pin sylvestre en forêt de Roumare (Seine-Maritime). *Environ. Pollut.*, ser. A, **24**: 245-262.

VITÉ J.-P., (1971). Silviculture and the management of bark beetle pests. *Proceedings Tall Timber Conference on Ecological Animal Control by Habitat Management*, **3**: 155-168.

VOUTE A.-D., (1964). Harmonious control of forest insects. *Int. Rev. Forest Research*, **1**: 325-383.

WARGO P. M., (1996). Consequences of environmental stress on oak: predisposition to pathogens. *Ann. Sci. For.*, **53**: 359-368.

WARGO P. M., PARKER J., HOUSTON D. R., (1972). Starch content in defoliated sugar maples. *For. Sci.*, **18**: 203-204.

WARING R. H., PITMAN G. B., (1985). Modifying lodgepole pine stands to change susceptibility to mountain pine beetle attack. *Ecology*, **66**: 889-897.

WATERS W. E., (1971). Ecological management of forest insect populations. *Proceedings Tall Timber Conference on Ecological Animal Control by Habitat Management*, **3**: 141-153.

WATT A. D., WHITTAKER J. B., DOCHERTY M. *et al.*, The impact of elevated atmospheric CO_2 on insect herbivores. *In:* R. HARRINGTON, N. E. STORK, (1996). *Insects in a changing environment.* 17th Symposium of the Royal Entomological Society. Academic Press, London, 197-217.

WAY M. J., BEVAN D., Dilemmas in forest pest and disease management. *In:* F.-H. PERRING, K. MELLANBY (1977). *Ecological effects of pesticides.* Academic Press, London, 95-110.

WELLENSTEIN G., (1965). Die Entwicklung der Waldameisen (*Formica rufa* Gruppe) auf die Biozönose Methoden und Ergebnisse. *Collada Verde*, **16**: 369-392.

WELCH R. C., (1981). The insect fauna of *Nothofagus*. *Institute of terrestrial Ecology. Annual report 1980*, 50-53.

WESELOH R. M., Host location by parasitoids. *In:* D. A. NORDLUND, R. L. JONES, W. J. LEWIS (1981). *Semiochemicals, their roles in pest control.* J. Wiley and Sons, New York, 79-91.

WHITE T. C. R., (1969). An index to measure weather-induced stress of trees associated with outbreaks of psyllids in Australia. *Ecology*, **50**: 905-909.

WHITE T. C. R., (1974). A hypothesis to explain outbreaks of looper caterpillars, with special reference to populations of *Selidosoma suavis* in a plantation of *Pinus radiata* in New Zealand. *Œcologia*, **16**: 279-301.

WHITHAM T. G., MOPPER S., (1985). Chronic herbivory: impacts on architecture and sex expression of pinyon pine. *Science*, **228**: 1,089-1,091.

WILLIAMS D.-W., LIEBHOLD M., (1995). Forest defoliators and climate change: potential changes in spatial distribution of outbreaks of western spruce budworm (Lepidoptera: Tortricidae) and gypsy moth (Lepidoptera: Lymantriidae). *Environ. Entomol.*, **24**: 1-9.

WORRELL R., (1983). Damage by the spruce bark beetle in south Norway 1970-1980. A survey and factors affecting its occurrence. *Medd. Norsk. Inst. Skogfors*, **38**: 1-34.

WRIGHT L. C., BERRYMAN A. A., WICKMAN B. E., (1984). Abundance of the fir engraver, *Scolytus ventralis*, and the Douglas-fir beetle *Dendroctonus pseudotsugae*, following tree defoliation by the Douglas-fir tussock moth, *Orgyia pseudotsugata. Canad. Ent.*, **116**: 293-305.

ZABORSKI E., TEAL P. E. A., LAING J. E., (1987). Kairomone-mediated host finding by spruce budworm egg parasite, *Trichogramma minutum. J. Chem. Ecol.*, **13**: 113-122.

ZVEREVA E.-L., KOZLOV M.-V., (1996). Avoidance of willows from moderately polluted areas by leaf beetle, *Melasoma lapponica*: effects of emission or induced resistance? *Ent. exp. appl.*, **79**: 355-362.

ZVEREVA E.-L., KOZLOV M.-V., NEUVONEN S., (1995). Decrease in feeding niche breath of *Melasoma lapponica* (Coleoptera: Chrysomelidae) with increase in pollution. *Œcologia*, **104**: 323-329.

7

Canopy insects

Canopy animals are those that live in crown foliage and the forest canopy. Research on these organisms, on the structure and dynamics of their communities and of their role in the functioning of the forest ecosystem was stimulated by work such as that done by Whittaker (1952), Southwood (1961), Ohmart & Voigt (1981), Southwood et al. (1982), Moran & Southwood (1982), and Erwin (1982, 1983). Research has been carried out in temperate as well as in tropical forests. Work done by Erwin on the canopy fauna of the Amazon forest was the basis of a prodigious development in research on biodiversity. Early estimates based on analysis of beetle populations of the canopy led Erwin to put the number of animal species at thirty million although the general consensus today is somewhat lower, at around ten million (cf. chapter 21).

For many arthropods, the canopy serves as habitat during only one part of their life cycle. Many caterpillars feed on foliage and drop to the ground to pupate. Adult weevils belonging to the genera *Phyllobius* and *Polydrusus* feed on leaves while their larvae eat the roots. Species distribution in the vertical plane is variable. Mining caterpillars of *Phyllonorycter* and *Stigmella*, and adult *Phyllobius argentatus* and *Rhynchaenus fagi* prefer to occupy positions beneath the tree crown. *Rhynchaenus fagi* galleries are particularly numerous at the periphery of populations (Nielsen & Ejlersen, 1977).

Canopy arthropods have very varied biology. They are also numerous and rich in species. These arthropods are mostly insects, spiders and mites, but there are also other, usually less numerous, representatives of other groups such as Apterygota, chilopods and diplopods. There are even other invertebrates such as a few molluscs. In the subtropical Everglades of Florida, for instance, gastropods of the genus *Liguus* are common on tree trunks.

Sampling canopy fauna is difficult, especially in forests in which the trees are very tall. In temperate forests this was long done by beating branches over a sheet, or by jarring tree trunks after tarpaulins had been spread on the ground to catch falling animals. A more recent method is the use of clouds of insecticides, known as "canopy fogging" or "smoking method", which is mostly used in the tropics but is also used in temperate forests (cf. chapter 21). A general account of methods of sampling canopy insects has been given by Stork *et al.* (1997).

1. GUILD STRUCTURE OF CANOPY ARTHROPODS

The biology of canopy arthropods are very diverse. These animals can be divided into a number of functional groups, or guilds, the main ones being as follows:

– Chewing herbivores: these are mostly Lepidoptera, a few sawflies and various Coleoptera which consume leaves from the outside.

– Mining herbivores: mainly Lepidoptera and a few Coleoptera which eat leaves from the inside.

– Gall animals: mites, Diptera and cynipid Hymenoptera.

– Sap-sucking herbivores: mainly Homoptera.

– Eaters of epiphytic cryptogams: the most notable elements are Psocoptera.

– Detritivores: a very heterogenous group including cockroaches, Psocoptera etc.

– Predators: also a very varied group, but dominated by Coleoptera.

– Parasitoids: Diptera and Hymenoptera.

To this list one must add ants, which are sometimes abundant in the canopy, and whose biology is unusual, as well as species which use foliage simply as a temporary shelter and which Southwood calls "tourists".

A study of the distribution of canopy arthropod guilds and their composition in terms of numbers of species, individuals and biomass has been made on six species of tree in Britain and in South Africa. Three trees, oak, false acacia (*Robinia*) and birch are common to both countries. Despite climatic differences in the two regions similarities in the guilds in each region are remarkable (figure 7.1). Phytophagous species constitute about one quarter of species, half the biomass, and two thirds of the number of individuals (Moran & Southwood, 1982).

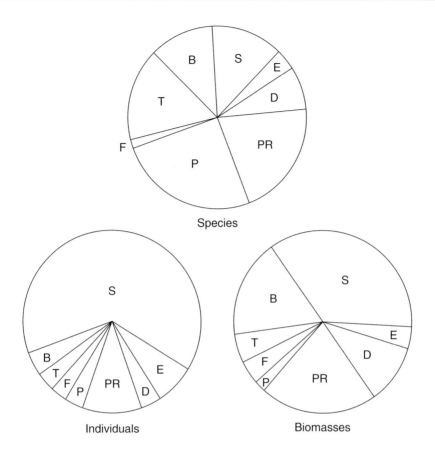

Figure 7.1. *Composition of populations of canopy arthropods. Average values of number of species, number of individuals and biomass of various guilds calculated from samples gathered on six species of tree in Britain and in South Africa. B: chewing species; S: sap-sucking species; E: feeders on epiphytes; D: detritivores; PR: predators; P: parasitoids; F: ants; T: tourists (after Moran & Southwood, 1982).*

2. SPECIES DIVERSITY AND POPULATION STRUCTURE OF CANOPY ARTHROPODS

Canopy arthropod population richness is high, even in temperate regions. A few examples will show the diversity of this fauna according to tree and region.

2.1. Canopy fauna of Aleppo pine

This fauna, surveyed in the Marseille area, is dominated by Homoptera and in particular aphids, as well as Psocoptera and spiders (Guillaumont,

1976). Numbers of individuals collected are divided among various systematic groups as follows:

Arachnids : 1,639
Myriapods: 5
Crustaceans : 2
Psocoptera : 2,140
Collembola: 321
Homoptera (except aphids): 1,883

Aphids: 1,165
Diptera : 280
Hymenoptera: 509
Coleoptera: 422
Thysanoptera: 490
Others: 157

2.2. Canopy fauna of Japanese red cedar (*Cryptomeria japonica*)

The fauna of this conifer has been studied by Hijii (1986). Dominant groups in terms of numbers on this tree are Collembola, followed by mites. Diplopods dominate in biomass, followed by Collembola and spiders (table 7.1). This dominance by mites and Collembola is remarkable, for these arthropods are usually rare in the canopy of other trees. There is a

	Average number of individuals per tree	Average biomass (in mg, dry weight)
Plant-eaters and detritivores		
Acari	3,302	37.55
Diplopoda	29	317.92
Collembola	12,754	146.74
Psocoptera	493	6.65
Hemiptera	148	26.15
Thysanoptera	1,416	6.72
Lepidoptera (caterpillars)	61	70.51
Diptera	298	56.78
Coleoptera	28	50.78
Hymenoptera	4	0.37
Others	9	14.18
Total	**18,542**	**734.35**
Predators and parasitoids		
Acari	1,113	2.21
Spiders	119	96.44
Chilopoda	2	2.20
Opiliones	22	13.75
Hemiptera	0.3	4.61
Coleoptera	14	4.97
Hymenoptera	81	6.42
Total	**1,351.03**	**130.60**
Others	**9**	**24.81**

Table 7.1.   *Abundance and biomass of arthropods in* Cryptomeria japonica *canopy sampled by fogging (Hijii, 1986).*

linear relationship of a logarithmic kind between abundance and biomass of arthropods and the size of host trees. Predators and parasitoids are less abundant than plant-eaters and detritivores. Considering the area occupied by a projection of the tree canopy on the ground, the average abundance of canopy arthropods is estimated at 3,755 individuals per square metre and their biomass at 165.87 mg (dry weight), which is a mean value. Values put forward by other authors vary between 74.3 mg for the fauna of *Larix leptolepis* in Japan, to 600.2 mg for the fauna of oak, and 1,402.5 mg for the fauna of birch in Britain.

2.3. Canopy fauna of a beech wood

An inventory of the canopy fauna of a beech wood of trees ninety years old was made in Denmark (Nielsen, 1975) using various sampling techniques. A total of 66,000 invertebrates was collected. Pterygotes accounted for about 70% of invertebrates (50,800 individuals) and spiders 20% (13,100 individuals). Groups of secondary importance (oligochaetes, gastropods, isopods, diplopods, chilopods, pseudoscorpions, opilionids, acarians and collembolans) amounted to only 1,700 individuals, but may have been underestimated owing to sampling techniques. Pterygote insects included representatives of 10 orders, 90 families and 256 determined species, although there were at least 350 species in all (table 7.2). The most notable were the following:

– Coleoptera. This order was represented by 33,000 individuals, 30 families and 160 species. Curculionids alone made up 85% of the beetles and were dominated by two species. *Rhynchaenus fagi* is an obligate feeder on beech, and *Phyllobius argentatus* is polyphagous. More than 100 species of Coleoptera were represented by fewer than ten individuals.

– Lepidoptera. This order accounted for 6,100 individuals and 34 species. About 90% of the moths belonged to two families, the geometrids including *Operophtera fagata* and the oecophorids including *Chimabacche fagella*. These two moths are not dependent on beech, but are common on it.

– Dermaptera. This order came in third place with 3 species and 3,300 individuals, of which more than 99% belonged to the ubiquitous species *Chelidurella acanthopygia* which may be classed in the tourist guild.

– Heteroptera. Fewer than 5% of invertebrates collected were Heteroptera. The five dominant species, which made up more than 50% of the Heteroptera, are *Troilus luridus* (Pentatomidae), *Psallus varians* (Miridae), *Anthocoris nemorum* and *A. confusus* (Anthocoridae) and *Lygus pratensis* (Lygaeidae).

– Homoptera. This order was represented mainly by *Phyllaphis fagi* which is specific to beech.

– Hymenoptera species were not determined. These parasitoid insects were braconids, ichneumonids and chalcids.

– Diptera. Syrphids were the most numerous of the flies.

Table 7.2 shows abundance of the commoner species which, according to sampling technique, amount to 87.1 to 91.2% of the insect population. There are great variations in abundance and species composition in the different samples, which shows the difficulty of such studies. Photoeclectors enable sampling of those invertebrates that move around on tree trunks. Samples obtained in this way are dominated by beetles, among which are species such as *Xyloterus domesticus* and *Hylecoetus dermestoides* which are associated with dead wood, *Dromius quadrimaculatus* and *Rhizophagus dispar* which are subcortical predators, and *Nebria brevicollis* which is a litter predator but sometimes wanders up trunks to hunt. The other two sampling methods mainly collect leaf-eaters, sap-suckers, Psocoptera which feed on algae or epiphytic lichens, and various predators (*Coccidula*, *Hemerobius*, *Malthodes*). Monophagous species dependent on beech, such as the aphid *Phyllaphis fagi*, the scale insect *Cryptococcus fagisuga* and the weevil *Rhynchaenus fagi*, amount to less than 5% of species. Tourists, occasional visitors such as the weevil *Phyllobius* or the earwig *Chelidurella*, constitute a not insignificant percentage of the fauna.

The canopy fauna of this beech wood is dominated by a group of 10 to 15 abundant species that amount to only 15 to 20% of the total number of species, but that make up 85 to 90% of the total number of insects present. This structure is found in all samples, whatever the method used, and appears to be a general one in all forests. From a trophic point of view, at species level 40 to 45% of insects are phytophagous, 25 to 30% are predators, 15 to 20% are saprophagous, and 5 to 10% are omnivorous. At the individual level 60 to 65% are phytophagous, 25 to 30% predators, 1 to 5% saprophagous, and 5% are omnivorous. Omnivorous and phytophagous insects are generally more abundant (200 to 300 individuals per species on average) than predators and saprophagous insects (10 to 40 individuals per species on average). The fauna is dominated by medium- to large-sized species measuring more than 3 mm. Saprophagous species are generally smaller and omnivorous species larger; herbivores and predators have intermediate body sizes.

2.4. Canopy fauna of cork oak and downy oak

This fauna has been studied in a location in Provence (Bigot & Ponel, 1985). The arthropod groups represented are very diverse (108 species), but, in terms of numbers of species and individuals, are dominated by Coleoptera (34 species), Psocoptera (11 species) and spiders (29 species). Maximum population diversity is in May, and minimum in February. Comparison of various canopy communities leads to the conclusion that the tree plays a

Species	A	B	C
Phyllobius argentatus (COL)	560	671	769
Rhynchaenus fagi (COL)	305	1,804	3,805
Operophtera fagata (LEP)	190	228	113
Chelidurella acanthopygia (DER)	103	21	262
Strophosomus melanogrammus (COL)	61	67	161
Rhizophagus dispar (COL)	46	0	0
Dromius quadrimaculatus (COL)	39	0	0
Strophosomus capitatus (COL)	38	146	85
Polydrusus mollis (COL)	35	0	0
Mesopsocus unipunctatus (PSO)	25	85	0
Nebria brevicollis (COL)	22	0	0
Chimabacche fagella (LEP)	21	27	124
Solenobia lichenella (LEP)	20	0	107
Hylecoetus dermestoides (COL)	17	0	0
Xyloterus domesticus (COL)	17	0	0
Psallus varians (HET)	0	76	254
Athous subfuscus (COL)	0	71	122
Coccinula 14-punctata (COL)	0	38	0
Phyllaphis fagi (HOM)	0	0	508
Typhlocyba cruenta (HOM)	0	0	158
Hemerobius micans (NEU)	0	0	145
Malthodes flaveolus (COL)	0	0	137
Total	1,499 (87.1 % of sample)	3,234 (90.2 % of sample)	6,592 (91.2 % of sample)

Table 7.2. *Abundance of commoner species in a Danish beech wood obtained by three different sampling techniques. A: "photoeclectors". B: beating. C: jarring trunks of trees over plastic sheeting (after Nielsen, 1975).*
The order to which species belong is abbreviated in brackets.
COL: *Coleoptera;* LEP: *Lepidoptera;* DER: *Dermaptera;*
PSO: *Psocoptera;* HET: *Heteroptera;* HOM: *Homoptera;*
NEU: *Neuroptera.*

secondary part in determining the species composition of canopy populations. The tree is above all a support on which species which depend on surrounding vegetation settle. The canopy fauna of cork oak in Morocco is dominated by spiders, which amount to 67% of samples; the remainder is composed of insects: Lepidoptera (8.5%), Collembola (7.8%), Hymenoptera (3.3%), Psocoptera (2.0%), Diptera (1.6%) and Heteroptera (0.5%) (Mahari, 1992). Insects that feed on cork oak leaves account for about one third of the fauna. The others are tourists, that is, phytophagous insects dependent on plants of the undergrowth. Most of them are Homoptera (jassids, cixiids and issids), tettigoniids, and beetles (scarabaeids, buprestids, chrysomelids, bruchids and curculionids).

3. EVOLUTION OF CANOPY FAUNA

One example will show that canopy fauna is a function of a tree's age and of the state of its foliage.

In a stand of young Douglas-fir in Oregon, arthropods are dominated by sap-sucking species of which the chermesid *Gilletteella cooleyi* accounts for 99% of the total number; moths are absent; there is only a single species of chrysomelid, a few ants, some parasitic wasps, and spiders (table 7.3). On the other hand diversity is much greater in old populations, and leaf-eating species are more abundant than sap-sucking ones in the latter (Schowalter, 1988; Schowalter *et al.*, 1988). Increase in the ratio of leaf-eaters / sap-suckers with the age of woods appears to be a general rule, at least in North American conifer forests.

4. EPIPHYTE FAUNA

Trees are often covered in micro-algae, epiphytic lichens and mosses. This cryptogam vegetation serves as food to about 20% of phytophagous canopy species in a Danish beech wood (Nielsen, 1975). Two geometrid moths belonging to the subfamily Lithosiinae, *Atolmis rubricollis* and *Lithosia deplana*, consume *Pleurococcus* algae which grow on oak, fir and beech, while *Miltochrista miniata* eats the lichen *Peltigera canina* on oak. Imagines of the stonefly *Nemoura cinerea*, the earwig *Chelidurella acanthopygia*, and the springtail *Entomobrya nivalis* also eat epiphytic algae or lichens, but the most numerous consumers of epiphytes are psocids. The particular structure of the mouthparts of these insects is probably related to such a diet. Their importance in the forest ecosystem is not insignificant, since primary production of epiphytes in temperate forest is comparable with that of the herbaceous stratum, and the biomass of psocids may be high. A density of 4,000 individuals per square metre of larch bark has been recorded in England, and a maximum biomass of 1.2 g per square metre of bark (corresponding to about 2 grams per square metre on the ground) for two species of *Mesopsocus* (Turner, 1975).

Some species of lichen which are obligate growers on trees are considered to be indicators of forest continuity. Such is the case of *Lobaria pulmonaria* in Finland and in Sweden (Nilsson et al., 1995; Kuusinen, 1996). This lichen is often associated, in the plots that were studied, with a number of other lichens dependent on old forest. For this reason these lichens risk extinction or considerable decline, and are on the red list of threatened species. There is a significant relationship between the numbers of threatened lichens and the numbers of beetles which are threatened because they are dependent on dead wood, hollow trees or lignicolous fungi.

Arthropod groups	Young *Pseudotsuga*	*Pseudotsuga* over 400 years old	*Tsuga* over 400 years old
Leaf-eaters			
Gilletteella	23,000	48	39
Cinara	100	0.4	0
Scale insects 4 spp.	2	51	110
Zeiraphera	0	2.8	0
Others 5 spp.	0	0.9	0.9
Flower- and seed-eaters			
Thrips 2 spp.	0	3.0	0.7
Kleodocerys	0	1.1	17
Others 4 spp.	0	0.4	0.9
Predators			
Camponotus	2	0.7	0.2
Aphids 3 spp.	7	1.5	0.6
Parasitic Diptera	0	0.2	0
Parasitic Hymenoptera > 6 spp.	6	1.5	1.7
Predatory Hymenoptera 3 spp.	4	2.4	0.2
Semidalis (Neuroptera)	0	1.3	0.7
Mites > 3 spp.	0	3.9	6.5
Spiders	12	10.6	15.5
Detritivores			
Collembola	0	0.7	0.2
Psocoptera	0	0.6	17
Camisia caroli (mite)	0	7.8	19
Tourists	4	0.6	0.7

Table 7.3. *Average abundance of canopy arthropods (number of individuals per kg of branches) in a population of* Pseudotsuga menziesii *and* Tsuga heterophylla *in Oregon. Young trees at the regeneration stage and trees over 400 years old in primary ("old growth") forest (Schowalter, 1988).*

In natural conifer forests, abundance of lichens is greater than in planted woodland. This enables a much richer arthropod fauna to occupy the crowns. Such increases affect all groups: the average number of insects per branch is 7.75 in planted forest and 36.5 in natural forest. The number of invertebrates over 2.5 mm long is consistently greater in natural forests. Moreover 2.5 mm is the minimum size of prey which insectivorous passerine birds can catch.

The decline of non-migratory passerines in northern Europe has been attributed to the destruction and fragmentation of forests. These studies show that it may also be due to reduction in available prey as a result of lower abundance of lichens (Nilsson *et al.*, 1995; Pettersson *et al.*, 1995).

Richness in Psocoptera is variable. Scots pine in England only harbours four species owing to paucity of *Protococcus* algae. Larch plantations in the Pennines harbour 24 species, for these trees are rich in lichens and algae belonging to the family Protococcaceae. Distribution of psocids on trees is not haphazard. *Caecilius burmeisteri* is confined to conifers, and *Stenopsocus stigmaticus* to broadleaf trees. Atmospheric pollution greatly alters the distribution of psocids through its effect on epiphytes. Some species vanish from polluted areas, particularly those that live on lichens, whereas those that live on micro-algae resistant to pollution survive. *Loensia variegata* occurs in high density in sites highly polluted by emissions of SO_2, as does *Cerobasis guestfalicus*. *Pseudopsocus rostocki* on the other hand is absent from areas where the air contains more than 80 g SO_2 m^{-3}. Psocids have proven to be good pollution indicators (Schneider, 1989).

Diet and food preferences in Psocoptera have been studied by Broadhead (1958) in species which live on larch. Living larch branches are mainly covered in unicellular algae of the genus *Pleurococcus*, and dead branches bear the lichen *Lecanora conizaeoides* in abundance. Some psocids prefer algae, but *Elipsocus maclachlani* and *Reuterella helvimaculata* prefer lichen. *Reuterella* consumes lichen entirely, whereas *Elipsocus* only eats certain parts, namely the fruiting bodies or apothecies. This difference in behaviour is explained by the lichen's chemical composition. A lichenic acid which is repellent to *Elipsocus* occurs in the lichen except in the apothecies. *Reuterella*, on the other hand, which eats all of the lichen, is insensitive to this lichenic acid.

Two very close species, *Mesopsocus immunis* and *M. unipunctatus*, coexist in larch populations. Cohabitation is made possible by the slight differences in biology. *M. immunis* prefers to lay in the basal and median parts of twigs on dormant lateral buds, while *M. unipunctatus* favours leaf scars of the distal areas in which to deposit its eggs (Broadhead & Wapshere, 1966).

5. A SPECIAL HABITAT: BARK

Bark is an important element of the forest ecosystem. The barks of trees are varied in structure, have very particular microclimates, and provide arthropods with very diverse micro-habitats. Bark is also a thoroughfare for species that migrate from soil or litter to tree crowns in the course of their life cycles. It has been established that 57% of spiders in a central European forest are found only on tree trunks. Many arthropods use trunks for mating, egg laying, larval growth and for rest or hibernation.

Microclimate varies according to the structure of bark (Nicolai, 1986). There are great differences in temperature between trees with thin, smooth bark such as beech and those with thick, fissured bark such as oak. In beech cambium, the temperature reaches 40 °C and that of the bark's surface is uniformly high. In oak cambium, the temperature does not exceed 30 °C and the temperature of the bark is higher on the raised parts than in crevices.

Tree trunk fauna is a heterogeneous collection of resident species (mainly feeders on epiphytic micro-plants) and predators. Spiders, Collembola, mites, psocids, a few beetles and a few Diptera are the dominant groups (Funke & Sammer, 1980). In Germany, bark fauna includes many oribatid mites whose distribution is governed by microclimate, particularly in the case of the dominant species, *Carabodes labyrinthicus*. Some spiders live only on bark, and one of them, *Drapetisca socialis*, is thought to be one of the major predators in European forests (Funke, 1973). Many Psocoptera live on bark. *Pseudopsocus rostocki* lives exclusively on bark (mainly on oak); it is parthenogenetic and has wingless females whose eggs overwinter on bark. Apterism (or wing reduction) in females is seen in half the species of bark psocids.

New trunks are colonised by wind-borne larvae. Parthenogenesis and apterism are seen in other Psocoptera and in Heteroptera, and these characters seem to provide an adaptive advantage. Bark Heteroptera are rare. *Empicoris vagabunda* (Reduviidae) lives exclusively on bark, mainly on oak and birch; *Loricula elegantula* (Microphysidae) mainly colonise beech.

Beech		Oak		Elm		Birch	
C. lab	88.1	C. lab	63.6	L. ror	36.9	C. lab	62.8
T. nub	5.3	E. pen	8.3	C. lab	15.8	D. soc	24.2
M. den	1.2	L. ror	5.2	M. exc	12.7	L. fas	3.1
L. ele	1.2	T. nub	3.7	K. bic	6.8	T. nub	2.9
D. soc	0.9	C. cym	2.6	M. jac	6.3	C. sch	2.3
Total	96.7	Total	83.5	Total	78.5	Total	95.3

Table 7.4. *Relative numbers and percentages of total number of individuals of 5 commonest species of arthropod (insects, mites, spiders) on trunks of 4 species of tree in Germany. Several species are common to different trees (Nicolai, 1989). C. lab: Carabodes labyrinthicus (oribatid); C. cym: Cymberemaeus cymba (spider); C. sch: Chamobates schuetzi (oribatid); D. soc: Drapetisca socialis (spider); E. pen: Entelecara penicillata (spider); K. bic: Kratochviliella bicapitata (spider); L. ele: Loricula elegantula (Heteroptera); L. fas: Loensia fasciata (psocid); L. ror: Lycia rorida (Diptera); M. den: Medetera dendrobaena (Diptera); M. exc: Medetera excellens (Diptera); M. jac: Medetera jacula (Diptera; T. nub: Tachypeza nubila (Diptera).*

Coleoptera such as the weevil *Strophosoma melanogramma* use bark as a corridor between soil and canopy. Empidids are the main flies (Diptera) in forest. Adult *Tachypeza nubila* occurs on the bark of many trees where it catches insects of all orders; its larvae live in the soil. The dolichopodid *Neurogona quadrifasciata* also frequents tree trunks. Species belonging to the family Lauxanidae also play a major role in this habitat (Nicolai, 1986).

Qualitative and quantitative differences between populations of various trees are slight; they are illustrated in table 7.4. The arthropods that settle on the bark of trees in a subtropical forest and in a savannah in Africa include many more species, and orders and families of insects such as Orthoptera and Coleoptera which are rare or absent in this habitat in temperate regions (Nicolai, 1989).

REFERENCES

BIGOT L., PONEL P., (1985). Étude d'un écosystème méditerranéen : la forêt domaniale mixte (chênaie à *Quercus ilex* L. et à *Q. pubescens* Willd.) de la Gardiole de Rians. Composition et structure du peuplement des Arthropodes frondicoles. *Bull. Ecol.*, **16**: 269-272.

BROADHEAD E., (1958). The psocid fauna of larch trees in Northern England. An ecological study of mixed populations exploiting a common resource. *J. Anim. Ecol.*, **27**: 217-263.

BROADHEAD E., WAPSHERE A.-J., (1966). *Mesopsocus* populations on larch in England. The distribution and dynamics of two closely-related coexisting species of Psocoptera sharing the same food resource. *Ecol. Monogr.*, **36**: 327-388.

FUNKE W., SAMMER G., (1980). Stammablauf und Stammanflug von Gliederfüssern in Laubwäldern (Arthropoda). *Entomol. Gen.*, **6**: 159-168.

GUILLAUMONT F., (1976). *Étude du peuplement et de la biologie des Psocoptères du Pin d'Alep dans la région méditerranéenne française.* Thesis, Université de Marseille.

HIJII N., (1986). Density, biomass, and guild structure of arboreal arthropods as related to their inhabited tree size in a *Cryptomeria japonica* plantation. *Ecological Research*, **1**: 97-118.

KUUSINEN M., (1996). Cyanobacterial macrolichens on *Populus tremula* as indicators of forest continuity in Finland. *Biol. Cons.*, **75**: 43-49.

MAHARI S., (1992). *Étude synécologique des niveaux de peuplement arthropodien de l'écosystème chêne-liège dans les subéraies de la Mamora et de Ben Slimane (Maroc).* Thesis, Faculté des Sciences Aix-Marseille.

MILLS N.-J., (1983). Possibilities for the biological control of *Choristoneura fumiferana* (Clemens) using natural enemies from Europe. *Commonwealth Institute of Biological Control,* **4**: 104-125.

MILLS N.-J., SCHÖNBERG F., (1985). Possibilities for the biological control of the Douglas-fir tussock moth *Orgyia pseudotsugata* (Lymantriidae), in Canada, using natural enemies from Europe. *Commonwealth Institute of Biological Control,* **6**: 7-18.

MORAN V.-C., SOUTHWOOD T.-R.-E., (1982). The guild composition of arthropod communities in trees. *J. Anim. Ecol.,* **51** : 289-306.

NICOLAI V., (1986). The bark of trees: thermal properties, microclimate and fauna. *Œcologia,* **69**: 148-160.

NICOLAI V., (1989). Thermal properties and fauna on the bark of trees in two different African ecosystems. *Œcologia,* **80**: 421-430.

NIELSEN B.-O., (1975). The species composition and community structure of the beech canopy fauna in Denmark. *Vidensk Meddel fra Dansk Natur Forening,* **138**: 137-170.

NIELSEN B.-O, EJLERSEN A., (1977). The distribution pattern of herbivory in a beech canopy. *Ecol. Ent.,* **2**: 293-299.

NILSSON S.-G, ARUP U., BARANOWSKI R., ECKMAN S., (1995). Tree dependent lichens and beetles as indicators in conservation forests. *Cons. Biol.,* **9**: 1,208-1,215.

OHMART C.-P, VOIGT W.-G., (1981). Arthropod communities in the crown of the natural and planted stands of *Pinus radiata* (Monterey pine) in *California. Canad. Ent.,* **113**: 673-684.

PETTERSSON R.-B, BALL R.-B, RENHORN K.-E, ESSEEN P.-A, SJÖBERG K., (1995). Invertebrate communities in boreal forest canopies as influenced by forestry and lichens with implications for passerine birds. *Biol. Cons.,* **74**: 57-63.

POPESCU C., BROADHEAD E., SHORROCKS B., (1978). Industrial melanism in *Mesopsocus unipunctatus* (Müll.) (Psocoptera) in northern England. *Ecol. Ent.,* **3**: 209-219.

SCHNEIDER N., (1989). Les Psocoptères du Grand Duché de Luxembourg. III. Faunistique et écologie des espèces urbaines. *Bull. Ann. Soc. R. Belge Entomol.,* **118**: 131-144.

SCHOWALTER T.-D., (1988). Canopy arthropod community structure and herbivory in old-growth and regenerating forests in western Oregon. *Canad. J. For. Res.,* **19**: 318-322.

SCHOWALTER T.-D, STAFFORD S.-G, SLAGLE R.-L., (1988). Arboreal arthropod community structure in an early successional coniferous forest ecosystem in western Oregon. *Great Basin Naturalist,* **48**: 327-333.

SOUTHWOOD T.-R.-E., (1961). The number of insects associated with various trees. *J. Anim. Ecol.,* **30**: 1-8.

Southwood T.-R.-E, Moran V.-C, Kennedy C.-E.-J., (1982). The richness, abundance and biomass of the arthropod communities on trees. *J. Anim. Ecol.*, **51**: 635-649.

Stork N.-E, Adis J., Didham R.-K., (1997). *Canopy arthropods.* Chapman & Hall, London.

Turner B.-D., (1975). Energy flow in arboreal epiphytic communities. An empirical model of net primary productivity in the alga *Pleurococcus* on larch trees. *Œcologia*, **20**: 179-180.

Whittaker R.-H., (1952). A study of summer foliage insect communities in the Great Smoky Mountains. *Ecol. Monog.*, **22**: 1-44.

◇ **8** ◇

Defoliating Lepidoptera

Defoliating insects, which are mostly moths, make up a major guild of forest insects. Most species are represented by low density populations of no economic importance. A few species can occur in major outbreaks and can cause spectacular damage.

Defoliating insects are primary pests which attack healthy trees, rather than those in a state of physiological deficiency. Choice of food is a function of tree species, and of the age, toughness and position of leaves on the tree. Defoliators play an important part in food chains by converting plant biomass into animal biomass, by providing food for many predators (birds, bats and other small mammals, insects), and by accelerating the recycling of mineral elements in the biomass within the forest ecosystem (cf. chapter 5). Their effects on tree growth may be considerable during outbreaks (cf. chapter 6). Caterpillars of defoliating moths belong to two guilds: species that consume leaves from the outside (true defoliators), and species that mine leaves from within in order to consume parenchyma.

1. CONSEQUENCES OF DEFOLIATION

The consequences of defoliation, which have been analysed in detail by Kulman (1971), are many and varied. It is however possible to outline a few general rules.

– Broadleaf trees (angiosperms) do not react in the same ways as conifers (gymnosperms). Angiosperms have greater starch reserves than gymnosperms, enabling them to produce new growth of leaves easily, and are therefore better able to withstand defoliation. Among the gymnosperms, pines are more resistant than spruces, and larches, being deciduous (an unusual feature in conifers), are fairly resistant.

– Mortality in defoliated trees varies greatly according to the tree's physiological condition. Trees in poor condition may have high mortality rates
(figure 8.1).

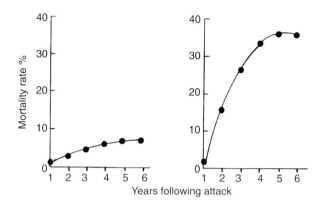

Figure 8.1.   *Mortality rates in various American oaks following severe
defoliation by* Lymantria dispar. *Left: trees in good physiological
condition. Right: trees in poor physiological condition
(Campbell & Sloan, 1977).*

– The time of year at which defoliation occurs is often of crucial importance.
Attacks on young leaves developed in spring are more harmful than those on
old leaves. The sawfly *Neodiprion rugifrons* usually eats old needles of
Pinus banksiana and causes little damage, but when a second generation of
sawflies is produced they consume the young leaves, and three quarters of
defoliated trees die (Wilkinson *et al.*, 1966).

– Repeated defoliation may suppress the dormancy of buds and alter the
time at which they open. One hypothesis (Heichel & Turner, 1976) is that
hormones such as gibberellins are more abundant in defoliated trees and
cause these changes.

– The necessity for synchronization of leaf and insect development has
often been demonstrated.

– A few examples showing effects of defoliation on tree growth and wood
production have been given (cf. chapter 6.3). Defoliation may have different
effects on spring wood and on summer wood. In Australia, the stick insect
Didymuria violescens feeds on *Eucalyptus delegatensis*, reducing summer
wood production during the year of defoliation, and spring wood in the
following year (Mazanec, 1968). In oak, defoliation by cockchafers results in
reduction in the amount of final (summer) wood and of the thickness of the
annual growth ring (Huber, 1982).

– Defoliation reduces cone and seed production. Spruce trees defoliated by
the black arches moth (*Lymantria monacha*) do not produce seeds.

Defoliation of *Pinus banksiana* by *Choristoneura pinus* leads to lowered production of male cones in the following years (Kulman *et al.*, 1963).

– Defoliation reduces root growth in many trees, and the remaining roots contain lower levels of starch than those of trees that have not been defoliated (Wargo *et al.*, 1972).

– Defoliation may alter the species composition of forests by reducing relative numbers of certain trees. Repeated defoliation by *Lymantria dispar* is detrimental to oaks and favours other trees such as pine and maple (figure 8.2).

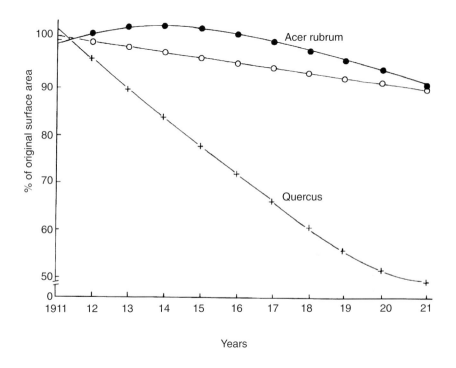

Figure 8.2. *Variations in relative numbers of various trees in a forest subjected to repeated defoliation by* Lymantria dispar. *Relative number of trees is the percentage of the basal area of each species, the value 100 being given to the year 1911. White circles correspond to the total number of trees (Campbell & Sloan, 1977).*

A well-studied case concerns a forest plot subjected to repeated infestations by *Diprion hercyniae* (Reecks & Barter, 1951). The number of trees in this plot fell from 331 in 1938 to 253 in 1946, and the volume of standing wood from 117 cubic metres to 76 cubic metres. The most remarkable change however was the increase in relative number of the fir *Abies balsamea* and decrease of that of *Picea glauca*, while *Picea mariana* maintained approximately the same percentage (table 8.1).

	Relative numbers in percentages		
	Picea glauca	*Picea mariana*	*Abies balsamea*
Number of trees in 1938	21.3	8.8	69.9
Number of trees in 1946	12.0	7.6	80.3
Volume of wood in 1938	35.0	7.3	57.7
Volume of wood in 1946	14.6	6.0	79.4

Table 8.1.   *Number of trees and volume of standing wood in a forest in 1938 and in 1946 following numerous defoliations by* Diprion hercyniae *(Reecks & Barter, 1951).*

Defence reactions of trees to phytophagous insects vary (cf. chapter 3). The main ones are as follows. (a) Low nutritional value. Young foliage, which is ephemeral, is rich in nitrogen compounds, whereas old leaves, which last longer and are therefore more likely to be eaten, are poorer in nitrogen compounds. (b) Accumulation of secondary repellent or toxic compounds. (c) Changes in chemical composition following depredation by insects. (d) Early leaf fall. (e) Growth of new foliage which is more active at photosynthesising, following partial defoliation, thus compensating losses.

2.   DIVERSITY OF DEFOLIATING LEPIDOPTERA

Temperate forests harbour several thousand species of Lepidoptera belonging to many families. A broadleaf forest in the Appalachians is home to about 475 species of macrolepidoptera, while microlepidoptera are much more numerous. A mixed conifer and broadleaf forest in Oregon has 452 species of macrolepidoptera. In Britain about one hundred species of Lepidoptera feed on oak leaves.

An inventory carried out in Fontainebleau forest shows the variety of defoliating moths. In Fontainebleau 14 species of identified microlepidoptera, plus a number of unidentified species belonging to the family Coleophoridae, may feed on the buds and young shoots of Scots pine. Eleven of these species are dependent on the genus *Pinus*, and three are polyphagous. Parts of the tree that are attacked vary according to species. Caterpillars of *Cedestis gysselinella*, *C. farinatella*, *Exoteleia dodecella* and *Rhyacionia buoliana* live by mining the needles; those of *Blasthestia posticana* and *Gravitarmata margarotana* devour the buds; those of *Rhyacionia pinicolana* and *Phycita mutatella* attack young shoots; those of *Petrova resinella* form resin galls which are then occupied by *Enarmonia cosmophorana* (Charles *et al.*, 1970).

Lepidoptera are much richer in species in tropical forests than in temperate ones. In a reserve in Costa Rica there are probably 200 species of hawk moth (Sphingidae) and 30 species of saturniids. This high biodiversity is also found in fruit- and seed-eaters. In the same Costa Rican forest there are at least 110 species of seed-eating beetle belonging to the families Bruchidae, Curculionidae and Cerambycidae (Janzen, 1981).

The main defoliating Lepidoptera belong to about twenty different families, the main ones being the following:

2.1. Eriocraniidae. These are small moths with wingspans under 13 mm and highly modified mouthparts. The rudimentary proboscis is devoid of lacinia and galea. Females have an ovipositor which they use to insert eggs into leaf parenchyma. The caterpillars are miners; *Eriocrania sparmanella* mines birch leaves.

2.2. Stigmellidae (= Nepticulidae). This family includes the smallest known moths (average wingspan 5 mm or less); they are often brightly coloured, and have long slender antennae. The wings have long fringes, and the venation is reduced. The larvae live by mining the leaves of various trees: *Stigmella aceris* bores galleries in maple leaves, *S. subbimaculella* in oak, *S. basalella* in beech, and *S. marginicolella* in elm. Mines of *S. basalella* are narrow and sinuous; those of *S. subbimaculella* are broader and form patches.

2.3. Tischeriidae (= Lyonetidae). This family contains small, pale, usually unicoloured moths. Their fringed wings give them a resemblance to gracillariids. Most species live by mining oak and chestnut leaves. *Tischeria ekebladella* (= *complanella*) is sometimes a pest of oak in eastern and southern Europe. The larvae bore patches of broad, irregular mines of a yellowish white colour (Jordan, 1995).

2.4. Coleophoridae (= Eupistidae). These small (6 to 15 mm) moths have narrow, elongate wings furnished with long fringes. *Coleophora laricella* mainly feeds in young populations of larch in lowlands and at medium altitudes. It occurs together with *Zeiraphera diniana* above 1,700 m in the Swiss Engadine, but there populations remain at low density. Young larvae mine the needles; from the third instar on they build shelters in which they live and which they carry with them; at this stage they feed on needles. The species is an occasional pest in France. *Coleophora serratella* is a common species in Europe, but is of no economic importance, although infestations have been recorded on birch in Sweden, on alder in Switzerland and on hornbeam in Czechoslovakia. Study of its parasite complex has revealed at least 25 species, including ichneumonids of the genus *Campoplex* and braconids of the genus *Apanteles*. This moth has been introduced to North America where it has become a major defoliator, particularly of the birch *Betula papyrifera*.

2.5. Yponomeutidae. These small moths hold their wings folded over the abdomen and their antennae adpressed to the body at rest. The green larva of

Ocnerostoma piniarella lives by mining pine needles. The moth, which appears in June to July, has narrow wings with a span of 10 mm. Eggs are laid at the tips of needles. Galleries begin at this end and continue towards the base, becoming gradually wider; pupation occurs between the needles. There is usually just one generation a year. *Cedestis gysselinella* and *Cedestis farinatella* are also miners of pine needles.

2.6. Gracillariidae (= Lithocolletidae). Representatives of this family, together with tischeriids and stigmellids, account for the majority of mining moths in broadleaf forests in the northern hemisphere. They are medium to small (often less than 6 mm) moths, often brightly coloured, with long slender antennae, wings furnished with fringes, and hind wings often reduced to a simple blade. The genus *Lithocolletis* (= *Phyllonorycter*) contains about fifty species in Europe which mainly live on oaks. Its larvae occur in two morphological types. During the first three or four instars the larva is of the "endophytic" type, resembling a minute white buprestid larva, with the body compressed dorso-ventrally and a very large prominent, triangular head and large and broad prothorax; legs are reduced or obsolescent (Grandi, 1933). This body shape enables the larva to move in its mines by reptation. In the last two instars the larva remains a miner, but it acquires an "ectophytic" shape, with thoracic and abdominal legs and a normal-shaped head. The mouthparts of *Lithocolletis* larvae are highly modified. The shape of the labrum appears to be designed to saw leaf parenchyma in order to detach it from the cuticle. *Lithocolletis* mines are bored, according to species, either in palisade parenchyma or in lacunary parenchyma, or in both. They form more or less strongly discoloured patches on leaves. A single oak leaf may be host to several species, and hundreds of mines of *Lithocolletis platani* are sometimes found on one plane tree leaf. Another species, *Lithocolletis messaniella*, attacks various oaks, chestnut, hornbeam and beech, but mainly *Quercus suber* and *Q. ilex*. The young larva begins by boring a flattened, linear gallery (known as phyllonome). This mine is then extended and becomes a patch (stigmatonome), and then a tentiform mine (ptychonome). The insect produces five or six generations per year. Damage is slight in Europe, but has become serious in New Zealand where various European oaks have been introduced as well as *L. messaniella* which has become polyphagous. Here it attacks not only oak but also endemic species of Lauraceae, Myrtaceae, Compositae and Fagaceae (Delucchi, 1958).

2.7. Oecophoridae. This family is heterogeneous. Its most notable representative is *Chimabacche fagella*, a moth with a wingspan of 20 to 25 mm which is strongly sexually dimorphic in the shape of the wings. The white caterpillar devours leaves of beech, oak, hornbeam and birch, which it gathers into a small nest. The third pair of legs are thickened and black. Reinforced abdominal pseudopods produce sounds by scraping leaf surfaces, which is the origin of the name "singer" given to this caterpillar in France. A few cases of total defoliation of trees have been recorded.

2.8. Gelechiidae. This family includes medium-sized (10 to 20 mm) dull-coloured moths with long, narrow forewings. *Exoteleia dodecella* lives as a miner in spring in pine needles, especially those of Scots pine, after which it penetrates buds.

2.9. Tortricidae. This family owes its name to the behaviour of the caterpillars of many species which roll up or twist the leaves they feed on, using silk threads. They are medium to small moths, with wingspans of 8 to 25 mm. Forewings are usually of a characteristic subrectangular or trapezoidal shape; hind wings are sub-trapezoidal and always uniform pale to dark grey. Many species of tortricid are defoliators of conifers. *Epinotia tedella*, a pest of spruce, has a wingspan of 13 to 14 mm, dark brown forewings with several transverse confluent silvery white lines and narrow pointed greyish-brown hind wings furnished with whitish fringes. The caterpillar is 8 to 9 mm long brownish-yellow with two reddish-brown or greenish dorsal stripes and a black head. Adults fly from May to August depending on the region. Eggs are laid singly on the upper surfaces of needles. After a fortnight spent wandering, the caterpillar bores a hole at the base of a needle which it then mines entirely, leaving the surface intact. Towards the end of summer more or less large piles of silk threads secreted by the caterpillars mixed with frass and detached dead needles can be seen. Hibernation and pupation take place in the ground. This moth is found in a large part of Europe up to 1800 m altitude. When outbreaks occur, infested spruce trees have a characteristic appearance, their needles more or less all turning a reddish-brown colour (Barbey, 1925; Baurant & Offergeld, 1972). *Rhycionia buoliana* infests pine shoots in a similar way.

Choristoneura murinana, the fir tortricid, is a monovoltine species (having only one generation a year) which grows in six larval instars. The first two instars do not feed, and a diapause during the second instar enables the larva to overwinter in a shelter in the tree. The four last instars present in April and May are those that cause the damage, as larvae eat the year's new growth of needles. The black pupae appear in May and adults in mid June. Each female may lay about a hundred eggs which are deposited in rows of about twenty on the upper surfaces of needles. Attacks by this species are usually concentrated on the terminal shoot of adult trees, after which all branches are colonised. During heavy outbreaks even young trees in the undergrowth are infested. Tree growth is greatly slowed, and trees have to be felled and removed to avoid infestations by secondary pests such as scolytids. An inventory of parasitoids has listed 14 species of ichneumonids, including *Cephaloglypta murinanae*, four tachinids and the braconid *Apanteles murinanae*. These parasites may play a part in controlling the pest's populations (Zwölfer, 1961). This species is a major pest of fir in central Europe where its gradations tend to last up to ten years (Franz, 1940; Zwölfer & Kraus, 1957; Zwölfer, 1961). In the last few years it has become a pest in France where it has invaded cedar plantations in the South-East and also populations of silver fir in the Alps, the Massif Central and even in Normandy and Brittany where it arrived in the wake of the trees (Du Merle & Cornic, 1989; Du Merle *et al.*, 1989).

Blasthestia turionella lives mainly on *Pinus sylvestris*. Adults, which appear in May-June, have a wingspan of 15 to 18 mm; they are recognisable by their forewings bearing alternate transverse browny-orange and grey stripes; the hind wings are yellowish-grey. Eggs are deposited singly on buds, and the larvae devour the buds entirely, causing large flows of sap. Pupation takes place in the hollowed out bud. Infested pines often acquire a characteristic bushy appearance. Damage has been reported in various regions of France. *Blasthestia posticana* is a close species which also feeds on pine needles. *Spilonota laricina*, *Exapate duratella* and *Ptycholomoides aeriferana* occur together with *Zeiraphera diniana* on larch in the Alps, and particularly in the Engadine.

The main tortricid on broadleaf trees is *Tortrix viridana*, a pest of oak. An economically less important species is *Archips crataegana* which is polyphagous but mainly frequents oak. *Gypsonoma aceriana* is a species which lives on all European poplars; the larvae bore into buds in spring and may cause much damage by hindering growth of annual shoots, especially on young trees, which become branched and bushy.

2.10. Pyralidae. A vast family of medium or small moths. The most important species in forest belong to the genus *Dioryctria*, and live on conifers. They are not defoliators, but destroyers of cones or xylophagous. *Acrobasis glaucella* defoliates cork oak in Morocco.

2.11. Geometridae. These moths are known as "loopers" owing to the method of locomotion of the caterpillars. The caterpillars' bodies are long and slender and they progress by alternately arching the body between the thoracic legs at the front of the body and the two pairs of false abdominal legs at the hind end, and stretching the body to its full length. The moths have wingspans of 15 to 40 mm, slender bodies, and wings that are often angular or dentate, in dull colours with numerous lines and transverse stripes. *Archiearis notha* lives on alder. *Ennomos quercinaria* is a polyphagous species that may defoliate hornbeam and beech; it totally defoliated these two trees in the Sarre between 1952 and 1954. This moth's parasite complex is rich in species, but in this case regression of the outbreak was due to a polyhedral viral disease of the genus *Borrelina*. The moth's main parasite was the tachinid *Ctenophocera pavida*, which lays microtype eggs, and preys on many species of Lepidoptera. *Operophtera fagata* lives mainly on beech. *Operophtera brumata*, the winter moth, lives on oak. *Erannis defoliaria*, known as the mottled umber, has wingless females; the male has a wingspan of 35 to 40 mm and is yellowish with two chocolate-brown stripes on the forewings. The caterpillar, which is about 30 mm long, is yellow to orange. This species is highly polyphagous and feeds mainly on oak and beech leaves which it gnaws around the edges. Fairly large invasions have been reported in Germany, Sweden and Poland. *Biston betularia*, the peppered moth, lives on oak, birch, elm and beech, and *Gonodontis bidentata* on oak. These moths, which rest on tree trunks, tend to melanism, pale forms being eliminated in industrial regions where atmospheric pollution blackens tree trunks. *Oporinia autumnata* is the main pest of birch in the subarctic forests of Scandinavia where its population fluctuation, known since 1862, has a nine- to ten-year cycle. In Finland this moth defoliated about 500,000 hectares in 1965. Weakened trees suffered infestations by secondary destroyers, pushing back the northernmost limits of forest. This species also lives on larch in the Engadine. Among other geometrids associated with conifers, *Eupithecia laricata, E. indigata, Erannis aurantiaria* and *Poecilopsis isabellae* live on larch; *Bupalus piniarius*, the pine looper moth, lives on pine.

2.12. Notodontidae. Species belonging to this family mostly live on broadleaf trees but they are not generally abundant and do not cause damage. Adults resemble noctuids but without the latters' characteristic spots on the wings. Caterpillars often have bizarre shapes. *Stauropus fagi* lives on oak and beech; its larvae are arched and hump-backed and the two pairs of hind legs are long and slender. *Phalera bucephala* has a wingspan of about 50 mm and is recognisable by its greyish-yellow colour and its forewings which bear a lighter spot internally ringed by a chocolate-brown stripe near the anterior angles. This species lives on broadleaf trees, especially oak; its caterpillars devour leaves, leaving only the midrib. *Cerura vinula* lives on poplar. *Drymonia ruficornis* may be a pest of cork oak and evergreen oak.

2.13. Thaumetopoeidae. This family includes four species of processionary caterpillars. *Thaumetopoea pityocampa*, the pine processionary caterpillar, is mainly distributed in southern France; *T. pinivora,* which feeds on pine, is a North European species which is rare in France; *T. processionea* is the oak processionary caterpillar; *T. wilkinsoni* is localised in the Near East (Cyprus and Israel); *T. solitaria* lives on pistachio trees in Iran.

2.14. Noctuidae. Moths of this family, which is very rich in species, have wingspans over 20 mm and robust hairy bodies. The triangular forewings bear characteristic markings, often in greys and browns. Older larvae often react by curling up when they are touched. *Panolis flammea*, the pine beauty moth, lives on pine. In central Europe the outbreak areas in which damage is frequent are limited to regions with a continental climate where rainfall does not exceed 400 to 600 mm and where July temperatures range between 17 and 19 °C. In Britain, outbreaks of *Panolis flammea* probably occur mainly in stands planted on deep and poorly-drained soils. This leads one to suppose that food quality affects larval development; this would provide a means of combating this defoliator by adapting cultivation methods (Stoakley, 1979).

The three species *Panolis flammea, Bupalus piniarius* and *Hyloicus pinastri* have been studied by Schwerdtfeger (1935), according to whom population drops following outbreaks are due to high levels of parasitism or to disease, and to scarcity of available food.

Species of the genus *Catocala* have brightly coloured, often red or blue, hind wings, while the forewings are dull brownish or grey. Adults emerge at the end of summer and feed readily on sap from injured trees. *Catocala nymphagoga* lives on oak which it may seriously defoliate. The larvae of *C. fraxini*, the Clifden nonpareil, live on ash during summer; they hibernate and in the following spring feed on various plants including plantain, honeysuckle and willow. Such changes in diet are rare in Lepidoptera. *Agrotis segetum* is a polyphagous noctuid; *Agrochola lychnidis* lives on ivy, and *Xanthia citrago* on lime. *Tethea or*, associated with poplar, belongs to the family Cymatophoridae which is close to the Noctuidae.

2.15. Lymantriidae. This small family includes species which are often very harmful. The nocturnal adults are whitish in colour or a more or less dark shade of brown. Males and often females too have strongly pectinate antennae, and the average size of species is quite large. *Lymantria dispar*, the gypsy moth, is a polyphagous species which feeds on many broadleaf trees. *Euproctis chrysorrhea* is entirely white except for the tip of the abdomen which varies from yellow to dark brown, hence its common name, brown tail. The male, which has a wingspan of 20 to 25 mm, is a little smaller than the female. Females lay in early July on the undersides of leaves near the tops of oaks, chestnut and various fruit trees; eggs are deposited in packets covered with brown scales from the tip of the female's abdomen. The gregarious young caterpillars gnaw the leaves of their host trees; leaves eaten by young caterpillars die and desiccate. The tops of infested trees then turn brown while foliage on the lower branches remains green. Around mid September, caterpillars congregate in a silky nest that they have woven together and in which they hibernate. This species only occurs in outbreaks in France at forest edges or in small copses and on trees planted in rows (such as by roadsides). In Corsica it may ravage chestnut woods in some years and wipe out the whole crop. The brilliant silky winter nests at the tops of branches stripped of foliage provide the best way of spotting the species in the event of an outbreak (Arevalo-Durup, 1993). The brown tail has been introduced in North America.

Females of *Orgyia antiqua*, the vapourer moth, are wingless. This polyphagous species, which occurs in Europe and in North America, feeds on broadleaf trees (oak, beech, elm, hornbeam and maple) as well as on conifers (pine, fir, spruce and larch). The female mates on its pupal cocoon, and deposits its eggs, numbering 100 to 300. Caterpillars are about 26 mm long, black and hairy, with four tufts of golden-yellow hairs on their backs and eight protuberances bearing yellow and black hairs. Other species of *Orgyia*, known as tussock moths, occur in North America where they may cause serious damage. The commonest is *Orgyia pseudotsugata* which feeds on firs and Douglas-fir (*Pseudotsuga menziesii*).

The pale tussock, *Dasychira pudibunda*, lives mainly on beech. Adults have a wingspan of 35 to 65 mm, are yellowish-grey with transverse brown stripes. The greenish-yellow caterpillar has four tufts of long yellow hairs on the metathorax and the first three segments, and a long brush of reddish hairs at the end of the body. The female lays about a hundred eggs, which are stuck side by side to form a greyish plaque, low down on tree trunks in May. The young larvae avidly devour foliage and drop many unconsumed leaf fragments to the ground. Pupation takes place in litter at the foot of host trees. This species rarely causes damage in France, but does so more often in central Europe. One outbreak occurred in 1926 in beech woods in Fontainebleau forest; trees on that occasion were almost entirely defoliated. The great majority of eggs were grouped on trunks at heights under two or three metres. Parasites were tachinids (*Compsilura concinnata* and *Carcelia* sp.), ichneumonids and a braconid (*Apanteles* sp.); predators were carabid beetles and the bug *Troilus luridus.*

Lymantria monacha, the black arches, is a polyphagous species ranging from Spain to Japan which caused great damage in central Europe during the nineteenth and early twentieth centuries in pure and often artificial populations of pines and spruce (Habermann, 1995). The moth is nocturnal and emerges in July to September. It is identifiable by its yellow forewings spotted with black and uniformly yellowish-grey hind wings. Males have clearly pectinate antennae and a wingspan of 40 to 50 mm. Eggs are deposited in piles under bark or lichens and hatch in April. The young, more or less gregarious, larvae live in heaps called "mirrors"; older caterpillars are solitary. They climb into trees to feed at night and often return to hide in floor litter during the day. Owing to these migrations the black arches moth can be combated by painting tree trunks with a glue to trap caterpillars.

Stilpnotia salicis is a defoliator of poplars, aspen and willows. The moth is almost entirely white except for its black legs. The caterpillars are hairy, variegated with white, blue and orange, and are very voracious. They destroy leaves and even buds, which may lead to desiccation and the death of entire trees.

2.16 Arctiidae. This family includes species of various sizes which are often brightly coloured in patterns of yellow, red, white and black. *Arctia caja,* the garden tiger moth, has a wingspan of 60 mm; the forewings are patterned in cream and chocolate-brown and hind wings are orange with black spots. It is a polyphagous species which lives on oak, lime, alder, birch etc. *Hyphantria cunea,* which measures 25 to 30 mm, is a black and white moth of North American origin which was introduced to Europe in 1940. Its highly poly-phagous larvae feed on many trees. This very fertile species occupies forest edge and does not penetrate the inner woods.

2.17. Saturniidae (Attacids). These often very large moths have strongly pectinate antennae. They are scarce and of little importance in European forests. *Graellsia isabellae* lives in Spain on *Pinus sylvestris* and *Pinus laricio*, and the subspecies *galliaegloria* lives in France on *Pinus sylvestris*. In France it is localised in a single pine forest in the Briançon region; this relict and protected species is the most remarkable lepidopteran of the European fauna. Adults have green wings veined and margined with reddish-brown. They fly at dusk and early in the night.

2.18. Lasiocampidae. This family includes a few species which are readily recognisable by their appearance, such as the lappet moth, *Gastropacha quercifolia*, and the oak eggar, *Lasiocampa quercus*. Another lasiocampid, the lackey moth (*Malacosoma neustria*), has a reddish-brown body and wings, with a darker transverse band on the forewings. In July the female lays eggs adpressed to each other in a layer that forms bands of 1 to 1.5 mm in diameter around tree twigs; both fruit trees and oaks are eaten. The same genus *Malacosoma* is represented in North America by several species known there as "tent caterpillars" because of the "tents" made of white silk which they construct in branches. Some species such as *Malacosoma disstria* are serious pests. *Dendrolimus pini* may totally defoliate pines and even destroy buds and bark during outbreaks. The larva, which measures up to 75 mm, is covered in short grey pubescence. Eggs are laid on needles in July and August. Caterpillars hibernate on the ground and resume feeding in spring before pupating towards the end of June.

The family Endromididae includes only one species, *Endromis versicolora*, known in Britain as the Kentish glory. It has a furry body and marked sexual dimorphism. The larger female, which reaches 70 mm, is buff coloured with simple antennae. The caterpillars' host plants are birch, alder, lime and a few other broadleaf trees.

2.19. Sphingidae. Hawk moths are few in forest. The pine hawk moth, *Hyloicus pinastri*, is very variable in abundance and may defoliate pines during outbreaks. It is a handsome grey moth with brown stripes which flies at dusk in summer to feed on nectar from flowers with deep corollas.

2.20. Rhopalocera. Butterflies include forest species belonging to several families. Of these, mention can be made of the pierids *Aporia crataegi*, the black-veined white, which lives on oak, hawthorn and walnut, and *Gonepteryx rhamni*, the brimstone butterfly, whose caterpillar feeds on buckthorn. In the family Nymphalidae, *Apatura iris*, the purple emperor, reaches a wingspan of 65 mm and has a bluish-purple metallic sheen. It often flies among the tree tops and only comes down to ground level when attracted by decomposing substances. *Limenitis populi* lives on aspen; *Ladoga camilla*, the white admiral, is dependent on honeysuckle, and *Nymphalis polychloros*, the large tortoiseshell, lives on various broadleaf trees. Most of these butterflies emerge between late spring and early summer. In the family Lycaenidae (the "blues" and "coppers") the holly blue, *Celastrina argiolus*, is dependent on ivy and is threatened by scarcity of its host plant which is systematically eliminated from forests.

Oaks are the trees that harbour the greatest number of species of Lepidoptera. There are several hundred of them in Europe, and at least one tenth of the French fauna live on various species of oak. Among the most typical oak wood Lepidoptera are *Apatura iris* and *Apatura ilia* (a slightly smaller relative of the purple emperor) and *Tortrix viridana* (the "oak leaf roller moth"). Beech woods harbour *Aglia tau* and *Operophtera fagata*. The great majority of Lepidoptera colonise all types of forest, but with a marked preference for edges and clearings, where they find their food plants and the sunny microclimate they need. Few Lepidoptera frequent the dark under-storey, and few species like *Lymantria monacha* and *Orgyia antiqua* feed both on broadleaf and coniferous trees.

3. SOME SPECIES ON BROADLEAF TREES

3.1. The oak leaf roller moth

The range of the oak leaf roller, *Tortrix viridana*, covers the whole of Europe and North Africa and extends eastward to Iran. It is one of the main defoliators of all the species of oak on which it depends. When there are out-breaks, gradations usually last two to three years, separated by recession periods of variable duration. In some cases other moths occur in outbreaks simultaneously. Damage may be considerable. Millions of hectares of ever-green oak in Spain are affected. Destruction of leaves and flower buds slows photosynthesis as well as wood production: it impedes acorn formation and natural regeneration.

The oak leaf roller produces one generation per year (figure 8.3). In the Paris region, eggs are laid in summer on the bark of young twigs and on leaf scars. These eggs overwinter, and in spring produce caterpillars that penetrate the buds to feed and then go on to feed on the leaves. The green, black-hea-

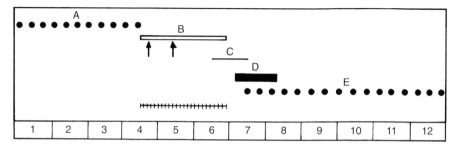

Figure 8.3. *Life cycle of* Tortrix viridana. *A: eggs; B: caterpillars; C: pupae; D: imagines; E: hibernating eggs in November-December. Figures 1 to 12 relate to months; the hatched line shows period of damage, and arrows periods of microbiological phytosanitary action.*

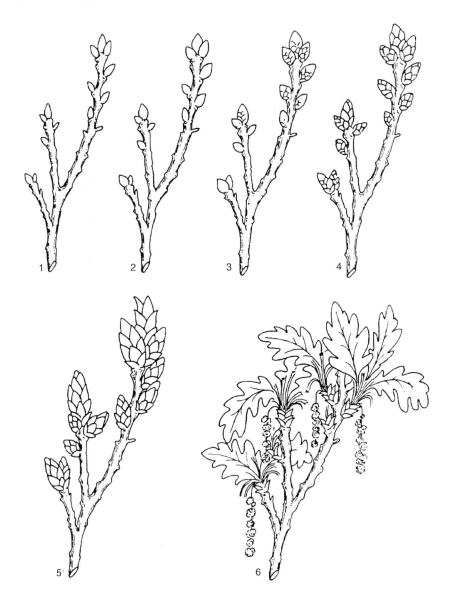

Figure 8.4.   *Various stages in the development of oak leaf buds. 1. entirely formed buds; 2. elongate buds with scales still unseparated; 3. tender buds with scales beginning to separate; 4. open buds with separated scales; 5. buds in full flush, often with one leaf already unfolding; 6. bent shoot with several leaves free and unfolded.*

ded caterpillars grow to 15 to 18 mm in their final instar. In late spring, groups of two or three leaves can be seen rolled into cornets, their edges bound by silken threads. These cornets contain the caterpillars and their faeces. Pupation takes place in June on leaves. The moths, which have a wingspan of 21 to 24 mm, emerge in June and live about three months. They are recognisable by their hind wings which bear white fringes.

Larval development of *T. viridana* is a function of temperature. A close coincidence is required between the emergence of young caterpillars and the unfolding of oak leaf buds, which must be at the third stage, in other words that at which scales on the tender bud begin to separate (figure 8.4).

In European populations of *Quercus robur* there is a polymorphism in the date at which leaves appear, early individuals being the majority. Late-budding individuals are resistant to attacks by the leaf roller because young larvae are unable to penetrate the bud scales in time and therefore die of starvation (Satchell, 1962).

In Portugal *T. viridana* lives on evergreen oak and cork oak. The latter develop leaves earlier, and the leaf roller's pupae are found on them at a time when eggs are only just hatching on evergreen oak. Altenkirch (1966) has shown that in such cases there are two physiological races of the moth which choose their host plants differently but which are morphologically indistinguishable. An analogous phenomenon has been discovered in southern France (cf. chapter 3.1.7).

3.1.1. The role of climatic factors

Research carried out in Ispina forest in Poland has shown that low temper-atures and rainfall at the time that larvae hatch considerably reduce the abundance of populations (Witkowski, 1975). Climatic factors that may explain outbreaks are, on the one hand, the sum of mean daily temperatures during April, and on the other the sum of rainfall during the same period. April was chosen because it is the hatching period for most of the cater-pillars; it thus corresponds to a critical period in the insects' life cycle. The following relationships were established:

Year	Sum of mean daily temperatures in April	Rainfall in mm during April	Outbreaks
1966	245.0	74.3	+
1967	238.4	58.9	+
1968	251.4	54.4	+
1969	*219.9*	*14.8*	−
1970	244.1	38.1	+
1971	242.2	24.8	+
1972	256.6	77.9	+
1973	*214.7*	*17.5*	−

It will be seen that there were no outbreaks of the leaf roller during the cold dry years 1969 and 1973.

3.1.2. The role of predators and parasitoids

Population growth of *Tortrix viridana* also depends on the impact of its numerous natural enemies. The oak leaf roller lives on oak together with several species of moth whose presence is important, for they may serve as replacement hosts for several of its parasitoids. In Germany more than 80 species have been listed, the most common being the following (Müller & Lotz, 1968):

Lymantriids	: *Dasychira pudibunda*	Tortricids	: *Tortricodes tortricellus*
Noctuids	: *Hylophila prasinana*		*Zeiraphera isertana*
Geometrids	: *Operophtera fagata*	Oecophorids	: *Chimabacche fagella*
	Operophtera brumata		*Teleia luculella*
	Erannis spp.	Gracillariids	: *Gracillaria alchimiella*

The main predators are birds and beetles (*Calosoma* sp. and *Cantharis* sp.), as well as scorpion flies. In Ispina forest 92% of oak defoliating caterpillars are eaten by secondary consumers, insectivorous birds accounting for 2.7 to 11% according to different cases. The role of insectivorous birds is therefore not inconsiderable. In Morocco the ant *Crematogaster scutellaris* consumes the oak leaf roller's eggs as well as those of the gypsy moth.

Parasitoids are numerous and varied. In Spain about forty species have been found, of which ichneumonids *Pimpla maculator, Ephialtes carbonarius, Phaeogenes stimulator* and various braconids are dominant. In Germany a detailed list of *T. viridana*'s parasitoids has been drawn up, some species of which also parasitise other moths of oak such as *Operophtera brumata* and *Zeiraphera isertana*, while others behave as hyperparasites. This list includes 60 ichneumonids, 32 braconids, 23 chalcids and 14 tachinids. In areas where the leaf roller populations are maintained permanently at low levels of abundance, rates of parasitism are higher, the parasitoid spectrum is different, and the connections between parasitoids and alternative hosts are stronger than in areas of regular outbreaks:

	Outbreak areas	Low abundance areas
% of parasitised caterpillars	5%	33%
Number of parasitoids	25 including 8 ichneumonids, 7 braconids, 4 chalcids, 6 tachinids	6 ichneumonids
Number of parasitoids that predate other species	8 predating 7 other species of moth	5 predating 6 other species of moth
Relationship of commonest parasitoids with alternative hosts	No alternative host	Several alternative hosts

In areas where population levels are low there is poor coincidence between hatching of larvae and the unfolding of oak leaf buds, and intense pressure from parasitoids. During gradation periods coincidence between hatching and unfolding of buds is perfected. At the same time parasitoids become less effective since they prey on a greater number of alternative hosts at the end of summer when there are no more leaf roller larvae. Alternative hosts are thus decimated, leaving the parasitoids with no chances of survival (Schwerdtfeger, 1971). The major role played by natural enemies prevents non-specific chemical intervention. Preparations based on *Bacillus thuringiensis* are effective. Treatment should be applied at the early larval stage before they roll leaves and settle in them.

3.2. The gypsy moth

The gypsy moth, *Lymantria* (= *Porthetria*) *dispar*, is one of the most studied of tree pests, and the one that has caused the most damage in the Old World as well as in the New. It is a pest of oak, sometimes of beech, hornbeam, elm, poplar and willow, and, accidentally, of conifers. This species, which originated in Japan, has spread to the entire northern hemisphere, and an Asiatic genotype has recently occupied California where the species did not previously occur.

3.2.1. Life cycle

In France the gypsy moth mainly infests cork oak. Damage is caused all over southern Europe. It is spectacular in Spain where this species is the second greatest defoliator after *Tortrix viridana*. Outbreaks have been recorded in Corsica, Morocco, Algeria and Tunisia. The English name "gypsy moth" refers to the caterpillar's showy colour: it is greyish-green with black markings and has six longitudinal rows of tubercles bearing long hairs; the tubercles on the thoracic segment and first two abdominal segments are blue, those on the following segments red. The adult female has a wingspan of 45 to 75 mm, while the male does not exceed 20 to 25 mm. The female is yellowish-white and the male brown with a complex pattern of lighter and black spots. This difference in colour accounts for the Latin epithet and for the French name "bombyx disparate".

Lymantria dispar has one generation a year. Females lay clusters of eggs covered with a buff felt on bark in summer. The egg clusters have a sponge-like appearance, giving rise to another vernacular name for the moth in France, "la spongieuse". Eggs overwinter and hatch in spring. Dispersal occurs during the first larval instar. The young caterpillars, which weigh less than 1 mg, are covered in long hairs that may be half as long as the whole body and shorter hairs called aerophores which have a globular basal swelling; these hairs increase the body's surface area and aid wind-borne dispersal. Larvae have been found at 600 m altitude, and larval migrations over 56 km have been recorded. Pupation begins in June, and the moths fly in July and August.

The cyclical character of outbreaks of the gypsy moth has been the subject of much research, but the largely unpredictable nature of this pest makes it difficult to study. Following an outbreak, relative stability in the number of defoliated trees is seen, and then the pest almost completely vanishes. During population recessions, egg-laying is usually reduced and egg clusters have fewer eggs. Population recession has been attributed to genetic factors, to variations in the caterpillars' sensitivity to polyhedrosis (a viral disease), or to nutrition.

3.2.2. Predators and parasitoids

Tits (*Parus* spp.) play a significant part by eating young *L. dispar* larvae. This is why nesting boxes are often provided in cork oak woods. The most notable predator is *Calosoma sycophanta*. This rather heavy, squat and brilliantly-coloured arboreal carabid, which reaches a length of 35 mm, may fly several kilometres in search of new prey. It predates many species; one adult may devour 235 to 336 gypsy moth caterpillars in one season.

Among other enemies of *Lymantria dispar* are insects which have been termed "dismantling predators" on account of their effect in dislocating egg batches, which plays a greater part in reducing numbers than predation itself. By burrowing into egg clusters these insects disperse the eggs which soon die. This phenomenon is of great importance in Mamora cork forest in Morocco where it was first observed (Villemant & Fraval, 1991). Thirteen species of Coleoptera and two species of Lepidoptera are known to practise this dismantling behaviour. The most remarkable are dermestid beetles (such as *Trogloderma versicolor* and *Dermestes lardarius*), an ostomatid, *Tenebrioides maroccanus*, a tenebrionid, *Akis tingitanus*, and a tineid moth, *Niditinea fuscipunctella*.

The two main parasites of eggs are *Ooencyrtus kuwanae*, which originated in Japan and was introduced to Europe in an attempt at biological control, and *Anastatus disparis*. The first larval instars are mainly preyed upon by ichneumonids of the genus *Apanteles* (figure 8.5). Pupae are parasitised by *Brachymeria intermedia*, *Pimpla instigator* and by numerous other ichneumonids (table 8.2).

3.2.3. The Lymantria dispar pheromone

A sex pheromone attractive to males is emitted by the females, whose voluminous abdomens make them unable to fly. This pheromone, which is known by the name disparlure, is *cis*-7,8 epoxy-2-methyloctadecane. It is

Parasites of eggs	Encyrtids Eupelmids	*Ooencyrtus kuwanae* *Anastatus disparis*	*Ooencyrtus kuwanae*
Parasites of larvae	Braconids	*Apanteles vitripennis* *Apanteles melanoscelus* *Apanteles porthetriae* *Hyposotes disparis*	*Apanteles solitarius* *Apanteles lacteicolor* *Apanteles porthetriae* *Meteorus pulchricornis*
	Tachinids	*Exorista fasciata* *Compsilura concinnata*	*Sturmia inconspicua* *Compsilura concinnata* *Exorista segregata* *Carcelia separata*
Parasites of pupae	Chalcids Torymids Tachinids	*Brachymeria intermedia* *Monodontomerus aereus* *Compsilura concinnata* *Exorista fasciata*	*Brachymeria intermedia* *Monodontomerus aereus*
	Ichneumonids	*Pimpla instigator*	*Pimpla instigator* *Pimpla turionellae*
Hyperparasites	Pteromalids	*Habrocystus poecilopus* *Dibrachys cavus*	*Habrocystus poecilopus* *Dibrachys cavus*
	Ichneumonids	*Hemiteles pulchellus* *Gelis provida* *Monodontomerus aereus*	*Gelis areator* *Monodontomerus aereus*
	Torymids Chalcids		*Brachymeria intermedia*
Predators	Carabids	*Calosoma sycophanta*	*Calosoma sycophanta*

Table 8.2. *Main elements of the parasite complex of* Lymantria dispar *in Spain (third column) and in Morocco (fourth column).*

attractive at doses as weak as 2.10^{-12} to 10^{-9} g. The closely related species *Lymantria monacha* has the same pheromone, but the two species do not hybridise because their rhythms of activity are different. *L. dispar* is diurnal and mates between 9.00 and about 16.00 hours, while *L. monacha* mates between 18.00 hours and one o'clock in the morning. Monitoring population abundance of *L. dispar* by observation and counting egg batches is at present the only way we have of taking preventive measures to limit damage. Treatment by means of *Bacillus thuringiensis* is effective.

Population regulation of *Lymantria dispar* has been much studied. Entomophagous insects do not appear to play an important part; population collapses are mainly caused by nuclear polyhedrosis, a viral disease that kills caterpillars. In Mamora forest in Morocco, diseases caused by pathogenic micro-organisms are unknown, and population regulation is mainly achieved by reactions of affected cork oak trees. When trees are stripped of their year's growth of leaves by an outbreak of the pest, caterpillars turn to other plants such as dwarf palm and to old cork oak leaves which are tough and indigestible, which causes most of the caterpillars to die of starvation.

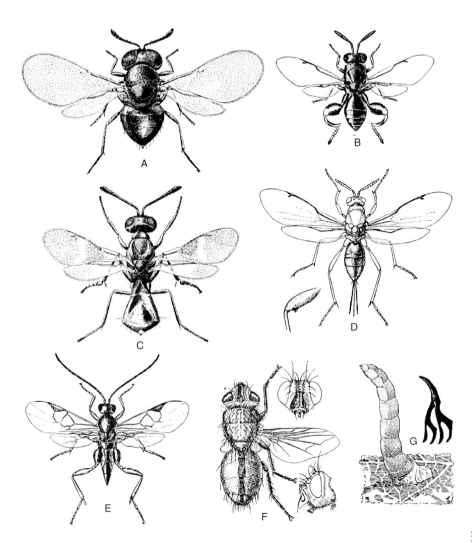

Figure 8.5.   *Some parasitoids of* Lymantria dispar. *A:* Ooencyrtus kuwanae
female; B: Brachymeria secundaria; *C:* Anastatus disparis *female;*
D: Monodontomerus aereus; *E:* Apanteles porthetriae; *F & G:*
Eudoromyia magnicornis *female and first instar larva on a leaf*
awaiting a caterpillar. Right, in black, the larva's buccal hook much
enlarged (Howard & Fiske, 1912).

Surviving caterpillars produce a small new generation of adults. As a reaction to the previous year's defoliation the cork oaks do not produce new leaves, but retain their old leaves, which prevents the caterpillars from growing, and gradation ceases. Population regulation may also be effected by high egg mortality due to the "dismantling" predators mentioned above. Biological control using the parasitoid *Ooencyrtus kuvanae* is not very effective. The best control method lies in treatments with *Bacillus thuringiensis*. A synthetic insecticide which interferes with moulting is also used.

3.3. The winter moth

The winter moth, *Operophtera brumata*, so called because the adults appear in early winter, and called "cheimatobie" in France, is a polyphagous geometrid which occurs on fruit trees such as apple as well as on forest trees, oak being a major host, and beech usually ignored. It is a potentially harmful species which is distributed from Scandinavia to Italy, and which ranges in altitude from sea level to the tree line. It has been introduced to Canada. Males have a wingspan of 25 to 30 mm, and normal wings, the forewings reddish-grey with transverse stripes; females have greenish vestigial wings which make them flightless. The larvae are yellowish-green with a darker dorsal stripe.

3.3.1. Life cycle

The winter moth has only one generation per year and its life cycle is regulated, as is that of the oak leaf roller, by seasonal changes in the chemical composition of oak leaves (cf. figure 3.3). Adults emerge from the ground in early winter. At dusk females climb trees, and males, which remain in the litter during the day, take to the wing; mating takes place on tree trunks. The females lay their eggs in bark crevices and on lichens, well above ground level. The larvae, which hatch at the time that buds unfold, finish feeding in May and then migrate to the ground. Pupation takes place in early summer in ground litter.

3.3.2. Predators and parasites

The winter moth's predators and parasites have been much studied (Wylie, 1960; Varley & Gradwell, 1968; Varley, 1971, etc.). Birds, tits in particular, are active predators. Hatching of great tit's eggs coincides more or less with the appearance of the winter moth's fourth instar larvae. Pupae in the litter are predated by beetles such as the staphylinid *Philonthus decorus*, the carabids *Pterostichus madidus, P. melanarius* and *Abax ater*, and the elaterid *Athous haemorrhoidalis*. Other predators have a lesser impact; they include small mammals such as shrews and voles.

The main parasitoids at the moth's egg stage are wasps such as the scelionid *Telenomus nitidulus* or chalcids of the genus *Trichogramma* such as *T. evanescens*. Among the parasitoids of larvae are some ten braconids of the genus *Apanteles* and several *Meteorus*, the ichneumonids *Agrypon flaveolatum*, *Campoplex pugillator*, *Lissonota biguttata* and *Phobocampa crassiuscula*, the eulophid *Eulophus larvarum*, and the tachinid *Cyzenis albicans*. Parasites of pupae are mostly ichneumonids, including *Aptesis abdominator, Pimpla turionellae* and *Cratichneumon culex*.

Cyzenis albicans lays microtype eggs which are deposited on the leaves of various plants. Clutches may reach 683 eggs in 12 days. As in other tachinids which lay microtype eggs, fertility depends on the female's body size. Eggs are ingested by caterpillars and hatch in the host's mesenteron. The parasite larva pierces the digestive tube and settles near the host's salivary glands. When the host pupates the *Cyzenis* larva develops a respiratory tube which opens outside the pupa. *Cyzenis* eggs are ingested by all defoliators that live on the same trees as *O. brumata*, and in particular by *Tortrix viridana* and *Erannis defoliaria*, but these uncustomary hosts do not allow the parasite to develop.

3.3.3. Population dynamics

Processes that affect population dynamics have been elucidated by several authors (Embrée, 1965, 1966, 1969; Varley & Gradwell, 1968; Varley, 1971). The name "key factor" has been given to those factors that can explain variations in abundance of populations. To discover them, variations in abundance observed over several years at each stage in the insect's life cycle are represented, together with the relative importance of various mortality factors that have been found. These values are shown on a graph in logarithmic coordinates, and a coefficient k equal to the ratio of population density before (A) and after (B) the effect of the mortality factor is defined. One thus has $k = \ln A/B$, or $k = \ln A - \ln B$. Total mortality for one generation is equal to $K = k_1 + k_2 + k_3 + ... + k_n$. A look at the graphs in figure 8.6 will show that winter mortality k_1 is the key factor, for it is the one that follows variations in factor K most closely.

Another form of graphic representation consists in studying variations in factors k as a function of population density. This reveals that the only factor directly dependent on density is predation of pupae buried in the ground in winter (figure 8.7). The three essential predators of pupae are the beetles *Abax ater, Pterostichus madidus* and *Philonthus decorus*.

In the winter moth's parasite complex the two main parasitoids are *Cyzenis albicans* and *Agrypon flaveolatum*. They are active at different times during gradation and therefore do not compete. This fact has been used to combat the winter moth in Canada where it is introduced.

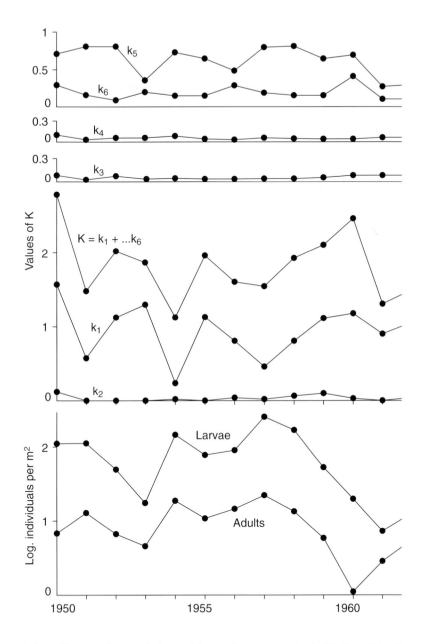

Figure 8.6. *Changes in populations of* Operophtera brumata *in Wytham Wood oak woods in England. Below, annual variations in abundance of larvae and adults. Above, annual variations in various mortality factors. K: total mortality; k_1: winter mortality; k_2: parasitism by Cyzenis; k_3: parasitism by other insects; k_4: Microsporidia larval diseases; k_5: predation upon pupae; k_6: parasitism of pupae by Cratichneumon spp. (Varley, 1971).*

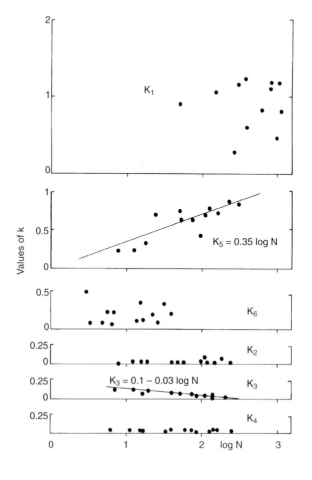

Figure 8.7. *Variations in mortality factor k as a function of population density N in the winter moth. It may be seen that k_1 and k_6 vary haphazardly and are independent of density; k_2 and k_4 are independent of density and almost constant; k_3 is inversely dependent on density; k_5 is the only factor directly dependent on density (Varley & Gradwell, 1968).*

4. SOME SPECIES ON CONIFERS

4.1. The larch bud moth

The larch bud moth *Zeiraphera diniana* is distributed in the Pyrénées and the whole arc of the Alps as far as the Tatras, and in the North reaches England and Scandinavia. This moth is dependent on conifers and feeds, in order of preference, on larch, Arolla pine, spruce, Scots pine and mountain pine. It ranges in altitude up to 3,100 m.

4.1.1. Life cycle

The life cycle of this univoltine (having only one generation a year) species varies according to altitude. At around 1,800 m in the Engadine, adults fly at dusk between 15 July and 15 September. Each female lays about 150 eggs in small clusters under the shelter of lichens that grow on branches. After one week's incubation in August at ambient temperature (ideal temperature being 11 °C) eggs go into diapause and can only be reactivated following at least 120 days at + 2 °C. These data make it possible to define the ideal climate for egg development, one which prevails in the Engadine and in most Alpine valleys in which larch flourishes. In the Engadine, diapause may last 210 days; in lower parts of the Alps at around 800 m, eggs develop at higher temperatures and over a longer period, since they are laid in June. Moreover they suffer massive destruction in winter; therefore in these areas the moth does not occur in outbreaks. Diapause plays a major part in synchronising the moth's life cycle with that of larch. Larvae must hatch at a time that coincides with the first growth of larch needles, the optimum size of the latter being 6 to 8 mm. The necessity for this coincidence explains why outbreaks of the larch bud moth only occur above 1,200 metres, and mostly between 1,700 and 2,000 metres.

The larval stage lasts from about 15 May to early July. The young caterpillar goes to the tip of a bud and penetrates it by insinuating itself between the needles which it gathers into a bundle by means of a silk thread before devouring them. Caterpillars change habitats three or four times before reaching the fourth instar during which they settle in a sort of funnel formed by the bases of eaten needles. In the final instar the caterpillar weaves a web along branches and migrates to the ground to pupate. The moth is grey with darker markings on the forewings, and has a wingspan of about 20 mm.

4.1.2. Chromatic polymorphism

Fifth instar larvae have two types of coloration. The form that lives on larch is almost black; the one that lives on mountain pine is orange. The two forms are inter-fertile and hybrids are highly polymorphic, which proves that several genes determine coloration. The larch form does not survive well on mountain pine, but the mountain pine form adapts easily to larch. The two forms have slightly different sex pheromones (cf. chapter 3.1.7).

4.1.3. Population dynamics

The larch bud moth has three types of fluctuation of abundance: (a) periodic fluctuations with cyclical gradations occurring at regular intervals of eight to ten years, without latency periods, in the sub-alpine larch and mountain pine form at around 1,800 metres; (b) temporary fluctuations with occasional outbreaks at lower altitudes around 1,400 metres; (c) more or less regular fluctuations but always remaining below the defoliation threshold, in larch plantations on the Swiss Plateau at the stage corresponding to oak/beech forest.

Periodical fluctuations which succeed each other without latency periods between two gradations last nine years on average, variations in abundance being as great as 1 to 30,000 between maximum and minimum abundance. During the 1957 to 1963 gradation a ratio of 1 to 104,000 between scarcity and abundance periods was recorded. Gradation has a progressive stage lasting 4 to 5 years followed by a culmination stage of 2 to 3 years during which damage appears in the form of spectacular defoliation of trees that turn brown over vast areas (figure 8.8). At the end of June an abundance of silk webs full of caterpillar exuvia and frass hangs from branches. Wood growth may be slowed by 30%, and seed production may cease almost entirely Z. diniana is the greatest pest of larch in the Alps.

In 1972, young mountain pines growing under cover of larch in the High Engadine were heavily infested by the form of Z. diniana that feeds on larch. In 1974, most of those trees had died because they had not been able to resist exploiters of weakness such as the weevil *Pissodes notatus* or the scolytids *Pityophthorus knotecki* and *Pityogenes bistridentatus*. This phenomenon, which occurs in some years, retards establishment of mountain pine climax forest because the tree is unable to take over from larch (Baltensweiler, 1975).

The regularity of the Z. diniana cycle has been known since 1850, in the Briançon region of France just as in the Engadine in Switzerland and in Italy and Tyrol. A dozen gradations were monitored in these regions. A model was established (Auer, 1968) with a view to predicting outbreaks of Z. diniana, and this model's validity has been widely tested. The following four factors are entered in the model in decreasing order of importance: available food, both in quantity and quality; the part played by entomophagous insects; diseases (bacteria and pathogenic viruses); temperature. The first three factors are dependent on density while the fourth is not.

When caterpillar abundance is very high a maximum biotic load is reached and caterpillars run out of food, which causes massive mortality. Mortality is further increased by intraspecific competition and there follows reduction in body weight of pupae and in the fertility of surviving imagos. Moreover, defoliation alters the physiology of larch trees; their needles, which appear later, are up to 30% shorter and have less nutritional value due to lower levels of proteins and higher levels of non-absorbable cellulose and lignin. Larch trees only recover their normal physiological condition three or four years after the damage. At low altitude the high mortality rate that prevails in the larval stage would not enable the moth to survive were it not for migrations facilitated by air currents from higher altitudes (cf. figure 2.6).

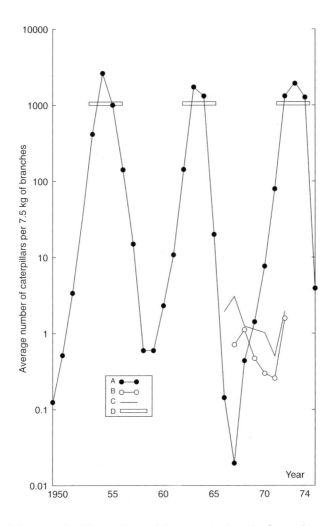

Figure 8.8. *Cyclical fluctuations of the larch bud moth. Curve A represents variations in abundance in the optimum zone in High Engadine; curves B and C correspond to latency periods on the Swiss Plateau; rectangles D show levels of abundance above which damage is visible. Total defoliation of trees occurs when a density of 1,250 caterpillars per 7.5 kg of sampled branches is reached (Auer & Servais, 1976).*

A few other species of Lepidoptera are associated with larch and show parallel variations in abundance to those of *Z. diniana*. These species are however always much less abundant, and as a rule are of no economic concern. These species are three tortricids, *Spilonota laricana*, *Exapate duratella* and *Ptycholomoides aeriferanus*; four geometrids, *Oporinia autumnata*, *Eupithecia indigata*, *Eupithecia laricata* and *Poecilopsis isabellae*; one gelechiid, *Teleia saltuum*; and one coleophorid, *Coleophora laricella*.

4.1.4. Predators and parasitoids

The larch bud moth is predated by a complex of 94 species of parasitoids and by many predators. The parasitoids are mainly ichneumonids, followed by braconids, chalcids and tachinids. The commonest parasites such as the ichneumonids *Phytodietus griseanae, Diadegma patens, Triclitsus pygmaeus* and *Phaeogenes osculator*, the braconid *Apanteles murinanae*, and the eulophids *Elachertus argissa* and *Sympiesis punctifrons* are virtually monophagous; thus the part played as parasite reservoirs for *Z. diniana* by other moths of larch is almost non-existent. Levels of parasitism, which are below 10% in periods when the larch bud moth is scarce, rise to a maximum of 70 to 80% two or three years after culmination. Ichneumonids dominate in the period of abundance whereas three species of eulophid take over during the regression and minimum abundance stages.

Predators of eggs may play a significant part due to the long period of time eggs are exposed on branches. These oophagous predators are mites such as *Balaustium murorum* (family Erythraeidae) and *Bdella vulgaris* (family Bdellidae), the bug *Deraeocoris annulipes*, and Neuroptera and Dermaptera larvae. Predation of eggs varies between 30 and 85% according to region (Delucchi *et al.*, 1974). Predation by wood ants and insectivorous birds has an insignificant impact (Bovey & Grison, 1975).

Diseases sometimes occur during culmination of gradation. Granulosis caused high larval mortality in the Engadine in 1953 to 1955, and then remained endemic without further outbreaks. Control of *Z. diniana* has been undertaken using *Bacillus thuringiensis*. Treatment was applied to third instar larvae just before damage appeared. A massive reduction of the population was achieved, which saved larches from serious damage. If treatment is applied earlier, at the minimum population stage, it may have adverse effects. This is because intraspecific competition plays an important part in population regulation, as does the nutritional quality of larch needles. Reduction of the caterpillar population allows surviving individuals to thrive in greater numbers owing to the lack of competition for food and space, and populations then increase to higher levels than in untreated, control plots. At the same time moths immigrate from untreated areas, drawn by foliage which has remained green (Martouret & Auer, 1977).

4.2. The pine looper moth

The pine looper, *Bupalus piniarius*, is a geometrid which occurs in the whole of Europe but only rarely causes damage in France. Invasions have been recorded periodically in Haguenau forest in Alsace and in Champagne. Infestations by the pine looper are numerous on the other hand in central Europe, where it is one of the main pests of pine. Young larvae only gnaw the tender parts of needles, leaving them entire but for channels cut into them; older larvae strip the needles, leaving only bunches of the liber and lignified parts. When populations rise above six caterpillars per square metre the trees are in danger. Damage generally only occurs in areas where annual rainfall is 500 to 600 mm or less and on sandy soils (Bevan & Brown, 1978).

4.2.1. Life cycle

The pine looper produces one generation a year. The moth has a wing-span of around 35 mm, and emerges between May and July. Males have dark brown wings with large yellow marks and strongly pectinate antennae; females are yellowish-brown with brown maculae and simple antennae. Eggs are laid in June to July; the greenish eggs, which are of the same dia-meter as pine needles, are deposited in rows on the needles. Larvae are active from June to October and reach a length of 3 centimetres; they are typical loopers, bluish-green in colour with paler longitudinal stripes. The caterpillars are protected from predators that hunt by sight by their colour pattern which mimics pine needles by optically dividing the body in two parts; they are sluggish and feed at night. Dispersal of eggs and larvae among the many needles is another protective device. The 150 to 200 eggs laid by each female are spread out in small clusters on needles. Larvae are solitary, which discourages predators (Kleinhout, 1968). These adaptations are inverse to those of species such as the sawflies *Diprion pini* and *Neodiprion sertifer* which are gregarious and have bright, aposematic colora-tion (Prop, 1960).

4.2.2. Population dynamics

The pine looper has been much studied (Schwerdtfeger, 1935; Klomp, 1966, 1968). In a Scots pine wood in the Netherlands several species of moth coexist with the looper and each of these has its own characteristic level of abundance. The pine looper has permanent fluctuations of abundance, its gradation periods being uninterrupted by stationary periods (figure 8.9).

Samples taken have been the basis for a mortality table drawn up each year for different stages in the life cycle. The search into key factors was done using the same methods as those used for the winter moth. Global mor-tality K during one generation depends first of all on larval mortality $k_2 - k_6$ and incidentally on mortality at the egg stage k_1 and of pupae $k_7 - k_9$. Small variations in sex ratio and in fertility have virtually no effect on total mortal-ity (figure 8.10).

The effects of climatic factors on larval mortality are probably the main key factor, but other factors may play a major role in some years, particularly auto-regulation due to a particular form of intraspecific competition which results in decrease in body weight of pupae in years in which larval density increases. Females that emerge from those pupae lay fewer eggs (figure 8.11).

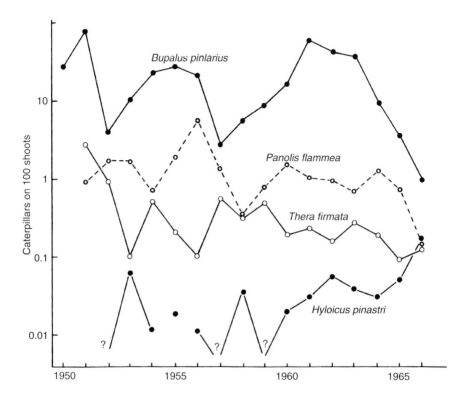

Figure 8.9.   Variations in abundance of four species of Lepidoptera living on Scots pine in the Netherlands. Bupalus piniarius *is always the most abundant species, and* Hyloicus pinastri *the rarest (Klomp, 1968).*

4.3. The pine shoot moth

The pine shoot moth, *Rhyacionia buoliana*, is distributed through the whole of Europe. It was introduced to Canada in 1925 where it soon became a pest. *R. buoliana* is a dangerous pest, especially in the Mediterranean region. In Spain it infests all pines except *Pinus uncinata* up to 1,500 metres in altitude. The moths have a wingspan of 20 mm, and orange forewings with seven transverse sinuate white stripes and brownish hind wings. They fly in summer and lay their eggs at the bases of needles and under bud scales. Larvae of the first two instars gnaw the bases of needles and then migrate to the buds on which they feed. Damage is very characteristic: growth is slowed and shoots are deformed, becoming forked or kinked (bayonet shaped). When all shoots are infested the trees take on the shape of bushes.

The impact of climatic factors on abundance of *R. buoliana* has been shown (Bejer-Petersen, 1972). In regions with a continental climate, cold winters cause severe mortality. Above average temperatures in March to April and low rainfall on the other hand favour the species' reproduction.

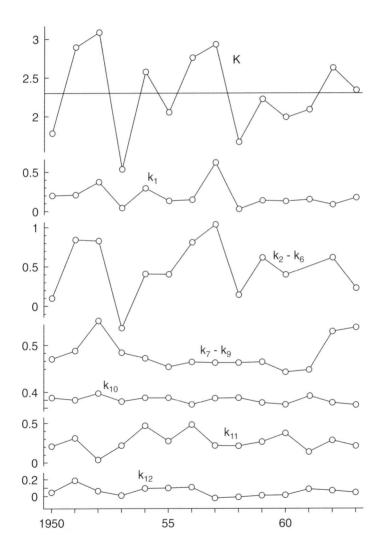

Figure 8.10. *Variations in different mortality factors in* Bupalus piniarius *from 1950 to 1963. K: total mortality; k_1: egg stage; $k_2 - k_6$: larval stages; $k_7 - k_9$: pupae; k_{10}: effects of sex ratio; k_{11}: adult mortality; k_{12}: lowered fertility (Klomp, 1966).*

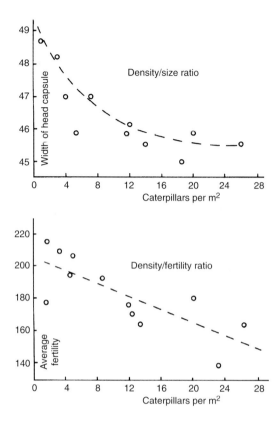

Figure 8.11. *Relationships in* Bupalus piniarius *between population density and size of old larvae, and between larval population density and fertility of resulting females (Klomp, 1966).*

REFERENCES

ALTENKIRCH W., (1966). Zum Vorkommen von *Tortrix viridana* L. in Portugal. *Zeit. ang. Ent.*, **53**: 403-415.

AREVALO-DURUP P., (1993). Le nid d'hiver du bombyx cul-brun, *Euproctis chrysorrhea* L. (Lepidoptera: Lymantriidae): description et structure interne. *Ann. Soc. Ent. Fr.*, **29**: 47-54.

AUER C., (1968). Erste Ergebnisse einfacher stochasticher Modelluntersuchungen über die Ursachen der Populationsbewegung des grauen Lärchenwicklers *Zeiraphera diniana* Gn. (= *Z. griseana* Hb.) im Oberengadin, 1949/66. *Zeit. ang. Ent.*, **62**: 202-235.

AUER C., SERVAIS B., (1976). Modalités de prévision de pullulation d'un ravageur forestier: la tordeuse du mélèze. *C. R. Ac. Agr. Fr.*, **16**: 1,172-1,179.

BALTENSWEILER W., (1975). Zur Bedeutung des grauen Lärchenwicklers (*Zeiraphera diniana*) für die Lebensgemeinschaft des Lärchen Arvenwaldes. *Bull. Soc. Ent. Suisse*, **48**: 5-12.

BALTENSWEILER W., BENZ G., BOVEY P., DELUCCHI V., (1977). Dynamics of larch bud moth populations. *Ann. Rev. Ent.*, **22**: 79-100.

BARBEY A., (1925). *Traité d'entomologie forestière*. Berger-Levrault, Paris, 2ᵉ édition.

BAURANT R., OFFERGELD J.-P., (1972). Observations sur un ravageur forestier: la tordeuse des aiguilles de l'épicéa. *Bull. Soc. Roy. For. Belgique*, **79**: 529-547.

BEJER-PETERSEN B., (1972). Relation of climate to the start of Danish outbreaks of the pine shoot moth (*Rhyacionia buoliana* Schiff.). *Det Forsog Danmark*, **33**: 41-50.

BETZ E., SCHWERDTFEGER F., (1971). Vergleichende Untersuchungen an der Kronenfauna der Eichenwicklers (*Tortrix viridana* L.). 2. Die Parasiten der Lepidopteren. *Zeit. ang. Ent.*, **67**: 149-170.

BEVAN D., BROWN R. M., (1978). Pine looper moth. *Forestry Commission, Forest Record* no. **119**: 1-13.

BOVEY P., GRISON P., (1975). La tordeuse grise (*Zeiraphera diniana* Gn.) important ravageur des mélézins alpins. *Trav. Sci. Parc Nat. Vanoise*, **6**: 115-139.

CAMPBELL R. W., SLOAN R. J., (1977). Forest stand responses to defoliation by the gypsy moth. *For. Sci. Monog.*, no. **19**.

CHARLES P.-J., DELOBEL A., RAIMBAULT J.-P., (1970). Microlepidopteres que atacan al pino silvestre en el bosque de Fontainebleau y primeros datas sobre los complejos parasitarios de *Petrova resinella* (L.) y de *Rhyacionia buoliana* (Schiff.). *Bol. Serv. Plagas For.*, **13**: 187-196.

DELUCCHI V., (1958). *Lithocolletis messaniella* Zeller (Lep. Gracillariidae): analysis of some mortality factors with particular reference to its parasite complex. *Entomophaga*, **3**: 203-270.

DELUCCHI V., RENFER A., AESCHLIMANN J.-P., (1974). Contribution à la connaissance des lépidoptères associés au mélèze en haute altitude et de leurs parasitoïdes. *La Recherche agronomique en Suisse*, **13**: 435-451.

DOANE C. C., Flight and mating behavior of the gypsy moth. *In*: J. E. ANDERSEN, H. K. KAYA (1976). *Perspectives in forest entomology*. Academic Press, New York, 127-136.

DU MERLE P., CORNIC J.-F., (1989). Répartition, niveaux de population et risques de pullulation de la tordeuse du sapin, *Choristoneura murinana* (Lepidoptera: Tortricidae) en France. Résultats d'une enquête par piègeage sexuel. *Ann. Soc. Ent. Fr.*, **25**: 265-288.

DU MERLE P., BRUNET S., CHAMBON J.-P., CORNIC J.-F., FABRE J.-P., (1989). Colonisation d'un végétal introduit (*Cedrus atlantica*) et de nouveaux milieux bioclimatiques par un insecte phytophage: *Choristoneura murinana* (Lep., Tortricidae). *Acta Œcol., Œcol. Applic.*, **10**: 289-301.

Embrée D.-G., (1965). The population dynamics of the winter moth in Nova Scotia 1954-1962. *Mém. Ent. Soc. Canada*, **46**: 1-57.

Embrée D.-G., (1966). The role of introduced parasites in the control of the winter moth in Nova Scotia. *Canad. Ent.*, **98**: 1,159-1,168.

Embrée D.-G.,. Biological control of the winter moth in eastern Canada by introduced parasites. *In*: C.-B. Huffaker (1969). *Biological control*. Plenum, New York, 217-226.

Franz J.-M., (1940). Die Tannentriebwickler *Cacœcia murinana* Hb. Beiträge zur Bionomie und Œkologie. *Zeit. ang. Ent.*, **27**: 345-407.

Grandi G., (1933). L'ipermetabolia nei Lepidotteri. *Mem. R. Acad. Sc. Ist. Bologna*, Cl. Sc. Fis., ser. 8, **10**: 115-121.

Habermann M., (1995). Zur Massenvermehrung der Nonne (*Lymantria monacha*) in Kiefernbestanden des nordöstlichen Niedersachsens. *Forst und Holz*, **50**: 558-564.

Heichel G.-H., Turner N. C., Phenology and leaf growth of defoliated hardwood trees. *In*: J. F. Anderson, H. K. Kaya (1976). *Perspectives in forest entomology*. Academic Press, New York, 31-40.

Huber F., (1982). Effet de défoliaisons des chênes par les hannetons sur la structure du bois. *Rev. For. Fr.*, **34**: 185-190.

Janzen D., (1981). Patterns of herbivory in a tropical deciduous forest. *Biotropica*, **13**: 271-282.

Jordan T., (1995). Biologie und Parasitoidenkomplex der Eichenminiermotte *Tischeria ekebladella* (Bjerkander, 1795) (Lep., Tischeriidae) in Nord-deutschland. *J. Appl. Ent.*, **119**: 447-454.

Kleinhout J., (1968). On visual adaptations against predators in the caterpillars of *Bupalus piniarius* L. (Lep.). *Zeit. ang. Ent.*, **61**: 154-156.

Klomp H., (1966). The dynamics of a field population of the pine looper, *Bupalus piniarius* L. (Lep. Geometridae). *Adv. Ecol. Res.*, **3**: 207-305.

Klomp H., A seventeen year study of the abundance of the pine looper *Bupalus piniarius* L. (Lep. Geometridae). *In* : T.-R.-E. Southwood (1968). *Insect abundance*. Blackwell, Oxford, 98-108.

Kulman H. M., (1971). Effects of insect defoliation on growth and mortality of trees. *Ann. Rev. Ent.*, **16**: 289-324.

Martouret D., Auer C., (1977). Effets de *Bacillus thuringiensis* chez une population de tordeuse grise *Zeiraphera diniana* (Lep. Tortricidae) en culmination gradologique. *Entomophaga*, **22**: 37-44.

Mazanec Z., (1968). Influence of defoliation by the phasmatid *Didymuria violescens* on seasonal diameter growth and the pattern of growth rings in alpine ash. *Austr. For.*, **32**: 3-14.

MÜLLER J., LOTZ G., (1968). Vergleichende Untersuchungen an den Kronen fauna der Eichen in Latenz und Gradationsgebieten des Eichen-wicklers (*Tortrix viridana* L.). 1. Die Lepidopteren. *Zeit. ang. Ent.*, **61**: 58-65.

PROP N., (1960). Protection against birds and parasites in some species of Tenthredinid larvae. *Arch. Néerland. Zool.*, **13**: 380-447.

REECKS W.-A., BARTER G. W., (1951). Growth reduction and mortality of spruce caused by the European spruce sawfly, *Gilpinia hercyniae* (Htg.) (Hymenoptera: Diprionidae). *For. Chron.*, **27**: 140-156.

RŒLOFS W.-L., CARDE R.-T., (1977). Responses of Lepidoptera to synthetic sex pheromone chemicals and their analogues. *Ann. Rev. Ent.*, **22**: 377-405.

ROMANYK N., RUPEREZ A., (1960). Principales parasites observados en los defoliadores de Espana con atencion particular de la *Lymantria dispar* L. *Entomophaga*, **5**: 229-236.

SATCHELL J. E., (1962). Resistance in oak (*Quercus* sp.) to defoliation by *Tortrix viridana* L. in Roudsea Wood National Nature Reserve. *Ann. Appl. Biol.*, **50**: 431-442.

SCHWERDTFEGER F., (1935). Studien uber den Massenwechsel eniger Forstschädlinge. I. Das Klima der Schadgebiete von *Bupalus piniarius* L. *Panolis flammea* Schiff. und *Dendrolimus pini* L. in Deutschland. *Z. Forst Jagdwes*, **67**: 15-38 et 95-104.- II. Uber die Populationsdichte von *Bupalus piniarius* L., *Panolis flammea* Schiff., *Dendrolimus pini* L., *Sphinx pinastri* L. und ihren zeitlichen Wechsel. *Idem*, **67**: 449-482 and 513-540.

SCHWERDTFEGER F., (1971). Vergleichende Untersuchungen an der Kronen-fauna der Eichen in Latenz und Gradationsgebieten des Eichenwicklers (*Tortrix viridana* L.). 3. Die Bedeeutung der Parasiten für den lokalen Fluktuationstyp des Eichenwicklers. *Zeit. ang. Ent.*, **67**: 296-304.

STOAKLEY J. T., (1979). Pine beauty moth. *Forestry Commission, Forest Record* no. **120**: 1-10.

VARLEY G. C., (1971). The effects of natural predators and parasites on winter moth populations in England. *Proceedings Tall Timbers Conference on Ecological Animal Control by Habitat Management*, **2**: 103-116.

VARLEY G. C., GRADWELL G. R., Population models for the winter moth. *In*: T. R. E. SOUTHWOOD (1968). *Insect abundance*. Blackwell, Oxford, 132-142.

VILLEMANT C., FRAVAL A., (1991). *La faune du chêne-liège*. Actes Éditions, Rabat.

WARGO P. M., PARKER J., HOUSTON D. R., (1972). Starch content in roots of defoliated sugar maple. *For. Sci.*, **18** : 203-204.

WILKINSON R. C., BECKER G. C., BENJAMIN D. M., (1966). The biology of *Neodiprion rugifrons*, a sawfly infesting jack pine in Wisconsin. *Ann. Ent. Soc. Amer.*, **59**: 786-792.

WITKOWSKI Z., (1975). Environment regulation of the population size of the oak leaf roller moth (*Tortrix viridana* L.) in the Niepolomice Forest. *Bull. Acad. Polon. Sci.*, sér. sci. biol., cl II, **23**: 513-519.

WYLIE H. G., (1960). Some factors that affect the annual cycle of the winter moth *Operophtera brumata* (L.) (Lepidoptera: Geometridae) in Western Europe. *Ent. Exp. Appl.*, **3**: 93-102.

ZWÖLFER W., (1961). A comparative analysis of the parasite complexes of the European fir budworm, *Choristoneura murinana* (Hub.) and the North American spruce budworm *C. fumiferana* (Clem.). *Tech. Bull. C. I. B. C.*, **1**: 1-162.

ZWÖLFER W., KRAUS M., (1957). Biocenotic studies on the parasites of two fir and two oak tortricids. *Entomophaga*, **2**: 173-196.

9

Processionary caterpillars

The name processionary caterpillars is given to certain gregarious arboreal species that build silk nests used as collective shelters and migrate head to tail in long columns on their way to pupate. These species belong to the family Thaumetopœidae, which contains only one genus with nine species, and is of considerable economic importance.

1. THE PINE PROCESSIONARY CATERPILLAR

The pine processionary caterpillar, *Thaumetopoea pityocampa*, feeds, in decreasing order of preference, on Austrian pine, laricio or Corsican pine, Salzmann's pine, maritime pine, Aleppo pine and cedars. Its distribution range is a function of its thermal requirements (cf. chapter 2.1.2). In France it is restricted to regions that get a minimum of 1,800 hours of sunlight a year and in which minimum January temperatures are not lower than − 4 °C. The present northward spread of this species' range is a sign of global warming. The processionary pine moth now occurs in the Barres arboretum in the Loiret Département, which coincides with an annual rise in temperature of 1.5 °C in thirty years (Demolin *et al.*, 1996).

1.1. Life cycle

On Mont Ventoux in South-east France females lay their eggs between 15 July and 15 August, and eggs hatch in August or September. The caterpillars feed at night and undergo four moults. The damage they cause becomes severe from the third instar on. In the fourth instar, in autumn, the caterpillars weave large silk bags that protect them in winter by acting as hothouses. Having grown to full size, in March or June according to region, the caterpillars migrate down from the trees in procession, bury themselves in the ground and weave a cocoon in which they pupate. Adult emergence in

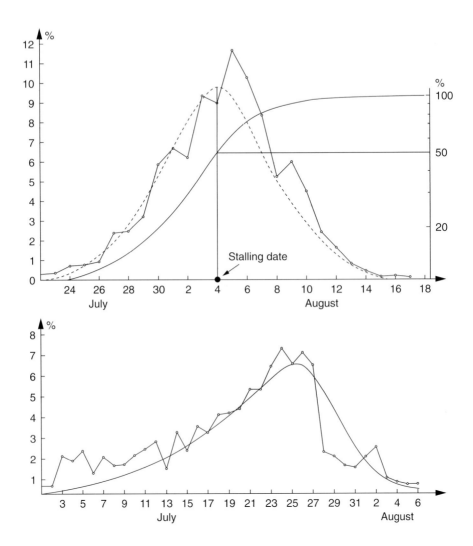

Figure 9.1. *Above, determination of the date at which the life cycle of the pine processionary moth can be stalled in a population with high growth potential. Emergence of adults is spread over 27 days with a daily maximum of 12%. The fine line curve represents percentage of daily catches in relation to total catch. The dashed line curve represents Gauss' theoretical curve adjusted to the real curve. The thick line curve corresponds to cumulative daily percentage. Dates in abscissa. Ordinates, on the right, show cumulative percentages of catches in probit scale. Stalling date corresponds to 50% of catches. Below, declining population. Emergence of adults is spread over 37 days with a daily maximum of 7% (Demolin, 1969).*

uplands in the South of France takes place between 15 July and 15 August. The moths mate and lay eggs within 48 hours and then die. Choice of tree by females and thermal preferences affect the distribution of infestations. Trees in sparsely planted stands, at forest edges and in mountains exposed to the south, are the most vulnerable to attack. Duration of the insect's life cycle varies according to region. On Mont Ventoux, where the caterpillars' habitat is limited by altitude at 900 metres, the life cycle takes one year. In the mountains of Corsica, where the species lives at up to 1,500 or 1,600 metres, the cycle lasts two years.

Monitoring adults with light traps has shown that, where a population has a high expansion potential, flight period remains fairly constant from one year to the next. Curves representing daily percentage of emergence as a function of the total, or of cumulative daily emergence, make it possible to determine of the date at which 50% of adult emergence occurred. This date enables one to "stall" the life cycle of the processionary moth in a given area (figure 9.1).

A graph drawn by Demolin (1969) reveals the pine processionary moth's life cycle when the stalling date has been established (figure 9.2). The graph reads in the following way. The point corresponding to the stalling date 15 August is situated in the middle of band number 1; this allows one to draw the horizontal **a-i** which bisects areas 1 to 8. It will be seen that: pre-winter growth **b-d** ceases in early November; the winter nest **d-e** is used from November to early January; post-winter growth **e-f** goes on until the end of February; procession **f-g** occurs in March; diapause **g-h** lasts from 1 April to 15 June; the new generation of imagoes **i** begins to emerge on 1 August.

1.1.1. Oviposition

Eggs are laid near the tips of branches the day following the emergence of females from the ground in which they pupated. Eggs are grouped in rings 4 to 5 centimetres long which are covered in pale buff scales which conceal them. The clutch may consist of 70 to 300 eggs per female.

1.1.2 Larval life

Caterpillars hatch 30 to 45 days after the adults emerge. There are five larval instars, during which the caterpillars change location as pine needles are consumed. Young caterpillars weave fine silk threads into pre-nests which are abandoned at each change of location. On low ground the average duration of each larval instar from L_1 to L_5 is as follows: L_1: 12 days; L_2: 14 days; L_3: 30 days; L_4 and L_5: 30 to 60 days each.

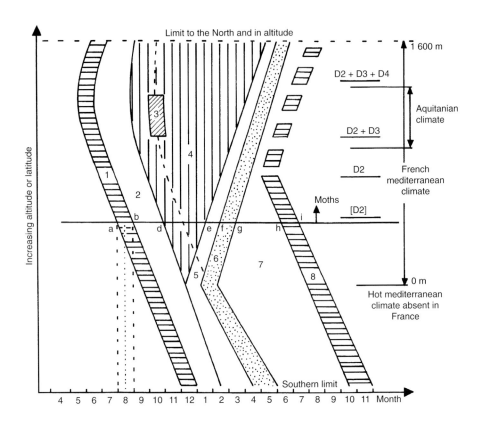

Figure 9.2.   *Graph showing seasonal periods of different stages in the life cycle of the pine processionary caterpillar (at different latitudes and altitudes) 1: Imagoes (horizontal hatching); 2: pre-winter growth of eggs and larvae (blank); 3: pre-winter processions (oblique hatching); 4: winter nest (vertical hatching); 5: post-winter growth (blank); 6: post-winter processions (dotted area); 7: pupal diapause (blank); 8: imago formation in pupae (horizontal hatching); D_2, D_3, D_4: prolonged diapauses in two, three and four year cycles. Vertical arrow shows adult emergence.*
Thaumetopœa pityocampa only lives in the French Mediterranean and Aquitanian climatic zones. In the hot Mediterranean zone it is replaced by T. wilkinsoni, *and north of its range it is replaced by* T. pinivora.

The winter nest is a voluminous silk bag up to 20 centimetres long in which the caterpillars spend the cold season. It is a shelter which ensures group cohesion and is also a heat accumulator in which air temperature may rise to 1.5 °C higher than the ambient temperature in one hour's insolation (figure 9.3).

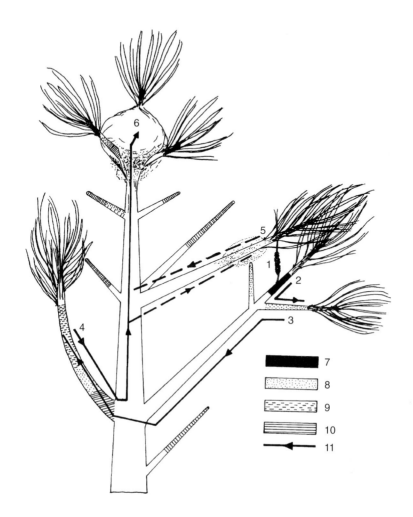

Figure 9.3. *Migrations of processionary pine caterpillars on a tree during their growth. 1: eggs; 2: location of L_1 caterpillars; 3: location of L_2 caterpillars; 4: location of young L_3 caterpillars; 5 optional pre-nest; 6: nest of old L_3 and of L_4 and L_5 caterpillars; 7: damage done by L_1 caterpillars; 8: damage done by L_2 caterpillars; 9: damage done by L_3; 10: damage done by L_4 and L_5 caterpillars; 11: migration routes.*

1.1.3. Pupation processions

These take place at the end of the larval stage from February to May. Processions are led by a caterpillar, usually a female, which heads for the most brightly lit and warmest area in the vicinity of the host tree. Processions only occur when ground temperature is between 10 and 22 °C.

1.1.4. Underground life

After burrowing 5 to 20 centimetres below ground, each caterpillar weaves a cocoon and pupates. A more or less lengthy diapause, depending on climatic conditions, ensues, during which all growth / metamorphosis ceases. This resumes only one month before the emergence of adults.

1.1.5. Adult life

Females emerge from the soil a little before sunset with the help of sclerified crests on their heads. They settle on a raised spot where they remain motionless as their wings unfold, and then, following two to four hours of inactivity, they evaginate their genitalia and become attractive to males. Copulation lasts about one hour. The female then deposits her eggs, starting at the bases of pine needles, each egg being covered with a small bundle of scales with which the tip of the ovipositing moth's abdomen is abundantly provided. A fertilised female only stops to lay when she encounters a pine tree, the only plant that satisfies her tactile requirements, these being needles of a diameter between 1.5 and 2 mm and having a rugose structure which the moth's tarsal claws can grip. In the absence of pines the moth may fly two kilometres or more in search of suitable laying sites, which explains why intact pine woods may be rapidly invaded. Adult moths do not live for much more than twenty four hours.

1.2. Processions and social life of caterpillars

Processionary caterpillars, as well as the larvae of a few other species of moth belonging to the families Lymantriidae (genera *Euproctis* and *Porthesia*) and Lasiocampidae (genera *Malacosoma* and *Eriogaster*), live gregariously, unlike most Lepidoptera. These societies have been termed organised groupings, or lower social groupings. They are characterised by coordination of individual activities into collective activity, as in the construction of nests. Moreover caterpillars spin silk trails that serve as guideposts to individuals, migrate collectively and feed together, which causes major changes in metabolism. This collective existence is only characteristic of larvae and disappears in the adult, unlike the behaviour of some other insects such as locusts whose bands also form organised groupings.

> Gregarious habits have produced group effects, including increases in metabolic and growth rates. Experiments carried out on pine processionary caterpillars reared alone or in groups of twenty showed that weight increases 2.5 to 3.8 times more rapidly in grouped caterpillars, and that their food consumption is greater. Respiration intensity is also stimulated by grouping. Some social activities such as nest weaving is however not vital to the caterpillars' survival; they are also able to weave individual nests.
>
> The spectacular nature of these caterpillars' life drew the attention of J.-H. Fabre as early as the last century. Mutual attraction of caterpillars shows itself from the

moment of hatching. Caterpillars disperse in order to feed, but remain in contact by means of a loose network of silk threads. A colony must migrate to the ground to find a suitable location for pupation. They undergo preconditioning exercises before setting off, the caterpillars gathering in a bundle and rubbing slowly against one another. The individual which in the course of this exercise acquires the greatest "collective memory" is usually a female, and she then leads the procession. The second caterpillar places itself between the setae at the end of the first one's abdomen and rests its head against the latter's anal scutellum. A third caterpillar takes place behind the second, and so on. When a caterpillar loses contact, the one which precedes it stops and then reverses; all the others do the same going up the line to the leader, thus making it possible to rescue the drop-out.

In the third instar, caterpillars develop cuticular structures known as "mirrors" which are furnished with irritant hairs. These may break off and be carried by the wind all around the colony. These hairs may cause itching, eye and respiratory problems and bouts of fever in man. Infested areas may become uninhabitable, even to birds and to small mammals, which suggests that the irritant hairs play a part in protection against predators.

1.3. Predators and parasitoids

An inventory of the processionary pine caterpillar's main enemies has been drawn up with a view to biological control. These are listed in table 9.1. Some, such as *Ephippiger*, are occasional predators; others, like *Meteorus versicolor*, are non-specific polyphagous parasitoids. The most promising are *Phryxe caudata* and *Villa brunnea*.

Predated stage of life cycle	Parasitoid, predator or disease	
Eggs	Predator	*Ephippiger* sp. (Orthoptera)
	Parasitoids	Chalcids
Larvae	Predator	*Xanthandrus comptus* (syrphid)
	Parasitoids	*Phryxe caudata* (tachinid) *Compsilura concinnata* (tachinid) *Erigorgus femoratus* (braconid)
	Disease	Viruses and bacteria
Pupae	Parasitoids	*Villa brunnea* (bombyliid) *Ichneumon rudis* (ichneumonid) *Conomorium eremita* (pteromalid)
	Disease	*Beauveria bassiana* (Fungus)
Imagos	Predators	*Vespa germanica* (Hymenoptera) Ants

Table 9.1. *Main elements of the processionay pine moth's parasite complex.*

1.3.1. Phryxe caudata

This tachinid fly has two generations a year and is a host-specific parasite of the processionary pine caterpillar. When it finds a processionary caterpillar, the female *P. caudata* settles next to it, bends its abdomen forward between its legs and extends its ovipositor which can stretch more than one centimetre. The tip of the ovipositor is used to feel the prey's integument and find the most suitable parts of the body on which to deposit its eggs. These are the inter-segmental membranes, the bases of pseudopods, and the ventral surface of the abdomen. As soon as the egg is laid, the young larva hatches and actively attacks the host's integument with the cutting edges of its mouth-parts. Within minutes it disappears inside its host which, in normal circumstances, is a third instar larva. The *Phryxe* larva remains free-moving within the host's general cavity for a certain period then, when the host reaches the fourth instar, moves to the area of the host's body situated between the last two abdominal stigmata, which is rich in tracheae. It then pierces a trachea and, by means of reactions of the host's tissues, forms a respiratory sheath for itself which almost entirely surrounds it. When it is fully-grown the parasite larva leaves its host to pupate in the ground or in the caterpillar's silk nest. Should the caterpillar die prematurely of a viral infection caused by *Smithiavirus*, the *P. caudata* larva can go on feeding on the dead body for several days. Flies resulting from such cases are smaller than normal, but fertile.

The use of *P. caudata* in biological control has been envisaged, so attempts to breed the parasite have been made (Grenier *et al.*, 1974). A possible substitute host on which the parasite might be reared is the 'wax moth' *Galleria mellonella*.

1.3.2. Villa brunnea

This bombyliid fly predates the underground stage of the pine processionary moth's life cycle (Biliotti *et al.*, 1965; Du Merle, 1971). Rates of parasitism vary from 5 to 75% according to location within the same locality, which explains the laying behaviour of females. The female's reproductive organs (figure 9.4) include a large perivaginal sack which, at rest, is closed ventrally by two lateral valves, the median margins of which are furnished with long brush-like setae. The vagina opens into this sack, and the anus is surrounded by a ring of stout spines. The female seeks sunny spots with powdery soil in which to lay. She inserts the tip of her abdomen in the soil and fills the perivaginal sack with very fine soil, the spines ringing the anus being used to scrape the soil which is then sifted by the long setae on the lateral valves.

At the next stage the female flies off at a height of 10 to 20 centimetres above the ground and releases an egg coated with a fine layer of soil. Eggs are always dropped in places away from direct sunlight, for instance at the foot of a stone. Up to ten eggs may be dropped in the same place. Egg laying behaviour is thus totally independent of the host, and depends only on the nature of the terrain. Later, processionary caterpillars tend to bury themselves in sunny areas, that is, in ones frequented by *Villa brunnea* larvae. Host / parasite coincidence is thus ensured by the two species' similar reaction to a micro-climatic factor. Where the terrain is varied, *Villa* tends to concentrate egg laying in certain suitable spots where the level of parasitism will, as a result, be high; where terrain is homogenous, eggs are distributed haphazardly and the level of parasitism will be about the same from one spot to another.

Villa brunnea eggs produce highly mobile planidium larvae that burrow into the ground in search of the processionary caterpillar pupae in which they will grow as endoparasites. During its life cycle *Villa brunnea* may

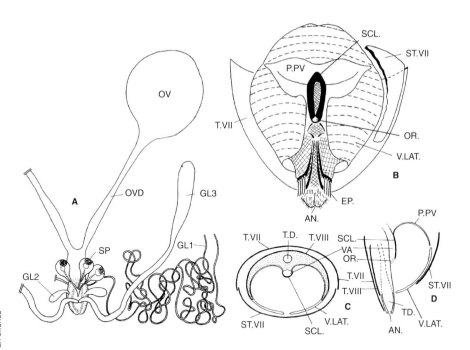

Figure 9.4. *Reproductive organs of* Villa brunnea. *A. Female reproductive organ. B. Schematic ventral view of apex of female abdomen with sternite VII partially removed and lateral valves spread open. C. Cross-section. D. longitudinal section. GL1, GL2, GL3: accessory glands; OV: ovary; OVD: oviduct; SP: spermatheca; AN: anus; EP: anal spines; OR: genital orifice; PPV: perivaginal sack; ST VII: sternite VII; SCL: sclerite surrounding vagina; T VII: tergite VII; TD: digestive tract; VA: vagina; VLAT: lateral valve (Biliotti et al., 1965).*

undergo optional diapauses, the first at the end of incubation, the second at the end of larval growth. For this reason the cycle from egg to adult may vary from two months to three years according to circumstances. This spread over time of adults emerging from a single clutch of eggs reduces the risk of starvation in periods where hosts may be scarce or absent.

1.4. Control of the pine processionary caterpillar

When caterpillars are abundant needles may be almost totally eaten. If these infestations last two to three years, which happens in young pine woods of trees under ten years old, tree mortality may be high. Old trees have better resistance, but their growth is slowed. On Mont Ventoux, populations of Austrian pine subjected to massive infestation every five to seven years lose 20 to 45% in volume every 50 years compared with healthy trees (Bouchon & Toth, 1971).

Action against the processionary caterpillar must be taken when the pest is most vulnerable, and without harming its parasite complex. Destruction of adults is difficult given their short life span. Trapping with light or pheromone traps has been tried. Destruction of eggs is also difficult. The most convenient time to act is between hatching and the building of winter nests. The ideal insecticides are those using *Bacillus thuringiensis*, which do not have the drawbacks of non-selective insecticides. The main virus that attacks the processionary caterpillar is *Smithiavirus pityocampae*. The disease causes lysis of the gut and kills the caterpillars. Attempts at biological control using this virus have been made on the slopes of Mont Ventoux, which are replanted with pines that suffer massive infestations. Reared larvae were used to prepare a virus-rich powder which was sprayed by helicopter on 320 hectares in 1958. Results were spectacular, the population being reduced to 4% of its former level, but the epidemic did not maintain itself, and the caterpillar population reconstituted itself in a few years (Grison *et al.*, 1959).

2. THE OAK PROCESSIONARY CATERPILLAR

The oak processionary moth, *Thaumetopœa processionea*, only lives on deciduous oaks in temperate western Europe. It occurs in the whole of France but only has major gradations in northern parts of the country. When trees are defoliated, oak processionary caterpillars may feed on hazel, chestnut, beech, hornbeam and birch. Mortality in these surrogate hosts is high.

2.1. Life cycle

The oak processionary moth has a single generation a year. In summer females lay eggs in groups forming regular plaques of a single layer of eggs on young twigs. Caterpillars hatch simultaneously with the unfolding of oak leaf buds in spring. All the individuals born of a single clutch form a colony, but several colonies may unite, especially during outbreaks. There are ten larval instars. The caterpillars migrate in procession; from the third instar they shelter each night in a collective silk nest situated under a large limb or on the trunk. Subsequent moults and pupation take place in the nest. Pupation takes place in July, and adults emerge in August; they are nocturnal and live no longer than a day or two.

Outbreaks of the oak processionary caterpillar are very irregular. Gradations extend over three to five years, after which a dramatic drop in the number of individuals is caused by natural enemies. There was a fifteen year period from 1952 to 1967 in the Paris region during which this species was very rare. Infested trees are mainly those in open and sparsely-wooded environments and on forest edge. Phenological coincidence between the unfolding of buds and hatching of larvae is a prerequisite: young caterpillars die in large numbers on late-budding trees. Association of oak processionary caterpillars with those of the gypsy moth is sometimes seen in the Paris region. The latter congregate during inactive periods, and to pupate, in the processionary caterpillars' nests. This unilateral interspecific attraction may be to the gypsy moth caterpillars' advantage, in protecting them from parasites or predators.

2.2. Predators and parasitoids

The oak processionary moth's parasite complex includes four categories of insect (Grison & Biliotti, 1952):

– Non-specific, polyphagous predators. These are mainly beetles. *Xylodrepa quadripunctata* is a yellow silphid with four black maculae on the elytra. *Calosoma sycophanta* and *C. inquisitor* are arboreal carabids which mainly feed on caterpillars. Pentatomid bugs of the subfamily Asopinae are also predators of *T. processionea*, especially the species *Picromerus bidens* and *Troilus luridus*, as are a few reduviid and, occasionally, mirid bugs. Larvae of the syrphid fly *Xanthandrus comptus*, which are highly polyphagous, also predate the caterpillars.

– Non-specific parasitoids. These are tachinid flies, and wasps. Tachinids may be species with microtype eggs that are ingested by the caterpillars, such as *Ctenophorocera pavida* and, more rarely, *Zenilia libatrix*; they may also be species with macrotype eggs deposited on the host's body such as *Phorocera agilis*. *Compsilura concinnata* is a peculiar tachinid which has an ovipositor enabling it to insert its larva inside the host's body. Wasps are chalcid parasites of eggs belonging to the genera *Trichogramma* and *Anastatus*, and ichneumonid parasites of larvae belonging to the genus *Phobocampe*, or of pupae (*Pimpla examinator, Theronia atalantae* etc.).

– Host-specific parasites. These appear to be few. Females of the tachinid *Carcelia processioneae* deposit their eggs when they are ready to hatch on the caterpillars' hairs or on silk threads of the nest, and the larvae actively penetrate the hosts' bodies.

– Hyperparasites. These predate the pupae of parasitic tachinids and the cocoons of ichneumonids.

Many predators are only attracted by the processionary caterpillars when the latter are very abundant. Some species may form close ties with their prey. The larvae of *Calosoma* occupy pupation nests where they prey on caterpillars, as do larvae of the syrphid *Xanthandrus comptus*.

The pupation nest constitutes a microhabitat occupied by various eaters of detritus: dermestid beetles (*Dermestes lardarius* and *D. laniarius*), phorids of the genus *Aphiochaeta*, anthomyiids, and muscids such as *Muscina stabulans*. These detritivores complete their entire life cycles in the pupation nests, but they are also found in other habitats. Some detritivores, such as the beetles *Dermestes aurichalceus* and *Micrambe perrisi*, have a much stricter association with processionary caterpillars.

2.3. Control of the oak processionary caterpillar

Knowledge of the processionary moth's life cycle is essential when using insecticide treatments if useful insects are not to be destroyed at the same time. Treatment should be applied before the caterpillars reach their third instar, at which stage most parasitic tachinids, and in particular, *Carcelia processioneae*, are active. A technique for predicting outbreaks of the oak processionary caterpillar has been described (Trouvelot *et al.*, 1953). In the Paris region treatment is applied in mid-April, between the hatching date (which varied from 7 to 14 April at Versailles over a period of five years) and the last larval moults.

REFERENCES

BILIOTTI E., DEMOLIN G., DU MERLE P., (1965). Parasitisme de la processionnaire du pin par *Villa quinquefasciata* Wied. apud Meig. (Dipt. Bombyliidae). Importance du comportement de ponte du parasite. *Ann. Epiphytes*, **16**: 279-288.

BOUCHON J., TOTH J., (1971). Étude préliminaire sur les pertes de production des pinèdes soumises aux attaques de la processionnaire du pin *Thaumetopœa pityocampa* Schiff. *Ann. Sci. For.*, **28**: 323-340.

DEMOLIN G., (1969). Bioecologia de la procesionaria del pino *Thaumetopœa pityocampa* Schiff. Incidencia de los factores climaticos. *Bol. Ser. Plagas For.*, **12**: 9-22.

DEMOLIN G., ABGRALL J.-F., BOUHOT-DELDUC L., (1996). Évolution de l'aire de la processionnaire du pin en France. *La Santé des Forêt*s (Ministère de l'Agriculture), **1**: 26-28.

DU MERLE P., (1971). Sur quelques facteurs qui régissent l'efficacité de *Villa brunnea* Beck (Dip. Bombyliidae) dans la régulation des populations de *Thaumetopœa pityocampa* Schiff. (Lep. Thaumetopœidae). *Ann. Zool. Écol. Anim.*, no. hors série: la lutte biologique en forêt: 57-66.

GRISON P., BILIOTTI E., (1952). Quelques aspects de la biocénose des chenilles processionnaires. *Ann. Sci. Nat. Zool.*, **14**: 423-432.

GRISON P., MAURY P., VAGO C., (1959). La lutte contre la processionnaire du pin (*Thaumetopœa pityocampa* Schiff.) dans le massif du Ventoux. Essai d'utilisation pratique d'un virus spécifique. *Rev. For. Fr.*, **5**: 353-367.

TROUVELOT B., GRISON P., BILIOTTI E., (1953). La prévision des infestations de processionnaires du chêne en vue de traitements chimiques. *C. R. Ac. Agric.*, **9**: 486-488.

◇
10
◇

Other defoliators:
sawflies, beetles, flies

Next to the Lepidoptera, which are the main defoliators of forest trees, mention must be made of species belonging to three other orders of insect: Hymenoptera, represented by sawflies, Coleoptera, represented mainly by weevils (Curculionidae), leaf beetles (Chrysomelidae) and chafers (Scarabaeoidea), and Diptera, represented by Cecidomyiidae.

1. SAWFLIES

Hymenoptera belonging to the sub-order Symphyta can be recognised by the abdomen being fused to the thorax, without the characteristic 'waist' of most wasps. There are three main groups: tenthredinid sawflies, which are almost all leaf-eaters in their larval stages, siricids, which are xylophagous, and orussids, which are rare parasites of buprestid beetle larvae. Sawflies are so called because females have a saw-like ovipositor which enables them to insert their eggs in leaf parenchyma. The larvae, which resemble those of Lepidoptera, are "false caterpillars" with a variable number of abdominal pseudopods, except in the family Pamphilidae, whose larvae have none. Sawflies are much less abundant than moths, but the group includes a few major forest species. Of the dozen or so families, the Tenthredinidae is the richest in species and the commonest on trees.

Sex pheromones have been discovered in several sawflies. In *Cephalcia laricifolia* the sex pheromone is emitted by virgin females, and its use as a lure to trap males has been considered in order to combat this pest species (Borden *et al.*, 1978). The existence of sex pheromones emitted by unmated females and attractive to males has also been demonstrated in *Diprion pini* and *Neodiprion sertifer*, as well as in several sawflies of conifers in North America. An alcohol, the 3,7-dimethylpentadecane-2-ol, is one constituent of this pheromone. The attractive power of sawfly pheromones varies greatly according to climatic conditions and the physiological condition of insects (Schönherr *et al.*, 1979).

1.1. Sawflies of broadleaf trees

Arge rustica (family Argidae) is distributed through the whole of Europe where it lives on oak. *Pamphilus silvarum* (family Pamphilidae) has larvae devoid of abdominal pseudopods, which live singly in a cornet formed by the rolled-up edge of an oak leaf. Other species belonging to the same genus have similar biology but live on poplar, hazel, or hornbeam. *Cimbex femoratus* (family Cimbicidae), which lives on oak, is recognisable by its large size (20 to 28 mm) and by its club-shaped antennae. *Caliroa annulipes* (family Tenthredinidae), which lives on oak and birch, has larvae covered with a black mucus which makes them resemble small, one-centimetre-long slugs.

Various species of *Periclista* (family Tenthredinidae) are pests of oak in Europe and North America. *P. andrei* and *P. dusneti* seriously defoliate cork oak in Portugal. *P. andrei* also occurs in Morocco; in Portugal its life cycle lasts one to three years according to the duration of the pre-pupal stage. *Fenusa pusilla* (family Tenthredinidae) is widely distributed in Europe on *Betula verrucosa* and *B. pubescens*. Eggs are laid in leaf palisade parenchyma, usually next to the midrib; larvae have five instars, and last instar larvae do not feed. Pupation takes place on the ground in a cocoon. One generation is produced in five to six weeks, and in Europe there may be two to four generations a year, depending on temperature. In Europe, infestations of this species are confined to trees in open environments, for instance trees planted along the sides of roads, and do not occur in forest. Since it was introduced to Canada and the United States this sawfly has become a major pest, and for this reason detailed studies of its parasite complex have been conducted in Europe (Eichhorn & Pschorn-Walcher, 1973). The two most effective parasites are the ichneumonids *Grypocentrus albipes* and *Lathrolestes nigricollis*.

1.2. Sawflies of conifers

Conifers are much more heavily attacked by sawflies than are broadleaf trees, and they harbour several pest species.

1.2.1. Spruce sawflies

The most important of these is *Pristiphora abietina*, which feeds on spruce in various parts of France and Europe. This sawfly mainly attacks young trees whose terminal shoots may be totally defoliated, giving trees a characteristic "broomstick" appearance. The young yellow and black adults are about 6 mm long, and their pale green larvae reach a length of 13 mm and are well concealed by their similarity in appearance to young spruce needles. Larval growth is rapid and may be completed in 15 days. Adults emerge in April and lay on young spruce needles that have just emerged from buds. Each females lays between 40 and 100 eggs. Successful completion of the insect's life cycle depends mainly on coincidence of egg-laying with the sensitive stage in the tree, that is, the time at which the buds unfold. Control of *P. abietina* using methods of cultivation is possible. These consist in selecting late-budding varieties of spruce which prevent the insects from laying eggs, and in encouraging rapid growth of young trees by means of fertilisers so that the crowns close in, thus creating an inhospitable environment for the pest.

Cephalcia abietis (family Pamphilidae) also lives on spruce. Adults have a black head and thorax with a few yellow spots and an orange abdomen. The larvae are yellowish or greenish. This species is widely distributed in Europe; it feeds mainly on old spruce needles. Young larvae shelter in a communal web full of their faeces and in which each larva has an individual cell with its own opening.

Gilpinia hercyniae (family Diprionidae), the only member of the genus that feeds on spruce, occurs in large parts of Europe. During outbreaks of this insect, growth of spruce trees is slowed, the tops of trees are defoliated, and some trees may die. This species has been introduced to Canada where it causes serious damage. Males are rare; females are recognisable by their black and yellow colour pattern. Fully grown larvae drop to the ground where they spin a pupation cocoon and overwinter. Wood mice and shrews may eat 30 to 70% of pupae but are unable to contain outbreaks, nor can parasitoids. *G. hercyniae* is very sensitive to a nuclear polyhedrosis virus which might be used in biological control (Billany, 1978).

1.2.2. Larch sawflies

Sawflies that live on larch have been studied in the Alps (Pschorn-Walcher & Zinnert, 1971). In the High Engadine several species of sawfly live on larch together with the larch bud moth *Zeiraphera diniana*. Abundance of these sawflies is always low (less than 1% of that of *Z. diniana*). The cyclical changes in the moth's population, which cause variations in the nutritional quality of larch needles, impose a cycle on sawfly populations characterised by a progression phase lasting five or six years followed by a rapid regression phase lasting three or four years, alternating with changes in the moth's population (Lovis, 1975).

The most important species is *Pristiphora erichsonii* which was introduced to North America in 1910 and has become a pest of two species of larch, *Larix occidentalis* and *L. laricina*. The larvae live in colonies and feed on needles. They defoliate trees which successive infestations may kill. In larch forests in the Alps, the years in which *P. erichsonii* abounds are those in which aphids are scarce, and vice versa. This is a result of predation by ants which are less attracted to trees when aphids are scarce.

The parasite complex of *Pristiphora erichsonii* includes eight ichneumonids and two tachinids. Predators are ants of the *Formica polyctena* group and the pentatomid bugs *Picromerus bidens* and *Pinthaeus sanguinipes*. Predation by ants is often the main cause of larval mortality in *Pristiphora erichsonii*, especially at high altitude where most trees are invaded by aphids and much frequented by ants in search of honeydew. This has been demonstrated by protecting trees from ants by means of adhesive bands placed on the trunks. Larval mortality in *P. erichsonii* is 24.7% on protected trees and 71.8% on non-protected ones.

1.2.3. Pine sawflies

Sawflies living on pine include most of the species of the genus *Acantholyda*, belonging to the family Pamphilidae (Charles & Chevin, 1977). *A. erythrocephala* prefers to feed on young trees aged ten to fifteen years, and *A. hieroglyphica* on saplings two to six years old. Adult *A. erythrocephala* are bluish-black with a metallic sheen on the thorax and a red head; larvae are grey-green with black spots on the thorax, and red legs. They live singly in a silk sheath. Adults are little seen and tend to be nocturnal. Larval growth is completed in five weeks. Infestations of *A. hieroglyphica* on young trees may almost totally strip trees of needles. The growth of trees is then greatly reduced, and repeated infestations may eliminate those trees. The two most harmful sawflies of pines belong to the family Diprionidae; they are *Diprion pini* and *Neodiprion sertifer*.

1.3. Diprionid sawflies

Sawflies of the family Diprionidae, sometimes also known on the continent by the old name "lophyres", all live on conifers – chiefly on pines. They are recognisable by their squat bodies and sexual dimorphism. Males are less broad, more often entirely black, with strongly pectinate antennae; females are lighter in colour with serrate antennae. Some species of diprionid are major pests, and for this reason have been much studied. The gregarious larvae of *Diprion pini* and *Neodiprion sertifer* display a particular behaviour, which consists in more or less rapid sideways thrashing of the front half of the body, while the hind half remains anchored to a pine needle by its false legs. Since these larvae are highly visible due to their colour, these movements are thought to be a defence mechanism designed to scare predators or parasites such as Hymenoptera, Diptera, and birds. Frequency of thrashing may increase twenty-fold when a potential enemy nears.

1.3.1. The pine sawfly, **Diprion pini**

The pine sawfly, which is distributed in Europe and North Africa, feeds on pine and occasionally on spruce, fir and Douglas-fir. The 16 mm-long males have strongly pectinate antennae, largely black bodies, and yellow legs; females are more extensively spotted with yellow and have simple antennae (figure 10.1). Young larvae have rufous heads and greenish-yellow bodies with series of black spots, and older larvae, which reach 25 mm, are green. These larvae are typical false caterpillars, with eight pairs of abdominal pseudopods.

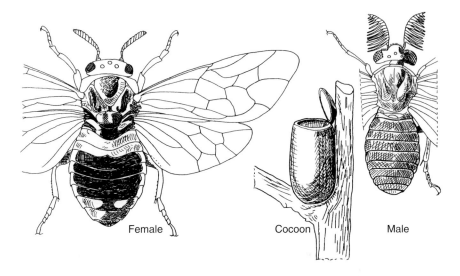

Female   Cocoon   Male

Figure 10.1. *The sawfly* Diprion pini. *Female, male, and cocoon placed at the base of a pine needle.*

– Life cycle.

There are usually two generations a year in France, but the event of a diapause may produce extra generations (Dussausoy, 1970; Dussausoy & Géri, 1971), which complicates the cycle (figure 10.2).

Eggs, which are deposited contiguously in rows on needles, are covered in a secretion. Eggs produced by the spring generation are laid on old needles, and those of the summer generation on needles of the current year's growth. Larvae are active during two periods, the first in May to June for the first generation and the other at the end of summer for the second generation. Larvae congregate in colonies consisting of all individuals born of eggs deposited on the tip of one branch. Gregarious behaviour in this species (as in *Neodiprion sertifer*) is due to reciprocal interaction between individuals. Absence of contact triggers search behaviour in isolated larvae which thus rejoin the group. When larvae migrate to new sources of food they appear to maintain cohesion by means of a scent trail.

Pupation takes place in a cocoon built among needles in the first generation, and in ground litter in the second generation. The ovoid cocoon is very
tough; the adult emerges after cutting out a lid (figure 10.1). Within the
cocoon the larva first goes through an eonymph stage, then through a pronymph stage, and does not turn into a true nymph, or pupa, until a fortnight

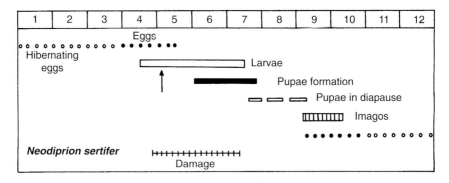

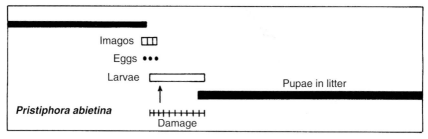

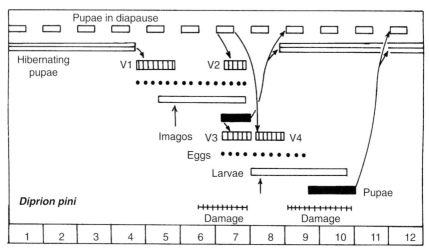

Figure 10.2. *Life cycles of three sawflies:* Neodiprion sertifer, Pristiphora
abietina *and* Diprion pini. *In the case of* Diprion pini V_1 *et* V_3
*indicate adults from pupae that have not undergone diapause,
and* V_2 *et* V_4 *individuals born of pupae with diapause. Vertical
arrows correspond to suitable times for insecticide treatment.*

before the adult emerges. It is during the eonymph stage that a more or less prolonged diapause may occur, giving various possibilities of later adaptations in the life cycle (figure 10.3). The same generation may produce up to six different waves of emergence, and thus six cohorts. This phenomenon plays a major part in population dynamics, as it disrupts coincidence between the sawfly and its parasites.

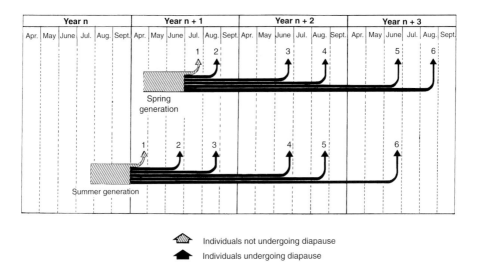

Individuals not undergoing diapause

Individuals undergoing diapause

Figure 10.3. *Theoretical schematic representation of waves of adult emergence of* Diprion pini *in Fontainebleau forest. Each generation may produce six different cohorts depending on duration of diapause in the eonymph stage (Dusaussoy & Géri, 1971).*

Foliage of young Scots pine trees is harmful to *Diprion pini*, for it contains large quantities of a resin acid called neoabietone. As needles age, the level of neoabietone falls, and they become palatable to the sawfly from the end of July onwards (Buratti *et al.*, 1987).

Invasions by *Diprion pini* have been reported on several occasions in France and elsewhere in Europe. Gradations are sudden and spectacular, but do not last much more than two years. Periodicity is very variable, and during latency periods the insect becomes so scarce that it goes virtually unnoticed. In 1963, invasions occurred in 25,000 hectares of the southern part of the Paris basin, in Fontainebleau and Orleans forests, in the Sologne and in the Pays d'Othe. Only Scots pine was infested; other species of pine were spared. Trees weakened during the 1963 outbreak were subsequently afflicted with ring disease caused by the fungus *Ungulina annosa*. In all cases outbreaks ceased spontaneously due to the effects of parasites, and without any phytosanitary measures having been taken (Grison & Jacquiot, 1964). The many studies conducted in Fontainebleau forest in 1963 and 1964 made it possible to draw up a mortality table in which various mortality factors were determined and their effects measured (table 10.1).

Stage x	Survivors Lx	Mortality factor	Deaths dx	Mortality quotient qx %
		First generation		
Eggs	94,231	Endogenous mortality	8,246	8.75
		Predators	3,770	4.00
		Parasites	13,920	14.75
		Other causes	5,890	6.25
		Total	31,826	33.75
L_1	62,405	Endogenous mortality	905	1.45
		Abiotic factors	3,601	5.75
		Total	4,506	7.20
L_2	57,899	Endogenous mortality	579	1.00
		Abiotic factors	637	1.10
		Total	1,216	2.10
L_3	56,683	Endogenous mortality	396	0.70
		Abiotic factors	170	0.30
		Total	566	1.00
L_4	56,117	Endogenous mortality	505	0.90
		Abiotic factors	393	0.70
		Total	898	1.60
L_5, L_6	55,219	Endogenous mortality	4,561	8.25
		Abiotic factors	2,744	4.95
		Total	7,305	13.20
Free eonymphs	47,914	Endogenous mortality	2,228	4.65
		Abiotic factors	1,342	2.80
		Parasitism and starvation	3,684	7.70
		Total	7,254	15.15
Pupae in cocoons	40,660	Diapause	4,391	10.80
		Physiological mortality	8,823	21.70
		Tachinids and *Exenterus*	24,354	59.90
		Various parasites	2,163	5.30
		Rodents	634	1.55
		Mortality in cocoons	41	0.10
		Total	40,406	93.35
Imagos	252	Migration, predators	138	54.75

Imagos active at laying time: 114
Sex ratio : 0.42
Fertile females: 16
Average fertility: 119 eggs per female

		Second generation		
Eggs	1,904	Parasites	1,497	78.60
		Physiological mortality	407	21.40
		Total	1,904	100.00
Population evolution index:		First generation : 1,904/94,231 = 2%		
		Second generation : 0 %		

Table 10.1. *Mortality table established for two generations de* Diprion pini *in Fontainebleau forest. Mortality quotient q_x is equal to the ratio, in percentages, of number of deaths d_x to the number of survivors L_x . This mortality table shows the large part played by parasites at the egg and cocoon stages (Géri & Dusaussoy, 1966).*

– Parasite complex

Diprion pini's parasite complex is a rich one, including more than a hundred species. An inventory of this complex, made in Fontainebleau forest, revealed the following species:

– Parasites of eggs. The two major parasites of eggs are chalcids of the family Eulophidae, *Achrysocharella ruforum* and *A. ovulorum*. Mortality at the egg stage due to *A. ruforum* may reach almost 100% in the culmination period and regress thereafter. *Dipriocampe diprioni* (family Tetracampidae) is also a parasite of eggs, in which it causes the chorion to turn brown.

– Parasites of larvae. The parasites of larvae are the tachinids *Drino inconspicua* and *D. gilva* and the ichneumonids *Holocremnus cothurnatus* and *Lamachus opthalmicus*. The relative importance of these different parasitoids varies greatly in different years and the total percentage of parasitism of larvae varies according to region.

– Parasites of nymphs and pupae. Parasitoids of free eonymphs and of pupae are more numerous in *Diprion* than in most other defoliating insects, which may be explained by the immobility of unprotected cocoons. Parasitoids of nymphs are the ichneumonids *Exenterus amictorius*, *Pleolophus basizonus* and *Gelis meigenii*. Rodents and insectivores (mammals) eat many pupation cocoons in ground litter; micro-organisms and climate also play a part in population regulation.

1.3.2. Neodiprion sertifer

This species, called the "rufous sawfly" (lophyre roux) in France, is distributed through the whole of Europe and extends to Japan. It has been introduced to North America. It feeds mainly on Austrian pine and, in the Alpine zone, on cembro and mountain pine up to an altitude of 2,000 metres.

– Life cycle

This sawfly produces one generation a year. Adults emerge in September - October. They lay immediately, and it is the eggs that overwinter. Eggs are deposited in a row in a crack made in a pine needle by the female's ovipositor; they hatch in April. Young larvae immediately begin to feed on the previous year's growth of needles. They remain grouped, and moult five or six times; when fully-grown they drop to the ground where they weave a pupation cocoon in which they remain in a state of summer diapause until September.

The effects of various climatic factors on *Neodiprion sertifer* have been studied in North America where this species is introduced. Pre-nymphal diapause is controlled by daylight length and temperature (Wallace & Sullivan, 1975). *Neodiprion sertifer* tolerates a great range of ecological conditions, which explains its vast distribution. It would appear that the physiological condition of the host tree is a major factor in outbreaks of this sawfly. Young trees in open environments are particularly prone to attack.

When they are disturbed, the larvae of *Neodiprion sertifer* expel a mixture of monoterpenes (α-pinene and β-pinene) obtained from food that originate in the pine tree on which they feed. This mixture acts as a repellent on many predators, and is stored in a diverticulum of the fore-gut.

– Parasite complex and population dynamics

Outbreaks of *Neodiprion sertifer* occur at irregular intervals in Europe, and just as suddenly as those of *Diprion pini*. The causes of these outbreaks are still little known. Their synchrony in different regions suggests climatic factors. Population collapse is due to three main factors: diseases, parasites and predators. The main disease is a polyhedric viral infection which has been used successfully in biological control in the United States. Predators, which are much less important, are the ant *Formica rufa*, the bug *Picromerus bidens*, spiders which prey on larvae, elaterid beetles, rodents and insectivorous mammals, as well as birds which eat the cocoons.

The parasite complex includes about thirty species which predate different stages in the life cycle. Two thirds of them are ichneumonids; the rest includes chalcids, tachinids and bombyliids. Only a dozen species are permanently associated with *N. sertifer*; the others play an insignificant role (table 10.2). Three ichneumonids belonging to the genus *Exenterus* live as ectoparasites. The eggs of *E. tricolor* are attached to the host's body by means of a pedicel inserted in the tegument, the base of which may end in a transverse bar that acts as an anchor. *Exenterus abruptorius*, the most abundant species and one that is almost exclusive to *N. sertifer*, deposits small eggs almost anywhere on the host larva's body. These eggs are inserted deep in the tegument and out of reach of the host's mandibles (figure 10.4). *Dipriocampe diprioni* is a chalcidoid which parasitises eggs and plays a

Degree of abundance	Species	Family	Stage predated
Constant and dominant parasites	*Exenterus abruptorius*	Ichneumonidae	Larva
	Pleolophus basizonus	Ichneumonidae	Pupa
	Lamachus eques	Ichneumonidae	Larva
	Lophyroplectus luteator	Ichneumonidae	Larva
Constant and subdominant parasites	*Dahlbominus fuscipennis*	Eulophidae	Pupa
	Exenterus amictorius	Ichneumonidae	Larva
	Drino inconspicua	Tachinidae	Larva
	Achrysocharella ruforum	Eulophidae	Egg
	Dipriocampe diprioni	Tetracampidae	Egg
Local and scarce parasites	*Synomelix scutulata*	Ichneumonidae	Larva
	Exenterus adspersus	Ichneumonidae	Larva
	Agrothereutes adustus	Ichneumonidae	Pupa

Table 10.2.   *Main species in* Neodiprion sertifer's *parasite complex in central Europe (Pschorn-Walcher, 1967).*

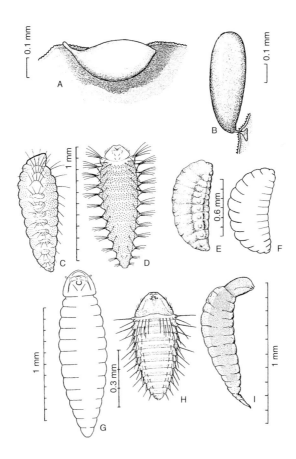

Figure 10.4. *Eggs and larvae of ichneumonids that parasitise diprionid sawflies. A: egg of* Exenterus abruptorius *deposited on the surface of the tegument. B: egg of* Exenterus tricolor *furnished with a peduncle attaching it to its host's tegument. C to F:* Exenterus abruptorius, *primary larva in lateral and dorsal views, old eonymph larva. G: primary larva of* Microcryptus basizonus. *H: primary larva of* Exenterus tricolor. *I: primary larva of* Holocremnus ratzeburgi.

major role directly by killing the eggs, and indirectly by reducing the number of larvae obtained from parasitised clutches. Moreover, feeding behaviour, which is regulated by group effect, is disrupted and mortality increases.

2. BEETLES (COLEOPTERA)

Beetles that eat forest tree leaves belong to three main families or super-families: Curculionidae, Chrysomelidae and Scarabaeoidea. Species are numerous but not usually abundant.

2.1. Leaf beetles (Chrysomelidae)

Many species of chrysomelid live in forest but few of them play important roles in the ecosystem. *Cryptocephalus pini*, recognisable by its fulvous
colour and its size (between 3.5 and 4 mm), lives on pine needles in which it
cuts long grooves. Its larva lives on the roots of herbaceous plants. A few
outbreaks of this species have been recorded on young pines. Another
species, *Luperus pinicola*, measures 3 to 4 mm, has a brown head and elytra
and a yellow pronotum with a black median spot. Adults feed on pine
needles as well as the bark of young shoots, while its larvae eat the roots.
Clytra quadripunctata is a cylindrical beetle 7 to 10 mm long with a shiny
black fore-body and reddish-yellow elytra with two black spots. The larvae
live in nests of the wood ant *Formica rufa* where they feed on vegetable
refuse. Adults devour the leaves of various trees such as oak, birch and
sallow. The most important chrysomelid is *Galerucella luteola* which
measures 6 to 7 mm and is a dirty yellow with a black band on each elytron.
This species is distributed all over Europe where it feeds on the leaves of
elm and may cause serious damage. Infestations weaken trees and make
them prone to later attack by scolytids which are vectors of the fungus that
causes Dutch elm disease.

2.2. Weevils (Curculionidae)

The huge family Curculionidae includes many species that feed on tree
leaves. Larval stages of some species live in the soil where they feed on
roots. Adults climb trees to feed on the leaves. Other species live in foliage
in their larval stage and spend part of their adult life underground. A few
weevils belonging to the genera *Phyllobius* and *Polydrusus* (such as
Phyllobius argentatus, P. oblongus and *Polydrusus mollis*) are the most common defoliators of birch in southern Finland (Rousi *et al.*, 1997).

2.2.1. Rhizophagous weevils of a beech wood

About 200 species of curculionid feed on roots in Europe. In Solling
beech wood in Germany the main species are *Phyllobius argentatus,*
Polydrusus undatus, Strophosomus melanogrammus, S. capitatus and
Otiorrhynchus singularis. These weevils account for 1.7% of all insects that
emerge from the ground, or 67 individuals per square metre and per year
(Funke, 1971; Schauermann, 1977). The energetics of these five species in
Solling beech wood have been worked out. The following figures were
obtained for one plot in that forest:

– Production of imagos in kg ha^{-1} year^{-1} (dry weight): 1.910

– Production of imagos in kcal ha^{-1} year^{-1} : 10.5 x 10^3

– Respiration in kcal ha^{-1} year^{-1} : 14.6 x 10^3

2.2.2. The beech weevil, Rhynchaenus fagi

The genus *Rhynchaenus* (= *Orchestes*) includes several species whose larvae live by mining the leaves of various plants. In forest *Rhynchaenus quercus* is found on the leaves of deciduous oaks. The best known species, *R. fagi*, the beech leaf miner weevil, is very common in beech woods almost everywhere in Europe. The black adult, which is 2.5 mm long, is readily recognised by its dilated, saltatorial hind legs. It is the larva that mines leaves; its body is flattened as in all leaf-mining insect species. A single infested leaf may be host to up to three larvae and have a third of its surface devoured. In addition to this damage the adults bite holes in the leaves to feed (figure 10.5). More than fifty species of wasp parasitise this weevil. The braconid *Triaspis pallidipes* preys on pupae and the chalcid *Eulophus pectinicornis* parasitises the larvae. Various predators are known to prey on the weevil, including the anthocorid bug *Anthocoris nemoralis*, as well as the larvae of *Chrysopa* and *Hemerobius*.

2.2.3. Weevils of cork oak

Weevils that defoliate cork oak have been surveyed in Morocco. Two species may destroy a large proportion of leaves by riddling them with small holes about 0.5 mm across. These are *Coeliodes ruber* and *C. conformis*. *Coeliodes* may also destroy flower buds and impede acorn formation. *Rhynchaenus erythropus* is a leaf-mining species, like its congener *R. fagi* which lives on beech. *Brachyderes pubescens* is a larger weevil, reaching 10 mm, which eats oak leaves but may also turn to *Eucalyptus*.

2.2.4. Leaf roller weevils

Leaf roller weevils, also known in France as "cigar rollers" are robustly built species with non-geniculate antennae and are often brilliantly coloured. They roll leaves to make cylinders or cigar-like tubes in which they deposit their eggs. Eggs are inserted in parenchyma or in leaf veins through incisions made by the female. Methods of rolling are particular to different species.

The metallic blue species *Byctiscus betulae* rolls leaves of birch, hazel, beech, alder and poplar. The entire leaf is rolled in its longitudinal axis without preliminary incisions. Adults emerge in mid May. They feed by chewing holes in leaves before mating and laying eggs in early June. The larva eats its rolled-up leaf from the inside of the tube and then pupates in the ground. *Deporaus betulae*, an entirely black species, lives on many species of broadleaf tree. The leaf is rolled after it has been cut in two by a transverse incision that leaves the midrib intact. The result is a downward opening cone suspended below the basal third or so of the leaf blade which remains intact. *Apoderus coryli* has a red body and black head and legs. This weevil uses

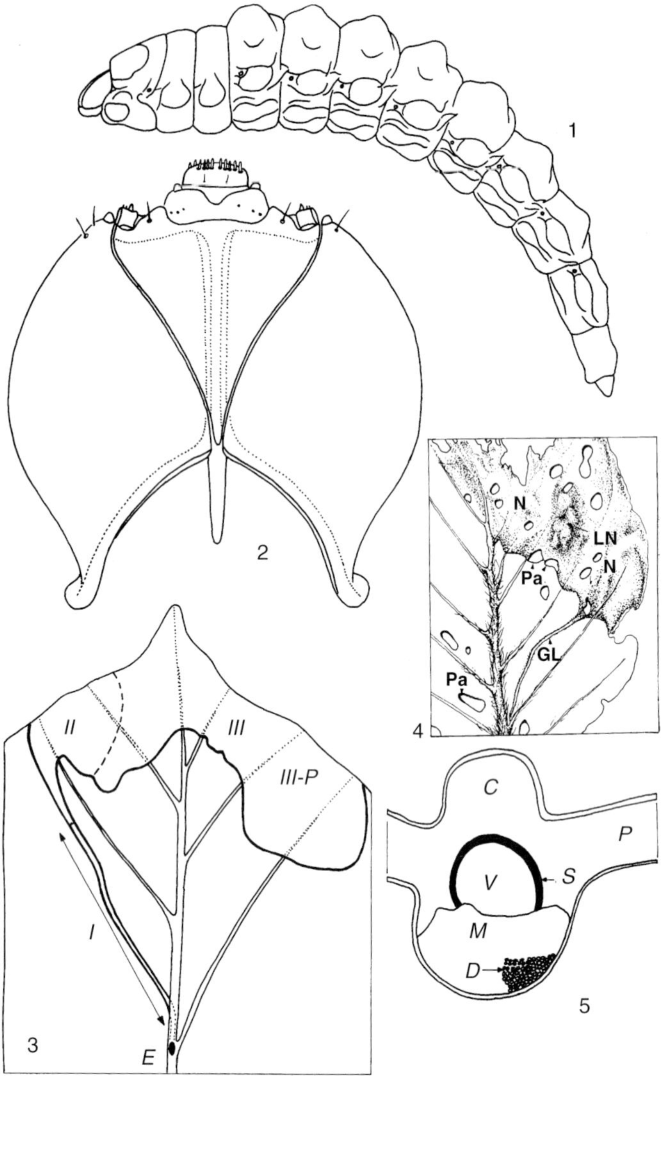

Figure 10.5. *The beech weevil* Rhynchaenus fagi. *1: Larva in lateral view.*
2: Head in dorsal view, mandibles omitted. 3: mine in a beech leaf.
The egg is in E; I, II, III and III-P indicate location of larval instars
and pupa. 4: Damage to a beech leaf. N: area suffering necrosis;
GL: larval gallery; Pa: Holes bitten by feeding adults; LN: Pupal
chamber. 5: Cross-section of a beech leaf vein. C: Collenchyma;
P: Mesophyll; S: Sclerenchyma; V: Veins; M: Larval mine;
D: Larval waste matter (Grandi, 1932, Nielsen, 1966; Chauvin et
al., 1976).

hazel, oak or beech leaves to roll the apical third of the leaf into a small cylinder after having severed it from the rest of the blade with a transverse incision that reaches the midrib. *Attelabus nitens* is red with black head, scutellum, underside and legs. This species lives on oak, beech, birch and chestnut. After having cut a leaf blade in two and softened the midrib by gnawing the entire length of the severed part, the female folds the latter longitudinally and rolls the folded blade into a cylinder which hangs perpendicular to the leaf (figure 10.6). Another species, *Lasiorhynchites sericeus*, may behave parasitically by laying its eggs in cylinders rolled by *Attelabus nitens*.

2.3. Chafers (Scarabaeidae)

A number of scarabaeid beetles related to the cockchafer (melolonthines) may defoliate trees. The common cockchafer, *Melolontha melolontha*, has now become rare. Its biology has been much studied because of the extensive

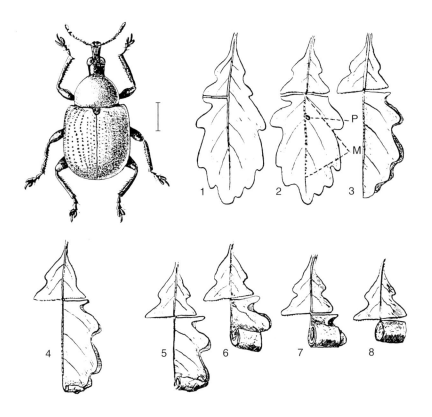

Figure 10.6. Attelabus nitens. *Adult appearance and stages in the rolling of an oak leaf by the female. P shows the location of eggs and M the part of the midrib softened by gnawing.*

damage it caused in the past. A closely related species, *Melolontha hippo-castani*, is more strictly confined to woodland where it lives on broadleaf trees and is still fairly common (Balachowsky, 1962; Paulian & Baraud, 1982). The pine chafer *Polyphylla fullo* is a larger insect, reaching 40 mm; it is dark brown, marbled with white. The male is distinguished by the last seven segments of the antennae which are much more strongly lamellate than in the female. This insect flies in June on sandy coasts, but may also occur inland up to an altitude of 1,000 metres. Adults eat the foliage and bark of pine trees; the larvae live underground on the roots of various plants. The life cycle lasts two or three years.

3. FLIES (DIPTERA)

Only one major family of defoliating Diptera need be mentioned: the cecido-myiids. Some of these flies, which belong to the suborder Nematocera, live by feeding on leaves and buds. Some cecidomyiids make galls, and others live on conifer cones.

3.1. *Agevillea abietis*

This species lives on silver fir. Adults are slender flies 3 mm long with a wingspan of 6.5 mm and usually orange-red in colour. The antennae have a characteristic structure, with knobbly segments inflated at their bases and tapering distally and furnished with verticils of long setae. Females lay their eggs between the half-opened leaf bud scales by means of their ovipositor. The young larva gradually penetrates the leaf parenchyma where it spends its whole life. The fully-grown larva is up to 3.5 mm long and possesses the T-shaped structure known as the sternal spatula that characterises the cecido-myiids. Damage to firs depends on the scale of infestation. Gnawed leaves turn yellow and fall; some branches are more or less totally stripped and atrophied terminal buds may die. Often branches infested one year are little or not at all infested the following year owing to delayed budding. Consequently, infested trees often have alternately defoliated and fully-leaved branches, the needles of the latter being more or less swollen and discoloured. This characteristic type of damage is readily recognisable from a distance. Invasions of *Agevillea abietis* are usually localised at an altitude of 300 to 500 metres. The main parasite, which plays a major part in grada-tion recession, is a wasp, *Platygaster manto*, that parasitises pupae and is host-specific to *A. abietis* (Hubault, 1945; Postner, 1962).

3.2. *Thecodiplosis brachyntera*

Larvae of this species live on young needles of Scots pine. Infested needles grow shorter and their bases swell, making them resemble galls, after which they turn yellow and fall. *T. brachyntera* may transmit the pathogenic fungus *Cenangium ferrugineum* to Scots pine. Periodic outbreaks occur in central Europe; these show a certain regularity, with intervals of 10 or 11 years, which corresponds to the cycles of other species such as *Bupalus piniarius*, *Dendrolimus pini* or *Hyloicus pinastri*. The most probable cause of outbreaks lies in coincidence of egg-laying and the growth of pine needles.

3.3. *Cecidomyia pilosa*

The eggs of this species are laid on young spruce branches. Presence of the larva is betrayed by formation of a small quantity of resin in which the larva sits, and by desiccation of the outer bud scales which turn dark brown. Resin seems to be the larva's only food. The bud that has sheltered the larva is atrophied, and needles in its vicinity fall. Repeated attacks may slow the tree's growth. In autumn the larva spins a hibernation cocoon, the shape and size of a spruce bud, which it attaches to needles (Gaumont, 1958).

REFERENCES

BALACHOWSKY A.-S., (1962). Entomologie appliquée à l'agriculture. Tome I, premier volume. *Masson, Paris.*

BILLANY D. J., (1978). *Gilpinia hercyniae*. A pest of spruce. *Forestry Commission, Forest Record*, **117**, 1-11.

BORDEN J.-H., BILLANY D.-J., BRADSHAW J.-W.-S. *et al.*, (1978). Pheromone response and sexual behaviour of *Cephalcia laricifolia* Wachlt (Hym. Pamphilidae). *Ecol. Ent.*, **3**: 13-23.

BURATTI L., GÉRI C., DELPLANQUE A., Scots pine foliage and *Diprion pini*. *In*: V. LABEYRIE, G. FABRES, D. LACHAISE (1987). Insects-Plants. *W. Junk*, Dordrecht, 379.

CHARLES P.-J., CHEVIN H., (1977). Note sur le genre *Acantholyda* (Hyménoptères-Symphytes-Pamphilidae), et plus particulièrement sur *Acantholyda hieroglyphica* (Christ.). *Rev. For. Fr.*, **29**: 22-26.

CHAUVIN G., GUEGUEN A., STRULLU D.-G., (1976). À propos d'une infestation des hêtres en Bretagne par l'*Orchestes fagi* L. (Coléoptère, Curculionide). *Rev. For. Fr.*, **38**: 343-348.

DUSSAUSOY G., (1970). Régulation des populations de *Diprion pini* L. (Hym. Tenthredinoidea) par la diapause éonymphale prolongée. *C. R. Ac. Sc.*, **270**: 1,692-1,694.

DUSSAUSOY G., GÉRI C., (1971). Étude des populations résiduelles de *Diprion pini* L. à Fontainebleau après la gradation de 1963-1964. *Ann. Sci. For.*, **28**: 297-322.

EICHHORN O., PSCHORN-WALCHER H., (1973). The parasites of the birch leaf miner sawfly *Fenusa pusilla*, Hym. Tenthredinidae in Central Europe. *Tech. Bull. C. I. C. B.*, **16**: 79-104.

FUNKE W., Food and energy turnover of leaf eating insects and their influence on primary production. *In*: H. ELLENBERG (1971). *Integrated Experimental Ecology*: 83-93.

GAUMONT R., (1958). La cécidomyie résinicole de l'épicéa (*Cecidomyiia pilosa* Bremi, 1847). *Arch. Zool. Exp. Gen.*, notes et revues, **96**: 1-26.

GÉRI C., DUSSAUSOY G., (1966). Étude d'une population de *Diprion pini* (Hym. Symphytes) en forêt de Fontainebleau. *Ann. Soc. Ent. Fr.*, **2**: 535-548.

GRANDI G., (1932). La morfologia delle larve endofite di due Coleotteri Curculionidi. *Bull. Lab. Ent. Bologna*, **5**: 93-103.

GRIMM R., (1973). Zum Energieumsatz phytophager Insekten im Buchenwald. I. Untersuchungen an Populationen des Russelkäfer (Curculionidae) *Rhynchaenus fagi* L., *Strophosomus* (Schönherr) und *Otiorrhynchus singularis* L. *Œcologia*, **11**: 187-252.

GRISON P., JACQUIOT C., (1964). Observations sur une invasion de *Diprion pini* en forêt de Fontainebleau. *C. R. Ac. Agr. Fr.*, 998-1 006.

HUBAULT E., (1945). Un parasite non encore signalé des aiguilles du sapin blanc (*Abies alba* Mill.). *Bull. Biol. France Belgique*, **19**: 17-30.

LOVIS C., (1975). Contribution à l'étude des Tenthrèdes du mélèze (Hym. Symphytes) en relation avec l'évolution dynamique des populations de *Zeiraphera diniana* Guénée (Lep. Tortricidae) en Haute Engadine. *Bull. Soc. Ent. Suisse*, **148**: 181-192.

NIELSEN B.-O., (1966). Studies on the fauna of beech foliage. 1. Contributions to the biology of the early stages of the beech weevil *Rhynchaenus (Orchestes) fagi* (Coleoptera Curculionidae). *Natura Jutlandica*, **12**: 162-181.

NIEMELÄ P., TOUMI J., MANNILA R., OJALA P., (1984). The effect of previous damage on the quality of Scots pine foliage as food for Diprionid sawflies. *Zeit. ang. Ent.*, **98**: 33-43.

PAULIAN R., BARAUD J., (1982). *Lucanoidea et Scarabaeoidea*. Faune des Coléoptères de France. Lechevalier, Paris.

POSTNER M., (1962). Zur Lebensweise und Bekämpfung der Larchenknospengallmücke *Dasyneura laricis* F Lw. (Cecid. Dipt.). *Anz. fur Schadlingkunde*, **35**: 169-174.

PSCHORN-WALCHER H., (1967). Biology of the Ichneumonid parasites of *Neodiprion sertifer* Geoffr. (Hym. Diprionidae) in Europe. *Tech. Bull. Commonwealth Biological Control*, **8**: 7-51.

PSCHORN-WALCHER H., ZINNERT K.-D., (1971). Zur Larvalsystematik, Verbreitung und Okologie der europaischer Lärchen-Blattenwespen. *Zeit. ang. Ent.*, **68**: 345-366.

ROUSI M., TAHVANAINEN J., HRENTTONEN J. *et al.*, (1997). Clonal variation in susceptibility of white birch (*Betula* spp.) to mammalian and insect herbivores. *Forest Science*, **43**: 396-402.

SCHAUERMANN J., (1977). Energy metabolism in rhizophagous insects and their role in ecosystems. *Ecol. Bull.*, **25**: 310-319.

SCHÖNHERR J., WEIK T., LANGGUTH C., REULECKE T., (1979). Factors influencing the female sex attractiveness in pine sawflies *Diprion pini* and *Neodiprion sertifer* (Hym. Diprionidae). *Bull. Soc. Ent. Suisse*, **52**: 217-222.

WALLACE D. R., SULLIVAN C. R., (1975). Effects of daylength over complete and a partial succeeding generation on development in *Neodiprion sertifer* (Hym. Diprionidae). *Ent. Exp. Apl.*, **18**: 399-411.

11

Sap-suckers: scale insects, aphids and bugs

Sap-sucking insects have mouthparts forming a rostrum adapted to piercing and sucking (figure 11.1). They belong to three orders:

– Homoptera. This suborder includes a great many pest species. Homopteran bugs fall into one of two main divisions, or series, according to the position of the rostrum. The Auchenorrhyncha are characterised by a more or less elongated rostrum which is borne on the posterior part of the head, and short antennae. They include cicadas, which are large insects, and leafhoppers, which are small insects belonging to several families, the main ones being Cercopidae, Jassidae and Membracidae. The Sternorrhyncha are much more numerous; they have a short rostrum borne further back, near the anterior coxae, and long antennae. They include aphids, scale insects, psyllids, aleyrodids, and chermesids.

– Heteroptera. True bugs have few major representatives in forests.

– Thysanoptera. These insects, also known as thrips, are mostly phytophagous. The only forest species of any importance is *Taeniothrips laricivorus*, which is mainly distributed in central Europe. Its bites lead to the death of terminal buds in larch and vegetative growth of buds situated lower down.

1. SCALE INSECTS

Scale insects, or coccids (= superfamily Coccoidea) are minute insects, many under one millimetre long. The life cycle is ametabolous (i.e. without metamorphoses) in the case of female larvae, whereas male larvae grow in the same way as holometabolous insects (those with metamorphoses). Female scale insects are always wingless; the larvae undergo little change and few moults (two in most cases). Male larvae always undergo one or two more moults than the females, and following their second moult resemble nymphs with the beginnings of wing development. Male scale insects have a single pair of wings and atrophied mouthparts. They are far less modified than the females. Three families of scale insects are recognised. Margarodids are primitive insects which retain compound eyes. The most familiar example is the orange tree scale insect *Iceria purchasi*. Females of some species of the family Lecanoidae have sack-shaped bodies covered in wax or lacquer. Diaspidids form the family with the greatest number of species. These are characterised by a wax shield that covers the female's body, hence the name "armoured scales" (Rosen, 1990).

1.1. Some common scale insects

The most spectacular scale insects of broadleaf trees are the cochineal insects belonging to the genus *Kermes,* which almost all live on oaks. The globular adults reach a length of 5 mm. Their tegument is hard, distended, opaque and brittle, and impregnated with lacquer. Legs and antennae are obsolescent or absent. They range in colour from brownish-yellow or reddish-brown to black. *Kermes* were used in ancient Egypt as a source of crimson dye for royal apparel. *Kermes ilicis*, which lives almost exclusively on evergreen oak and is found in southern France, Italy, Spain and North Africa, was formerly harvested in the South of France for the manufacture of kermes dye (Balachowsky, 1950).

The genus *Asterolecanium* is represented by several species which live on oaks. *A. variolosum* appears as a small oval brown or yellow scale 1.5 mm long under which the insect shelters. *Physokermes piceae* lives on spruce. Females have a globular reddish-brown shield 3 to 6 mm long; males are rare. This scale insect produces a honeydew in which the *Fumago* fungus grows and which attracts ants, bees and wasps.

Pines are hosts to various species of *Leucaspis* whose females are sheltered by a shield shaped like a minute mussel shell 2 to 3 mm long. The larvae settle on the inner face of pine needles which turn yellow and fall. The pine needle scale, *Leucaspis pini*, has a pure white shield and measures 2.2 to 2.8 mm. Both young and old individuals always settle in the fold of a needle which sometimes makes them difficult to see. Females lay in July. Infested needles turn yellow and are shed prematurely. *Lepidosaphes ulmi*, the oyster-shell scale, is also shell-shaped (figure 11.1). This species lives on many trees including elm and beech, and may settle on smaller branches as well as on petioles of on leaf midribs. Toxic substances in this species' saliva cause necrosis in leaves, which begins on the blade margins and gradually spreads to the midrib (De Groat, 1967).

The two most notoriously harmful scale insects are *Matsucoccus feytaudi*, which lives on maritime pine, and *Cryptococcus fagisuga*, which lives on beech.

1.2. *Matsucoccus feytaudi* and maritime pine wilt

Maritime pine in Provence was subject to serious withering and death, the first major manifestations of which were reported in the Bormes and

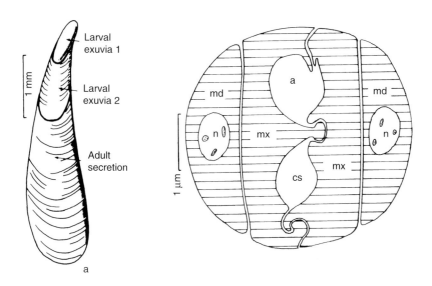

Figure 11.1. *Left:* Lepidosaphes ulmi, *female shield. Right: schematic cross-section of a scale insect's rostrum. The rostrum, which is circular in cross-section, is composed of four fine elongate styles a few µm in diameter. On the outside are two mandibular styles (md), which are modified mandibles, and on the inside two maxillary styles (mx), which are modified maxillae. Inside each mandibular style a cavity contains three neurones (n) which connect with mechano-receptors. The maxillary styles are joined by close coaptations and delimit a salivary duct (cs) and an alimentary canal (a).*

Saint Tropez region in 1957. This affliction spread rapidly since by 1972 damage extended to 120,000 hectares, and more than 100 million trees were destroyed or had to be exploited prematurely. By 1973, trees in the Alpes-Maritimes were affected, and those in Liguria and the Apennines threatened, as well as those in other regions such as Corsica. Only maritime pine was infested, other pines in the region remaining unharmed. The first signs of wilt consist of more or less heavy exudations of resin. Next, foliage on branches turns yellow; this gradually spreads to the whole tree which defoliates and dies. This process lasts from a few months in young trees to several years in old ones.

1.2.1. Causes of wilting

The causes of wilt in maritime pine were discovered gradually. An early hypothesis held that trees enfeebled by the severe cold weather in the winter of 1956 had been easy prey to pests that exploit weakness, such as the scolytid *Tomicus piniperda*. Then in 1961 the discovery of the scale insect *Matsucoccus feytaudi* led to the conclusion that this insect bore "the entire responsibility for the wilting phenomena" (Carle, 1971). The scale insect is the primary pest, and the first link in a vast complex of secondary pests that follow in its wake. The fundamental role of *M. feytaudi*, which had been accidentally introduced from the Landes (SW France), is now established. This however does not exclude other factors, for otherwise it would be difficult to understand why damage does not occur in the Landes and in other areas in which the scale is endemic. There would appear to be differences of a genetic nature between the bark structure of maritime pines in the Landes and those in Provence.

1.2.2. Biology of **Matsucoccus feytaudi**

This scale insect, in the family Margarodidae, belongs to a genus of some twenty species which all live on species of pine. *M. feytaudi*, which is host-specific on maritime pine, has one generation a year. It is known to occur throughout South-west France as far North as the Vendée, in a large part of Spain, in northern Portugal and in Morocco. These populations are endemic. It is only in South-east France, where the species is introduced, that outbreaks are seen. Young larvae may settle on several species of pine, but they die in their second instar unless they are on maritime pine.

Eggs produce mobile hexapod larvae (first instar larvae, or L_1) which settle by implanting their piercing styles two days at the latest after hatching. Several months later they moult and final instar larvae, or L_2 larvae, legless and sack-shaped, appear. Females issue directly from these, whereas males go through two further stages, a mobile hexapod pronymph stage, and a nymphal stage, the latter being spent in a whitish ovoid cocoon. Adult males are winged. Earliest emergence dates in the Maures and Esterel region of Provence are as follows. Emergence of adult males and females: first half of February; beginning of hatching of L_1 larvae: end of March to early April; moulting of L_2 larvae: early September; emergence of male pronymphs: early December; emergence of male nymphs: early January.

Mating takes place soon after adults emerge. Females lay about 300 eggs contained in a silk ootheca. Larval stage L_1 lasts 140 days in normal conditions. High temperatures in July and August trigger aestivation towards the end of stage L_1, and temperatures between 10 and 15 °C are necessary to allow transition to stage L_2. *Matsucoccus feytaudi* mainly implants itself on tree trunks and large limbs where bark is sufficiently rough and fissured for the insect's styles to reach the phloem. Settling is helped by pronounced thigmotaxis in larvae, and in females, which prefer to lay their eggs in the numerous fissures of maritime pine bark. Dispersal is ensured by non-settled forms, and mainly by newly-hatched larvae which can survive wind-borne dispersal and travel 5 to 7 kilometres in the case of the Maures - Esterel woods. A large proportion of larvae settle in the vicinity of the tree in which they hatched, with a tendency to climb upward, while those that arrive by air and land on needles or small branches tend to move downward towards the trunk.

1.2.3. Factors regulating populations

These factors may be climatic or biotic. Climate may have an effect in the form of dew condensing, or rain trickling, on hatching eggs. In the Landes, 90% of young larvae do not find their way out of the ootheca after hatching and die there. In the Maures nearly all larvae leave the ootheca and manage to implant. Climate may also affect bark structure, which is smoother and less fissured in maritime pines in the Landes, thus providing the insects with fewer suitable spots in which to implant. The essential factor regulating populations of *M. feytaudi* thus lies in high mortality of larvae in their early stages due to unfavourable climatic factors; this in itself is enough to explain differences observed in populations of the Landes and of Provence.

Predators appear to play only a minor role, or at any rate to act too late to halt the scale insect in its period of expansion. The main predator is an anthocorid bug, *Elatophilus nigricornis* (figure 11.2).

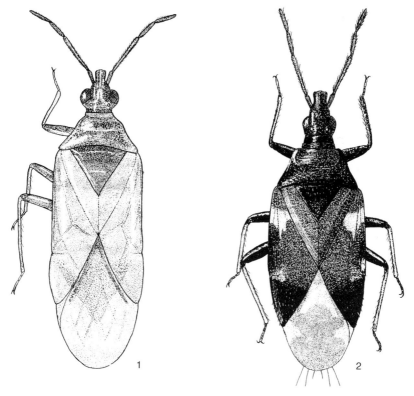

Figure 11.2. *Two anthocorid Hemiptera which predate scale insects. Left,* Elatophilus nigricornis. *Right,* Temnostethus pusillus. Elatophilus nigricornis *lives on various species of pine. The female inserts her eggs in pine needles. Both larvae and adults are predators of scale insects.* Temnostethus pusillus *lives among lichens on the branches of many trees, but mainly broadleaf trees. Anthocorids predate many insects, but especially Homoptera.*

Elatophilus however is unable to stem growth of the scale insect's population, because population growth of the two species does not coincide. *E. nigricornis* only becomes abundant several years after an infestation, when the damage is already done. Other, secondary or occasional predators are various ladybirds, lacewings, mites, and perhaps birds: in the case of *Matsucoccus pini* which lives on Scots pine in Britain it has been shown that tits may feed on the mobile stages of the scale insect.

A certain auto-regulation of *M. feytaudi* populations takes place. The maximum level of abundance, which is about ten L_2 larvae per square centimetre of fissured trunk, is not maintained for long. Rapid decrease in abundance occurs and populations stabilise at a level of around 0.0001 individuals per square centimetre, or 10,000 times less than the maximum. This phenomenon is largely due to lesions on the tree caused by perforations made by the insects, for necrosis renders tissue unfit for implanting by the insect for a certain length of time. This leads to competition for space and food among the larvae, which explains why populations decline.

1.2.4. Harmfulness of Matsucoccus feytaudi

The harm this insect causes is due to its method of feeding by piercing and sucking. When the scales are few, isolated perforations only produce benign lesions in phloem. The harm threshold is estimated to be reached, in Provence, when there are 6 to 8 L_1 larvae per square centimetre of fissures on the trunk. At this point minimum reactions can be seen in the form of light exudations of resin. At the L_2 stage harm becomes apparent at a density of two larvae per square centimetre. In years of prolonged drought (as in 1970), one sees renewed outbreaks of wilt and cases of mortality in maritime pine; the opposite effect is produced by cool humid summers like that of 1972. A tree's characteristics also affect its sensitivity to attack by the scale insect. Lesions develop more quickly when the phloem layer is reduced; such is the case in the main terminal shoot and at the bases of larger branches in young trees which are, for this reason, rapidly infested by another pest, the weevil *Pissodes notatus*. This explains why populations of young trees are rapidly exterminated in Provence. Similarly, wounds on bark, trauma of all kinds, and malformations caused by cankers produced by the ascomycete fungus *Calliciopsis pinea* all favour infestation by the scale insect.

Lesions caused by *M. feytaudi* bites are characterised by inhibition of growth of tissue situated under the larva, and, further down, hyperplasia of cells touched by the insect's rostrum. Unlike most Homoptera, which leave traces of the styles' sheath, *M. feytaudi* leaves few traces. Premature ageing of surface cells transforms them into cork. At this stage damage is reversible thanks to normal shedding of rhytidome and to the activity of the cambium. It is only when intensive contamination continues into the second year that secondary lesions are formed, together with reduction in woody tissue and slowing of growth. Hypertrophied cells lose their lipid content and become filled with resinous substances which are released outside or accumulate in resin pockets. This change in cells towards patholgical production of resin is characteristic of perforations by various species of *Matsucoccus*. The cause of these changes is certainly due to a toxic agent in the scale insect's saliva; the symptoms can be reproduced by applying a preparation of mashed scale insect bodies to a pine tree. The active substance, which has not yet been isolated, is perhaps an amino acid. Extensive lesions of pine branches blocks conduction in the phloem and causes the needles to turn yellow. Simultaneously, growth and lignification of wood cells ceases, leading to reduced growth of needles and formation of points of least resistance and to branches being broken off by wind (Carle *et al.*, 1970).

1.2.5. Insect successions following infestation by **Matsucoccus feytaudi**

Wilting of maritime pine in Provence was followed by heavy increases in numbers of xylophagous pests that exploit weakness, most of which are beetles. More than seventy species were counted, of which about ten play a major part. The main species are as follows:

Coleoptera	Curculionids	*Pissodes notatus*
	Scolytids	*Tomicus destruens; Orthotomicus erosus;*
		Pityogenes bidentatus; Pityoceragenes calcaratus;
		Hylastes sp.; *Hylurgus* sp.; *Crypturgus* sp.
	Cerambycids	*Monochamus galloprovincialis;*
		Acanthocinus aedilis; Rhagium sp.
	Buprestids	*Phaenops cyanea; Buprestis* sp.
Hymenoptera	Siricids	*Paururus juvencus*
Lepidoptera	Phycitids	*Dioryctria splendidella*

Attacks by xylophagous insects usually occur when the scale insect's population reaches a level at which their perforations induce subcortical lesions. Invasions by these secondary pests most often occur in the following order:

– *Pissodes notatus* and *Tomicus destruens* arrive more or less simultaneously during the first two years of wilting.

– Then, in areas in which wilting begins to be general, *Tomicus destruens* disappears, to be almost entirely replaced by *Pissodes notatus*, whose population maintains itself for five or six years. *Orthotomicus erosus* invades; *Phaenops cyanea* and *Arhopalus syriacus* may in some cases play a major part. In coastal areas *Monochamus galloprovincialis* may replace the two latter species on young trees a few years later.

– After ten years, in residual areas of invasion, populations of *Pissodes notatus* decline sharply and those of *Orthotomicus erosus* rise to finish off most of the stricken trees.

1.2.6. Biology of some secondary pests

– *Pissodes notatus. Pissodes* species, which belong to the family Curculionidae, are black or brown weevils covered with scales that form patterns of yellow or white spots. Their larvae live under conifer bark, with the exception of those of *Pissodes validirostris*, which live in pine cones. The commonest species in Europe is *P. notatus*, which lives on various pines (figure 11.3). This weevil infests weakened trees by settling on the collars of

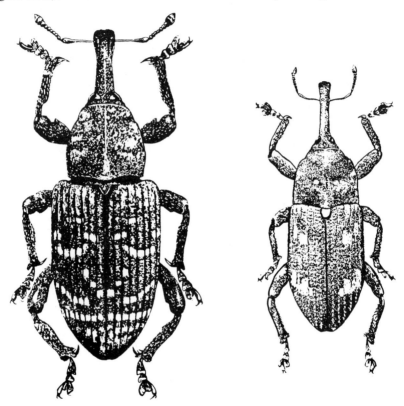

Figure 11.3. *Two curculionid pests of pines. Left,* Hylobius abietis. *Right,* Pissodes notatus. *The antennae of* Pissodes *are inserted near the middle of the rostrum, whereas those of* Hylobius *are inserted near the tip, which makes the two genera readily distinguishable.*

young trees or on the trunks of older trees. It also invades trees infested by *M. feytaudi* (Carle, 1968). This beetle sometimes acts as primary pest. In such cases it prefers to lay eggs in that area of the trunk where the bark is between 0.5 and 0.8 cm thick, an area which thus varies in height with the age of the tree (Alauzet, 1973). In Provence it is invariably the first exploiter of weakness to appear on the scene of stricken maritime pines. Its effect is crucial, for its larvae accelerate the withering of trees irreversibly, and open the way to increased attacks by other xylophagous insects. The behaviour of *Pissodes notatus* following infestations by *M. feytaudi* differs from its normal behaviour. It may settle at any height on a tree trunk, but choice is determined by the density of *M. feytaudi* larvae and by the number of lesions they have caused. In August there must be at least 10 *Matsucoccus* larvae per square centimetre, and around 6.6 per square centimetre in September (Carle, 1968). It is probable that female *Pissodes* are attracted by terpene vapours (and particularly α- and β-pinene) emanating from lesions caused by *Matsucoccus*. Once they are established, *Pissodes* larvae cause secondary degradation of the tree, so that other parts of the tree poor in *Matsucoccus* in turn become attractive.

In Provence emergence of adults occurs in two different seasons, the first from May to July, the second from September to December. Young adults settle on the current year's growth of shoots and feed by gnawing bark and phloem. This feeding and maturation stage lasts until they mate. The females perforate the bark and lay one to three eggs in the cavity they have created. Each female has one or more laying cycles, each of which lasts forty days on average, and each cycle produces about 200 eggs. Larvae grow under the bark. Insects born from eggs laid in May and June emerge before winter, and those from eggs laid in July to September emerge in April of the following year.

– *Phaenops cyanea*. This blue or green, 7 to 12 mm long buprestid is distributed through the whole of Europe (except the British Isles). It lives on various species of pine. Adults emerge in late May or early June. Eggs are laid during a 40 to 60 day period, and larvae live in large tree trunks. The life cycle may last one or two years, depending on temperature.

– *Orthotomicus erosus*. This polygamous scolytid bores galleries having three or four radiating branches five to ten centimetres long, each of which is occupied by one female, while the male remains stationed in the entrance vestibule (figure 11.4). It is the male that begins by boring the vestibule, after which he is joined by females and, after mating, each of them bores a branch gallery. Swarming starts when maximum temperatures exceed 18 °C and continues more or less constantly during the entire warm season, due to overlapping of generations. There may be as many as four generations a year in the Mediterranean region. Each generation may produce up to six sibling generations in spring.

– *Tomicus destruens*. This scolytid, which is closely related to *T. piniperda*, is localised in coastal areas of Mediterranean Europe. Adult emergence begins in April and may go on for more than 80 days. Adults that fly off settle on branches grown in the current year and feed in galleries that they bore at the bases of terminal shoots. Infested branches dry out and drop off. When these adults are sexually mature they mate and lay, usually on un-

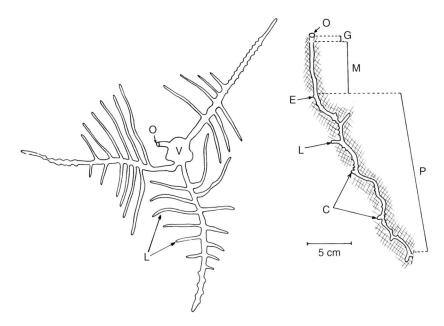

Figure 11.4. *Left, sub-cortical gallery system bored in maritime pine by*
Orthotomicus erosus. *O: entrance orifice; V: vestibule, where
mating takes place; L: Larval galleries. Right, sub-cortical galleries
bored by* Hylurgus ligniperda; *O: entrance orifice;
G: access gallery; M: maturation gallery; P: incubation gallery;
E: egg niches; L: long lateral galleries reserved for feeding males;
C: short lateral galleries used as turning space for females.*

healthy trees infested with *Pissodes* larvae. Each female bores a linear gallery
in an upward direction in the tree's axis; in it she lays an average of 70 eggs.

– *Hylurgus ligniperda.* This scolytid, which lives solely on pines, is common in maritime pine infested by *M. feytaudi*. Adults, which emerge in
spring when the temperature reaches 15 to 16 °C, may fly several kilometres
some ten metres above tree tops in search of wilting trees or areas being
felled. They then turn to tree collars and stumps, on which they settle. The
female bores a short access gallery to the cambium and then a maturation
gallery parallel to the surface and finally an incubation gallery provided with
egg niches. In the Maures forests, adult *Hylurgus ligniperda* swarm in
spring, lay, fly off again, and produce a second batch of eggs. The two
sibling generations grow practically at the same time to produce adults that
emerge in late summer and produce two further sibling generations in warm
climates or a single generation in areas where winter comes earlier (Fabre &
Carle, 1975).

– *Arhopalus syriacus.* This cerambycid lives on various pines. Adults live
only a few days and do not feed. Females lay on large trunks with bark 2 to

6 cm thick. Larvae go through 12 to 15 stages over a period of one year in areas with mild winters, or up to three years in harsher climates.

1.3. *Matsucoccus pini*

Mastucoccus pini lives on pines on which it is often associated with the chermesid *Pineus pini*. It has no known parasites; its predators are mostly those of *Pineus*, in other words mites (*Hemisarcoptes malus* and *Cheyletia flabellifera*), lacewings (chrysopids and hemerobiids), raphidids, carabids of the genus *Dromius*, ladybirds, spiders of the genus *Theridion*, etc. *M. pini* is a parthenogenetic species which has two larval stages. It occurs up to 1,400 metres on Mont Ventoux, where it infests Scots pine, Aleppo pine and mountain pine but spares stone pine and maritime pine. This scale insect is also known to occur in central Europe, from Germany to Russia. On Scots pine the female settles in bark fissures of trees at least fifteen years old. Massive attacks followed by damage have been recorded (Boratynski, 1952; Rieux, 1975; Siewniak, 1976).

1.4. *Cryptococcus fagisuga* **and beech wilt**

A disease caused by the combined effects of the beech scale insect *Cryptococcus fagisuga* and the ascomycete fungus *Nectria coccinea* produces wilting of the beeches *Fagus silvatica* in Europe and *Fagus grandifolia* in North America. Wilt has been reported in almost the whole of western Europe. In France the most heavily affected were beech woods in upper Normandy and to the north of Paris in the 1970s. This disease made it necessary to fell and remove many trees, and jeopardises the future of beech woods should it persist and spread.

Symptoms of infestation by the beech scale insect vary. A single stand of beech trees will often be seen to contain both healthy trees and heavily infested ones. Infested trees are usually older ones (between 60 and 140 years old), although trees as young as 25 years old may be stricken. Early symptoms are minute white spots, usually within fissures of the bark, which are the females' waxy secretion. When numbers increase, beech trunks may be coated with the white substance in summer and autumn. Later, blackish fluxes rich in bacteria appear, bark necroses and the scales, which are no longer able to feed, disappear. The fungus *Nectria coccinea* then becomes apparent by turning sub-cortical tissue an orange colour. Foliage becomes chlorotic, leaves fall prematurely in autumn, and budding is delayed in spring. Occasionally the fungus' peritheciae, looking like little bright red balls, develop on dead bark. Deterioration is compounded by invasion by other lignivorous fungi, peeling bark, and attacks by other xylophagous insects whose galleries mine the wood.

1.4.1. Biology of Cryptococcus fagisuga

The genus *Cryptococcus* includes five species distributed worldwide. Males and alate forms are unknown in *Cryptococcus fagisuga*, which produces a single generation a year. The female lays up to fifty eggs in June or July and then dies. Eggs are deposited in groups of 5 to 8 and coated with a white, waxy secretion. Young larvae, which are the only mobile stage, hatch in September. They are orange-yellow, in size about 0.3 x 0.16 mm, have three pairs of well-developed legs and 5-segmented antennae. They feed by piercing the bark with their styles. Before attaching themselves they may colonise other trees by means of wind-borne dispersal over distances of more than a hundred metres (cf. chapter 2, figure 2.5). Rain on tree trunks washes away many larvae. Survivors that overwinter moult a second time in April. They turn into larger, legless larvae which moult again in May to produce female adults which are scarcely differentiated from the larvae. Their styles reach a length of about 2 mm, or about twice body length. The adult female, which may live for three months, is lemon-yellow, measures 0.5 to 1 mm and has rudimentary legs. She shelters under a waxy shield secreted by numerous glands on the dorsal surface. Females die immediately following egg-laying.

Dispersal of mobile larvae is mainly determined by prevailing winds; the stronger the wind, the further larvae may travel. Larval density diminishes rapidly away from 'source' trees. Maximum dispersal distance may exceed 200 metres, but most larvae are carried to trees within a radius of twenty metres. This is why heavily infested trees are found in small groups (Augustin, 1986).

1.4.2. Perforations made by Cryptococcus fagisuga and their effect on beech

The scale insect's rostrum penetrates the phelloderm and may reach the phloem. It is probable that, as in many scale insects, the saliva contains phytotoxic substances that damage affected cells. There are sclerified phloem rays in beech that penetrate the bark and give it a rigid structure. Rhytidome is lacking, and only the periderm protects the bark (figure 11.5). Following bites by scale insects a periderm interrupted by the phloem rays is formed as well as fissures caused by growth in diameter. This allows necrosis to spread. Invasion by *Nectria coccinea*, then by other lignivorous fungi, is made possible by the fissure and necrosed areas (figure 11.6).

Beech trees infested by the scale have a defence reaction which consists of raised levels of tannins of the procyanidin group, and in changes in metabolism (Lunderstädt & Eisenwiener, 1989). Wilt is due to combined effects of the fungus and the scale insect, but exceptional climatic conditions, such as very dry summers or very cold winters, must be taken into account to

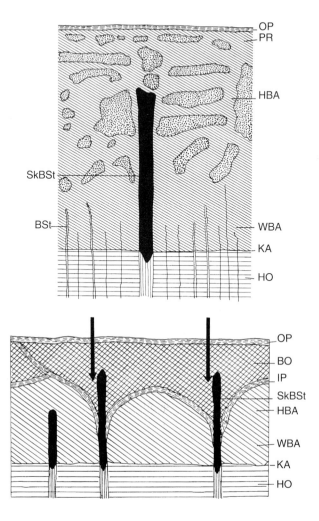

Figure 11.5.

Above, structure of bark 5 mm thick on a 30 year old beech tree. Below, schematic illustration of changes in the bark caused by a lesion and leading to formation of pathological bark. A layer of scar tissue IP, or internal periderm, is formed, broken by the sclerified phloem rays. Bark fissures occur at the points indicated by black arrows as a result of tension produced by growth in thickness. OP: external periderm; PR: primary bark; HBA: parenchyma; WBA: phloem; KA: cambium; SkBSt: sclerified phloem ray; HO: wood; BO: pathological bark. Sclerenchyma is indicated by dashes (Brauns, 1976).

explain the advent of necrosis of the bark. According to this hypothesis the fungus and the scale are merely secondary organisms taking advantage of these climatic conditions (Zycha, 1960). A dry summer causes an imbalance in the tree's water levels, and lack of rainfall to wash the trunk enables many larvae to settle. In Denmark an invasion of *Cryptococcus fagisuga* in 1930 coincided with a succession of hot summers, and its regression was attributed to cold winters from 1939 on. An identical hypothesis may explain the outbreak of the scale in Normandy (Malphettes, 1977). This notion is supported by the fact that in Lyons forest (Normandy) scale insects settle mainly on the eastern sides of trunks, that is on the sides sheltered from prevailing winds which carry rain that washes the trunks. According to Kunkel (1968) survival of the insect is linked to the possibilities of implanting its rostrum in areas where it can reach living tissue, in other words to bark thickness. Bark is thinnest towards the base of the trunk. This hypothesis

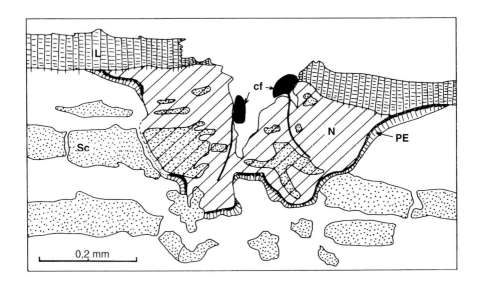

Figure 11.6. *Schematic section of beech bark showing necrosis of tissues caused by perforations made by* Cryptococcus fagisuga.
L: cork; Sc: sclerenchyma; N: necrosed tissue (oblique hatching); PE: newly-formed periderm; cf: scale insects (Kunkel, 1968).

may explain the fact that dominant, thick-barked trees in dense woodland remain unharmed. Changes in beech wood known as "T disease" may correspond to early infestation by the scale (Jacquiot, 1961).

Beech wilt may be detected by aerial photography and the use of emulsions giving "false colours", by which trees appear in different shades of colour according to their condition. Dead trees show up in blue, healthy trees in red, and intermediate, pinkish shades indicate deficient or badly affected trees (Perrin, 1977).

1.4.3. Fauna associated with **Cryptococcus fagisuga**

No parasitoids of this scale insect are known, and predators are few. The commonest ones are the ladybirds *Exochomus quadripustulatus* and *Chilochorus renipustulatus*. Most authors agree in estimating that predators, and ladybirds in particular, are unable to control the scale's population. They may at most have an effect on the remains of heavy infestations already reduced by climatic factors. Other predators include a cecidomyiid fly belonging to the genus *Lestodiplosis*, the anthocorid bug *Temnostethus gracilis*, the lacewings *Chrysopa flavifrons* and *Sympherobius elegans*, and the larvae of an earwig of the genus *Forficula* (Baylac, 1980). Stricken trees are invaded by secondary destroyers. These include lignivorous fungi such as *Hypoxylon fragiforme*, *Stereum hirsutum*, *Fomes fomentarius*, *Pleurotus ostreatus* and

Armillaria mellea. Xylophagous insects also attack moribund trees. One of the commonest of these is the scolytid *Trypodendron domesticum*, which is very common in beech woods in the Paris basin, and carries spores of the fungus *Bjerkandera adusta* which causes white rot. *Hylecoetus dermestoides* invades next and bores its larval galleries right to the heart of the tree (Thomsen *et al.*, 1949). The insect successions that follow this are those normally associated with dead beech.

2. APHIDS

Aphids ("greenfly", "plant lice" etc.) are remarkable for their large number of species and for their many biological peculiarities, such as complex life cycles, often involving alternate sexual and parthenogenic generations, and frequent dependence on two host plants. They are of great economic importance owing to the many outbreaks that occur on cultivated plants as well as on trees.

Aphids belong to one of four families: aphidids, or aphids proper, (cinarids are often isolated in a family of their own); eriosomatids or pemphigids, which include gall aphids (cf. chapter 12) and woolly aphids; phylloxerids; and chermesids (or adelgids) which are gall aphids of conifers (cf. chapter 12). Aphids bear a pair of more or less strongly developed cylindrical appendages, called cornicles, on the sixth abdominal tergite, the function of which is poorly known (figure 11.7). In cinarids the cornicles are minute, whereas in other aphids they may be longer than the body itself. A valve situated at the tip of the cornicle and moved by muscles controls emission of substances which these organs secrete and whose function is little known. In many species these substances act as defences against predators. An alarm pheromone emitted by molested individuals has been shown to exist in a few species (Dixon & Stewart, 1975). The alarm substance contains triglycerides which "glue" and may immobilise an enemy, as well as *trans-β*-farnesene which incites neighbouring aphids to detach themselves from the plant and fall to the ground. The alarm substance may also alert ants which live symbiotically with the aphids and come to their rescue by attacking the enemy. In such cases the aphids do not disperse but remain fairly grouped (Nault *et al.*, 1976).

Aphids release honeydew from their anuses. When aphids feed on elaborated sap that circulates in phloem, they absorb a complex mixture of sugars, amino acids, starches, proteins, mineral salts and vitamins. This mixture is modified as it passes through the digestive tract, as some of its constituents are absorbed and others are added. Nitrogen compounds form 0.2 to 1.8% of honeydew (dry weight) and 70 to 95% of these nitrogen compounds are amino acids and starches. Most (90 to 95%) of the honeydew consists of

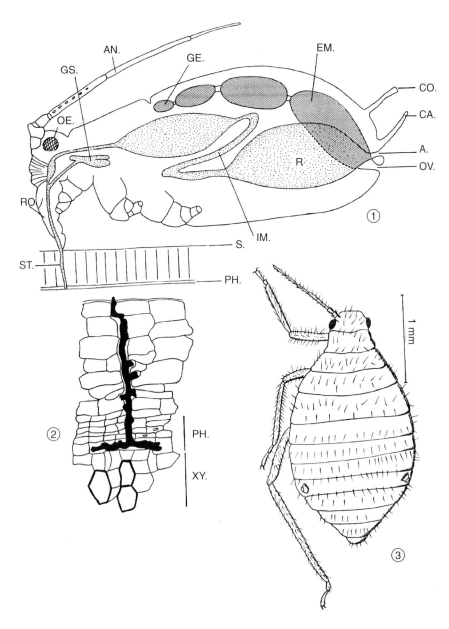

Figure 11.7. *Above (1), schematic illustration of a feeding aphid. Below (2), section of a plant showing the path taken by the insects' styles marked by residual sheaths, and (3) appearance of the aphid* Schizolachnus tomentosus, *which lives on pines. A: anus; AN: antenna; CA: cauda; CO: cornicle; EM: embryo; GE: germarium; GS: salivary gland; IM: middle intestine; OE: eye; OV: oviduct; PH: phloem; R: rectum; RO: rostrum; S: surface of plant; ST: styles of rostrum; XY: xylem.*

sugars such as glucose, saccharose, fructose, trehalose and various polysaccharides. The amount of honeydew produced by an aphid may be considerable. In *Tuberolachnus salignis* it equals 133% of the insect's body weight every hour. Honeydew is usually ejected in the form of a pendent droplet; some aphids such as *Drepanosiphum platanoides* eject it in the form of a vaporised spray. Honeydew is sought by honey-producing Hymenoptera, and particularly by bees, which can use it to increase honey production considerably. Honeydew also increases fertility in useful parasitoid wasps. Honeydew may however also drip onto leaves and dry out leaving a characteristic shiny coating which burns the leaves by dehydrating cells owing to its high osmotic pressure. Fungi take advantage of this nutritious medium and their mycelia cover the surface with a sooty coating called fumago. Honeydew attracts ants which consume it avidly and protect the aphids that produce it. A true symbiotic relationship exists between certain aphids and ants, the commonest of which belong to the genera *Lasius*, *Dendrolasius* and *Formica*. The most remarkable example is without doubt that of the aphids *Stomaphis quercus* and *Stomaphis longirostris* which live under the bark of oaks and whose colonies the ants protect by constructing domes of dried mud. Oviparous females and founders of *Stomaphis* and *Dryaphis* species are cared for by ants which carry them to safety on tree roots; in spring they move the young founding females to the upper parts of the tree.

Symbiotic relationships between ants and aphids, and honeydew exchange, are accompanied by characteristic behaviour which has been described by various authors (Hölldobler & Wilson, 1990). It was known that ants from a same nest exchange food, a phenomenon known as trophallaxis. The offer of a drop of honeydew by the aphid *Lachnus roboris* is preceded by an exchange of mechanical signals between the aphid and the ant *Formica polyctena*. The aphid raises and waves its hind legs before releasing the honeydew. This posture makes the hind part of the aphid's body resemble the head of a worker ant about to exchange food, the cornicles simulating open mandibles and the hind legs simulating the movements of an ant's antennae.

The major forest aphids are cynarids many of which live on bark. This family includes some large (up to 8 mm long) monophagous (host-specific) species. *Cinara*, mostly dark or black species, and often covered with a waxy coat, live on conifers. They are to be found in large colonies on smaller branches. Massive infestations cause foliage to turn yellow, especially in young trees. The abundant honeydew they produce attracts ants such as *Formica* and *Camponotus* spp. The grey pine aphid *Cinara pinea* measures 3 to 5 mm in length. It mainly colonises the bases of the current year's growth of needles, as well as twigs. Females lay their eggs in rows on needles in autumn. Hibernation takes place at the egg stage. On Arolla pine, *Cinara cembrae* may weaken the tree and make it prone to attack by wood-eating *Pissodes* weevils. *Cinara pini* is a large bronze-coloured aphid that lives on Scots pine, colonising the bark of young shoots in spring, then moving to older branches. This species, which is a big honeydew producer, is not known to harm trees, and it even appears that its bites stimulate the tree's growth. *Cinara cedri*, which lives on cedar, originated in Morocco and has been introduced in cedar plantations in the South of France, where it may become a pest.

Schizolachnus pineti, sometimes known as the "grey needle aphid" is common on pines throughout Europe. It is recognisable by the long fine pubescence that covers the whole body, including appendages. Colonies of this aphid appear in spring. Individuals prefer to group themselves on needles of the current or of the preceding year's growth. The reasons for forming groups are poorly understood; it may be a method of searching for the best feeding places, interaction between individuals, or simply the result of inertia of insects that do not move far from the location in which they hatch before settling. The main predators of *Schizolachnus pineti* are syrphids. Analysis of predation shows that the number of aphids which escape the syrphids is proportional to the number of aphids grouped in a colony. Grouping is thus an effective form of protection against predators. This phenomenon is also known in larvae of the sawfly *Diprion swainei* (cf. chapter 3.3).

The aphid *Lachnus roboris* is common on oaks, on which it forms large colonies and produces abundant honeydew. *Stomaphis quercus*, which also lives on oak, is remarkable for the length of its rostrum, which is greater than that of the body (figure 12.12).

2.1. The woolly beech aphid, *Phyllaphis fagi*

The aphid *Phyllaphis fagi*, which lives exclusively on the genus *Fagus*, occurs in Europe and in North America. It is characterised by an abundant waxy coating which gives the leaves a bluish or glaucous hue. Its life cycle, as in all aphids, is complex, and may serve as a general example. The cycle includes morphologically distinct generations produced by parthenogenesis with the exception of one, the sexual generation. Winter eggs, which are deposited in the crowns of old beech trees produce wingless founding females in spring. These produce the first generation of viriginparous individuals whose adults form colonies of a woolly white colour on the undersides of young beech leaves in the lower strata of the crown. The progeny of the virginiparous generation are winged females which return to the tree crowns in June. In September a sexuparous generation appears which gives birth to winged males and wingless females. These sexual individuals produce the fertilised winter eggs which are laid in bark fissures of branches at the tops of beeches.

Variation in abundance of this species throughout the year is considerable. As in other aphids, this is linked to the nutritional quality of leaves. A first maximum in abundance occurs in spring when larvae hatched from winter eggs migrate to the leaves of young beeches in the undergrowth. A second maximum occurs in late September when alates have settled in the crowns of tall trees. Infestations of the aphid are serious when they are on young trees, and they may jeopardise natural regeneration. Bites destroy leaves which turn brown and desiccate. Damage may be compounded by another piercing/sucking homopteran, the beech leafhopper *Typhlocyba cruenta*; this 4 mm long yellowish-white to reddish-brown hopper belongs to the family Typhlocybidae. In order to avoid infestations by the woolly aphid, nurseries should be planted at a distance from adult trees. If insecticide treatments are applied, it should be done early in order to avoid killing the aphid's many predators (lacewings, ladybirds, syrphids) and parasitoids (braconids), which play a major part in the collapse of the aphid's population in late summer.

2.2. The spruce aphid, *Elatobium abietinum*

The green spruce aphid *Elatobium abietinum* lives chiefly on two spruces, *Picea sitchensis* and *Picea excelsa*, and also sometimes on pines (Bejer-Petersen, 1962; Scheller, 1963; Carter, 1972). It is a gregarious species, small-sized and of the colour of the needles, and thus difficult to spot. Low, shady parts of the foliage are those most heavily infested. This species' life cycle is complex, for it varies according to region. Sexual generations only appear in regions with a continental climate where overwintering takes place at the egg stage. In regions with milder winters it is parthenogenetic females that overwinter. Production of winged forms depends both on the amino acid content of needles and on the density of aphid populations (Parry, 1977).

Levels of soluble nitrogen compounds in Sitka spruce needles decrease regularly from March until August (figure 11.8). Fertility in *Elatobium abietinum* changes accordingly; it remains more or less constant and then decreases rapidly (Parry, 1974).

2.3. The sycamore aphid, *Drepanosiphum platanoides*

Drepanosiphum platanoides is a species close to *Phyllaphis fagi* that lives on *Acer pseudoplatanus*; it is unusual in that all its generations are fully winged. Its main predators and parasitoids have been listed (Dixon, 1970; Dixon & Russel, 1972) and are as follows:

– Predators: *Adalia bipunctata* (coccinellid beetle); *Anthocoris nemorum* and *A. confusus* (anthocorid bugs); *Chrysopa carnea* and *C. ciliata* (lacewings); *Syrphus vitripennis* and *S. balteatus* (syrphid flies); *Tachydromia arrogans* (empidid fly).

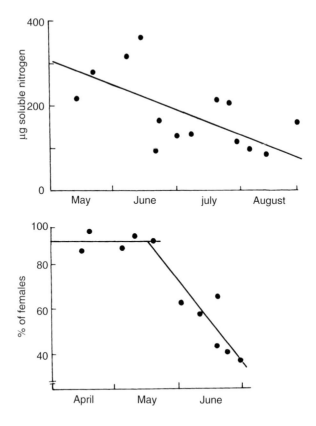

Figure 11.8. *Above, seasonal variations in levels of soluble nitrogen compounds in Sitka spruce needles. Below, seasonal variations in
fertility in the aphid* Elatobium abietinum *expressed as percentage
of females bearing embryos (Parry, 1974).*

– Parasitoid wasps: *Monoctonus pseudoplatani, Trioxys cirsii, Dyscritulus
planiceps, Aphelinus flavus.*

These natural enemies do not appear to be able to control the aphid's
population. Effects of various aphids, including *Drepanosiphum
platanoides*, on tree growth have been proved (cf. chapter 6.3).

3. OTHER HOMOPTERA

Other Homoptera belong to the division Auchenorrhyncha. The family
Cicadidae includes cicadas, which are readily recognised by their large size
and the presence of three ocelli. These insects, which are distributed in all the
warm parts of the world, are remarkable for the complex male sound apparatus situated in the first abdominal segment. *Cicada* (= *Lyristes*) *plebeja*

is a common cicada of pine woods in the Mediterranean region. The sun-loving adults are found on the bark of trees from which they suck sap; larvae live underground where they burrow with their highly modified forelegs.

Members of the family Cercopidae are commonly known as "spittle bugs" or "frog hoppers". *Cercopis* species, which measure up to 10 mm, have a black head and prothorax, wings patterned in red and black, and a black abdomen with the posterior tergites margined in red (figure 11.9). Like

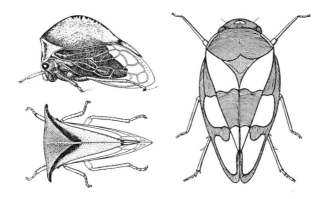

Figure 11.9. *Two homopterans. Left, male* Ceresa bubalus *in lateral and dorsal views. Right,* Cercopis vulnerata *in dorsal view.*

all spittle bugs, *Cercopis* hop vigorously and fly; they are common in oak woods. These insects, which are also called aphrophores, live in the larval stage concealed in a sort of froth called "cuckoo spit" secreted by the anus from alimentary fluids. The Saratoga spittle bug, *Aphrophora saratogensis*, is a major pest of young pines (particularly of red pine, *Pinus resinosa*) in the eastern United States, where its biology has been much studied (Ewan, 1961; McAloney & Wilson, 1971; Linnane & Osgood, 1976). This spittle bug is common on pines growing on poor, sandy soils. The insects can suck large quantities of fluid and dehydrate phloem. Their styles penetrate the cambium, and the bites produce necrosed areas and pockets filled with resin. Although the tree's cambium can regenerate, permanent blockages of sap flow remain. *Aphrophora saratogensis* feeds primarily on pine, but has a number of alternative food plants, a fern and a bramble in particular. One method of controlling the spittle bug is to avoid planting plant pines in areas where these alternative food plants exist, or to weed the latter out.

Membracids have a pronotum that usually bears bizarrely shaped out-growths. *Stictocephala bisonia* (= *Ceresa bubalus*), which originated in America, is easily recognisable by its voluminous prothorax extended into spikes at the sides and back. This bug lays its eggs under the bark of young branches of various fruit trees, lime, hazel and dogwood. The incisions it

makes impede sap flow, and open the wilting branches to attacks by xylophagous insects.

4.   TRUE BUGS (HETEROPTERA)

Heteropterous bugs do not play a large part in forest and few species are
truly harmful. The most characteristic species belong to the family
Aradidae. These are sombrely coloured, blackish-brown insects with flat
tened bodies, living under bark where they suck the mycelium of lignicolous
fungi (figure 11.10). One common species, *Aradus cinnamomeus*, has a quite

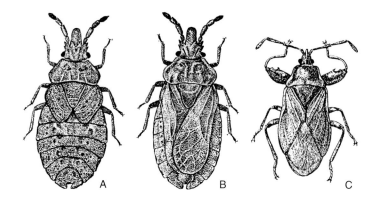

Figure 11.10. *Two species of Heteroptera.* Aradus cinnamomeus, *A: brachypterous female; B: macropterous female. C:* Gastrodes abietum.

different diet and may be harmful to Scots pine, on which it sucks the outer
bark layers and the phloem. Resistance of pine trees to attacks by this species
appears to be a function of concentrations in the tree of various terpenes such
as α-pinene, β-pinene and Δ-3 carene (Doom, 1974). *A. cinnamomeus*' life cycle lasts two years. Some individuals are normally
winged, and others are brachypterous. Damage takes the form of yellowing
and shortening of the needles, which may fall, and slowing of vertical growth.
Bark peels and the phloem develops black spots of necrosis; in some cases
the tree may be killed. Serious infestations have been recorded in central
Europe. Up to 1,000 bugs may be found on a single trunk 30 cm in diameter.
Dry soil, hot summers and sparsely-planted trees are all factors favouring
invasion by *A. cinnamomeus*. Weakened trees are more prone to attacks by
other pests such as *Pissodes*. Enemies of this bug have been found; they
include fungi of the genus *Beauveria*, predators such as mites of the genus
Allothrombidium, and the coccinellid *Scymnus suturalis*. Chemical control is
difficult, as the bugs lie protected in bark crevices, and bugs return to treated
trees two years later. Systemic insecticides give better results than contact

insecticides. In Finland there are two distinct geographic areas. In one area adult *Aradus* appear in even years and in the other they appear in odd years. A probable explanation for this is intraspecific competition between larvae and adults – the larvae, being the stronger competitors, manage to overcome the adults (Heliövaara & Väisänen, 1986).

Many bugs belonging to the family Anthocoridae are found on trees. Some are predators while others have a mixed animal and vegetable diet. Although they are polyphagous, anthocorids show preferences for certain trees: *Elatophilus* species always live on pines. Among the lygaeids, *Gastrodes* species live on conifers. These bugs have oval, flattened bodies, are blackish-brown, and have three-segmented tarsi and long antennae. *Gastrodes grossipes* lives on Scots pine and sometimes on spruce. Eggs are laid in rows along needles, and the adults hide between the scales of pine cones or wander on branches. Another species, *Gastrodes abietum*, is associated with spruce, and the adults hibernate in the cones.

The family Miridae includes a number of forest bugs, among which those that are more or less closely associated with oak or with Scots pine should be mentioned (Ehanno, 1965). *Deraeocoris lutescens* lays on oak; eggs are inserted deep inside young shoots; the bug overwinters as an adult. This bug has a mixed diet: it preys on small-sized organisms but also sucks from leaves. *Allaeotomus gothicus*, which belongs to a closely related genus, is associated with Scots pine.

5.   A SPECIAL HABITAT: SAP FLUX

Sap that flows from wounded tree trunks is a scarce and little studied micro-habitat, and one with a special fauna. Insects that frequent sap fluxes have mainly been studied in the United States, where it has been shown that these insects propagate spores of the fungus *Ceratocystis fagacearum* which causes the disease known as "oak canker" or "oak wilt" (Dorsey *et al.*, 1953, 1956).

Sap flux on oak may result from various injuries caused either by man or by natural phenomena such as bark fissured by frost or galleries bored by goat moth (*Cossus*) larvae. Such flows of sap are home to a number of Coleoptera which have been listed in the Italian Alps (Ratti, 1978), but also live in many other parts of Europe. Many of them are hardly ever found in other habitats. The most remarkable of these species is the nosodendrid beetle *Nosodendron fasciculare*. This insect is often associated with ceratopogonid fly larvae on which it may feed. Other characteristic beetles of this habitat are various nitidulids, also known as "sap beetles" such as *Epuraea* spp., *Soronia grisea*, *Librodor hortensis*, *Cryptarcha strigata*, and many staphylinids, including *Thamiaraea hospita* which seems to be a predator of mites found in this environment.

The fauna of sap flux from elms in England has been described by Robinson (1953). Primary consumers of sap are the mite *Hericia hericia*, all of whose stages are found in this habitat, as well as larvae of the flies *Forcypomyia* spp., *Dasyhelea obscura*, *Aulacigaster leucopeza* and *Brachyopa insensilis*. Primary carnivores are two species of fly of the genus *Systenus* whose larvae prey on those of *Dasyhelea*, and the beetle *Thamiaraea hospita*, both the larvae and adults of which prey mainly on the mite *Hericia*. Secondary carnivores are represented by flies of the genus *Phaonia*.

REFERENCES

ALAUZET C., (1973). Étude de la zone préférentielle de ponte de *Pissodes notatus* (F.) (Coléoptère, Curculionidae) sur *Pinus maritima* en l'absence de tout autre parasite. *Bull. Écol.*, **4**: 144-150.

ALAUZET C., (1977). Cycle biologique de *Pissodes notatus* (Coleoptera, Curculionidae) dans la région toulousaine (France). *Canad. Ent.*, **109**: 597-603.

AUGUSTIN S., (1986). Étude de la dispersion des larves de premier stade de *Cryptococcus fagisuga* Lind. (Hom. Coccoidea) dans une hêtraie. *J. Appl. Ent.*, **102**: 178-194.

BALACHOWSKY A.-S., (1950). Les *Kermes* (Hom. Coccoidea) des chênes en Europe et dans le bassin méditerranéen. *Proc. 8th Int. Congr. Ent. 1948*, 739-754.

BAYLAC M., (1980). Faune associée à *Cryptococcus fagi* (Baer.) (Homoptera : Coccoidea) dans quelques hêtraies du nord de la France. *Acta Œcol.,Œcol. Applic.*, **1**: 199-208.

BEJER-PETERSEN B., (1962). Peak years and regulation of numbers in the aphid *Neomyza abietina* Walker. *Oikos*, **13**: 155-168.

BORATYNSKI K.-L., (1952). *Matsucoccus pini* (Green, 1925) (Homoptera, Coccoidea: Margarodidae): Bionomics and external anatomy with reference to the variability of some taxonomic characters. *Trans. R. Ent. Soc. London*, **103**: 285-326.

BRAUNS H.-J., (1976). Das Rindensterben der Buche *Fagus sylvatica* L., verursacht durch die Buchenwollschildlaus *Cryptococcus fagi* Bär. I. Die Anatomie der Buchenrinde als Basis-Ursache. II. Ablauf der Krankeit. *European Journ. Forest Pathology*, **6** : 136-146 et **7**: 76-93.

CARLE P., (1968). Attractivité des lésions provoquées par *Matsucoccus feytaudi* Duc. (Hom. Coccoidea) sur le pin maritime à l'égard de *Pissodes notatus* (Col. Curculionidae). *Ann. Sci. For.*, **25**: 25-33.

CARLE P., (1971). Les phénomènes présidant aux successions d'insectes dans le dépérissement du pin maritime du Var. *Ann. Zool. Écol. Anim.*, no. h.s.: *La lutte biologique en forêt*: 177-192.

CARLE P., CARDE J.-P., BOULAY C., (1970). Comportement de piqûre de *Matsucoccus feytaudi* Duc. (Coc. Margarodidae). Caractérisation histologique et histochimique des désorganisations engendrées dans le végétal hôte (*Pinus pinaster* Aït. var. *mesogeensis*). *Ann. Sci. For.*, **27**: 89-104.

CARTER C. I., (1972). Winter temperatures and survival of the green spruce aphid. *Forestry Commission, Forest Record*, **84**, 10 pp.

DE GROAT R. C., (1967). Twig and branch mortality of American beech infested with oystershell scale. *For. Sci.*, **13**: 448-455.

DIXON A. F. G., Quality and availability of food for a sycamore aphid population. *In:* A. Watson (1970). *Animal populations in relation to their food resources.* Blackwell, Oxford, 271-287.

DIXON A. F. G., RUSSEL R. J., (1972). The effectiveness of *Anthocoris nemorum* and *A. confusus* (Hemiptera Anthocoridae) as predators of the sycamore aphid *Drepanosiphum platanoides*. II. Searching behaviour and the incidence of predation in the field. *Ent. Exp. Appl.*, **15**: 35-50.

DIXON A. F. G., STEWART W. A., (1975). Function of the siphunculi in aphid with particular reference to the sycamore aphid *Drepanosiphum platanoides*. *J. Zool. London*, **175**: 279-289.

DOOM M., (1974). The pine bark bug *Aradus cinnamomeus*, a common and dangerous pest of Scots pine. *Nederl. Bosbow. Tijdschr.*, **46** : 136-137.

DORSEY C. K., JEWELL F. F., LEACH J. G., TRUE R.-P., (1953). Experimental transmission of oak wilt by four species of Nitidulidae. *Plant Disease Rep.*, **37**: 419-420.

DORSEY C. K., LEACH J. G., (1956). The bionomics of certain insects associated with oak wilt with particular reference to the Nitidulidae. *J. Econ. Ent.*, **49**: 219-230.

EHANNO B., (1965). Notes écologiques sur les Mirides (Insectes Heteroptera), observés en Bretagne sur le chêne. *Vie et Milieu*, **16**, sér. C: 517-533.

EWAN H. G., (1961). The Saratoga spittlebug. A destructive pest in red pine plantations. *USDA For. Serv. Tech. Bull.* no. 1,250.

FABRE J.-P. , CARLE P., (1975). Contribution à l'étude biologique d'*Hylurgops confusus* (Col. Scolytidae) dans le sud-est de la France. *Ann. Sci. For.*, **32**: 55-71.

HELIÖVAARA K., VÄISÄNEN R., (1986). Bugs in bags: intraspecific competition affects the biogeography of the alternate-year populations of *Aradus cinnamomeus*. *Oikos*, **47**: 327-334.

HÖLLDOBLER B., WILSON E.-O., (1990). *The ants.* Springer, Berlin.

JACQUIOT C., (1961). Note préliminaire sur une maladie du bois de hêtre dans l'est de la France. *Rev. For. Fr.*, **13**: 167-170.

KUNKEL H., (1968). Untersuchungen über die Buchenwollschildlaus *Cryptococcus fagi* Bär. (Insecta, Coccoidea), einen Vertreter der Rindenparenchymsauger. *Zeit. ang. Ent.*, **61**: 373-380.

LINNANE J. P., OSGOOD E. A., (1976). Controlling the Saratoga spittlebug in young red pine plantations by the removal of alternate hosts. *Univ. Maine Agric. Exp. Stn., Tech. Bull.* no. 84.

LUNDERSTÄDT VON J., EISENWIENER U., (1989). Zur physiologischen Grundlage der Populationsdynamik der Buchenwollschildlaus, *Cryptococcus fagisuga* Lind. (Coccidae, Coccina) in Buchen - (*Fagus sylvatica*) beständen. *J. Appl. Ent.*, **107**: 248-260.

MALPHETTES C.-B., (1977). *Cryptococcus fagi* (Bär.) et dépérissement du hêtre en forêt domaniale de Lyons (départements de l'Eure et de Seine-Maritime) (France). *Ann. Sci. For.*, **34**: 159-173.

McALONEY H.-J., WILSON L.-F., (1971). The Saratoga spittlebug. *US Dept. Agric., For. Pest. Leafl.*, **3**, 6 pp.

NAULT L. R., MONTGOMERY M. E., BOWERS N. S., (1976). Ant-aphid association: role of aphid alarm pheromone. *Science*, **192**: 1,349-1,351.

PARRY W. H., (1974). The effects of nitrogen levels in Sitka needles on *Elatobium abietinum* (Walker) populations in north-eastern Scotland. *Œcologia*, **15**: 305-320.

PARRY W. H., (1977). The effects of nutrition and density on the production of alate *Elatobium abietinum* on Sitka spruce. *Œcologia*, **30**: 367-375.

PERRIN R., (1977). Le dépérissement du hêtre. *Rev. For. Fr.*, **29** : 101-126.

RATTI E., (1978). La Colleotterofauna delle ferite di *Quercus robur* L. nelle Prealpi varesine. *Atti. Conv. Ecol. Prealpi Or*, 295-325.

RIEUX R., (1975). La spécificité alimentaire dans le genre *Matsucoccus* (Homoptera: Margarodidae) avec référence spéciale aux plantes hôtes de *M. pini* Green. Classement des *Matsucoccus* d'après leurs hôtes. *Ann. Sci. For.*, **32**: 157-168.

ROBINSON I., (1953). On the fauna of a brown flux of an elm tree (*Ulmus procera*) Salisb. *J. Anim. Ecol.*, **22**: 144-153.

ROSEN D., (1990). *Armored scale insects. Their biology, natural enemies and control*. Elsevier, Amsterdam.

SCHELLER H.-D., (1963). Zur Biologie und Schädwirkung der Nadelholzspinmilbe *Oligonychus ununguis* Jacobi und der Sitkafichtenlaus *Liosomaphis abietina* Walker. *Zeit. ang. Ent.*, **51**: 258-284.

SIEWNIAK M., (1976). Zur Morphologie und Bionomie der Kieferborkenschildlaus *Matsucoccus pini* (Green) (Hom. Coccoidea, Margarodidae). *Zeit. ang. Ent.*, **81**: 337-362.

THOMSEN M., BUCHWALD N.-F., HAUBERG P.-A., (1949). Attack of *Crypto-coccus fagi*, *Nectria galligena* and other parasites on beech in Denmark 1939-43. *Det. Forst. Forsog*, **18**: 97-326, 43 pl.

ZYCHA H., (1960). Die kranken Buchen: Ursachen und Folgerungen. *Holz-Zentralblatt*, **86**, no. 146: 2,061-2,063.

12

Galls and gall insects

Galls, or cecidia, are abnormal plant structures formed by association with a parasite. These structures, which may be either defined or diffuse, are part of the host plant and composed entirely of the latter's tissues. Gall formation may be due to various organisms: bacteria, fungi or animals. Gall mites and insects, which are the main animals responsible for gall formation, are called cecidozoans.

Mechanical interference alone (bites, oviposition by piercing) is not enough to form a gall: chemical action triggered by the parasite's secretions or excretions is necessary. Much research has pointed to a secretion of the salivary glands producing a substance analogous to plant hormones, or auxins, such as indol-acetic acid. Gall formation by hyperauxinia has been shown in the case of aphids. In other cases substances acting like hormones may be produced by the plant as a rejection reaction to stimulants of insects such as the buprestid *Agrilus biguttatus*. In the cecidomyiid *Contarinia medicaginis*, which causes galls on lucerne, a factor originating in the saliva stimulates synthesis of substances which trigger the cellular reproduction and hypertrophy characteristic of the gall (Strebler, 1975).

Galls are found on a great many plants. Nearly 50% of galls known in the northern hemisphere are formed on trees of the family Fagaceae, particularly oaks and beeches. Different parts of a plant are affected with varying frequency. In oaks, on a global scale, 2% of galls are on flowers, 4% on acorns, 5% on roots, 22% on buds, and about 63% on leaves.

Insects responsible for the greatest number of galls are cynipid and tenthredinid wasps, cecidomyiid flies and chermesid bugs. Other orders play only a secondary role. The two major families of gall mites are the Tarsonemidae and Eriophyidae. Gall insects and mites are of particular interest because of their peculiar biology and their very complex life cycles. A few species are pests.

1. CYNIPID GALL WASPS OF OAKS

The family Cynipidae includes many phytophagous species, most of them gall makers; they are particularly abundant on oaks. More than 200 kinds of cynipid galls are known on *Quercus robur* alone. Cynipids of oak are nearly always heterogonous, in other words there is a regular alternation of a sexual generation including males and females with an agamous generation devoid of males and reproducing by parthenogenesis. Each generation produces a particular type of gall, and individuals often differ one from another, which explains why representatives of each generation were long given different specific and even different generic names.

1.1. *Biorrhiza pallida*

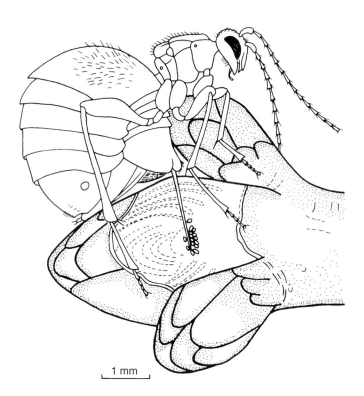

1 mm

Figure 12.1. *Female* Biorrhiza aptera *laying in an oak leaf bud. The bud is partially opened to show the ovipositor and the eggs. The female perforates the tough outer scales, then bores a small cavity at the heart of the bud in which she deposits her eggs together with a colourless fluid produced by glands next to the reproductive apparatus.*

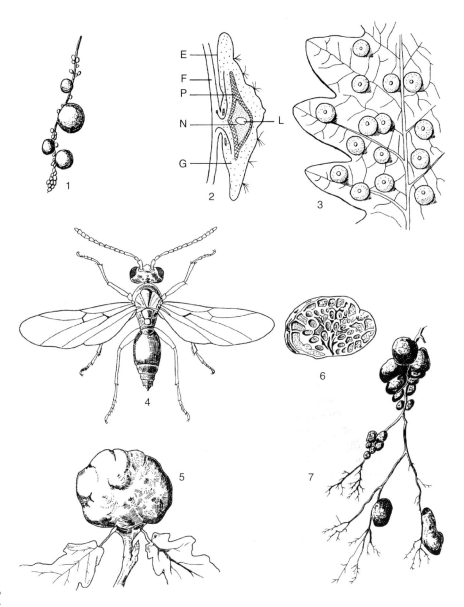

Figure 12.2. *Cynipid gall wasps of oak.* Neuroterus lenticularis. *1: gall of the form* baccarum *on male flowers; 2 and 3: gall on a leaf, general appearance and cross-section. E: epidermis; F: leaf; G: gall parenchyma; L: larval cell; N: food tissue; P: protective tissue.* Biorrhiza aptera. *4: sexual form* pallida; *5 and 6: gall of the sexual form, entire and in cross-section; 7: gall of agamous form on rootlets.*

The name *Biorrhiza aptera* designates the agamous generation of the cynipid *Biorrhiza pallida*. Adults emerge in winter from galls grown either on large oak roots or on rootlets as much as one metre below the surface. These yellow galls turn dark brown and woody after the insect emerges; they form agglomerations about the size of a walnut. Parthenogenetic females can live and lay eggs even when the temperature is below 0 °C. They climb the trunks and lay in young buds which turn into galls of characteristic appearance called "oak apples" which are yellow flushed with red and range in size from that of a walnut to that of a small apple. From each gall emerge up to 200 insects which are the adult, sexual, generation known by the name *Biorrhiza pallida* (figure 12.2). After mating, females lay their eggs underground in the bark of oak roots. The life cycle lasts two years. *Biorrhiza pallida* is exceptional in that the galls of most cynipid gall wasps of oak contain only a single larva.

1.2. *Neuroterus lenticularis*

This cynipid produces lenticular galls a few millimetres in diameter on the undersides of oak leaves; the upper surface of the gall is broadly conical, whitish flushed with yellow or red, and bears small clusters of radiating brown hairs. The insects that emerge from these galls are individuals of the agamous generation which go on to lay their eggs on young oak shoots which then develop spherical galls resembling gooseberries. From these galls emerge sexual male and female insects named *Neuroterus quercusbaccarum* which lay on young leaves which develop galls that come to maturity in September. Females of the agamous generation belong to one of three types: androphorus females producing only males; gynephorus females producing only females; a few rare females may have progeny of both sexes. Males are haploid, as is the general rule in Hymenoptera; the chromosome number is 2 n = 20 (figure 12.3).

There are other common *Neuroterus* species:

– *Neuroterus numismalis* has disc-shaped brownish-yellow agamous generation galls 3 mm in diameter furnished with a median depression covered in dense hairs. The sexual generation known as *Neuroterus vesicator* produces light-bulb-shaped galls 3 mm long on leaf margins.

– *Neuroterus tricolor* has greenish-yellow disc-shaped agamous generation galls 3 mm in diameter with inflated edges and sparsely covered in hairs. Galls of the sexual generation are spherical, about 6 mm in diameter, more or less hairy, and grow on the undersides of leaf veins.

– *Neuroterus albipes* has cream-coloured, cup-shaped agamous generation galls 4 mm in diameter; sexual generation galls are oval, 2 mm long, and green, and develop on leaf margins (figure 12.4).

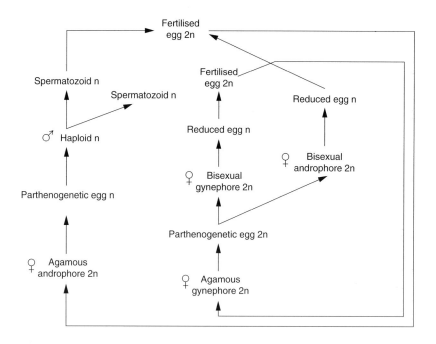

Figure 12.3. *Life cycle of* Neuroterus lenticularis.

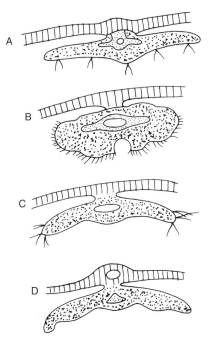

Figure 12.4. *Schematic cross-section of agamous generation galls of four species of* Neuroterus. *A:* N. lenticularis; *B:* N. numismalis; *C:* N. tricolor; *D:* N. albipes.

1.3. Distribution of *Neuroterus* galls

Neuroterus galls are particularly abundant on young trees at forest edges. Galls of several species may be found on the same leaf blade. In some cases 80 to 90% of a leaf's surface is covered with galls; such density explains the development of intra- and interspecific competition. Water plays an important part in gall formation; the first galls to form deprive later ones of water, thus preventing them from growing normally. The offspring of females that lay the first eggs have an advantage over those of females that lay later. A study of four sympatric species of *Neuroterus* has shown that *N. albipes* always lays its eggs first, and *N. tricolor* lays last, about six weeks later. *N. numismalis* and *N. quercusbaccarum* lay at some time between these two. Competition between these species is partly avoided by their different preferences in laying sites on leaves and on other parts of an oak tree (cf. chapter 3.3).

1.4. *Andricus kollari*

Andricus kollari (= *Cynips kollari*) has two forms of agamous female. The first form produces offspring of both sexes known by the name *Andricus circulans* which live on Turkey oak (*Quercus cerris*) in galls which are relatively few in number. These galls, which are situated on axillary buds, are ovoid, 2.5 mm long, brownish or reddish-yellow, and lie embedded half way in a bud which may be host to up to eight galls. The other form of agamous female only has permanently agamous offspring that may reproduce outside the range of *Quercus cerris* on the leaves of other deciduous oaks, on which they produce spherical, yellowish-brown galls 12 to 23 m in diameter which are completely smooth, very hard, and often plentiful on twigs near the bases of buds.

The gall produced by *Andricus kollari*, like those of nearly all cynipids, has a complex structure. It is composed, going outwards from the core, of food tissue which is consumed by the larva, woody protective tissue formed of cells with thickened walls, a spongy parenchyma which constitutes the main bulk of the gall, and finally of a surface epidermis (figure 12.5). When the cynipid larva has exhausted the supply of food tissue, the cell walls of the protective layer produce a lysate which provides additional food. Cinipid galls are very rich in tannins (up to 65% of weight in some species), and have been used for tanning leather, dying fabrics, and manufacturing ink.

1.5. *Andricus fecundator*

Andricus fecundator has agamous females that develop in galls resulting from oviposition in buds. These galls reach 2 cm, and look somewhat like an artichoke or a hop strobilus; they are covered with enlarged imbricate

green- to rust-coloured scales surrounding a central cavity in which lies the
gall proper, which harbours the larva (cf. figure 12.13). Sexual individuals

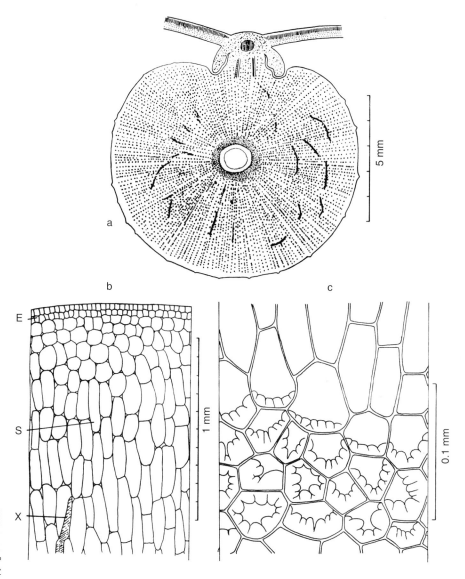

Figure 12.5. Andricus kollari *gall on an oak leaf. a: cross section of the gall
showing the larval chamber at the centre and the various
surrounding tissues. b: detail of outer area showing epidermis E,
spongy tissue S and conductive tissue X. c: detail of protective
tissue with thickening of inner and lateral cell walls.*

develop in galls growing on the male flowers of deciduous oaks. The galls are oval, 3 mm long, hairy and green, later turning brown; these galls are those of the form named *Andricus pilosus*.

Another species of the same genus, *Andricus quercuscalicis,* has irregularly-shaped galls that grow on the bases of acorns on *Quercus robur*. In some years this may cause the destruction of up to 90% of acorns, which is in excess of destruction by rodents (Crawley, 1987). Galls fall to the ground in autumn, and in the following spring females that emerge from them lay on the male flowers of *Quercus cerris*, which has been introduced here and there in Europe. Males and females of the wasp develop in small galls formed on these flowers, and females return to lay on pedunculate (= English) oak.

1.6. Cynipid galls as micro-ecosystems

Cynipid wasp galls are microhabitats which often harbour occupants other than the cecidogeneous larva. One hundred and seventy seven species of "tenants" in cynipid galls on oaks have been listed, of which 139 are parasites belonging to groups as varied as nematodes, mites, Thysanoptera, Coleoptera, Lepidoptera, Diptera and Hymenoptera. Parasitic chalcid wasps are particularly well represented; they may account for 90% of animals that emerge from the galls. These occupants fall into four categories.

– Inquilines or commensals which feed on the gall's tissues while the cecidogeneous host is still in situ. Inquilines are either unable to cause gall formation themselves, or are only feebly cecidogeneous. About twenty species of inquiline were found in galls of *Biorrhiza pallida*, among which were several cynipids of the genus *Synergus*. Some of these species, such as *Synergus melanopus*, occupy *Andricus kollari* galls where they live in a cavity they form in the parenchyma; others, such as *Synergus reinhardi*, which also occurs in *Andricus kollari* galls, kill the host larva and take its place. Cecidomyiid flies of the genus *Clinodiplosis* as well as mites are also inquilines. The line between an inquiline and a cecidogene may be a fine one, as an inquiline's presence may by itself cause hypertrophy of the gall. A transitional stage from gall producer to inquiline exists in *Andricus curvator* which prefers to lay its eggs in galls in which another species belonging to the same genus, *A. fecundator*, has already laid its own. *A. curvator*'s small galls grow on the much larger galls of *A. fecundator*. An inquiline may thus be a gall-forming species that gradually specialises in living in another species' galls. This view is lent support by the existence of Thysanoptera and of mites of the genus *Eriophyes* which may, according to species, be either cecidogeneous or inquilines.

– Eaters or destroyers of galls are of various kinds. They include birds, ants such as *Camponotus ligniperda* which bores its galleries into the "oak apples" of *Biorrhiza aptera*, about ten species of caterpillar, and saprophytic or parasitic fungi many of which are host-specific on cynipid galls of oaks.

– Successors occupy a gall after the gall-former and inquilines have left. They include a great variety of non-social arthropods (about a hundred species are already known) and a few species of ant (Torossian, 1971, 1972).

– Predators and parasitoids are dominated by Hymenoptera. Some are primary parasitoids of the gall wasp; others are hyperparasites or parasitoids of inquilines. *Eurytoma brunniventris* is a chalcid wasp which parasitises 49 species of cynipid as well as their inquilines and parasites.

Because species diversity is so high, trophic networks centred on galls are very complex. They may include up to seven or eight different trophic levels, which is exceptional in other ecosystems (figure 12.6).

2. CHERMESID APHIDS

Within the superfamily Aphidoidea, chermesids (or adelgids) are gall aphids that are confined to conifers. They are characterised (in alate forms) by their wings which are held roof-wise over the abdomen, and by their five-segmented antennae, of which the last three segments each bear a large sensory organ called a rhinarium.

Chermesids are remarkable by virtue of their polymorphism and their complex life cycles. The typical cycle always includes at least five different generations of which only one is sexual, with both males and females, the other generations being composed solely of parthenogenetic females. The cycle normally lasts two years and involves two species of conifer: a primary host, a spruce, on which the sexual generation depends, and a secondary host belonging to a different genus (pine, fir or larch) which feeds the partheno-genetic generations known as exiles. Passage from primary to secondary host is achieved by two alate generations: winged emigrants or gallicoles travel from spruce to the secondary host, and the winged sexuparous insects travel from the secondary host back to spruce trees on which they produce sexual offspring (figure 12.7). Apterous exiles which undergo a diapause in their first larval instar are called sistens (plural sistentes). Holocyclic species complete their cycle on two trees in the manner that has been outlined. Anholocyclic species derive from the former and complete their cycle on the secondary host.

Because of the great variability of species, classification of chermesids is difficult. The old genus *Chermes* is now divided into two groups: the genus *Pineus*, which has four pairs of abdominal stigmata and does not have sistens forms, and the genera *Chermes, Dreyfusia, Cnaphalodes, Sacchiphantes, Gilletteella* and *Cholodkowskya*, which have five pairs of abdominal stigmata and a sistens form. Common species are few (table 12.1). Their rapid proliferation explains why chermesids may cause consid-erable damage to trees in the form of deformed terminal shoots bearing galls,

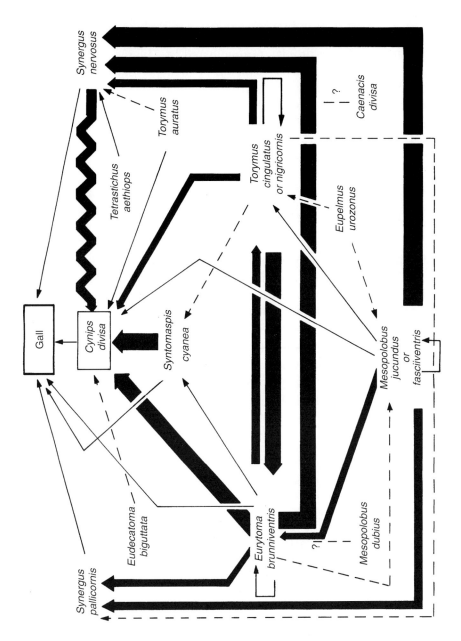

Figure 12.6. *Structure of the trophic network around* Cynips divisa *galls in England. Arrows point to food sources or to a parasitised species; their thickness is a function of frequency of the relationship. The wavy line indicates that* Synergus nervosus *attacks* Cynips divisa *but does not eat it (Askew, 1961).*

and exhaustion of the tree due to loss of sap and weakened and bent needles.

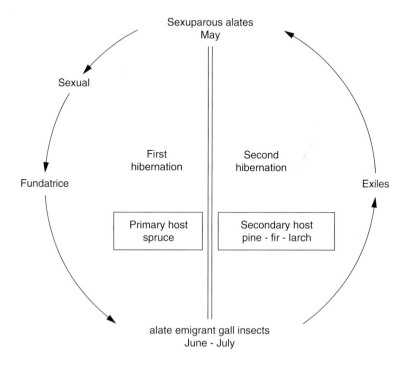

Figure 12.7. *Theoretical schematic illustration of the life cycle of an holocyclic chermesid.*

2.1. Species of the genus *Dreyfusia*

In western Europe the firs *Abies alba* and *Abies nordmanniana* (the latter being an introduced species) are host to four species of *Dreyfusia*, which are difficult to tell apart, of which *D. nordmannianae* and *D. piceae* are the commonest. Like all chermesids, *Dreyfusia* are minute insects covered with an abundant white waxy secretion which inspired the name "woolly aphids"(figure 12.8).

2.1.1. Dreyfusia nordmannianae

This species (formerly known as *D. nuesslini*) occurs in France in the Jura and the Vosges, but it is mainly a central European species. It sometimes infests young trees whose needles become deformed as a result of the insect's bites. Repeated attacks cause the main terminal shoot to wilt, and it

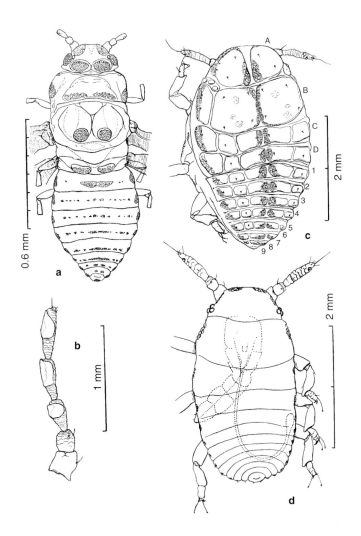

Figure 12.8. Dreyfusia nordmannianae. *a: emigrant alate with wings removed. b: antenna (note the sensory organs called rhynaria on the last three segments). c: founding female in slightly oblique dorsal view. Pineus pini. d: primary larva produced by an exiled alate, dorsal view. The rostrum and curvature of the styles are seen in transparency (Marchal, 1913).*

turns brown and sheds its needles. *D. nordmannianae*'s life cycle is one of the most complex found in chermesids. Overwintering is by founding larvae hatched from fertilised eggs; these larvae settle at the bases of spruce buds. The wingless adults they turn into produce gall larvae in the following spring; these larvae produce "pineapple" galls in bud scales. The species' switch to fir trees is effected by winged females which emerge from these galls in June. Females hibernate on the fir and in the following spring produce a sexuparous generation whose females return to spruce. This species also has parthenogenetic "exiled" generations which remain on fir in the absence of the primary host, as is often the case in France.

2.1.2. Dreyfusia piceae

This parthenogenetic species lives only on fir. Its range extends to the Massif Central and the Normandy fir plantations. It mainly colonises the bark of trunks and branches of older trees and causes malformations of smaller branches, which jeopardises the trees' growth. This and the preceding species of *Dreyfusia* have become pests of fir in southern England, where two predatory beetles, *Laricobius erichsoni* and *Scymnus impexus* have been introduced from the Black Forest in an attempt at biological control. Heat and drought are conducive to outbreaks of *D. piceae*. Outbreaks and invasions occurred in Austria and in central Europe in 1945 and the following few years. The critical factor is rainfall in March to June.

Scientific names	Primary host	Secondary host
Pineus orientalis (= *Chermes pini*)	*Picea orientalis*	*Pinus sylvestris*
Pineus pini (= *Chermes pini*)	0	*Pinus sylvestris*
Dreyfusia nordmannianae (= *Chermes nuesslini*)	*Picea orientalis*	*Abies nordmanniana* and *A. alba*
Dreyfusia piceae (= *Chermes piceae*)	0	*Abies nordmanniana* and *A. alba*
Gilletteella cooleyi (=*Chermes cooleyi*)	*Picea sitchensis*	*Pseudotsuga menziesii*
Adelges laricis (= *Chermes strobilobius* = *Cnaphalodes strobilobius*)	*Picea excelsa*	*Larix*
Adelges tardus (= *Chermes tardus*)	*Picea excelsa*	0
Sacchiphantes viridis (= *Chermes viridis*)	*Picea excelsa*	*Larix*
Sacchiphantes abietis (= *Chermes abietis*)	*Picea excelsa*	0
Cholodkowskya viridana (= *Chermes viridanus*)	0	*Larix leptolepis*

Table 12.1. *Commonest chermesids of the French fauna, and primary and secondary hosts. The first name is the current scientific name; former names are in brackets.*

The chermesid *Dreyfusia piceae* was introduced to western Canada and the United States, where it is known as the "balsam woolly aphid", and may, during outbreaks, cause damage, particularly to *Abies balsamea* which is a major resource for the manufacture of paper. Infestations of this aphid cause characteristically shaped- galls, called "gouty galls" on terminal shoots and twigs; affected buds do not open or produce short shoots; longitudinal and radial growth ceases; the tree's starch reserves decrease and the tree loses vigour. Xylem cells form abnormal wood known as "rotholz", or red wood, in German-speaking countries. These cells have thickened walls and narrower apertures than normal cells, which slows sap circulation (Davies, 1968; Puritch & Petty, 1971). Trees may sometimes recover following severe infestation.

2.1.3. Predators and parasites of Dreyfusia

Chermesids are well protected by their galls in the larval stage, and parasites are few. Predators of adults are more numerous. In central Europe predators of *D. picea* were found to fall into three categories (Delucchi,

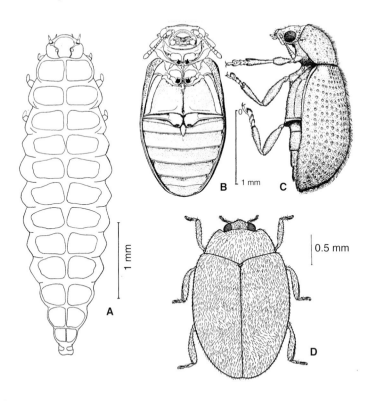

Figure 12.9. *Two predators of* Dreyfusia piceae. Laricobius erichsoni. *A: fourth instar larva; B and C: adult in ventral and in lateral views.* D: Pullus impexus.

1954; Pschorn-Walcher & Zwölfwer, 1956; Eichhorn, 1969). Dominant species are present regularly and are usually abundant; these are two beetles, *Laricobius erichsoni* (Derodontidae) and *Pullus impexus* (Coccinellidae), and four flies, *Aphidoletes thompsoni* (Cecidomyiidae), *Neocnemodon dreyfusiae* (Syrphidae), *Cremifania nigrocellulata* and *Leucopis obscura* (Chamaemyiidae). Sub-dominant species include coccinellids, syrphids, chrysopids and mites of the thrombidiid family. Occasional species are coccnellids, syrphids and chrysopids (figure 12.9).

The morphology and biology of *Laricobius erichsoni* have been studied with a view to possible practical use of this predator against *Dreyfusia piceae* in North America (Franz, 1958). *L. erichsoni* preys on several species of chermesid, but *D. piceae* is its favourite. The beetle's larvae eat the eggs of *D. piceae* while adult beetles hunt the adult aphids. Prey may be detected over great distances by the adult beetles which are strong fliers and are able to disperse to colonise areas newly planted with fir outside the beetle's original range of distribution. In Canada and in the western United States biological control of *Dreyfusia* was tried using the following predators imported from Europe: *Laricobius erichsoni* and the two coccinellids *Pullus impexus* and *Aphidecta obliterata*; the Diptera *Cremifania nigrocellulata*, *Leucopis* sp. and *Aphidoletes thompsoni*. These species proved successful at controlling *Dreyfusia piceae* (Clark & Brown, 1958).

2.2. Species of the genus *Sacchiphantes*

In this genus the two most important species are *Sacchiphantes viridis* and *S. abietis*. The life cycle of *S. viridis* is somewhat similar to that of *Gilletteella cooleyi*. Its primary host is spruce and its secondary host is larch. Both tree species are alternately colonized by five generations within two years. Pineapple-shaped galls on spruce impede the growth of shoots and cause malformation of the young trees grown in nurseries to be sold as Christmas trees. On larch, the perforations may cause an early fall of the needles. The galls of *S. viridis* are easy to identify. Their green colour and the red-bordered -hem of their scales make them easily recognizable. The best method of control is prevention: it consists in interrupting the insect's life cycle by not growing spruce and larch side by side. Chemical control in nurseries may consist of spraying organophosphorous insecticides. *Sacchiphantes abietis* has a complete life cycle on spruce on which it may often form a large number of galls.

In Switzerland two predators have been found on these two *Sacchiphantes*. The former is *Aphidoletes abietis* (Diptera, Cecidomyiidae) whose female lays its eggs on the gall surface. The young larva of this predatory Diptera bores a hole and enters the gall. There may be up to five larvae in the same hole. When the larvae of *Aphidoletes* reach their last instar they cause the gall to open up prematurely and the larvae fall to the ground where they pupate. The latter predator is a Syrphidae of the genus *Cnemodon*. Its larva bores itself into a young gall and its entire development is synchronized with that of the gall. The pupation of *Cnemodon* also occurs in the ground (Mitchell and Maksymov, 1977).

2.3. *Gilletteella cooleyi*

The chermesid aphid *Gilletteella cooleyi* is the Sitka spruce gall aphid. This North American insect lives alternately on two host species: Sitka spruce (*Picea sitchensis*), the primary host on which the sexual generation develops, and Douglas-fir (*Pseudotsuga menziesii*) on which only partheno-genetic generations live. The aphid spread to Europe with the introduction of Douglas-fir, and has been recorded from various parts of France. Its life cycle lasts two years and involves several generations. In France only those forms that can survive on Douglas-fir by successive generations of parthenogenetic females are regularly found. The sexual generation, as well as founding gener-ations and those that live in galls, live on Sitka spruce; they are rare in France. *G. cooleyi* becomes obvious in summer on Douglas-fir needles in the form of a white felting which covers both larvae and adults, and by yellowish spots on the needles caused by bites. Needles become deformed and have an untidy appearance. Infestations occur chiefly in nurseries. On Sitka spruce damage is more serious. Elongate galls appear at the tips of branches; they impede growth and sometimes destroy the tree's main terminal shoot.

In America, and in countries such as Britain where *G. cooleyi* is intro-duced, its main predators are coccinellids, syrphids, anthocorids, chrysopids, hemerobiids, spiders and mites. A study conducted in Scotland showed that the coccinellid *Aphidecta obliterata* is probably the chief predator.

2.4. Species of the genus *Pineus*

The chermesid aphid *Pineus orientalis* is thought to have originated in the Caucasus where it has a complete life cycle on oriental spruce and on Scots pine. In France the species maintains itself by indefinite parthenogenesis on pine, with possible migration to the primary host where this is planted. There is also an indigenous French race, considered by some to be a distinct species, called *Pineus pini*, which lives entirely on pine and is unable to reproduce sexually on oriental spruce. Predators of *Pineus pini* include five main species which are close to or identical with those that predate other chermesids. These are the two Diptera *Leucopis obscura* and *Lestodiplosis pini*, the Neuroptera *Hemerobius sigma* and *Wesmaelius concinnus*, and the coccinellid beetle *Exochomus quadripustulatus* (Wilson, 1938).

2.5. Formation and structure of chermesid galls

Chermesid galls result from changes that take place in the branches of parasitised trees. These bud galls are of the type in which the bases of hypertrophied needles become imbricated like the bracts of an artichoke. A more or less spherical larval cavity is formed at the base between each needle (figure 12.10). Subsequent growth of parts affected by cecidogenesis

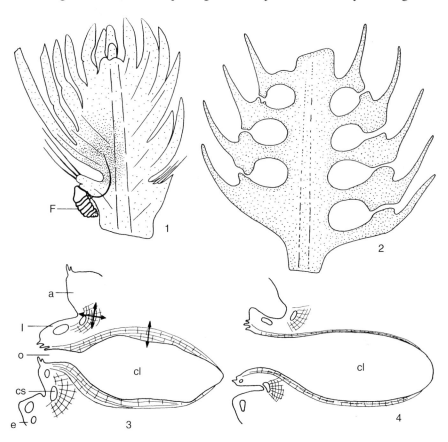

Figure 12.10. *A "pineapple" gall of* Sacchiphantes abietis *on spruce. 1. Cross-section of a spruce bud showing a founding female F with her styles embedded in the tissue in which cell division, indicated by more or less dense stippling, is induced. 2. Longitudinal section through a ripe gall showing swollen needle bases and larval cavities. 3 and 4. Section of a larval cavity, first closed, then open. Arrows point to cellular deformations caused by water loss from the pectic intercellular matter. cl: larval cavity; cs: secretory canals; e: scutellum; l: lip; a: needle; o: ostiole (Rohfritsch, 1966).*

causes the bases of the needles to fuse. The finished galls measure between 20 and 30 mm in the case of *Sacchiphantes abietis*. Gall development is due to the effects on spruce buds of two successive generations of insects. The founding female plays a determining role: she begins formation of larval chambers by the process of oviposition. Gall insects play a part in differentiation which can be shown experimentally (Rohfritsch, 1966, 1967). Transplantation of *Adelges laricis* gall larvae to a gall induced by a founding female *Sacchiphantes abietis* whose own larvae have been removed shows that primordial inductions are not specific, since the transplanted species is able to go on growing in a spruce bud prepared by another species. However, maturation of the gall, characteriscd by lignification and yellowing, as well as by opening of the cavities, does not take place if the gall insects are killed. Production of mixed galls (figure 12.11) shows that gall insects play a major part in gall formation and maturation. Gall insects may modify growth and induce specific morphological characteristics (such as colour and the time of maturation and dehiscence) as well as histological and cytological characteristics (Gaumont, 1966).

Studies of *Sacchiphantes abietis* galls have shown that the cavities in each gall almost all open at the same time. The opening mechanism results from water loss from the pectic intercellular matter in the area that bounds the orifice of the larval cavities. The active parenchyma cells are so shaped and disposed that their desiccation causes the gall to open (figure 12.10).

3. OTHER GALL HOMOPTERANS

3.1. Phylloxerids

Gall insects are met with in other groups of aphids. Members of the family Phylloxeridae are recognisable by the way in which the wings are borne flat on the body. All generations are oviparous. *Acanthochermes quercus* is a species devoid of alate forms and has a simple life cycle. The founding female forms a lenticular gall on the underside of a deciduous oak leaf. She lays eggs that produce a sexual generation which in turn produce winter eggs. *Phylloxera glabra*, a species that also lives on oak, has a more complex life cycle. The founding female forms a small gall on a leaf margin. Her offspring are viriniparous and wingless, and disperse to other parts of the same leaf on which their bites cause necrosis of the leaf parenchyma, causing the leaf to become covered with yellow spots. After producing two or three generations during the summer the virginiparous insects produce winged sexuparous offspring which in turn produce a sexual generation. Each fertilised female lays a single winter egg. *Periphyllus testudinaceus*, which lives on the undersides of maple leaves, is remarkable for the leaf-like hairs on its body (figure 12.12).

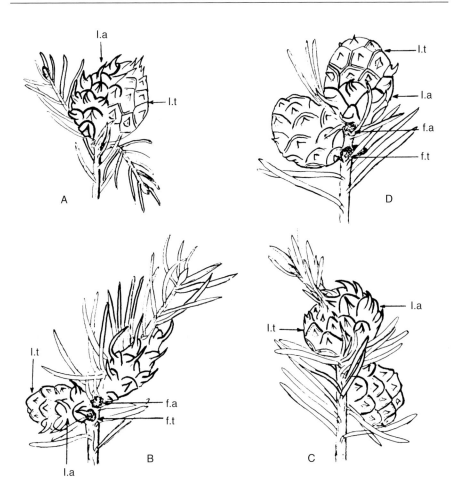

Figure 12.11. *Mixed galls of* Sacchiphantes abietis *and* Adelges tardus.
A: bipartite gall. A founding female S. abietis *and a founding female* A. tardus *settled at the base of the same bud. The insects each took a share of the chambers.*
B: crossed gall. Founding females of S. abietis *and* A. tardus *each settled on neighbouring buds. A number of* S. abietis *gall larvae invaded the gall induced by the* A. tardus *founding female and determined cavities of the* abietis *type.*
C and D: crossed galls. A. tardus *gall larvae have invaded cavities induced by the founding female* S. abietis. *f.a.: founding female* S. abietis; *f.t.: founding female* A. tardus; *l.a.:* abietis *type cavities; l.t.:* tardus *type cavities (Gaumont, 1966).*

3.2. Pemphigids

Members of the family Pemphigidae (or Eriosomatidae) are characterised by their reduced cornicles and numerous wax glands. In America the woolly

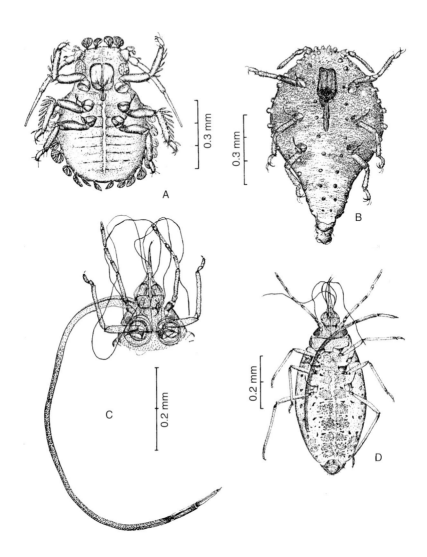

Figure 12.12. *A few aphids. A:* Periphyllus testudinaceus, *female, summer form.*
B: Acanthochermes quercus, *wingless founding female.*
C: Stomaphis quercus, *head, prothorax and rostrum of*
parthenogenetic female. D: Stomaphis quercus *wingless*
parthenogenetic female
(cf. chapter 11.2) (Gaumont, 1923).

aphid *Eriosoma lanuginosum* causes curling of the leaves of the American elm *Ulmus americana*; the curled leaves then become closed galls containing the founding female and her progeny. In winter, winged emigrants colonise apple tree roots which serve as the secondary host. In Europe, where American elm does not occur, the life cycle is simplified: the aphid lives only on apple on which it reproduces by parthenogenesis.

Various species belonging to the genus *Pemphigus* form galls on poplar petioles or leaf blades; *Pemphigus* complete their cycles on two plants: poplar, and an herbaceous plant that varies according to the species of aphid.

Pemphigus betae does not form its galls haphazardly on the leaves of its host, *Populus angustifolia*. The leaves that bear the greatest numbers of galls are those that are potentially the largest, and the poorest in phenol compounds. This explains the territorial behaviour of females, which defend their chosen laying sites against intruders (Zucker, 1982). The curious behaviour of another species of *Pemphigus* has been described by Witham (1980, 1986). The female chooses to settle on a young poplar leaf, on the basal part of which she feeds; this creates a depression which becomes a gall in which the larva will grow. If another female tries to settle in the same place the two females engage in a battle by pushing each other. This struggle may last up to two days; while it goes on the females are unable to initiate gall formation, and the gall that later develops is smaller than usual. One of the two females may even die during the battle. What then is the purpose of this behaviour? A female that has a whole leaf to herself has a greater rate of reproductive success than one that has to share with another female, as is shown by the figures in table 12.2.

Number of galls per leaf	Average size of leaf (cm)	Average number of offspring		
		One female	Two females	Three females
1	10.2	80	–	–
2	12.3	95	74	–
3	14.6	138	–	29

Table 12.2. *Number of offspring per gall for a species of* Pemphigus. *Figures indicate numbers of offspring of the first female to settle, followed by the second and the third. The first female to settle always produces the greatest number of offspring.*

4.   GALL MIDGES

The most remarkable gall-producing members of the true flies (Diptera) belong to the family Cecidomyiidae. They are found on oaks and beech and less frequently on birch. *Macrodiplosis dryobia* forms characteristic galls on the margins of oak leaves characterised by downward folding of the leaf lobes. The larva lives between the two leaf lobes, the histology of which is scarcely changed. *Macrodiplosis volvens* produces similar galls, but the curled portions of the leaf margin are narrower and situated between the lobes (figure 12.13).

Figure 12.13. *Some galls on oak and beech.* Hartigiola annulipes *on beech.*
A: galls on a beech leaf; B, C, D, E: four stages of development in
cross-section: F: Macrodiplosis volvens *on oak;*
G: Macrodiplosis dryobia *on oak.* Andricus fecundator.
H: artichoke-shaped galls of agamous generation; I: galls of sexual
generation on oak male flowers.

The most widespread cecidomyiid of beech is *Mikiola fagi*. The pear-shaped galls of this species are up to one centimetre long; their distribution on beech leaves is not haphazard, but of a negative binomial type. This may be explained by supposing that a female may lay several eggs on the same leaf and that she has to find parts of the leaf where the latter's growth stage is receptive to the egg (Legay, 1963). The hollow galls grow on the upper surface of leaves. The larva moves about actively in the chamber, stopping at regularly spaced points of the chamber's surface, stimulating even growth of the gall (Boysen-Jensen, 1952). The wall of the gall consists of two layers: the internal layer is composed of food cells and the outer layer of sclerenchyma; conductive tissue lies between the two. An abscission area allows the gall to break off and fall to the ground, where the larva pupates at the end of winter (figure 12.14).

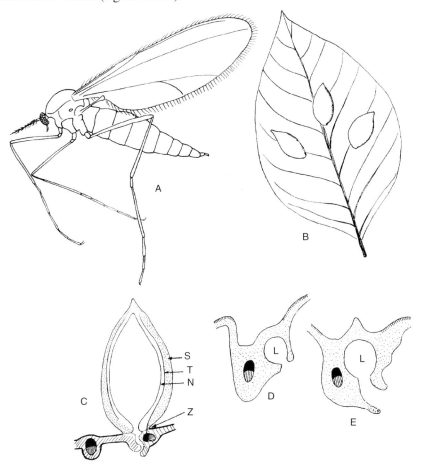

Figure 12.14. *The midge* Mikiola fagi *and its gall. A: adult; B: galls on a beech leaf; C: cross section of a mature gall (T: conductive tissue; S: sclerenchyma; N: food tissue; Z: abscission area); D and E: two stages in gall growth (L: larval chamber).*

The morphology and biology of some cecidomyiid gall midges of birch have been studied by Askew & Ruse (1974). *Massalongia rubra* is a common species which may parasitise 20% of a birch tree's leaves. Its galls take the form of swellings one centimetre long situated along the midrib. The larva is about 4 mm long and yellow to orange. It has the appearance of a typical cecidomyiid larva, but its sternal spatula is of small size. *M rubra*'s parasite complex includes mainly chalcids of the genera *Torymus* and *Tetrastichus*.

5. OTHER GALL INSECTS

The order Coleoptera includes few gall-forming species on trees. The most noteworthy is the buprestid *Agrilus biguttatus* which causes tissue formations resembling galls on oak (cf. chapter 16).

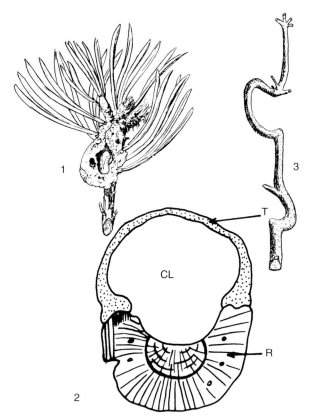

Figure 12.15. *Resin gall of* Petrova resinella. *1: gall on a pine twig; 2: cross-section showing the larval cavity CL and what is left of the twig T; 3: typical bayonet shape of a pine branch, a deformation caused by the caterpillar of* Ryacionia buoliana.

Gall Lepidoptera include the tortricid moth *Petrova resinella* which produces a structure known as a "resin gall" on pine. The female deposits her eggs on the surface of and near the tips of branches of various pines. The caterpillar spins a shelter between the twig and a few needles; it then gnaws the bark, which causes a few drops of resin to flow. A sort of larval lodge is thus formed, the roof of which is made of dried resin mixed with frass. The gall properly speaking is a composite of bark tissue and un-gnawed wood which grows considerably and swells laterally until it reaches the size of a cherry (figure 12.15). This moth's life cycle lasts two years. Moths emerge from May to June.

REFERENCES

ASKEW R. R., (1961). On the biology of the inhabitants of oak galls of Cynipidae (Hymenoptera) in Britain. *Trans. Soc. Brit. Ent.*, **14**: 237-268.

CLARK R. C., BROWN N. R., (1958). Studies on the predators of the balsam woolly aphid, *Adelges piceae* (Ratz.) (Homoptera : Adelgidae). V. *Laricobius erichsoni* Rosen (Coleoptera: Derodontidae), an introduced predator in eastern Canada. *Canad. Ent.*, **90**: 657-672.

CRAWLEY M. J., The effects of insect herbivores on the growth and reproductive performance of English oak. *In*: V. LABEYRIE, G. FABRE, D. LACHAISE (1987). *Insects-Plants*. W. Junk, Dordrecht, 307-311.

DAVIES J. M., (1968). Adelgids attacking spruce and other conifers. *Forestry Commission Leaflet*, **7**: 1-12.

DELUCCHI V., (1954). *Pullus impexus* (Muls.) (Coleoptera, Coccinellidae), a predator of *Adelges piceae* Ratz., with notes on its parasites. *Bull. Ent. Res.*, **45**: 243-278.

EICHHORN O., (1969). Investigations on woolly aphid of the genus *Adelges* An. (Hemipt., Adelgidae) and their predators in Turkey. *Tech. Bull. CICB*, **12**: 83-103.

FRANZ J.-M., (1958). Studies on *Laricobius erichsoni* Rosenh. (Coleoptera, Derodontidae) a predator on Chermesids. *Entomophaga*, **3**: 109-198.

GAUMONT R., (1966). Sur les galles mixtes de Chermésides (= Adelgidés). *Marcellia*, **33**: 159-164.

GAUMONT R., (1978). Tableaux pratiques de détermination des principales formes de Chermésides (= Adelgidés) de France. *Rev. For. Fr.*, **30**: 21-36.

MARCHAL P., (1913). Contribution à l'étude de la biologie des chermes. *Ann. Sci. Nat. Zool.*, **18**: 153-385.

MITCHELL R. G., MAKSYMOV J.-K., (1977). Observations of predation on spruce gall aphids within the gall. *Entomophaga*, **22**: 179-186.

PSCHORN-WALCHER H., ZWÖLFER H., (1956). The predator complex of white fir woolly aphids (genus *Dreyfusia*, Adelgidae). *Zeit. ang. Ent.*, **39**: 63-75.

PURITCH G. S., PETTY J. A., (1971). Effects of balsam woolly aphid *Adelges piceae* (Ratz.) infestation on the xylem of *Abies grandis* (Dougl.) Lindl. *J. Exp. Biol.*, **22**: 946-952.

ROHFRITSCH O., (1966). Rôles respectifs de la fondatrice et des gallicoles dans le développement et la maturation de deux galles de Chermesidae: *Adelges abietis* Kalt., *Adelges strobilobius* Kalt. *Marcellia*, **33**: 209-221.

ROHFRITSCH O., (1967). Conditions histo-cytologiques nécessaires à la base des aiguilles d'épicéa pour permettre le développement des larves de Chermesidae. *C. R. Ac. Sc.*, **205**: 1,905-1,908.

STREBLER G., (1975). Mécanismes de formation des zoocécidies. *Ann. Zool. Écol. Anim.*, **7**: 273-293.

TOROSSIAN C., (1971). La faune secondaire des galles de Cynipides. I. Étude systématique des fourmis et des principaux arthropodes récoltés dans les galles. *Insectes Sociaux*, **18**: 133-154.

TOROSSIAN C., (1972). Étude biologique des fourmis forestières peuplant des galles de Cynipidae des chênes. Rôle et importance numérique des femelles fondatrices. *Insectes Sociaux*, **19**: 25-38.

WHITHAM T. G., (1980). The theory of habitat selection: examined and extended using *Pemphigus* aphids. *Am. Nat.*, **115**: 449-466.

WHITHAM T. G., (1986). Costs and benefits of territoriality: behavioral and reproductive release by competing aphids. *Ecology*, **67**: 139-147.

WILSON F., (1938). Notes on the insect enemies of Chermes, with particular reference to *Pineus pini* Koch and *P. strobi* Hartig. *Bull. Ent. Res.*, **29**: 373-389.

ZUCKER W.-V., (1982). How aphids choose leaves: the roles of phenolics in host selection by a galling aphid. *Ecology*, **63**: 972-981.

◇ **13** ◇

Insects on flowers, fruits and seeds

In the forest environment the flowers of trees are eaten by some insects, while the fruits and seeds are a rich and sometimes major source of food that insects share with birds and mammals. Here we will consider only fruits and seeds of the more important trees, both broadleaf and conifer. The insects of conifer cones in France have been the subject of a very comprehensive study by Roques (1983, 1988).

1. IMPORTANCE OF FRUITS AND SEEDS IN THE FOREST ENVIRONMENT

Biomass and productivity of fruits and seeds have been evaluated in various types of forest. In a plantation of forty year old pine trees in Britain the biomass of cones is 0.4 tons per hectare (dry weight). These cones are almost entirely produced by dominant trees, which, in a normal year, are the only ones able to accumulate amounts of products of photosynthesis in excess of those needed for maintenance and growth. In an oak / hazel wood in Belgium the biomass of acorns is 2,000 kg per hectare and that of hazel nuts 260 kg per hectare. In some Belgian forests productivity of the whole of fruits, flowers and inflorescence bracts is 0.8 to 1.2 tons per hectare and per year for a *Querceto-Carpinetum*, 0.8 tons for a *Fagetum*, and 0.06 tons for a *Piceetum* (Duvigneaud, 1974; Duvigneaud *et al.*, 1971). In Carpathian forests productivity is 1 ton per hectare of acorns, 2 tons of beechnuts and 1.2 tons of pine seeds.

The chemical composition of fruits and seeds is very different from that of a tree's other organs: they have low water content (5 to 18%), are rich in easily absorbable organic substances such as glucids, lipids, and in some cases protids, and are poor in minerals. Calorific value of seeds is high: 407 cal g^{-1} of dry matter in the case of acorns, and 6,700 cal g^{-1} in the case of beechnuts. These properties make them choice forage for many wild animals and sometimes as food for man (pine nuts, chestnuts) or fodder for domestic animals. Acorns were formerly a major source of food for pigs which foraged freely in forest.

1.1. Variability in fruit and seed production

Many trees go through periods of abundant fruiting, known as masting, followed by years of poor production. In Denmark beechnut production varied between 0 and 53.5g m^{-2} (dry weight) between 1967 and 1975 (Nielsen, 1977). Inter-annual variability in cone and seed production in various species of conifer is well known (figure 13.1). These trees have great

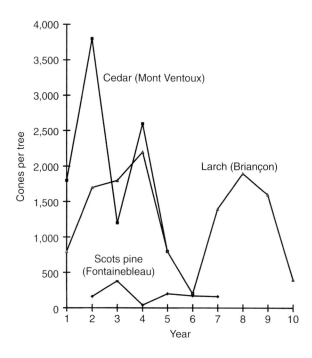

Figure 13.1. *Inter-annual variability in cone production in three species of conifer (Roques, 1988).*

reproductive aptitude. A sequoia may produce 10^9 to 10^{10} seeds during its lifetime. These observations have led some to accept the "prodigal parent" theory, according to which the great number of seeds produced means that at least one of them is bound to germinate and grow into a tree. This has led to minimising the destructive part played by insects, although Janzen (1971) brought forward many arguments against this theory. Difficulties in propagating larch were long attributed to the nature of soils or to competition between seedlings and herbaceous plants. It is only recently that Roques *et al.* (1984) have shown that insects reduce the number of viable seeds by 60 to 80%. In the case of beech seeds, those that fall to the ground are often of very poor quality. Five to 20% are sterile and 7 to 55% are eaten by *Laspeyresia fagiglandana* caterpillars. Germinating plants are attacked by the weevil *Phyllobius argentatus* and the aphid *Phyllaphis fagi* as well as by geometrid moths and the Homopteran *Typhlocyba cruenta*. Beechnuts are also attacked by the larvae of *Athous* (Elateridae) (Watt, 1923; Le Tacon & Malphettes, 1974; Nielsen, 1977).

Masting is interpreted as a tactical response by trees to destruction of their reproductive organs. Periods of low fruit and seed production correspond with low numbers of seed predators. The latter do not have time to rapidly increase their numbers during masting and thus the number of seeds that escape destruction and may produce seedlings remains high.

Fruit and seed mortality due to insects and other animals such as rodents may be high. In an English oak wood the probability of an acorn germinating and producing a one year old seedling is lower than 0.0006 (Shaw, 1968). In Virginia the blue jay (*Cyanocitta cristata*) carries away and hides 54% of the acorns of *Quercus palustris*; about 20% of acorns are eaten on the spot and what is left under the trees is parasitised by weevil larvae (Darley-Hill & Johnson, 1981). Fruits attacked by insects often fall prematurely. This has been shown to be the case in *Carya glabra* fruits which fall in autumn when they are viable and in summer when they have been attacked by insects (Boucher & Stork, 1979).

1.2. Prolonged diapause

For many insects diapause is a fundamental stage in life which enables them to survive an unpropitious period. Some insects specialised in exploiting forest tree fruits or seeds may undergo prolonged (up to seven years) diapause at the larval or pupal stage in accordance with annual fluctuations in abundance of fruits and seeds. (Prolonged diapause is not to be confused with the very slow growth of some xylophagous insects such as the American buprestid *Buprestis aurulenta* whose life cycle may last 20 to 50 years.) Studies of population dynamics of these species show that they rapidly regain their former numbers when abundance of fruits and seeds increases again. One mechanism that regulates population is prolonged diapause, which has been observed in at least 55% of species in the European fauna that live on conifer cones. In contrast to these, most generalist feeders do not undergo prolonged diapause (Debouzie & Menu, 1990; Roques, 1990).

> The adaptive value of prolonged diapause has been established for the chestnut weevil *Curculio elephas*, 32 to 56% of individuals of which go into prolonged diapause. This beetle's larvae bury themselves in the soil in autumn. Some have a short diapause, and pupate and produce adults in the following summer. Others undergo diapauses of one, two or three years, spreading emergence of adults over a period of

several years. The species is thus protected against unforeseeable changes in abundance of chestnuts. Differences in diapause time appear to be determined genetically. Computer simulation shows that genotypes that produced larvae with a fixed diapause period die out in less than 100 generations, while a population of mixed genotypes which prolong diapause in 50% of larvae maintains itself easily.

Lasiomma melania (Diptera, Anthomyiidae) is the greatest destroyer of larch seeds in France. After consuming the seeds of a cone, the larva of this species migrates to the ground where it builds a puparium in June or July, depending on altitude. Pupation takes place in the following spring, or the larva may undergo a diapause lasting up to three years. The proportion of adults that go into diapause varies from 5 to 50% in some years, and rises as cone production drops. It would seem that the cones contain as yet unidentified substances that stimulate physiological changes that prepare this species for diapause. The various species of torymid wasp belonging to the cosmopolitan genus *Megastigmus* live on species of conifer. Only those that live on conifer species whose seed production is very variable tend to undergo diapause. *Megastigmus strobilobius* feeds on spruce, which has fruiting peaks on average every three years. The larva lives its whole life in the seed on which it feeds. Pupation takes place either in the following spring or after a prolonged diapause which may last from one to five years. *Megastigmus amicorum*, on the other hand, lives on *Juniperus oxycedrus* which produces few fruits but does so regularly every year, and the wasp's larvae very rarely undergo diapause, and then only for one year.

Optional diapause, providing protection to a proportion of individuals against temporary unfavourable conditions, occurs in insects other than seed-eaters. It has been recorded in *Villa brunnea*, a bombyliid fly whose larvae are endoparasites of the processionary pine caterpillar. Diapause may take place at the egg stage or as the fully-grown larva inside the host's pupa. Adults that emerge in the following summer thus stem from two different sources, which spreads risk in the face of absence or scarcity of the host (Du Merle, 1969).

Many other adaptive characters are met with in insects that live on cones. In the genus *Megastigmus* there is a relationship between ovipositor length and thickness of cones, a character that represents the distance between seeds and the surface of the cone (figure 13.2). This adaptation affects the nature of tree species that may be attacked in the case of introduced species. In Finland *Megastigmus specularis*, which is introduced from the United States, has a short ovipositor and only attacks cones under 10 cm long (such as those of *Abies sibirica*), whereas *M. suspectus*, which is also introduced from the same source, can lay in cones between 10 and 20 cm long by means of its ovipositor which is about 1.5 times longer (Annila, 1970).

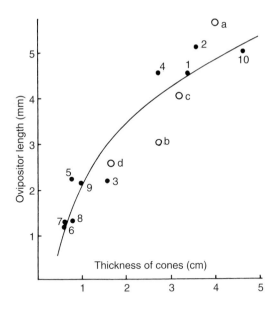

Figure 13.2. *Relationship between conifer cone thickness and ovipositor length in various species of* Megastigmus. *Native species.*
1: Megastigmus strobilobius *on* Picea abies;
2: M. suspectus *on* Abies alba;
3: M. pictus *on* Larix decidua;
4: M. wachtli *on* Cupressus sempervirens;
5: M. amicorum *on* Juniperus oxycedrus;
6: M. bipunctatus *on* Juniperus communis;
7: M. bipunctatus *on* Juniperus sabina;
8: M. bipunctatus *on* J. thurifera;
9: M. amicorum *on* Juniperus phoenicea;
10: M. suspectus *var.* pinsapinis *on* Cedrus atlantica.
The following species introduced from North America are not taken into account on the regression curve. a: M. pinus *on* Abies nordmanniana; *b:* M. specularis *on* Abies sibirica; *c:* M. spermotrophus *on* Pseudotsuga menziesii; *d:* M. atedius *on* Picea orientalis *(Roques, 1988).*

1.3. Economic importance of pests of fruits and seeds

The economic importance of insects and other organisms that destroy fruits and seeds of forest trees has only recently been taken into consideration, following changes in forestry techniques. In the past, natural regeneration of forest was the rule, and thus losses of seeds might go unnoticed. These days cultivation of seedlings from seed and planting young trees is becoming the rule, particularly in areas that were originally un-wooded or forested with other species. For this reason damage to seeds has become more apparent. In Norway harvesting spruce cones is not economically profitable in some years (Bakke, 1963). In parts of the Carpathians seed loss in pines may reach 84%. In Normandy beech woods there has been no natural regeneration of beech for many years due to destruction of beechnuts by rodents and insects. Proliferation of pests of coniferous trees is aided by planting in regions in which soil or climate are unsuitable. This explains the importance given to an insect such as *Megastigmus spermotrophus,* which lives on Douglas-fir seeds in North America, and was introduced to Europe where it has become a major pest of that tree, particularly in seed orchards. The economic importance of fruit- and seed-eating insects accounts for the many studies undertaken on these insects, particularly in regard to conifers, in the past twenty years, both in France and Europe and in North America.

2.   FLOWER-EATING INSECTS

Of the factors that may reduce numbers of flowers and infloresences of trees, insects play a part that has long been known but has been little studied. Our knowledge of this subject is largely based on oak inflorescences. Cynipid gall wasps attack the reproductive organs of these trees and impede acorn production. The most remarkable case is that of *Andricus quercuscalicis* which may cause the destruction of 80% of acorns and jeopardise tree regeneration (Scutareanu & Roques, 1993).

Observation of oak flowers from the moment they appear until they are fertilised shows massive reduction in numbers of flowers in the period preceding fertilisation. Reduction reaches 91% in *Quercus robur.* Two phenomena account for this: microclimate and insects. Late frosts at night cause dessication and the fall of 45 to 50% of flowers. The insects in question are the caterpillars of *Tortrix viridana* and *Operophtera brumata* in France and *Lymantria dispar* in Romania. Cecidomyiid larvae such as *Contarinaria amenti* grow inside the flowers' ovaries. Moreover thirteen species of weevil previously known to eat the leaves have been found to also attack flowers or flower stalks, causing them to wilt and drop. The most noteworthy of these weevils are four species of *Polydrusus,* two species of *Strophosomus, Brachyderes incanus, Phyllobius piri,* and especially *Coeliodes ilicis* (Rougon *et al.,* 1995). Later consequences, in years of low acorn production, are all the more serious the greater the loss of flowers has been.

3.   CONIFER CONE INSECTS

3.1. Biological diversity

Cone-dwellers, which have been given the name conobionts, have been classified in various categories. (a) Stenoconobionts are associated exclusively with cones. Those whose entire life cycle is completed in a cone are called conophiles. Where part of a species' life cycle takes place outside the cone (for instance if it pupates in the soil) the species is known as a conoxene. (b) Heteroconobionts are able to complete a part or all of their life cycle on other organs such as buds, leaves or bark. Those species that attack intact cones are primary heteroconobionts while those that can only settle in cones that have already been attacked by others are secondary heteroconobionts. The number of phytophagous insects that exploit cones in western Europe has been evaluated at 59 species belonging to 30 genera, 13 families and 4 orders, plus one species of mite (figure 13.3). (c) Added to this group of phytophagous species are parasites and predators, some of which are host-specific on stenoconobionts while others are polyphagous. Parasites and predators have so far only been listed for 38 species of cone-eaters. They are mostly wasps, and belong to the following families:

Hymenoptera

Ichneumonidae: 43 species	Braconidae: 28 species
Pteromalidae: 22 species	Torymidae: 5 species
Eulophidae: 4 species	Eupelmidae: 3 species
Tetrastichidae: 2 species	Platygasteridae: 1 species
Eurytomidae: 1 species	Encyrtidae: 1 species
Trichogrammatidae: 1 species	Cynipidae: 1 species

Diptera

Tachinidae: 5 species	Lonchaeidae: 2 species (predators)

Neuroptera

Chrysopidae: 1 species (predator)	Rhaphidioptera: 2 species (predators)

The number of parasite species associated with each species of conobiont is limited and is usually fewer than ten. The cryptic life style of most conobionts, which limits their accessibility to parasites, appears to be the cause of this.

3.2. The concept of merocenose

Cones produced by female inflorescences are plant units having their own characteristic morphological and physiological properties. Within the tree a cone forms a particular microhabitat which has variously been termed "structural element" ("Strukturteil") or "merotope" (Tischler, 1955). Insects and other organisms that inhabit them form merocenoses. These are not totally independent of other parts of the tree: they are tied to them in particular by the presence of heteroconobionts as well as by numerous non-specific predators and parasites of conobionts that also prey on many other species. The merocenose is a relatively autonomous group of species but is unable to regulate itself and is under the control of the pine tree's overall biocenose. Conobiont population regulation is largely due to control coming from various parts of the tree.

Figure 13.3.
Diversity of orders and families of insects and Acarians that exploit cones in western Europe.
L: Lepidoptera;
H: Hymenoptera;
D: Diptera;
C: Coleoptera;
A: Acari.
The asterisk indicates a detritivorous species (Roques, 1988).

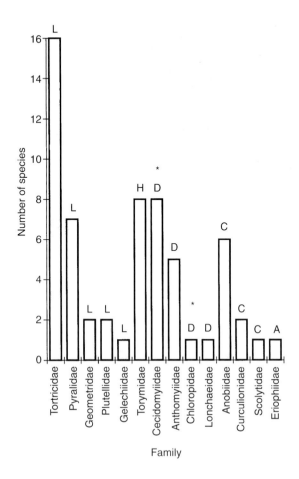

These relationships may be illustrated with the example of Scots pine (figure 13.4). Relationships exist between the parasite complexes of cones, those of *Petrova resinella* galls, and those of the moths *Rhyacionia buoliana* and *Blasthestia turionella*. The following are three examples: (a) Adults of the ichneumonid *Scambus sudeticus* which as larvae parasitised *Pissodes validirostris* or *Petrova resinella* are unable in spring to find *Pissodes* larvae or *Petrova* caterpillars. This ichneumon wasp then predates caterpillars of *Rhyacionia buoliana* which are the only available hosts at that time of year. Adults resulting from these eggs emerge in July, when *Pissodes* and *Petrova* are available as hosts. The weevil *Pissodes validirostris* thus serves in this case as an intermediate host for the parasite *Scambus sudeticus*. Any reduction of the *Pissodes* population will therefore be an inhibiting factor in the *Scambus* population. (b) The tachinid fly *Actia nudibasis* emerges in mid July from old *Rhyacionia buoliana* caterpillars which its parents have parasitised. At this time of year it can lay its eggs in the caterpillars of either *Rhyacionia*, *Dioryctria* or *Petrova*. In the latter case the cone pests provide a host reserve for a parasite of pests other than cone pests. (c) In the case of

Plate 1

Above: view in a non-exploited forest, la Massane beech wood in the Pyrénées-Orientales. A clearing has been made by the fall of very old beech trees, one of which remains standing and bears numerous bracket fungi, in the centre of the photograph. In the background are young trees which are part of the process of natural regeneration of the forest free from any human interference. In this relict forest saproxylic beetles are numerous and include many rare species. Species of cerambycid include *Rosalia alpina*, *Morimus asper* and *Aegosoma scabricorne*; anobiids include *Anobium inexpectatum* and *Xestobium rufovillosum*; scolytids include *Scolytus koenigi*, and the elaterids include *Melanotus tenebrosus*, *Ampedus praeustus* and, most notably, *Ampedus quadrisignatus* for which this forest is the only remaining French locality and which may by now be extinct. The mycetophagous beetle fauna living in bracket fungi is also very rich and includes the tenebrionids *Bolitophagus reticulatus* and *Diaperis boleti*.

Below: a nest mound belonging to the wood ant *Formica polyctena* in the Crouzet forest in the Caroux hills (Hérault, southern France). The forest is a beech wood at about 1000 m altitude containing many introduced pines. The nest, which is made mostly of pine needles, was about one metre high. It contained numerous commensal arthropods such as mites, a psocid, Heteroptera and the beetle *Myrmecoxenus vaporarium*.

An emergence trap placed on a fallen Scots pine trunk. This system monitors the life cycles of pine scolytids such as *Ips sexdentatus* as well as those of their predators and parasitoids.

A window trap set in woodland. Flyir insects strike the glass pane which the cannot see and fall into the containe below it which are filled with water which a wetting agent is added to preve the insects from escaping.

Plate 2 -
Methods of sampling forest insects

A pheromone trap placed on the trunk of a western yellow pine (*Pinus ponderosa*) tree. Pheromones are placed in the trap with the addition of glue to trap insects attracted to them. This gives indications of variations in abundance and emergence periods of various pests such as scolytids and is used in trials of biological control.

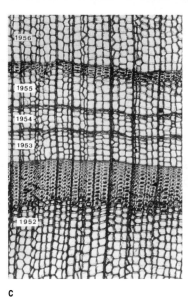

b   c

Plate 3 - The larch bud moth,
Zeiraphera diniana

e

: damage caused by fifth instar larva on a twig. Note the
mount of food wasted by these highly voracious cater-
llars. Uneaten needles are gathered in bunches by a silky
etwork of threads to which frass adheres. These needles
ry up and give larch trees a characteristic brown appear-
nce.

: remains of needles grouped in "funnels" by fourth or fifth
star larvae. Needle tips have been eaten. Each funnel is
rmed of the central needles produced by a bud and
athered together by silk threads.

. part of a section of a larch tree trunk from the High
ngadine, damaged in 1953-1954. Thickening of wood is
reatly reduced in years in which outbreaks of the larch bud
aterpillar occur.

: the moth

: larch tree completely stripped of needles and with its
unk bearing webs spun by starving caterpillars in search
f food.

Plate 4

Above, left: typical "bayonet" shape of a pine shoot caused by *Rhyacionia buoliana*.

Above, right: *Rhyacionia buoliana* caterpillars hibernating in the terminal shoot of a pine tree.

Below, left: *Lymantria dispar* moths and pupal exuvia. On the bottom left of the photograph are a pair of mating moths; the male is darker than the female and has strongly pectinate antennae.

Below, right: gregarious false caterpillars; these are larvae of the sawfly *Neodiprion sertifer* gathered in characteristic clusters on a pine that they are busy devouring.

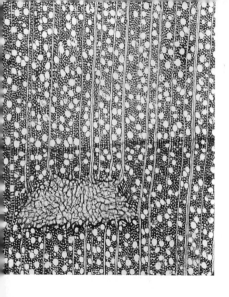

b

d

e

f

ate 5

cross-section of wood of a tree of the genus *Sorbus*
owing disruption of the xylem caused by fly larvae belonging
the family Cecidomyiidae (magnified × 26).

larval galleries in spruce of the siricid wasp *Xeris spectrum*.
the right are two galleries leading away from a common
netration orifice and to the left are three other galleries.

the cerambycid beetle *Cerambyx cerdo* on a wilting oak tree
nk in which its larvae live. This dark brown species, which
ches a length of 54 mm, is one of the largest longhorn
etles in the European fauna and is still common in many
eas. The larvae live three or four years.

and **e**: a gall of the chermesid gall aphid *Chermes tardus* on
ruce. **d** shows the open gall from which the insects (alate
igrants) have flown. In **e** the gall is still young and closed,
owing the shape of the scales.

ir branch with arrows indicating galls formed as a reaction
occupation by hibernating adults of the scolytid beetle
yphalus piceae.

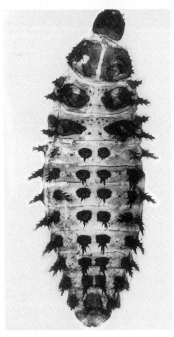

a

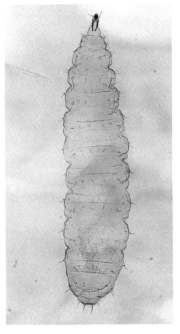

b

c

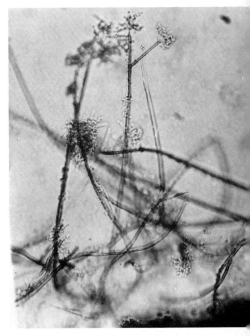

d

Plate 6

a: fourth instar larva of the ladybird *Exochomus quadripustulatus*, a predator of the beech sca Cryptococcus fagisuga.

b: (? second instar) larva of a cecidomyiid fly of the genus *Lestodiplosis*, another predator of the scale inse

c: inside of a larval gallery of the mycetophagous scolytid *Xyleborus monographus* showing the symbic fungus of the "Ambrosia" type.

d: detail of the same fungus enlarged × 290

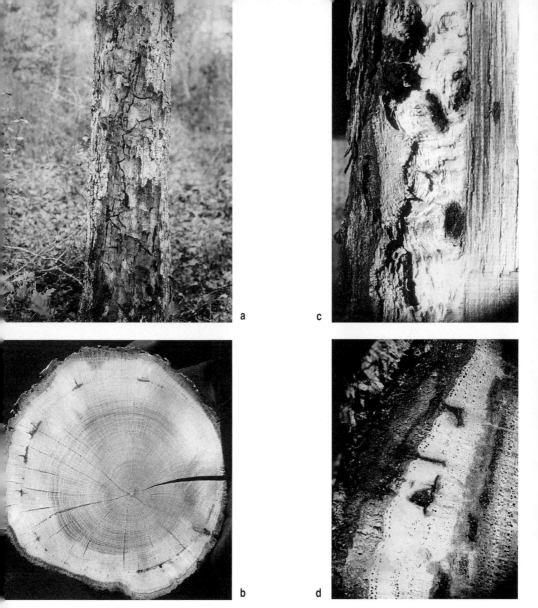

a

c

b

d

ate 7 - Effects of the beetle *Agrilus biguttatus* on oak

downy oak (*Quercus pubescens*) infested by the buprestid's larvae in Fontainebleau forest. Larval gal-
ies are clearly visible in the form of black lines where the bark has been removed.

cross-section of an infested branch. The attack occurred two years before the tree fell. Each gallery is an
and of dead cambium and phloem dyed black by tannins. Around each gallery cambial activity has been
mulated and annual growth rings are broader.

outer part of a branch in radial section. Normal growth areas (to the right in the photograph) show longi-
dinal fibres which contrast with the left side where fibres are diverted following attack by *Agrilus* larvae.
eir galleries are seen in cross-section (magnified x 4).

cross-section in the outer part of an infested branch in the vicinity of *Agrilus* galleries. From right to left
e the last growth ring formed before the attack, the dead cambium and the phloem, and a new growth area
h two annual growth rings in which there are more of the beetle's galleries. Note the reduced diameter of
ssels, making the wood resemble ivory. A canal opening outward allows a blackish fluid to drain to the
ter surface of the branch (magnified x 4).

Plate 8

Above: a recently burned area in Yosemite National Park in California. Trees are mostly conifers, especially Douglas-fir (*Pseudotsuga menziesii*) and pines (*Pinus* sp.). Among the insects attracted to the fire were various carabid beetles, including *Sericoda bogemanni*, *Sericoda bembidioides* and *Nomius pygmaeus*, the former two species being very abundant. *Rhysodes hamatus* and a platypodid belonging to the genus *Platypus* were found under bark. Many buprestids settled on scorched trees.

Centre: gallery systems of the scolytid beetle *Cryphalus piceae*. On the left, are systems of two opposite galleries and H-shaped systems with four galleries. On the right, is a three maternal gallery system with galleries perpendicular to the grain of the wood. Egg laying notches are clearly visible on the sides of galleries.

Below, left: dorsal surface of the posterior part of the prothorax of a female *Platypus cylindrus* showing the mid-longitudinal furrow, and the mycangium which consists of numerous (about 400) depressions which contain the symbiotic fungus. In the male these depressions are less numerous, sparser and larger (magnified × 120).

Below, right: the click beetle *Alaus melanops* belonging to the family Elateridae. Species of the genus *Alaus* are over 25 mm long and are recognisable by the two eye-like dark spots on the prothorax. This species is common in California. Its larvae live mainly in the dead wood of pines and *Pseudotsuga*. They predate other insects and also feed on dead wood.

Plate 9

Above, left: view in Gros Fouteau Biological Reserve in Fontainebleau forest. In the right foreground, is an oak tree aged about 400 years whose bent trunk is due to growing conditions in a dense stand of trees. The curvature of its trunk, produced by phototropism, has enabled it to exploit a breach in the canopy to reach the dominant layer. In the left middle-ground, another oak tree about 200 years old has also avoided the shade cast by the beech tree which has a straight trunk. This stand is composed of trees of all ages and includes thickets and coppice in which beech dominates. A high incidence of bent trunks is a characteristic of natural forest, whereas in exploited forests cleared areas allow vertical growth of selected trees.

Below, left: the Basidiomycete fungus *Dryodon coralloides* on a beech tree trunk in an advanced stage of decomposition in la Tillaie Biological Reserve in Fontainebleau forest. This spectacular species (its creamy-white fruiting body may grow to more than 40 cm) is becoming very rare, for it only grows on fallen dead trees of large diameter. This environment is gradually being lost in exploited forest.

Above, right: *Trypodendron lineatum*, a scolytid beetle that bores deep galleries in various conifers, especially fir, but also spruce, and more rarely in Scots pine and larch. The photograph shows a maternal gallery and the beginnings of larval galleries. The wood of the gallery walls is died black by an "Ambrosia" type symbiotic fungus.

Bottom, right: the asilid fly *Andrenosoma bayardi*. Like all asilids, this species has a robust body thickly covered in hairs. The photograph shows an individual settled on pine bark blackened by fire, and blending almost perfectly with the background.

Zeuzera pyrina (Cossidae)

Sesia (= *Aegeria*) *apiformis* (Sesiidae)

Bupalus piniarius male (Geometridae)

Panolis flammea (Noctuidae)

Dendrolimus pini (Lasiocampidae)

Lymantria dispar female (Lymantriidae)

Plate 10 - A few common moth

nnomos quercinaria (Geometridae)

Erannis defoliaria male (Geometridae)

uproctis chrysorrhea (Lymantriidae)

Thaumetopoea pityocampa (Thaumetopoeidae)

ymantria dispar male

Hyloicus pinastri (Sphingidae)

f the forest environment

Plate 11

Plate 11

Above: the Malaise trap, named after its inventor, has been used intensively in many areas. It consists of a large tent-shaped cage made of fine netting into which insects fly through a gap. Insects are trapped in a bottle (seen on the top of the trap) owing to their behaviour which induces them to fly upward when they encounter the vertical wall of netting. This trap is efficient for catching Diptera and Hymenoptera, but less so for other insect orders. Combining a Malaise trap with a light trap makes it possible to catch insects both during daytime and at night.

Below: larva of the cerambycid beetle *Ergates faber* in pine (southern France).

Plate 12

Above: a tree used as trap. This consists of a tree (*Pinus ponderosa* in this case) which has been felled and stripped of branches and on which the arrival of xylophagous insects is monitored (Flagstaff, Arizona, July).

Centre: a male of the cerambycid *Monochamus scutellatus*. Many beetles belonging to this species were attracted to the trap tree on which they came to mate and lay. As in most cerambycids, the male is recognisable by its antennae which are longer than the body.

Below, left: larval galleries of the buprestid *Chrysobothris affinis* under beech bark.

Below, right: an adult *Cucujus clavipes*, a beetle of the family Cucujidae, under pine bark in California. This species is still common in North America. In Europe the two related species *Cucujus cinnabarinus* and *Cucujus haematodes* are relicts headed for extinction. *Cucujus haematodes* has not been recorded in the last hundred years.

Plate 12

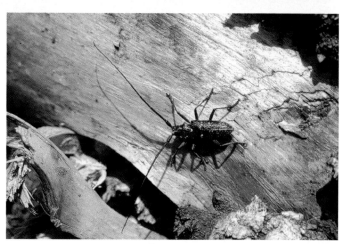

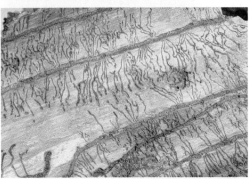

Plate 13

Above, left: a stand of spruce trees that have died following infestation by the scolytid *Dendroctonus mica* (Bor forest, Lozère, southern France).

Above, right: gregarious larvae of *Dendroctonus micans* massed under bark.

Centre, left: subcortical galleries of the scolytid *Dendroctonus brevicornis* on *Pinus jeffreyi* in California.

Centre, right: characteristic resin flux forming at a *Dendroctonus* penetration orifice in pine bark.

Below, left: larva of a buprestid of the genus *Chrysobothris*.

Below, right: larval galleries of the beetle *Hyleoetus dermestoides*, belonging to the family Lymexylonid in beech. This beetle's galleries are of two types. The first penetrate the wood and are disposed radially (th are visible in cross-section). The others are sub-cortical and seen in longitudinal section. Blackening of galleries is due to the symbiotic fungus on which the larvae feed.

ate 14

ove, right: a bristle cone pine (*Pinus longaeva*) of the White Mountains of California. These trees, which
ow at high altitude (3,000 metres and above) in harsh (arid) climatic conditions and on poor soils, live to
very old (4,600 years); as their growth is very slow and their wood very hard they show no traces of dam-
e by insects or fungi.

ove, left: a gall known as "oak apple" formed by the cynipid *Biorrhiza pallida* on oak.

ntre, left: a newly emerged *Uleochaetes leoninus*, a beetle of the family Cerambycidae, still in its pupal
amber under pine bark in California. This beetle has abbreviated elytra and is patterned in yellow and
ick, which makes it resemble a wasp. This may be a case of mimicry evolved to protect the beetle against
ssible predators.

ntre, right: *Cynips quercusfolii* gall on oak.

low, left: the cerambycid *Tragosoma depsarium*, which lives on conifers. This is a rare species localised
France in mountains above 1,500 metres.

low, right: "pineapple" gall of the chermesid *Adelges tardus* on spruce.

Plate 15

Above, left: nests of the processionary pine caterpillar on Austrian pine on Mont Ventoux (SE France).

Above, right: the dorso-ventrally flattened larva of *Pyrochroa coccinea* which lives under the detached ba of dead trees, and two larvae of *Cucujus clavipes* under conifer bark. Although they belong to two differe families the larvae of these two species are very similar in appearance; this is a case of morphologic convergence brought about by living in the same habitat.

Centre, left: eggs of the sawfly *Pristiphora abietina* laid on the undersides of spruce needles.

Centre, right: base of a beech tree trunk covered in white spots which are the shields of females of the sca insect *Cryptococcus fagisuga* (Villers-Cotterêts forest, Aisne, France).

Below, left: the woolly beech aphid *Phyllaphis fagi*. The leaves that are attacked wilt and are covered with waxy coating secreted by the aphid.

Below, right: the carabid beetle *Carabus intricatus* is a typical species of forest soil.

the two wasps *Bracon* and *Exeristes*, a second, intermediate host (marked X1 and X2 in figure 13.4) is required between the cone pest and the other pest.

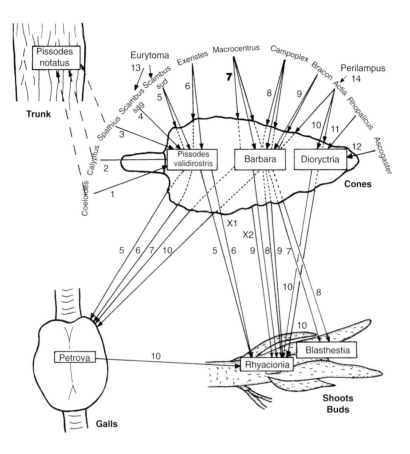

Figure 13.4. *Schematic representation of relationships between the insects of Scots pine cones, those of shoots and buds, those of* Petrova resinella *galls, and those of the tree trunk (Roques, 1975). Parasitic species of the three main phytophagous species are numbered 1 to 12.*

3.3. Variations in species-richness

Species richness in populations of cone insects varies according to the species of conifer and to geographical location. In Europe the fauna of spruce cones includes 11 to 20 phytophagous species and 9 to 41 predators and parasites, according to region.

There is a link between species-richness and the present extent of the distribution ranges of various species of conifer (figure 13.5). This is but one particular aspect of a very general law governing plant / insect relationships (cf. chapter 4).

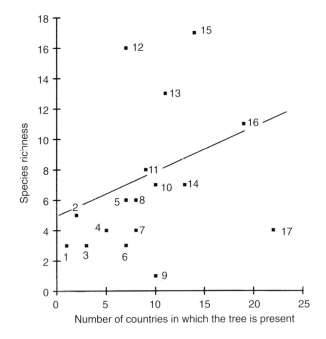

Figure 13.5. *Species richness of insects living in the cones of 17 species of European conifer as a function of the distribution ranges of host trees represented by the number of countries in which they occur (Roques, 1988). Conifers studied are as follows:*
1: Pinus omorica; 2: Cupressus sempervirens; 3: Juniperus thurifera; 4: Pinus pinaster; 5: Pinus uncinata; 6: Pinus cembra; 7: Pinus pinea; 8: Pinus halepensis; 9: Juniperus sabina; 10: Juniperus phoenicea; 11: Pinus nigra; 12: Larix decidua; 13: Abies alba; 14: Juniperus oxycedrus; 15: Picea abies; 16: Pinus sylvestris; 17: Juniperus communis.

Variation in numbers of insect species within a single tree species appears to be linked to the age of the population. In the case of spruce, local species richness varies according to a gradient which follows remoteness of the date at which the tree first appeared in the Quaternary; curves are similar for the stenoconobionts and heteroconobionts (figure 13.6). Their asymptotic shape shows that the population is virtually definitive after the first 2,000 years.

Colonisation of introduced and taxonomically isolated tree species (for example belonging to the genera *Pseudotsuga*, *Tsuga*, and *Sequoia*, which do not occur in Europe) is very slow. In the case of conifer cones the seeds are the last part of the tree to be colonised, unless the pest is introduced together with the host. Douglas-fir is a typical case: while its associated fauna is relatively rich in its native land, the fauna that has adapted to it in Europe is still poor in species, but it was accompanied from America by the seed-eater *Megastigmus spermotrophus* (table 13.1).

North America	France
Seed-eaters *Megastigmus spermotrophus* (Torymidae) *Contarinia oregonensis* (Cecidomyiidae)	*Megastigmus spermotrophus* (Torymidae)
Cone- and seed-eaters *Choristoneura occidentalis* (Tortricidae) *Dioryctria abietivorella* (Pyralidae) *Dioryctria pseudotsugella* (Pyralidae) *Eupithecia albicapitata* (Geometridae) *Earomyia barbara* (Lonchaeidae) *Earomyia aquilonia* (Lonchaeidae) *Henricus fuscodorana* (Cochylidae)	*Assara terebrella* (Pyralidae)
Cone-eaters *Contarinia washingtonensis* (Cecidomyiidae) *Ernobius punctatulus* (Anobiidae) *Holcocerca augusti* (Blastobasidae) *Holcocerca immaculella* (Blastobasidae)	*Asynapta* sp. (Cecidomyiidae) *Ernobius mollis* (Anobiidae) *Dasyneura* sp. (Cecidomyiidae) *Hapleginella laevifrons* (Chloropidae)

Table 13.1. *Fauna associated with Douglas-fir cones in its North American homeland and in France, where it has been introduced.*

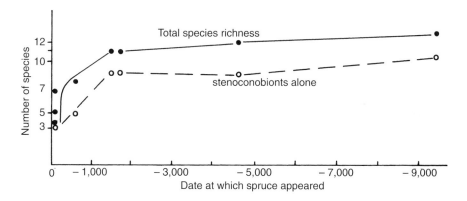

Figure 13.6. *Local species-richness of insects associated with spruce cones as a function of the first appearance of the tree during the Quaternary era in different parts of Europe (Roques, 1988).*

3.4. Merocenose of Scots pine cones

3.4.1. Composition and evolution

Four stages in the growth of a Scots pine cone may be seen. (a) Fertilisation of the female inflorescence leads to the formation of a conelet which, in April of the following year, rapidly increases in volume as does its water content, while its colour changes from brown to light green. (b) A pre-elongation stage is seen towards the end of April, together with an increase in dry weight and in fresh weight and a slight increase in water content. (c) An elongation stage begins from mid May on; water content increases heavily. The very elongate cone contains up to 445% water and is light green in colour. (d) A lignification stage occurs from mid June to July; dry weight increases and water content falls.

Merocenose growth is characterised by coincidence between the emergence of various species and changes in the physicochemical characters of cones, and particularly in water content (figure 13.7).

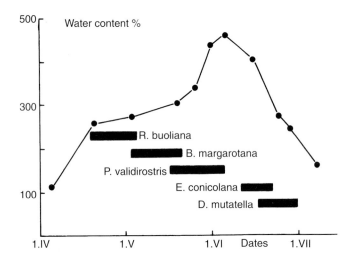

Figure 13.7. *Relationships between infestation periods by various pests of Scots pine cones in Fontainebleau forest and water content of cones (Roques, 1988).*

The five main pests infest trees at slightly different periods. Data gathered in Fontainebleau forest have, ignoring species which were only found on single occasions, provided a list of 25 species of conobionts belonging to the following categories:

– Stenoconobionts: two moths, *Barbara margarotana* and *Enarmonia conicolina*, and a beetle, *Pissodes validirostris*.

– Heteroconobionts: three moths, *Rhyacionia buoliana*, *Rhyacionia pinicolana* and *Dioryctria mutatella*.

– Detritivores: two flies, *Camptomyia pinicolana* and *Hapleginella laevifrons*.

– Predators: two lacewings, *Hemerobius pini* and *Raphidia* sp.

– Parasites: thirteen wasps and one fly. None of these parasites is host-specific on cone pests, but may live on other species. The parasites of *Pissodes validirostris* also predate *Pissodes notatus*, *Rhyacionia buoliana* and *Petrova resinella*. The main parasites found are as follows:

Parasites of *Pissodes validirostris*: three braconids, *Calyptus atricornis*, *Coeloides melanostigma* and *Spathius rubidus*; two ichneumonids, *Scambus* sp. and *Exeristes ruficollis*; one eurytomid, *Eurytoma wachtli*.

Parasites of *Barbara margarotana*: two braconids, *Macrocentrus resinellae* and *Bracon praecox*; one ichneumonid, *Campoplex ramidula*.

Parasites of *Dioryctria mutatella*: one braconid, *Ascogaster armatus*; one perilampid, *Perilampus tristis*; one pteromalid, *Rhopalicus guttatus*; one tachinid, *Actia nudibasis*.

3.4.2. Biology of the main species

Pissodes validirostris is a weevil found throughout Europe that lives on the cones of various species of pine. It has also colonised introduced *Pseudotsuga*. It is the main pest of *Pinus pinea* cones in central Spain, where the percentage of stricken trees exceeds 70% and sometimes reaches 100%, causing heavy economic losses, since pine nuts are an economic food crop. The consequences of infestation by *P. validirostris* are mainly due to reaction by the cones which secrete resin which glues the bracts to each other and prevents even intact cones from opening to release their seeds. Adult weevils gnaw cones as they enter the elongation stage, in other words when they are richest in nutritive substances. Oviposition time depends on changes in water content of cones (figure 13.7).

Barbara margarotana is a tortricid moth that lives on pine, fir and spruce cones. Larvae penetrate cones at the small cones' pre-elongation stage which is the time at which water content rises and pollen is released. The moth produces a single generation a year and the species overwinters in the pupa stage.

Dioryctria mutatella is a pyralid moth which occurs throughout Europe. It feeds on shoots and cones of various species of pine and is particularly associated with the cones. Eggs are laid randomly on trees and the young larvae must migrate to the cones in which they feed.

3.5. Insects of spruce, fir and larch cones

The insects of spruce cones have been studied in central Europe and in Scandinavia (Bakke, 1963). A list of 44 species made in Bavaria contains the following five categories:

– Insects that attack seeds. These are the chalcid *Megastigmus abietis*, the cecidomyiid *Plemeliella abietina* and the anobiid *Ernobius abietis*.

– Insects that attack cones. *Dioryctria abietella* settles before cones have fallen to the ground. Later arrivals include another moth, *Laspeyresia strobilella* and the cecidomyiid flies *Kaltenbachiola strobi*, *Camptomyia strobi* and *Clinodiplosis piceae* as well as the anobiid beetle *Ernobius piceae*.

– Parasites, which are all Hymenoptera.

– A bug, *Gastrodes abietum*, which hibernates in the cones.

– A detritiphagous beetle of the genus *Corticaria* and a predatory species of the genus *Raphidia*.

Kaltenbachiola strobi is the main cecidomyiid of spruce seeds and is widely distributed in the tree's entire geographical range. Its larvae predate the seeds and then pupate between the bracts.

The insects of pine cones are the same as or very close to those of spruce cones. The main ones are *Laspeyresia strobilella*, *Kaltenbachiola strobi* and *Ernobius abietis*.

Insects that attack larch cones are the moths *Petrova perangustana* and *Dioryctria abietella*, the cecidomyiid flies *Asynapta laricis* and *Resseliella skuhravyorum*, and the wasp *Megastigmus pictus*.

4. INSECTS ON ACORNS, BEECHNUTS AND OTHER BROADLEAF TREE SEEDS

The acorns of various species of oak and beechnuts form, as do conifer cones, a distinct microhabitat occupied by a particular merocenose which changes in composition as the acorns and beechnuts deteriorate until they are finally incorporated into forest floor humus.

4.1. Faunal successions in decomposing acorns

The different stages of this deterioration have been described in the case of American red oak by Winston (1956). Infestation begins with destruction of the embryo by various insects, the main one being the weevil *Balaninus rectus*. As they penetrate the acorn these insects introduce micro-organisms such as fungi which contribute to decomposition. In the next stage the embryo enclosed in the acorn is completely destroyed by the weevil and by the larvae of two moths, *Valentinia glanduella* and *Melissopus latiferreanus*. Several species of fungus occupy the acorn and serve as food for mites such as *Tyrophagus* sp. and *Rhizoglyphus* sp., for the sciarid fly *Sciara coprophila* and the drosophilid *Chymomyza amoena*. Then the outer covering or peri-carp is attacked by fungi able to decompose cellulose or lignin. Collembola such as *Xenylla* sp. and *Tullbergia* sp. contribute in accelerating decay. Where acorns fall on humid soil mites are numerous. In the last stage fungal infestation becomes intense. Many soil invertebrates become involved, parti-cularly ants, collembolans, myriapods and enchytraeids. The acorn finally falls to pieces and is incorporated in the humus.

It is likely that analogous successions exist in Europe where the main insects of acorns are weevils of the genus *Balaninus*, the moths *Laspeyresia splendana* and *Pammene fasciana* as well as various cecidomyiid flies.

The enemies of cork oak acorns have been studied in Mamora forest in Morocco. They appear to be few and close to, if not identical with, those that occur in Europe. They include the moth *Cydia fagiglandana*, two tineids, and about ten cecidomyiids which seem mostly to behave as mycetophages, feeding on moulds that grow in the acorns. In the Middle Atlas *Cydia fagiglandana* destroys 9 to 13% of evergreen oak acorns; *Cydia splendana* is also present, as is the weevil *Balaninus glandium*. Finally many scarabaeoid beetle larvae may impede natural regeneration by eating the roots of seedlings (Villemant & Fraval, 1991).

4.2. Beechnuts

In a Danish beech wood beechnuts have a biomass varying between 2 and 54 g m^{-2} year^{-1}. These beechnuts are of very poor quality since 5 to 20% are empty and non-viable, and 7 to 55% are predated by caterpillars of the tortricid *Cydia fagiglandana*. Density on the ground is on average 287 nuts per square metre in autumn. Among other despoilers of beechnuts are insects that attack germinating seeds, the commonest of which are the weevil *Phyllobius argentatus*, the aphid *Phyllaphis fagi*, the elaterid *Athous subfuscus* and the hopper *Typhlocyba cruenta* (Le Tacon & Malphettes, 1974). The effects of these insects, and particularly of *Cydia fagiglandana*, are sufficiently important to jeopardise natural regeneration of beech.

4.3. Some insects on fruits and seeds of broadleaf trees

Larvae of weevils of the genus *Balaninus* live in the fruits of various forest trees such as hazel and oaks. One exception is *Balaninus villosus* which lives in galls of the cynipid *Biorrhiza pallida*. Acorns harbour *Balaninus elephas*, *B. pellitus* and *B. glandium*. Hazelnuts are home to *B. elephas* and *B. nucum*, while *B. elephas* also exploits chestnuts. *Balaninus* species are remarkable for their long thin rostrum which may exceed body length in females, and for the position of their mandibles which are in the vertical plane, a unique characteristic among weevils (figure 13.8).

The female *Balaninus elephas* bores a hole in an acorn by means of its rostrum. Boring the hole is a lengthy process given the hardness of the shell. This operation was described by J.-H. Fabre in characteristically lyrical language: "The mother, more heavily built than the male and armed with longer drilling equipment, inspects her acorn, doubtless with a view to laying her eggs. She paces it from stem to stern, above and below... Her choice is made; the acorn is seen to be of good quality. She has now to bore her probe. The pale, due to its excessive length, is unwieldy. In order to obtain the best mechanical effect the instrument must be brought to bear at a right angle to the piece's convexity and the cumbersome tool, which outside working hours is borne pointing forward, must be lowered under the craftsman's body. To

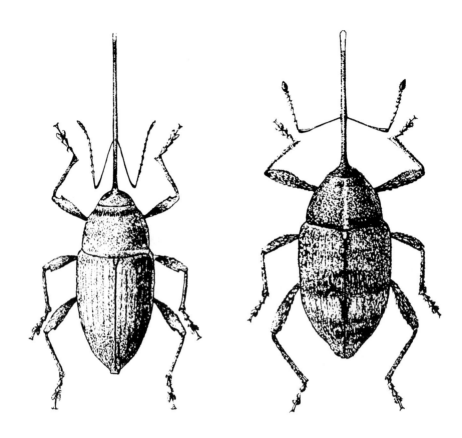

Figure 13.8. *Two species of* Balaninus. *Left,* B. elephas; *right,* B. villosus.

achieve this the animal braces itself with its hind legs and raises its body on the tripod formed of its hind tarsi and the tips of its elytra. Boring begins. Very slowly the insect turns, from right to left, then from left to right, over and over again... and goes on twisting, rests, starts again. An hour, two hours elapse...". When the hole is bored the egg is introduced and put in place by means of an ovipositor that is almost as long as the rostrum, and which at rest fits entirely inside the body.

The tortricid moth genus *Laspeyresia* is rich in species that attack the fruits of various plants, and particularly those of oaks, beech, chestnut and hazel. *Laspeyresia splendana* lives in acorns and chestnuts. Eggs are laid in July on leaves situated near acorns or chestnuts. The young larvae wander freely for a day or two and then penetrate the acorn through the base of a bract of the cup and enter the shell through its insertion point on the calyx. Infested acorns turn brown, lose weight, and fall prematurely. Other species of the same genus are found in the fruits of Fagaceae. *L. amplana* lives in acorns and in beechnuts; *L. fagiglanda* attacks beechnuts as well as acorns and chestnuts.

Among Diptera that live in acorns are various cecidomyids such as *Dasyneura squamosa* which occupies the embryo, whereas the larvae of *Contarinia* spp. settle between the acorn and the cup. It is probable that various cecidomyiids of the genus *Clinodiplosis* predate *Contarinia* (Barnes, 1955).

REFERENCES

BAKKE A., (1963). Studies on the spruce-cone insects *Laspeyresia strobilella* (L.) (Lep. Tortricidae), *Kaltenbachiola strobi* (Winn.) (Diptera, Ithonidae) and their parasites (Hymenoptera) in Norway. *Medd. det Norske Skogsfor.*, **19**:1-150.

BARNES H. F., (1955). Gall midges reared from acorns and acorn cups. *Ent. Mon. Mag.*, **91**: 86-87.

BOUCHER D. H., STORK V. L., (1979). Early drop of nuts in response to insect infestation. *Oikos*, **33**: 440-443.

DARLEY-HILL S., JOHNSON W. C., (1981). Acorn dispersal by the blue jay (*Cyanocitta cristata*). *Œcologia*, **50**: 231-232.

DEBOUZIE D., MENU F., Diapause prolongée et démographie chez le balanin de la châtaigne, *Curculio elephas*. Gyll. *In*: P. FERRON, J. MISSONNIER, B. MAUCHAMP (1990). *Régulation des cycles saisonniers chez les invertébrés*. INRA, Paris, 97-100.

DU MERLE P., (1969). Existence de deux diapauses facultatives au cours du cycle biologique de *Villa brunnea* Beck. (Dipt. Bombyliidae). *C. R. Ac. Sc.*, **268**: 2,433-2,435.

DUVIGNEAUD P., DENAEYER DE SMETS S., *et al.*, (1971). Biomasse, productivité et cycle des éléments biogènes dans l'écosystème "chênaie caducifoliée". Essai de phytogéochimie forestière. *Mém. Inst. R. Sc. Nat. Belgique*, **102**: 339-354.

JANZEN D.-H., (1971). Seed predation by animals. *Ann. Rev.. Ecol. Syst.*, **2**: 465-492.

LE TACON F., MALPHETTES C.-B., (1974). Germination et comportement de semis de hêtre sur six stations de la forêt domaniale de Villers-Cotterêts. *Rev. For. Fr.*, **26**: 111-123.

NIELSEN B.-O., (1977). Beech seeds as an ecosystem component. *Oikos*, **29**: 268-274.

ROQUES A., (1975). *Étude de la mérocénose des cônes de pin sylvestre en forêt de Fontainebleau*. Thesis Université Paris VI.

ROQUES A., (1983). *Les insectes ravageurs des cônes et graines de conifères en France.* INRA, Paris.

ROQUES A., (1988). *La spécificité des relations entre cônes de Conifères et insectes inféodés en Europe occidentale: un exemple d'étude des interactions plantes-insectes.* Thesis, Université de Pau et des pays de l'Adour, 242 pages.

ROQUES A., Comment s'ajustent des populations d'insectes phytophages confrontées à des fluctuations saisonnières acycliques d'abondance de leurs hôtes? *In:* P. FERRON, J. MISSONNIER, B. MAUCHAMP (1990). *Régulation des cycles saisonniers chez les invertébrés.* INRA, Paris, 113-116.

ROQUES A., RAIMBAULT J.-P., DELPLANQUE A., (1984). Les Diptères Anthomyiidae du genre *Lasiomma* Stein, ravageurs des cônes et graines du mélèze d'Europe (*Larix decidua* Mill.) en France. II. Cycles biologiques et dégâts. *Zeit. Angew. Ent.*, **97**: 350-357.

ROUGON C., ROQUES A., ROUGON D., LEVIEUX J., (1995). Impact des insectes sur les potentialités de régénération des Chênes (*Quercus* spp.) en France. I. Mise en évidence d'une action méconnue des curculionides phyllophages sur les organes reproducteurs avant la fécondation. *J. Appl. Ent.*, **119**: 455-463.

SCUTAREANU P., ROQUES A., (1993). L'entomofaune nuisible aux structures reproductrices mâles et femelles des chênes en Roumanie. *J. Appl. Ent.*, **115**: 321-328.

SHAW M.-W., (1968). Factors affecting the natural regeneration of sessile oak (*Quercus petraea*) in North Wales. II. Acorn losses and germination under field conditions. *J. Ecol.*, **56** : 647-660.

TISCHLER W., (1955). *Synökologie der Landtiere.* Gustav Fischer, Stuttgart.

VILLEMANT C., FRAVAL A., (1991). La faune du chêne-liège. Éditions Actes, Rabat.

WATT A. S., (1923). On the ecology of British beechwoods with special reference to their regeneration. *J. Ecol.*, **11**: 1-48.

WINSTON P. W., (1956). The acorn microsere, with special reference to arthropods. *Ecology*, **37**: 120-132.

14

Wood and its use
by xylophagous organisms

Wood amounts to a major reserve of organic material and mineral elements. Wood decomposes much more slowly than do leaves. A major part played by some forest invertebrates lies in acceleration of the process of decomposition and mineralisation of wood. Xylophagous insects are the most important of these invertebrates. Many fungi of the kind described as lignicolous operate simultaneously with insects in the process of decomposition. Analysis of five major mineral elements shows two stages in wood decomposition (table 14.1). The first stage, in which fungal attacks dominate, is characterised by little loss of mineral elements except potassium, which is probably lost even before branches fall. Accumulations of calcium and nitrogen are difficult to account for. There may be nitrogen fixation by fungi which also accumulate calcium. The second stage is that in which xylophagous animals invade; it is characterised by loss of minerals, which are removed by the animals.

Nature of wood	Mineral elements (in $\mu g\ cm^{-3}$)				
	N	P	K	Ca	Mg
Living wood	1,057	69.5	767	1,935	146
Wood invaded by fungi	1,304*	54.4 ns	141**	3,871**	118 ns
Wood invaded by animals and fungi	710***	27.8**	20.9**	1,777***	58.3***

Table 14.1. *Main mineral elements present in branches of oak, ash and birch fallen to the ground following attacks by fungi alone and following attacks by both fungi and animals. Composition of living wood is given for comparison. ns: non-significant variation; ***, **, * correspond to significant variations for probabilities P = 0.001, 0.01 and 0.05 respectively (Swift, 1977).*

1. WOOD: CHEMICAL CHARACTERISTICS

A cross-section of a tree trunk shows the main tissues (figure 14.1). These are, from the outside inward, bark or rhytidome formed by cork and the phelloderm, between which lies the first generating cell layer; primary phloem and secondary phloem or liber; the generating cell layer or cambium; primary xylem and secondary xylem, the whole of which is the wood proper. This wood is divided into outer wood or sapwood and inner wood or heartwood. In each annual ring spring wood and summer wood can be distinguished, the latter having conductive elements which are more numerous and of lesser diameter than those of the spring wood.

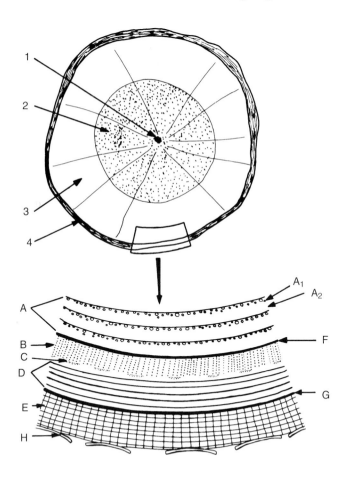

Figure 14.1. *Schematic diagram of a cross-section of a tree trunk (above) and detail of tissues of the outer areas (below). 1: pith and primary xylem; 2: heartwood; 3: sapwood; 4: bark; A: secondary xylem; A_1: spring wood; A_2: summer wood; B: secondary phloem; C: primary phloem; D: phelloderm; E: cork; F and G: generating cell layers; H: epidermis.*

Wood is a very unusual material in terms of its hardness and chemical composition. This affects the morphology and physiology of insects that feed on or grow in it. Cellulose, hemicelluloses and lignin are its main constituents.

The chemical composition of wood varies according to whether one is dealing with bark, phloem, cambium, sapwood or heartwood, which explains why xylophagous insects prefer one or another area. Cerambycid and scolytid larvae, which lack cellulase, occupy the peripheral area which is richer in starch, soluble sugars and absorbable proteins (table 14.2).

	Bark	Phloem	Cambium	Sapwood	Heartwood
Cellulose	20 - 30	21 - 37	20 - 43	40 - 45	43 - 44
Hemicelluloses	?	9 - 30	30 - 45	23 - 35	25 - 34
Lignin	27 - 58	4 - 53	2 - 20	21 - 30	23 - 30
Pectin	?	3 - 18	3 - 22	1 - 4	0.3 - 1
Suberin	2 - 40	0	0	0	0
Starch and sugars	?	4 - 23	3 - 37	1 - 5	0
Nitrogen compounds	0.2 - 0.6	0.2 - 2	1 - 5	0.05 - 0.3	0.03 - 0.1
Lipids	1 - 38	0.3 - 4	?	0.1 - 7	0.2 - 0.8
Minerals	0.2 - 3	1 - 10	3 - 22	0.2 - 0.7	0.2 - 0.8

Table 14.2. *Chemical composition of different parts of wood. Values obtained for trees in temperate regions in % of total biomass.*

1.1. Cellulose

Cellulose accounts for 40 to 62% of the dry weight of wood according to tree species. Cellulose is a d-glucose monopolymer formed by association of n molecules of the carbohydrate with elimination of $n - 1$ water molecules (figure 14.2). There may be as many as 3,000 glucose molecules united in one cellulose molecule. Cellulose can only be hydrolysed by fungi, bacteria, a very few marine invertebrates (such as sea urchins, lamellibranchs such as the shipworm, isopod crustaceans of the genus *Limnoria* and amphipods of the genus *Chelura*) and by insects. There are also a few other terrestrial arthropods, such as isopods of the genus *Hemilepistus* and diplopods, which feed on leaf litter, that are able to digest cellulose (Crawford, 1988).

Wood cell walls contain a primary membrane in which cellulose molecules have no preferred orientation, and a secondary membrane. In the latter, cellulose takes the form of microfibrils which are oval in cross-section and measure about 10 x 4 nm situated in the middle of an envelope formed of

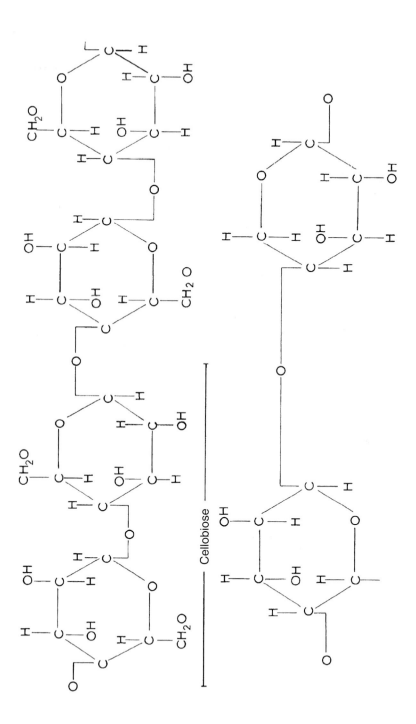

Figure 14.2. *Structure of a cellulose molecule and of an arabinose hemicellulose.*

polysaccharides and water about 100 µm wide. The cellulose molecules, which may reach a length of 7 µm, lie parallel to each other in areas that constitute crystallites and are bi-refractive. Between the crystallites lies the amorphous cellulose formed of non-parallel chains interspersed with hemi-celluloses (figure 14.3). Crystalline cellulose is more resistant to hydrolysis than amorphous cellulose.

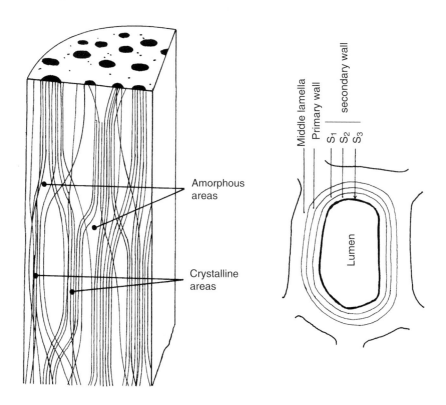

Figure 14.3. *Left: schematic illustration of the layout of cellulose microfibrils in an area of amorphous cellulose and an area of crystalline cellulose. Right: cross-section of a xylem cell showing the primary wall, the middle lamella, the secondary wall of three layers S_1, S_2 and S_3, and the central lumen.*

1.2. Hemicelluloses

This inaccurate but time-honoured term covers a group of heterogeneous elements whose polymerisation degree of about 200 is lower than that of cellulose. Hydrolysed hemicelluloses give simple sugars, which makes them classifiable into pentosans (such as arabans and xylans) and hexosans (such as fructans and galactans). Among the better known hemicelluloses is a

galactan extracted from the alga *Gelidium corneum* used in the manufacture
of gelose. Hemicelluloses are degraded by enzymes called hemicellulases.
Different chemical composition of hemicelluloses of broadleaf trees and of
conifers, as well as the presence of terpenes in conifer resin, may explain
preferences of many xylophagous insects for one or the other category of
tree. Hemicellulose content varies from 14.6% in *Pinus ponderosa* wood to
31.68% in that of birch.

1.3. Lignin

The word lignin is a generic term covering complex and very diverse substances. Lignin accounts for 18 to 38% of matter in wood. It is formed
essentially by polymerisation of monomers, aromatic alcohols such as coniferylic, sinapylic and coumarylic alcohol (figure 14.4). Lignin and cellulose
are united by true chemical bonds to form lignocelluloses. This structure is
proven by the fact that lignin protects cellulose against attacks by organisms
which only possess cellulolytic enzymes. Many bacteria and fungi are
known that are able to hydrolyse cellulose, but no known bacteria are able to
attack cellulose combined with lignin. Only highly specialised fungi such as
various basidiomycetes and a few ascomycetes that possess both cellulase
and enzymes that attack lignin and decompose wood can be considered true
lignivores.

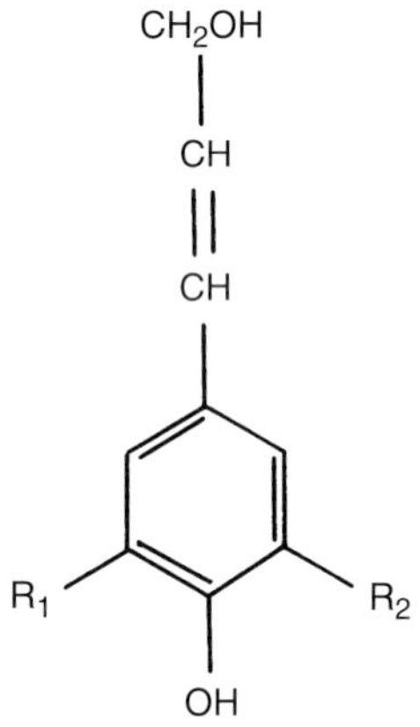

Figure 14.4. *Alcohol components of lignin.*
Coumarylic alcohol ($R_1 = R_2 = H$);
coniferylic alcohol ($R_1 = OCH_3$, $R_2 = H$);
sinapylic alcohol ($R_1 = R_2 = OCH_3$).

1.4. Secondary components

Secondary components, whose weight rarely exceeds 2 to 3% of the total,
play an important part in the lives of xylophagous insects for they provide

various carbohydrates such as starch and soluble sugars, the main ones being saccharose, levulose, glucose and melezitose (table 14.2). Levels in proteic and non-proteic nitrogen are very low: 0.045 to 0.227% in gymnosperms and 0.057 to 0.104% in broadleaf trees (Cowling & Merrill, 1966). Tannins are found mainly in the heartwood and in bark. Lipids also occur, especially in winter. Conifer oleoresins are very varied, often specific, and play a major part in regulating the colonisation process, especially by scolytids.

Wood is a body of tissues relatively poor in mineral elements. Calcium, potassium and phosphorus are the most abundant, as are nitrogen compounds. Despite this relative paucity in mineral elements, the great biomass it forms immobilises large quantities which must be mineralised by means of decomposers so that they can later be reincorporated in biogeochemical cycles. In a beech wood in Fontainebleau forest the biomass of wood is around 300 tons per hectare (Lemée, 1978). This biomass of wood contains 23% of the ecosystem's potassium, 20.8% of its calcium, and 16% of its nitrogen. Reserves in the soil are however significantly greater than those immobilised in the biomass (table 14.3). This is true of temperate regions. In tropical forests wood biomass may reach and even exceed 500 tons per hectare, and the soil is very poor in mineral elements, most of which are found in the biomass. More wood decay is therefore necessary for the maintenance of these forests; this is done either by organisms such as fungi that are much more active than in temperate regions, or such as termites which are almost absent from temperate forests.

	K	Ca	N
Incorporated in biomass of wood	498	1,829	1,450
Incorporated in deciduous parts (leaves, twigs, flowers)	49	61	58
Incorporated in litter	97.8	180	215
Incorporated in soil	1,560	6,670	7,480

Table 14.3. *Mass of the three main elements (in kg per hectare) in a beech wood in Fontainebleau forest. Total biomass of wood is around 300 tons per hectare.*

2. STRUCTURE OF THE XYLEM CELL

Wood cells have complex walls that include a primary membrane heavily laden with lignin (71% of total mass), and a secondary membrane formed after cell differentiation that contains less lignin and is thicker than the primary membrane, since it forms on average 76% of the cell's total volume. The middle layer which unites cells has a heavy lignin content, like the primary wall. The secondary wall has three distinct layers. The outer layer S_1

and the inner layer S_3 contain cellulose whose microfibrils form loose helices; the middle layer S_2 contains cellulose with microfibrils disposed in tight helices and in concentric areas (figure 14.3).

Wood, like many cellulose material, has microscopic capillary spaces 0.5 to 4 nm wide, the average width being 1.6 nm. These microspaces are important to organisms involved in wood decomposition since they provide channels for penetration and allow the spread of oxygen and water. They do not however allow the spread of cellulolytic enzymes, which are protein molecules between 5 and 9 nm in diameter. Cellulolytic enzymes therefore attack wood in the immediate vicinity of the fungal hyphae that secrete them or they spread when, as in the case of brown rot, deterioration of cell walls begins and creates large enough spaces for these molecules.

3. IMPORTANCE OF WOOD IN THE FOREST ECOSYSTEM

Plant biomass is high in forest. The greater part of this biomass exists in the form of wood. In natural, non-exploited forest, the biomass of dead trees may reach very high levels. In managed and exploited woodland most trees are removed before they die, but a certain quantity of wood remains in the form of stumps and fallen branches. The few following examples give some idea of levels of biomass.

In a beech wood in Fontainebleau 14 to 22 trees per hectare were found that had died standing and whose biomass could not be evaluated (Lemée, 1978). Fallen branches and twigs alone amounted to 2,000 to 3,270 kg per hectare. Details of biomasses are as follow (in tons per hectare):

– Tree stratum: total wood: 284, of which trunks: 195; branches: 48; roots: 41; leaves and bracts: 3.2; reproductive organs: 0.9.

– Shrub stratum: trunks and branches: 4.8; leaves: 0.7.

– Herbaceous stratum: parts above ground: 1.82; parts underground: 1.92

Duvigneaud & Denaeyer-De Smets (1970) measured the following biomasses in Virelles oak / hornbeam wood in Belgium:

– Biomass above ground: 120 tons per hectare, of which woody parts: 112.1 tons per hectare; buds and leaves: 8.9 tons per hectare.

– Biomass below ground (roots): 35 tons per hectare.

– Dead wood: 1,410 kg per hectare.

In broadleaf forests of western Europe wood biomass is between 150 and 300 tons per hectare. Amounts of dead wood, whether in the form of standing dead trees or of fallen trunks and branches, varies. It amounts to only 1,410 kg per hectare in the exploited forest at Virelles, but reaches 6 tons per hectare in the non-exploited beech wood of la Massane in the Pyrénées-Orientales (Dajoz, 1974). In an uncultivated beech and oak coppice in England biomass of dead fallen wood varies from 3,110 to 4,400 kg per hectare (Swift *et al.*, 1984). In different broadleaf forests of temperate Europe and North America the quantity of dead wood varies from 2,280 kg to 9,060 kg per hectare, with a tendency to increases in mass of dead wood the older the forest, and an average of 5 tons per hectare where the average age of trees reaches 150 to 200 years (Christensen, 1977).

Much higher biomasses have been established in old conifer forests in the western United States. In an Oregon forest composed mainly of *Pseudotsuga menziesii, Thuya plicata* and *Tsuga heterophylla* with an average age of 450 years, biomass above ground is between 492 and 976 tons per hectare, and that of large sized woody debris between 143 and 215 tons per hectare (Spies *et al.*, 1988).

4. DIGESTION OF WOOD BY INSECTS

4.1. Digestion of cellulose

Cellulose is of great biological and economic importance, yet study of cellulases was long neglected. Until the 1950s it was assumed that the name cellulase described a single enzyme. The discovery that some organisms that are unable to live on pure cellulose (such as cotton) are able to grow on modified forms of water-soluble cellulose such as carboxymethylcellulose (or CMC) led to the present view that recognises two distinct enzymes that act in turning cellulose into cellobiose, the latter substance being then converted to glucose by a third enzyme. This model seems to be particularly common in fungi and in micro-organisms and insects able to perform cellulolysis, and may be general.

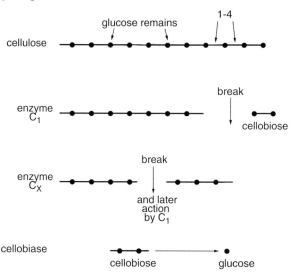

Figure 14.5. *Schematic representation of the cellulose molecule and of the action of three enzymes that convert it to glucose.*

Insects digest cellulose, as do fungi, by means of a complex of three hydrolases (figure 14.5). The enzyme called C_x is a β-endoglucanase which degrades linear chains of cellulose only in the amorphous areas, while cellulose is in part "crystalline". The cellulose is attacked in chemical bonds 1 – 4, creating areas suitable for action by the following enzyme. Enzyme C_x acts on CMC and other cellulose derivatives such as cellulose swollen with phosphoric acid and has no effect on normal cellulose.

Enzyme C_1 or cellobiohydrolase is an exoglucanase which produces oligosaccharides or cellobiose from non-reductive chains liberated by enzyme C_x, and at the same time creates new areas of attack for enzyme C_x. Enzyme C_1 acts on pure micro-crystalline cellulose such as cotton and has virtually no effect on CMC.

Synergistic action of C_1 and C_x ends in the complete disruption of cellulose and its conversion into cellobiose.

As cellobiose, which is a potential inhibitor of enzymes C_1 and C_x, is produced it is converted into glucose by cellobiase, which is a β1 – 4 glucosidase.

Since the three enzymes act synergistically, cellulose decomposition can only take place when all three enzymes are present. Presence of the whole enzyme complex in an insect is shown by the latter's gut content's ability to hydrolyse pure cellulose in such forms as cotton or filter paper. The presence of the C_x enzyme alone is demonstrated by degradation only of CMC, or of some form of amorphous cellulose.

4.1.1. Insects able to digest pure cellulose

Few insects are able to degrade cellulose totally. Many xylophagous insects that live on bark, on phloem or on wood do not digest cellulose. Examples are bostrychid, curculionid and lyctid beetles as well as many buprestids and cerambycids, which only possess the C_x enzyme. The list of insects known to digest cellulose completely is a short one, including in all about fifty species belonging to the following orders:

– Thysanura. The silverfish *Ctenolepisma lineata*, although not a xylophagous species, is able to digest cellulose (Lasker & Giese, 1956).

– Dictyoptera. Two cockroaches are able to digest cellulose: *Cryptocercus punctulatus*, a xylophagous species that lives in dead wood (Cleveland, 1934), and *Periplaneta americana*, which is not xylophagous (Bignell, 1977).

– Isoptera. Of the many species of termite, a few species belonging to five families are able to digest cellulose. These are *Mastotermes darwiniensis* (Mastotermitidae), *Kalotermes flavicollis* and *Neotermes bosei* (Kalotermitidae); *Zootermopsis angusticollis* (Hodotermitidae); *Coptotermes formosanus, Heterotermes indicola, Reticulitermes lucifugus* and *R. speratus* (Rhinotermitidae); *Macrotermes natalensis, M. subhyalinus, Nasutitermes ephratae* and *Trinervitermes trinervoides* (Termitidae).

– Hymenoptera. Three siricid wasps are known for their ability to digest cellulose. These are *Sirex cyaneus, S. gigas* and *S. phantoma.*

– Coleoptera. This order contains the largest number of species that digest cellulose. They are the following species belonging to four families:

Anobiidae: *Anobium punctatum, A. striatum, Ernobius mollis, Ptilinus pectinicornis, Xestobium rufovillosum.*

Buprestidae: *Capnodis* spp., *Chalcophora mariana.*

Cerambycidae: *Acanthocinus aedilis, Aegosoma scabricorne, Cerambyx cerdo, Ergates faber, Gracilia minuta, Hoplocerambyx spinicornis, Hylotrupes bajulus, Leptura rubra, Macrotoma palmata, Morimus funereus, Oxymirus cursor, Phymatodes testaceus, Plagionotus detritus, Rhagium bifasciatum, R. inquisitor, R. mordax, Saperda populnea, Smodicum cucujiforme, Stromatium barbatum, S. fulvum, Xylotrechus rusticus.*

Scarabaeidae: *Oryctes nasicornis, Xylotrupes gideon, Sericesthis geminata, Anomala polita, Oxycetonia albopunctata.*

A few aquatic xylophagous Coleoptera belonging to the family Elmidae, such as *Lara avara,* do not possess cellulase, but absorb nutrients produced by microbial attacks on submerged wood.

This list has been drawn up using four different techniques:

(a) Comparison of the cellulose content of food and faeces, giving an idea of approximate percentages of cellulose digested. This is 74 to 91% in *Kalotermes flavicollis* (Seifert & Becker, 1965), 31% in *Anobium punctatum,* 22% in *Sirex gigas* (Muller, 1934), and 14 to 47% in *Macrotoma palmata* (Mansour & Mansour-Bek, 1934).

(b) Demonstration of cellulose digestion by means of cellulose marked with ^{14}C and search for $^{14}CO_2$ in the insect's metabolism. This has been done in particular with *Ctenolepisma lineata* (Lasker & Giese, 1956), *Oryctes nasicornis* (Bayon & Mathelin, 1980), *Reticulitermes flavipes* (Mauldin, 1977) and *Monochamus marmorator* (Kukor & Martin, 1986a).

(c) Demonstration of an ability of fluids in the digestive tract to degrade filter paper, cotton, or any other form of pure cellulose. This has been done with *Macrotermes natalensis* (Martin & Martin, 1978), *Chalcophora mariana* (Schlottke, 1945), *Ernobius mollis* and *Ptilinus pectinicornis* (Parkin, 1940) as well as with three species of *Rhagium* (Deschamps, 1954) and with *Sirex cyaneus* (Kukor & Martin, 1983).

(d) Demonstration of an insect's ability to survive in an environment composed of pure cellulose. This has been proven in the cases of the cockroach *Cryptocercus punctulatus* (Cleveland, 1934), the silverfish *Ctenolepisma lineata* (Lasker & Giese, 1956) and the termite *Reticulitermes flavipes* (Mauldin, 1977).

Further research would certainly expand this list and extend it to other taxonomic groups such as Diptera larvae which are highly diversified in dead wood. Cellulase is absent from those leaf-eating insects (Lepidoptera, Coleoptera, Orthoptera, Diptera) that have been tested.

4.1.2. Species possessing only the C_x enzyme

Species that possess the C_x enzyme and are unable to degrade CMC are more numerous than those that possess both the C_1 and C_x enzymes. They include some fifteen species of cockroach, about thirty species of aphid, three stoneflies (Plecoptera), a dozen caddis flies with aquatic larvae and, among the Coleoptera, all the scolytids that have been studied, numerous cerambycids, a few curculionids (*Cryptorrhynchus lapathi, Hylobius abietis, Pissodes hercyniae* and *P. notatus*), one psocid (Psocoptera) and three bugs (Heteroptera).

4.1.3. Acquired enzymes

Many invertebrates that feed on plant debris ingest a complex mix of bacteria, protozoans and fungi together with their food. These micro-organisms are sources of enzymes which, once freed, remain in the animal's digestive tract thereby increasing the latter's ability to digest substrates such as cellulose, hemicelluloses, pectin, lignin and other molecules that are not easily degradable (Martin & Kukor, 1984). The part played by acquired enzymes was first shown in *Macrotermes natalensis* (Martin & Martin, 1978); this termite, which has no symbiotic protozoans, lives in association with various fungi including species of *Termitomyces*. The latter are ingested and provide C_1 enzymes, while C_x enzymes are brought by other fungi. Similarly *Sirex cyaneus* possesses acquired enzymes able to hydrolyse wood polysaccharides; these enzymes come from the symbiotic fungus *Amylostereum chailletii* which lives in larval galleries (Kukor & Martin, 1983).

Monochamus marmorator is a cerambycid beetle whose larva lives in the wood of fir. The midgut of larvae fed on wood invaded by the mycelium of the fungus *Trichoderma harzianum* contains an enzyme complex which, when it was purified and analysed by chromatography, proved to be identical to that produced by the fungus (figure 14.6). Larvae reared in an artificial environment devoid of the fungus lose their ability to digest cellulose and can only attack CMC and hemicelluloses, which shows that the insect itself only produces the C_x enzyme. *Monochamus* larval galleries are invaded by various fungi able to attack cellulose, particularly *Trichoderma harzianum*, which is the commonest, but also *Stereum sanguinolentum, Amylostereum chailletii* and *Hirschioporus abietinus*. Another cerambycid, *Saperda calcarata*, is unable to exploit cellulose under normal conditions. This species lives in the cambium and sapwood of living *Populus tremuloides* branches where there are no fungi. Digestive fluids of its midgut act only on starch, pectin, a hemicellulose, a polysaccharide, callose, and CMC, which points to the presence of the C_x enzyme and absence of C_1. Non-digestion of cellulose was confirmed by addition of cellulose marked with ^{14}C in food. No trace of it was found in carbon dioxide emitted by respiration. The *Saperda* larva does however digest cellulose when a cellulase solution extracted from

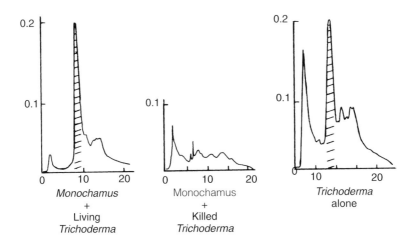

Figure 14.6. *Chromotographic study of the digestive enzymes of the cerambycid* Monochamus marmorator. *Left, contents of the midgut of* Monochamus *fed on wood infected with the living fungus* Trichoderma harzianum. *Middle, contents of the midgut of* Monochamus *fed on wood and* Trichoderma *killed by heat treatment. Right, fluid from a pure culture of* Trichoderma. *The hatched area of the graph represents maximum activity on crystalline cellulose. This aera is absent in* Monochamus *reared in the presence of* Trichoderma *that had been killed by sterilisation (Kukor & Martin, 1986a).*

Penicillium funiculosum culture is added to its food (Kukor & Martin, 1986a). This case is experimental verification of an insect's ability to acquire and use enzymes produced by a fungus.

Sustained activity of ingested enzymes is made possible by the pH of insects' digestive tract, which is close to neutral (usually between 6.8 and 7.2). Within this pH range enzymes remain stable and active. Furthermore the insects' proteolytic enzyme activity, which might destroy acquired enzymes, appears to be low in the anterior part of the midgut, which is where cellulolytic activity is most intense.

Acquired enzymes are not confined to insects. In the isopod crustacean *Tracheoniscus rathkei* (common woodlouse), which feeds on plant debris, addition of a commercial cellulase preparation extracted from *Penicillium funiculosum* enables it to digest cellulose, which is not the case under natural conditions (Kukor & Martin, 1986b). In the earthworm *Pontoscolex corethrurus,* cellulase comes from micro-organisms ingested together with soil (Zhang *et al.*, 1993).

The termite *Macrotermes mulleri* has in its gut a β-glucosidase which is different from that of the symbiotic fungus *Termitomyces*. Decomposition of cellulose would appear to be initiated by the fungus and continued in the termite's digestive tract by means of an exocellulase and a β-glucosidase produced in the insect's salivary glands. These enzymes would therefore appear not to be ingested, unlike the situation outlined above. Ingestion of enzymes coming from symbiotic fungi would then merely improve cellulose hydrolysis and contribute to maintaining the termite's nutritional balance (Rouland, 1986; Rouland *et al.*, 1986).

4.2   Digestion of lignin

Lignin does not seem to be degraded by enzymes secreted by animals, or at any rate by insects. It is however decomposed by fungi and by bacteria, and perhaps even by protozoans, several of which live symbiotically in insects' digestive tracts. It has been established that these symbiotic relationships have enabled some insects to exploit lignin.

The larva of the cerambycid beetle *Hoplocerambyx spinicornis* ingests 584 mg of wood (dry weight) per gram of body weight per day and digests 40.4% of it. It assimilates cellulose, hemicelluloses, starch and a few hexosans as well as a part of the lignin; this appears to be aided by micro-organisms in the digestive tract. Moreover this larva appears to be able to fix atmospheric nitrogen by means of the micro-organisms. Digestion results partly in the formation of more or less polymerized humic compounds of low molecular weight (Mishra, 1985).

Termites digest cellulose and hemicelluloses, and at least some of them, such as *Kalotermes flavicollis*, also exploit lignin. Hypermastigine flagellates and bacteria are found in the termite's rectal caecum where they ensure digestion of cellulose. Digestion of lignin by these protozoans has been confirmed by Lavette (1964). The use of lignin marked with ^{14}C has shown that *Nasutitermes exitiosus* decomposes 5 to 6% of lignin in heartwood. Decomposition rates determined for several other species range from 0.33 to 83% (Cookson, 1987). Bacteria whose action on lignin has been confirmed include the genera *Nocardia, Pseudomonas* and *Streptomyces*.

Changes in the composition of healthy poplar wood caused by the termite *Microcerotermes edentatus* are characterised by attack on lignin, with a reduction in its quantity, and a marked decrease in cellulose content. Attacks on the same wood by the lignivorous fungi *Ganoderma applanatum* and *Trametes trabea* take the form of large loss of mass owing to decomposition of lignin and cellulose. Termites prefer to feed on decayed rather than on healthy wood. They are attracted by wood whose constituents are mainly cellulose and pentosans (Kovoor, 1964).

4.3.   Digestion of hemicelluloses and other components of wood

Many xylophagous insect enzymes hydrolyse parts of wood other than cellulose and lignin. Hemicelluloses have been recorded in many species, for instance among the Cerambycidae. There are also various osidases such as amylase, saccharase, maltase and trehalase, and also proteases and lipases (table 14.4). Distribution of these enzymes in different parts of the digestive tract varies according to species.

Species	Enzymes						
	Amylase	Saccharase	Maltase	Lactase	Cellulase	Hemicellulase	Proteases
Chalcophora mariana (Buprestidae)	+				+	+	+
Pissodes notatus (Curculionidae)	+	+	+	+	+	+	+
Bostrychoplites cornutus (Bostrychidae)	4	4	3	1	0	0	3
Lyctus brunneus (Lyctidae)	3	3	3	3	0	2	3
Anobium punctatum (Anobiidae)	2	1	1	2	1	4	3
Xestobium rufovillosum (Anobiidae)	3	3	3	3	1	4	3
Cerambyx cerdo (Cerambycidae)	+				+		+
Ergates faber (Cerambycidae)	+				+		+

Table 14.4. *Enzyme arsenals of a few species of xylophagous beetle. + shows that the enzyme is present but without indication of the intensity of attack on the substrate. Figures 1, 2, 3 and 4 show that the enzyme is present and indicate intensity of attack.*

In larvae of the house longhorn (*Hylotrupes bajulus*) growth is more rapid and mortality lower in sapwood than in heartwood. This species can only live on conifers. The larva only uses 12.3% of cellulose present and 2% of hemicelluloses; in this group it only attacks hexosans and cannot digest pentasans. When wood is leached of sugars, starch and proteins by means of hot water, growth is inhibited. The same occurs when wood is treated with sulphuric acid which destroys proteins. The limiting factor of growth in *Hylotrupes* is protein levels in wood. Addition of sugars does little or nothing to improve growth, but when amino acids, peptones or corpses of other *Hylotrupes* are added growth is 10 to 15 times faster. Larval growth may in this way be speeded from three years to one. At the same time the larva consumes less wood, as it no longer needs to search for the nitrogenous substances it needs (Becker, 1942).

Research has shown that cerambycids fall into one of two ecological groups. The first contains species which possess cellulase and live in non-decomposed heartwood poor in soluble sugars and starch; their only source of carbohydrates is thus cellulose. *Cerambyx cerdo* belongs to this category. The second group is composed of species (such as *Callidium* spp.) which feed on phloem or on wood that has already been decomposed by fungi, and which depend on soluble sugars since they do not possess cellulase.

5. THE ROLE OF SYMBIOTIC MICRO-ORGANISMS

Many insects, and particularly xylophagous ones, live in association with symbiotic organisms most of which are fungi, but also include protozoans and bacteria in the case of organisms that live in the digestive tract.

5.1. Symbiotes of the digestive tract

These are present in termites (cf. above) as well as in a few moths, scarabaeid beetles, cockroaches and flies (Diptera). The larva of the wood-boring goat moth (*Cossus cossus*) does not secrete cellulase or amylase and would be able to absorb only soluble sugars were it not for symbiotic bacteria and fungi in its gut that convert cellulose to glucose and glycuronic acid that can be used by the insect. The bacteria also produce hydrogen, methane, fatty acids and lactic acid. The fungi produce alcohol, carbon dioxide, acetone and lactic acid. Healthy wood never contains these micro-organisms which are only ever found in the caterpillars' gut or in by-products of their scrapings in the galleries. These then are true symbiotes.

Several species of cockroach digest wood. The earliest known example is that of *Cryptocercus punctulatus* which was studied in the United States by Cleveland (1934). This cockroach lives in dead tree trunks; protozoans, in particular flagellates of the genera *Trichonympha* and *Barbulonympha*, and polymastigines, live symbiotically in its hind gut. They secrete cellulases and cellobiases which the cockroach does not possess; this insect is unable to survive if it is deprived of its flagellates.

Various scarabaeid beetles feed on wood that has been more or less decomposed by other insects, fungi, and bacteria. All xylophagous scarabaeids such as *Dorcus parallelipipedus*, *Oryctes nasicornis*, *Osmoderma eremita* and *Potosia cuprea* have a fermentation chamber formed by a strongly dilated part of the hind gut that fills with bacteria and wood fragments. *Potosia cuprea*'s fermentation chamber is inhabited by the bacterium *Bacillus cellulosae fermentans* which converts cellulose to carbon dioxide and hydrogen and fixes atmospheric nitrogen (Werner, 1926). *Osmoderma* and *Oryctes* harbour various bacteria that decompose wood, fix nitrogen, and are later digested by the larva (Wiedermann, 1930). Crane-fly larvae (*Tipula* spp.) also have cellulolytic bacteria in their digestive tract, but these bacteria, which also live free in dead wood, are not true symbiotes (Pochon, 1939).

In *Oryctes nasicornis*, which feeds on wood that has already begun to decompose, passage of food through the gut lasts on average 18 1/4 hours. Food undergoes continuous brewing in the proctodeum, which is a pocket-shaped swelling of part of the hindgut. Sustained anaerobic fermentation is carried out in the proctodeum by bacteria. Cellulose is decomposed into various products including fatty acids, carbon dioxide and methane. The fatty acids may pass through the gut wall into the haemolymph; methane is partly or wholly discharged from the body (Bayon, 1980; Bayon & Mathelin, 1980). This type of anaerobic digestion resembles that found in termites and ruminants. It is characterised by discharge of methane, volumes of which may be considerable from abundant insects such as termites. Methane emission by termites (xylophogous, fungus-rearing and humivorous), has been estimated at 27.10^6 tons per year (Rouland, 1994). Methane production by workers of several African termite species is as follows (average values in micromoles of methane per gram of termite per hour):

– four xylophagous species: 0.13 to 0.21; average: 0.16
– four fungus-rearing species: 0.38 to 0.88; average: 0.52
– four humivorous species: 0.64 to 1.09; average: 0.88.

5.2. Ectosymbiotic organisms

Ectosymbiotic organisms that attack wood and are later eaten by the symbiotic insects are known to exist in a few families of Coleoptera and in siricid wasps.

5.2.1. Symbiosis in siricid wasps

Siricids of the genera *Sirex*, *Tremex* and *Urocerus* live in association with basidiomycete fungi belonging to the genus *Amylostereum*, the commonest species being *A. chailletti*. When female *Sirex juvencus* lay their eggs in wood no offspring are produced unless the fungus is present; when it is present fertilised females produce both male and female offspring, whereas unfertilised females produce only males, which shows arrhenotocous parthenogenesis in this species. The fungus serves as food for the young larvae which are unable to bore in wood. Propagation of the fungus is by means of two spore pouches, or mycetanges, situated in the female's abominal cavity and opening at the base of the ovipositor. Spores are ejected together with eggs as they are laid in wood. Mycetanges are also present in female larvae. During the final larval moult the pupal tegument is contaminated by the spores, thus ensuring the fungus' transmission to the adult female (Francke-Grosmann, 1957, 1967).

5.2.2. Symbiosis in lymexylonid beetles

The small family Lymexylonidae includes one widely-distributed species, *Hylecoetus dermestoides*. The Ascomycete fungus *Ascoidea hylecoeti* grows inside the beetle's larval galleries which may penetrate wood to a depth of 20 cm. This fungus is a true symbiote which is only found in *Hylecoetus* galleries and which provides the beetle with indispensable elements by attacking the wood. Transmission of the fungus in larval galleries is by means of pouches containing numerous spores situated near the tip of the ovipositor (figure 14.7).

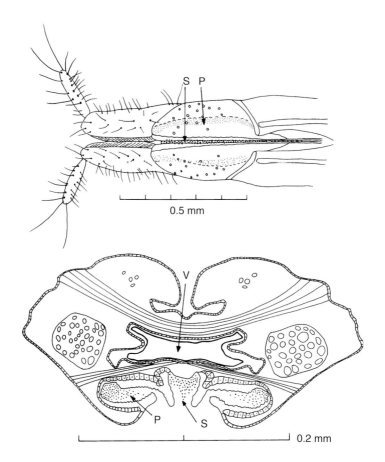

Figure 14.7. *The beetle* Hylecoetus dermestoides. *Above, tip of the female's ovipositor in ventral view; below: in cross-section. P: pouches containing spores of the symbiotic fungus; S: median furrow; V: vagina (Francke-Grosmann, 1967).*

5.2.3. Symbiosis in scolytid and platypodid beetles

Scolytids and platypodids contain two groups of species. Species of the first group bore galleries deep into wood. These galleries are kept clean and free of faeces and their walls are lined with the mycelium of a fungus known as ambrosia. Insects that cultivate it are called ambrosia beetles. Species of the other group live under bark in the cambial area and do not have ambrosia.

Most, perhaps even all scolytids, are associated with symbiotic fungi, and particularly with pathogenic fungi of the genus *Ceratocystis* which play a major role in the process of attacking trees (cf. chapter 3). Transmission to wood of spores is not a simple mechanical process. Spores are often incapable of germinating until after they have undergone a period of maturation and after proliferation in specialised organs called mycangia. These are invaginations of the tegument into which specialised glandular cells discharge alimentary secretions. Adult *Dendroctonus* reared from an axenic larval culture have empty mycangia, whereas those that have grown normally have the fungus' spores (Paine & Birch, 1983). The presence of secretory glands in the mycangium wall suggests that secretions play a part in controlling the fungus and in feeding it (Schneider & Rudinsky, 1969). In *Anisandrus dispar* the symbiotic fungus has seasonal variations which are synchronous with those of ovarian development. Development of oocytes and egg-laying only take place after the female has ingested the symbiotic fungus (French & Roeper, 1975). In the Scolytidae mycangia are only present in the sex that first attacks the host tree, this, in most cases, being the female. In the Platypodidae both sexes are furnished with mycangia. Presence of the fungus is vital to ambrosia beetles, for their larvae are unable to eat wood and they feed on the fungus.

Several types of mycangia may be distinguished (figure 14.8).

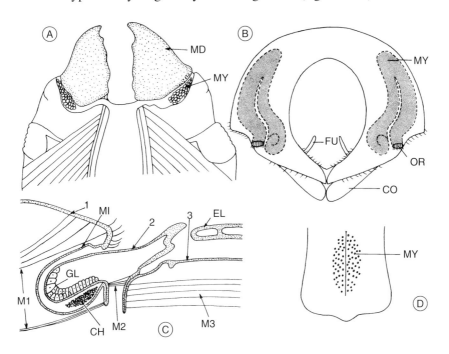

Figure 14.8. *Mycangium structure of female scolytids. A: frontal section of the head of* Ips acuminatus. *B: Posterior view of the prothorax of* Platypus cylindrus. *C: sagittal section through the body of* Anisandrus dispar. *D: dorsal view of the prothorax of a female* Platypus oxyurus. *MD: mandible; MY: mycangium; FU: furca; CO: coxa; MY: mycangium and its opening OR; 1, 2 and 3: thoracic segments; MI: intersegmentary membrane; M1, M2 and M3: longitudinal muscles; CH: symbiotic fungus; GL: glandular cells; EL: elytron (Francke-Grosmann, 1967).*

– Buccal mycangia situated at the base of mandibles and opening into the buccal cavity, as in *Xyleborus affinis* or *Dryocoetes confusus*.

– Pronotal mycangia situated on the pronotum, as in *Platypus cylindrus*.

– Pleuro-prothoracic mycangia which are invaginations of the propleural area, as in *Trypodendron lineatum* or *Dendroctonus brevicomis*.

– Prosternal-subcoxal mycangia that open into the highly dilated coxal cavity, as in *Gnathotrichus sulcatus*.

– Elytral mycangia formed in a cavity situated in the anterior margin of the elytron, as in *Xyleborus saxaseni*.

5.3. The role of endosymbiotes

Many insects have endosymbiotic organisms localised in different organs, according to genus. In anobiid larvae and adults they live in diverticula of the midgut where they may be almost independent. In cerambycids it is little-developed side pockets of the midgut that house the endosymbiotes (figure 14.9). Transmission of these symbiotes from one generation to the next is done in various ways. In female cerambycids and anobiids interseg-mental sacs filled with yeasts open into the vagina. The systematic position of these symbiotes is difficult to establish; many of them are yeasts of the genus *Candida*.

The role of endosymbiotes has been much disputed. They do not appear to play any part in digestion, but they do provide a number of nutrients that the insects cannot do without, particularly vitamins of the B complex, which they are able to synthesise, and amino acids.

6. THE ROLE OF LIGNIVOROUS FUNGI

Wood is a resource which very often requires a preliminary attack by fungi before it can be invaded by insects; for this reason the effects of fungal attack on wood must be understood. The heartwood of many trees contains substances (alkaloids, terpenes, phenols, etc.) which make them unaccep-table to insects. Fungi may neutralise these compounds by means of their well-developed enzymatic arsenal. The African termite *Coptotermes sjost-edti* only consumes the wood of the tree *Australella congolensis* following fungal infestation (Becker, 1975). *Pinus caribaea* wood contains high quan-tities of resin and is only acceptable to the termite *Coptotermes niger* after it has been attacked by the fungus *Lentinus pallidus* (Williams, 1965).

Fungi enrich the substrate by concentrating scarce nutrient elements such as nitrogen, phosphorus and potassium in their mycelium. An insect has to consume 36.2 g of wood in order to obtain a quantity of nitrogen equal to that found in 3.7 g of mycelium or in 1 g of spores of the lignicolous fungus

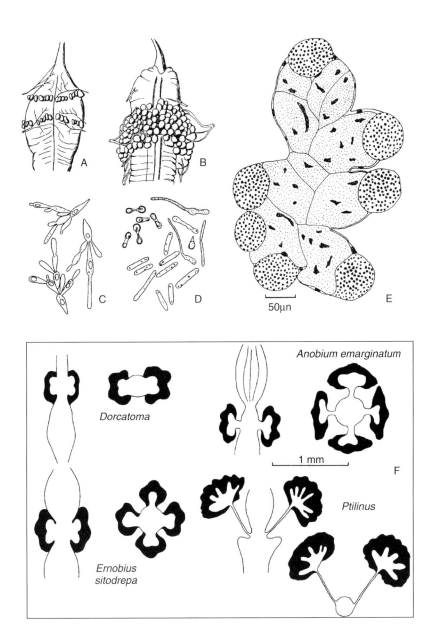

Figure 14.9. *Endosymbiotes of xylophagous Coleoptera. A and B: side pockets of the midgut in larvae of* Leptura rubra *and* Oxymirus cursor. *C and D: yeast type symbiotes of* Leptura rubra *and* Spondylis buprestoides; *some cells are sporulating. E: cross-section of the mycetome containing symbiotes of* Lyctus linearis. *F: various types of pocket containing endosymbiotic organisms in anobiids (Buchner, 1953; Koch, 1967).*

Ganoderma applanatum (Merrill & Cowling, 1966). Growth of cerambycid and anobiid larvae on wood that has undergone fungal attack is more rapid as a result of this enrichment (Campbell & Bryant, 1940; Bletchly, 1953; Becker, 1977). Growth of the anobiid *Xestobium rufovillosum* is a function of the state of wood decomposition by fungi. In intact wood, larval life lasts 50 months; in wood infested by fungi it may be reduced to as little as 16 months (Fisher, 1940). Accelerated growth may also be due to physical changes in the wood, the larva being better able to feed on softer wood.

Nitrogen levels in the mycelium of lignicolous fungi vary between 0.23 and 3.27% according to species. These levels are significantly higher than those found in wood. This nitrogen cannot have come solely from wood invaded by the mycelium; it is drawn from bark, or from fixation of atmospheric nitrogen. Hyphomycete fungi also play a part in the decomposition of wood that falls into water and is then attacked by various aquatic insects.

6.1. Enzymes of fungi

The enzymes of fungi are numerous and active. Many species – perhaps as many as several thousand – of basidiomycete attack lignin, as the use of lignin marked with ^{14}C has shown unequivocally (Kirk & Fenn, 1982). Phenolases attack lignin by oxidation, leaving by-products such as vanillin and vanillic acid. Fungal cellulase has the same structure and acts in the same way as that of insects (Reese, 1977; Montgomery, 1982). Cellulase is present in basidiomycetes and also in ascomycetes and lower fungi. Fungal attacks on wood are rapid: *Stereum hirsutum* decomposes 27% of lignin in oak wood in thirteen weeks, and *Coriolus versicolor* decomposes 35% of beech lignin in ten weeks.

6.2. Different types of attack on wood

Ability to decompose components of wood cell walls has evolved in many different families of fungi, the most important of which are basidiomycetes. There are about 1,700 species able to rot wood in North America, and probably as many in Europe. The four main families are the Polyporaceae and Corticiaceae in the order Aphyllophorales, and the Cortinariaceae and Tricholomataceae in the order Agaricales. These fungi often have vast ranges over large parts of the northern hemisphere. Many of them show a marked preference for old forests in which there is an abundance of dead trees with large trunks and in stages of advanced decomposition. They become much rarer in exploited forests (Bader *et al.*, 1995).

Five different types of attack on wood by fungi may be distinguished. The deterioration caused by fungi is known as rot.

IMPORTANCE OF FUNGI IN THE FOREST ECOSYSTEM

Fungi play a major role in the forest ecosystem:

– Through species richness (more than 1,000 species in some forests). They are a major element of biodiversity, coming second only to insects in number of species.

– Many fungi harbour a rich fauna of mycetophagous insects, this being another element of species biodiversity in forests.

– Some fungi, particularly basidiomycetes such as *Amanita*, *Boletus*, and *Russula*, form true symbiotic relationships with trees and grow mycorrhiza on roots, which provides the tree with vital nutritive elements.

– Other species such as bracket fungi are parasites which may kill living trees which they may penetrate following wounding. Some of these such as *Ceratocystis* may be carried by scolytid beetles.

– Many species are agents of various rots that exploit dead wood and accelerate recycling of mineral elements.

A FEW COMMON LIGNIVOROUS BASIDIOMYCETES

– Species growing on standing trees:

Hypholoma (= *Nematoloma*) *fasciculare*. Cubic rot on broadleaf trees

Pleurotus ostreatus. White rot, especially on beech

Armillaria mellea. White rot on broadleaf trees

Phaeolus schweinitzii. Cubic rot on conifers

Heterobasidion (= *Ungulina*) *annosa*. Alveolar rot of conifers; agent of "ring disease"

Fomes fomentarius. White rot on broadleaf trees

Polyporus squamosus. White rot on various broadleaf trees

Polyporus (= *Laetiporus*) *sulfureus*. Cubic rot on oak and chesnut

Fistulina hepatica. Cubic rot, especially on oak

Ganoderma applanatum. White rot, especially on beech

– Species growing on fallen trees and stumps:

Coniophora puteana. White rot on all wood, including timber and furniture

Coriolus versicolor. White rot on all broadleaf trees

Daedalea quercina. Brown dry rot on oak and timber exposed to rain

Trametes gibbosa. White rot on broadleaf trees

Stereum hirsutum and *Stereum purpureum*. White rot on all broadleaf trees

Stereum sanguinolentum. White rot on conifers

6.2.1. Cubic rot

In this type of rot, also known as brown rot, cellulose is destroyed, whereas lignin is not. Wood turns brown and fissures along three perpendicular planes, so that it crumbles into cubic pieces. Fungal hyphae that reach xylem cells settle on wall S_3 which remains little affected, but they cause profound changes in other layers whose cellulose is destroyed. Cubic rot is the work of *Polyporus sulphureus, Fistulina hepatica* and *Gyrophana lacrymans* which often attacks timber.

6.2.2. White rot

In this type of rot, also known as fibrous rot, lignin is attacked and what remains of cell walls turns white and takes on a fibrous consistency; wood is greatly softened. White rot fungi cause gradual lysis of the S_3 layer, followed by the S_2 layer of xylem cells. They include *Coriolus versicolor, Coniophora puteana* and the bracket fungus *Fomes fomentarius*. Phenolases degrade lignin by oxidation, giving off by-products such as vanillin and vanillic acid.

6.2.3. Alveolar rot

This type of rot is characterised by destruction of cellulose and lignin in tubular portions of wood aligned with the grain, creating alveoli lined with white mycelium. It is caused by species such as *Stereum sanguinolentum, Ungulina annosa, Schizophyllum* etc.

6.2.4 Soft rot

Soft rot is caused by fungi which are unable to attack the primary, lignin-rich membrane, and station themselves in the secondary membrane. Hyphae grow in this membrane where they follow the route of helicoid cellulose fibres. Affected wood is softened and when it dries fissures in rectangular planes. Soft rot fungi attack damp or even saturated wood; they are often the first to colonise wood lying on the ground. They are ascomycetes such as *Chaetomium globosum* or micromycetes such as *Alternaria* spp. and *Phialophora* sp.

6.2.5. Blue stain fungi

Blue stain fungi, so called because they cause wood to turn a characteristic bluish colour, are ascomycetes, particularly species of *Ceratocystis*, and also micromycetes that mainly attack conifers and live in sapwood at the expense of living cells. Affected wood is discoloured but its physical properties are little changed.

REFERENCES

BADER P., JANSSON S., JONSSON B.-G., (1995). Wood-inhabiting fungi and substratum decline in selectively logged boreal spruce forests. *Biol. Cons.*, **72**: 355-362.

BAYON C., (1980). Transit des aliments et fermentations continues dans le tube digestif d'une larve xylophage d'insecte: *Oryctes nasicornis* (Coleoptera, Scarabaeidae). *C. R. Ac. Sc.*, **290**: 1,145-1,148.

BAYON C., (1980). Volatile fatty acids and methane production in relation to anaerobic carbohydrate fermentation in *Oryctes nasicornis* larvae (Coleoptera: Scarabaeidae). *J. Insect. Physiol.*, **26**: 819-828.

BAYON C., MATHELIN J., (1980). Carbohydrate fermentation and by-product absorption studied with labelled cellulose in *Oryctes nasicornis* larvae (Coleoptera: Scarabaeidae). *J. Insect. Physiol.*, **26**: 833-840.

BECKER G., (1942). Untersuchungen über die Ernährungsphysiologie der Hausbock Käferlarven. *Zeit. Vergl. Physiol.*, **29**: 315-326.

BECKER G., (1975). Coptotermes in the heartwood of living trees in Central and West Africa. *Mater. Org.*, **10**: 149-154.

BECKER G., (1977). Ecology and physiology of wood destroying Coleoptera in structural timber. *Mater. Org.*, **12**: 141-161.

BIGNELL D. E., (1977). An experimental study of cellulose and hemicellulose degradation in the alimentary canal of the American cockroach. *Canad. J. Zool.*, **55**: 579-589.

BLETCHLY J. D., (1953). The influence of decay in timber on susceptibility to attack by the common furniture beetle, *Anobium punctatum* Deg. *Ann. Appl. Biol.*, **40**: 218-221.

BUCHNER P., (1953). Endosymbiose der Tiere mit pflanzischen Mikro-organismen. Birkhäuser, Basel.

CAMPBELL W. G., BRYANT S. A., (1940). A chemical study of the bearing of decay by *Phellinus cryptarum* Karst. and other fungi on the destruction of wood by the death-watch beetle (*Xestobium rufovillosum* Deg.). *Biochem. J.*, **34**: 1,404-1,414.

CHRISTENSEN O., (1977). Estimation of standing crop and turnover of dead wood in a Danish oak forest. *Oikos*, **26**: 187-195.

CLEVELAND L. R., (1934). The wood feeding roach *Cryptocercus*, its Protozoa and the symbiosis between Protozoa and roach. *Mem. Amer. Ac. Arts Sc.*, **17**: 185-242.

COOKSON L. J., (1987). ^{14}C-lignin degradation by three Australian termite species. *Wood Sci. Technol.*, **21**: 11-25.

COOKSON L. J., (1987). Influence of laboratory maintenance, relative humidity and coprophagy on [14C] lignin degradation by *Nasutitermes exitiosus*. *J. Insect. Physiol.*, **33**: 683-687.

COWLING E. B., MERRILL W., (1966). Nitrogen in wood and its role in wood deterioration. *Canad. J. Botany*, **44**: 1,539-1,554.

CRAWFORD C. S., (1988). Nutrition and habitat selection in desert invertebrates. *J. Arid. Environ.*, **14**: 111-121.

DAJOZ R., Les insectes xylophages et leur rôle dans la dégradation du bois mort. *In*: P. PESSON (1974). *Écologie forestière*. Gauthier-Villars, Paris, 257-287.

DESCHAMPS P., (1954). Contribution à l'étude de la xylophagie. La nutrition des larves de Coléoptères Cerambycidae. *Ann. Sc. Nat., Zool.*, **11**ᵉ série: 449-533.

DUVIGNEAUD P., (1974). *La synthèse écologique*. Doin, Paris.

FISHER R. C., (1940). Studies of the biology of the death watch beetle *Xestobium rufovillosum* De Geer. III. Fungal decay in timber in relation to the occurrence and rate of development of the insect. *Ann. Appl. Biol.*, **27**: 545-552.

FRANCKE-GROSMANN H., (1957). Über das Schicksal der Siricidenpilze während der Metamorphose. *Ber. Deut. Entom.*, **8**: 37-43.

FRANCKE-GROSMANN H., Ectosymbiosis in wood inhabiting insects. *In*: M. HENRY, *Symbiosis*, vol. 2 (1967). Academic Press, New York, 141-205.

FRENCH J.-R., RŒPER R.-A., (1975). Studies on the biology of the Ambrosia beetle: *Xyleborus dispar* (F.) (Coleoptera Scolytidae). *Zeit. ang. Ent.*, 78: 241-247.

KIRK T. K., FENN P., Formation and action of the ligninolytic system in basidiomycetes. *In:* J. C. FRANKLAND, J. N. HEDGER, M. J. SWIFT, (1982). *Decomposer basidiomycetes: their biology and ecology*. Cambridge University Press, 67-90.

KOVOOR J., (1964). Modifications chimiques provoquées par un Termitide (*Microcerotermes edentatus*, Was.) dans du bois de peuplier sain ou partiellement dégradé par des champignons. *Bull. Biol. France Belgique*, **48**: 491-510.

KUKOR J. J., (1986). The transformation of *Saperda calcarata* into a cellulase digester through the inclusion of fungal enzymes in its diet. *Œcologia*, **71**: 138-141.

KUKOR J. J., MARTIN M. M., (1983). Acquisition of digestive enzymes by siricid woodwasps from their fungal symbiont. *Science*, **220**: 1,161-1,163.

KUKOR J. J., MARTIN M. M., (1986a). Cellulose digestion in *Monochamus marmorator*: role of acquired fungal enzymes. *J. Chem. Ecol.*, **12**: 1,057-1,070.

KUKOR J. J., MARTIN M. M., (1986b). The effect of acquired microbial enzymes on assimilation efficiency in the common woodlouse, *Tracheoniscus rathkei*. *Œcologia*, **69**: 360-366.

LASKER R., GIESE A. C., (1956). Cellulose digestion by the silverfish *Ctenolepisma lineata. J. Exp. Biol.*, **33**: 542-553.

LAVETTE A., (1964). La digestion du bois par les Flagellés symbiotiques des termites: cellulose et lignine. *C. R. Ac. Sc.*, **258**: 2,211-2,213.

LEMÉE G., La hêtraie naturelle de Fontainebleau. *In*: M. LAMOTTE, F. BOURLIERE (1978). *Problèmes d'Écologie, Écosystèmes terrestres.* Masson, Paris, 75-128.

MARTIN M. M., (1983). Cellulose digestion in insects. *Comp. Biochem. Physiol.*, **75**: 313-324.

MARTIN M. M., KUKOR J. J., Role of mycophagy and bacteriophagy in invertebrate nutrition. *In*: M. J. KLUG, C. A. REDDY (1984). *Current perspectives in microbial ecology.* American Society for microbiology, Washington, 257-263.

MARTIN M. M., MARTIN J. S., (1978). Cellulose digestion in the midgut of the fungus-growing termite *Macrotermes natalensis*: the role of acquired enzymes. *Science*, **199**: 1,453-1,455.

MAULDIN J. K., (1977). Cellulose catabolism and lipid synthesis by normally and abnormally faunated termites, *Reticulitermes flavipes. Insect. Biochem.*, **7**: 27-35.

MERRILL W., COWLING E. B., (1966). Role of nitrogen in wood deterioration: amount and distribution of nitrogen in fungi. *Phytopathol*, **56**: 1,083-1,090.

MISHRA S.-C., (1985). Chemical changes in wood during the digestive process in larvae of *Hoplocerambyx spinicornis. Mater. Org.*, **20**: 53-64.

MISHRA S.-C., SEN-SARMA P.-K., (1985). Carbohydrases in xylophagous coleopterous larvae (Cerambycidae and Scarabaeidac) and their evolutionary significance. *Mater. Org.*, **20**: 221-230.

MONTGOMERY R. A. P., The role of polysaccharidase enzymes in the decay of wood by basidiomycetes. *In:* J. C. FRANKLAND, J. N. HEDGER, M. J. SWIFT (1982). *Decomposer basidiomycetes: their biology and ecology.* Cambridge Univ. Press, 51-65.

MULLER W., (1934). Untersuchungen uber das Symbiose von Tieren mit Pilzen und Bakterien. III. Über die Pilzsymbiose holzfressender Insektenlarven. *Zeit. Mikrobiol.*, **5**: 84-147.

PAINE T. D., BIRCH M. C., (1983). Acquisition and maintenance of mycangial fungi by *Dendroctonus brevicomis* LeConte (Coleoptera: Scolytidae). *Envir. Ent.*, **12**: 1,384-1,386.

PARKIN E. A., (1940). The digestive enzymes of some wood boring beetle larvae. *J. Exp. Biol.*, **23**: 369-400.

POCHON J., (1939). Flore bactérienne cellulolytique du tube digestif de larves xylophages. *C. R. Ac. Sc.*, **208**: 402-404.

REESE E. T., (1955). Enzymatic hydrolysis of cellulose. *Appl. Microbiol.*, **4**: 39-45.

REESE E. T., (1977). Degradation of polymeric carbohydrates by microbial enzymes. *Recent Adv. Phytochem.*, **11**: 311-368.

ROULAND C., (1986). *Contribution à l'étude des osidases digestives de plusieurs espèces de termites africains.* Thesis, Université Paris Val-de-Marne.

ROULAND C., (1994). Les mécanismes de production du méthane par les termites en forêt tropicale. *Le Courrier de l'environnement de l'INRA*, **23**: 57-62.

ROULAND C., MORA P. et al., (1986). Étude comparative entre la β-glucosidase présente dans le tube digestif du termite *Macrotermes mulleri* (Termitidae, Macrotermitinae) et la β-glucosidase du champignon symbiotique *Termitomyces* sp. *Actes Coll. Insectes Sociaux*, **3**: 109-118.

SCHLOTTKE E., (1945).Ueber die Verdauungsfermente im Holzfressender Käferlarven. *Zool. Jahrb., Zool. Physiol.*, **61**: 88-140.

SCHNEIDER I., RUDINSKY J.-A., (1969). Mycetangial glands and their seasonal changes in *Gnathotrichus retusus* and *G. sulcatus. Ann. Ent. Soc. Amer.*, **62**: 39-43.

SEIFERT K., BECKER G., (1965). Der chemische Abbau von Laub und Nadelholzarten durch verschiedene Termiten. *Holzforschung*, **19**: 105-111.

SPIES T. A., FRANKLIN J. F., THOMAS T. D., (1988). Coarse woody debris in Douglas-fir forests of western Oregon and Washington. *Ecology*, **69**: 1,689-1,702.

SWIFT M. J., (1977). The ecology of wood decomposition. *Sci. Prog.*, **64**: 179-203.

SWIFT M. J., BODDY L., HEALEY J. N., (1984). Wood decomposition in an abandoned beech and oak coppices woodland in SE England. II. The standing crop of wood on the forest floor with particular reference to its invasion by *Tipula flavolineata* and other animals. *Holarctic Ecology*, **7**: 218-228.

WERNER E., (1926). Die Ernährung der Larve von *Potosia cuprea. Z. Morph. Okol. Tiere*, **19**: 226-256.

WIEDEMANN K., (1930). Die Zelluloserverdauung bei Lamellicornierlarven. *Z. Morph. Okol. Tiere*, **19**: 228-258.

WILLIAMS R. M. C., (1965). *Termite infestation of pines in British Honduras.* Overseas Res. Pub. no. 11. HMSO, London.

ZHANG B.-G., ROULAND C., LATTAUD C., LAVELLE P., (1993). Activity and origin of digestive enzymes in gut of the tropical earthworm *Pontoscolex corethrurus. Eur. J. Soil Biol.*, **29**: 7-11.

15

Bark beetles (Scolytidae) and their associated fauna

Beetles of the family Scolytidae are of great importance as pests of conifers, and incidentally of some broadleaf trees. Their unusual biology has been much researched, especially with a view to developing methods of protecting forests from their depredations. The closely related family Platypodidae contains fewer species and is mainly represented in the tropics. A few species in the family Curculionidae have a biology similar to that of scolytids. A great number of insects, mites and other invertebrates with very diverse diets live in the company of scolytids in their habitat under bark.

General accounts of the Scolytidae may be found in Balachowsky (1949) and in Wood (1982), and for species associated with conifers in France, in Chararas (1962). A general bibliography relating to scolytids and platypodids has been compiled by Wood (1987).

1. MORPHOLOGY AND ANATOMY OF SCOLYTIDS

The family Scolytidae are small beetles including more than 6,000 known species. Many are under 2 mm long; *Ips sexdentatus* reaches 8 mm, and *Dendroctonus*, the giants of the group, reach almost 10 mm. Most species have cylindrical bodies and denticulations on the declivitous apical part of the elytra (figure 15.1). Their last three antennal segments form a club that, with a very few exceptions, is fused, which distinguishes them from curculionids whose antennal club segments are not fused. Curculionids have a well-developed rostrum which is lacking in scolytids, except in a few genera such as *Hylastes* in which a prolongation of the head simulates a rudimentary rostrum (figure 15.2). Platypodids, with their very elongated cylindrical bodies, are chiefly distinguished by their extended first tarsomeres. Scolytid larvae, like those of curculionids, have soft white teguments; they are subcylindrical and hunched, and lack legs. There are as a rule five larval instars except in a few American species of *Dendroctonus*.

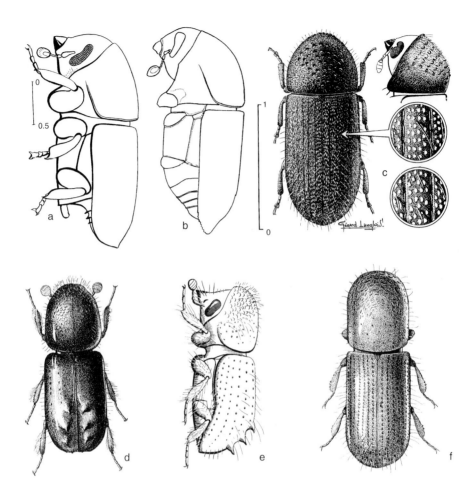

Figure 15.1. *Some scolytids. a:* Scolytus multistriatus, *male in lateral view;*
b: Scolytus intricatus, *female in lateral view; c:* Cryphalus abietis;
d and e: Pityogenes chalcographus, *dorsal and lateral views;*
f: Xyleborus dryographus.

The scolytid alimentary tract has a relatively homogenous structure.
Adults have a proventriculus furnished with sclerified denticles that act as
filters by preventing excessively large fragments of wood from entering the
midgut; the larval proventriculus is devoid of sclerified pieces. The midgut
and hindgut are moderately long and much less developed than in other
xylophagous beetles such as cerambycids. Crypts shaped like the fingers of a
glove lead off the midgut (figure 15.3). There is a great similarity between
the scolytid digestive tract and that of curculionids of the tribe Cossonini.
Scolytids are sometimes considered to be a subfamily of the Curculionidae
owing to major structural similarities both in the larvae and the adults. The
hypothesis has also been put forward that characteristics of the head, the
shape of the body, and the digestive tract which are found in scolytids, in

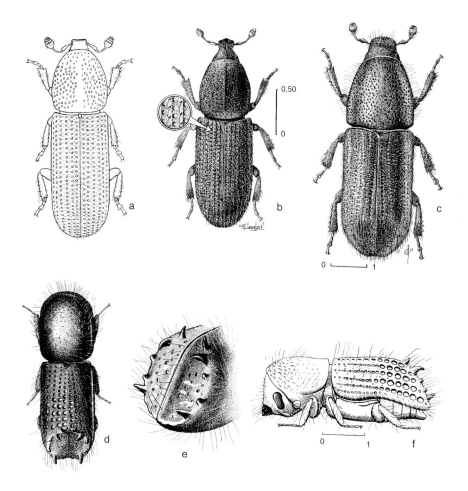

Figure 15.2. *Some scolytids. a:* Hylastes ater*; b:* Hylastes attenuatus*;
c: Hylurgus ligniperda*; d:* Pityokteines curvidens, *male;
*e and f: apical declivity of the elytra, and the same species in
lateral view.*

curculionids of the tribe Cossonini as well as in bostrychids and some ciids
are merely the result of convergent evolution owing to their common xylophagous habits. Scolytid mandibles act on a quasi horizontal plane, which
likens them to anthribids, and sets them apart from curculionids whose mandibles move in the vertical plane (Morimoto, 1962).

Scolytids have a great variety of mating systems (Kirkendall, 1983). The
most unusual is seen in tropical species and in *Dendroctonus micans*. A system of inbreeding exists whereby sibling beetles mate in the gallery system
in which they developed. In these scolytids the sex ratio is aberrant and
unbalanced in favour of females. Many platypodids have dwarf, wingless
males that do not leave the galleries. Scolytids belonging to the tribes

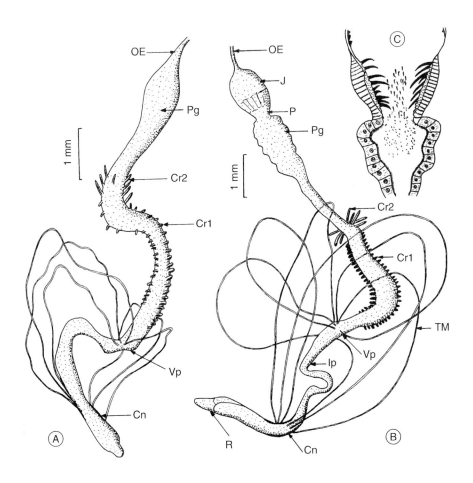

Figure 15.3. *Alimentary tract of* Ips typographus. *A: larva; B: adult; C: schematic longitudinal section through the proventriculus showing denticles that filter fragments of wood; OE: oesophagus; Pg: gastric pouch; J: crop; P: proventriculus; Ip: hindgut; Cr1 and Cr2: short and long intestinal crypts; Vp: pyloric valvule; Cn: cryptonephridism area; TM: Malpighian tubes; R: rectum.*

Pityophthorini (genera *Pityophthorus, Pityogenes, Pityokteines, Orthotomicus* and *Ips*) and Polygraphini (genera *Carphoborus* and *Polygraphus*) are polygamous and bore two to seven maternal galleries per family system. The galleries are vertical in *Ips*, horizontal in *Pityokteines*, and radial in *Pityogenes* and *Pityophthorus* species. Monogamous scolytids belong to the tribes Hylastini (genera *Hylastes, Hylurgus, Tomicus* and *Dendroctonus*), Hylesini (genus *Xylechinus*), Crypturgini (genera *Crypturgus* and *Dryocoetus*), and Cryphalini (genus *Cryphalus*). They build family systems in which there is a single maternal gallery which may be vertical (in *Hylastes* and *Dryocoetes*, and in *Tomicus piniperda*), or horizontal (in

Tomicus minor and *Xylechinus*), or else have an expanded shape (in *Dendroctonus micans*). European species of scolytid may be classed in three subfamilies (Balachowsky, 1949). The subfamily Scolytinae contains robust species two to two and a half times as long as broad in which the elytra do not entirely cover the abdomen. *Scolytus*, all European members of which live on broadleaf trees, is the only genus in the subfamily. The subfamily Hylesinae includes beetles bearing a row of prominent granules distinct from the general surface rugosity on the anterior margins of the elytra. Some genera such as *Hylastes* and *Hylurgops* have the head prolonged into a short rostrum and live on conifers; they prefer trees whose wood is already slightly decomposed. The genera *Dendroctonus, Tomicus* and *Hylurgops* also belong to this subfamily, as do *Leperisinus, Phloesinus, Carphoborus* and *Polygraphus*. The subfamily Ipinae is composed of species without granules on the anterior margin of the elytra. Most live on conifers. The main genera are *Ips, Crypturgus, Orthotomicus, Xyleborus* and *Trypodendron*.

1.1. Diet

All scolytids are phytophagous and most live on woody plants; only a few species live on herbaceous plants. *Thamnurgus kaltenbachi* lives in Labiatae, and *Coccotrypes dactyliperda* attacks the seeds of various palm trees. There is a clear division between the fauna of broadleaf trees and that of conifers, species that live on both being very few indeed. Conifers suffer most attacks, and above all the genus *Pinus*. Eighty one species of scolytid are known to depend on conifers in Europe, of which 39 live on pines, 23 on spruce, 6 on fir and 2 on larch; eleven species live on two or more genera of conifer.

Primary pests attack perfectly healthy trees whose cambium and phloem are functional and rich in starch and proteins. Secondary pests seek more or less decomposed trees. Exclusively primary pests are few, but when outbreaks occur many other species such as *Ips typographus, Tomicus piniperda, Dendroctonus micans* and *Platypus cylindrus* are capable of becoming so. Many scolytid secondary pests cause irreversible changes in the trees they infest, such damage sometimes being indirect, as in the transmission of diseases such as Dutch elm disease. Some scolytids such as species of *Ips, Pityokteines, Orthotomicus, Pityophthorus, Hylesinus,* and *Leperisinus* seek trees suffering from physiological deficiencies caused by drought, damage by defoliating caterpillars, fire or competition. The phloem and cambium of such trees retain their mechanical properties, but osmotic pressure is lower than normal, as are levels in starch and proteins. Other scolytids go for fallen trees or those still standing but moribund, in which the cambium and phloem are discoloured but still retain their mechanical properties. A final category is that of scolytids that invade trees in a much more advanced stage of decay, whose cambium and phloem have begun to ferment, are discoloured and have an acid pH, and in which levels of starch and proteins are low and levels of water high. Beetles in this category include species of *Dryocoetes* and *Hylurgops*.

Several species of scolytid show a preference for one or another part of a tree. Some choose large branches, others smaller branches, twigs or bark of a specified thickness. Among the scolytids of conifers, 10 species settle at the base of roots, 31 species opt for the trunk and 40 species choose branches or buds. *Hylastes ater* lives at the base of trunks and on exposed roots of pines and more rarely of spruce; *Dendroctonus micans* may nest anywhere on spruce trunks but prefers the bases of roots; *Ips acuminatus* attacks branches and the upper parts of pine trunks where the bark is thin; *Pityokteines curvidens* seeks the thin bark of fir trees.

Most scolytids of temperate areas live in the sub-cortical area in the vicinity of phloem, and are known as phloeophages. A few species such as *Trypodendron lineatum*, *Anisandrus dispar* and *Platypus* spp. bore galleries deep into the wood and live in association with fungi. These are xylomycetophages, also known as ambrosia beetles.

1.2. Different types of gallery

Scolytid larvae live in family gallery systems which are of characteristic design in each species (figure 15.4). Three main elements are found in a gallery system. The entrance orifice, or penetration corridor, may be bored by one or other sex; it is short in the case of mycophagous species. Egg-laying, or maternal galleries, are bored by adults who keep them free of detritus, and they are used by females to lay their eggs, which are deposited in lateral notches. Maternal galleries are sometimes furnished with structures that have been called ventilation shafts but which are in fact mating chambers. Larval galleries, which are bored by the larvae themselves, lead off the maternal galleries and end in more or less widened dead ends in which the larvae pupate, and where the adults often remain in a state of diapause before emerging.

The main types of scolytid gallery are as follows:
– Simple longitudinal galleries: the maternal gallery is aligned from bottom to top in the tree trunk's axis; they have no initial, or central vestibule. Examples: *Scolytus scolytus; Tomicus piniperda.*
– Double longitudinal galleries: the maternal gallery is divided in two by a mating chamber consisting of a widening of the gallery excavated by the male. Example: *Ips typographus.*
– Simple transverse galleries: these have no mating chamber. Example: *Scolytus intricatus.*
– Transverse galleries with adjoining mating chamber: the maternal gallery has a lateral mating chamber. Example: *Leperesinus fraxini.*
– Stellate transverse galleries: The mating chamber is in the centre and is excavated by the male; egg-laying corridors, each of which is occupied by

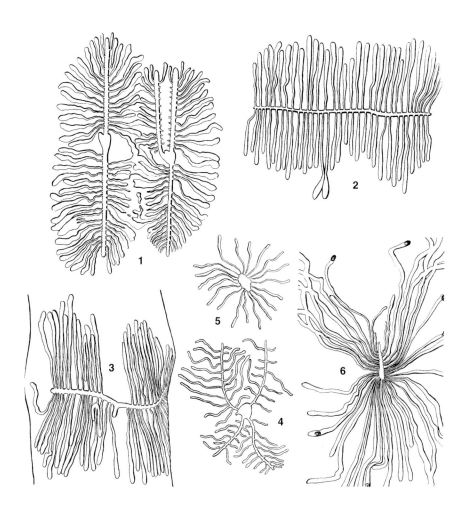

Figure 15.4. *Layouts of some scolytid gallery systems. 1: double and triple longitudinal gallery with central vestibule of* Ips typographus; *2: transverse double gallery of* Pteleobius kraatzi, *a species that lives on elm; 3: double transverse gallery with central vestibule adjoining the main gallery of* Leperesinus fraxini, *which lives on ash; 4: stellate gallery with central vestibule of* Pityophthorus pityographus *which lives on spruce; 5: "false" stellate gallery of* Cryphalus piceae, *a species that lives on spruce; 6: simple transverse gallery of* Scolytus intricatus, *which lives on oak.*

one female, lead off it. This system is seen in polygamous species such as *Cryphalus piceae, Pityogenes chalcographus* and *Ips acuminatus*.

Species of the genus *Dendroctonus* have unusually-shaped galleries that do not fit into this system of classification.

2. COLONISATION OF TREES

Initial attraction of scolytids to the host tree was foreseen as early as 1931 in a study of *Dendroctonus brevicomis* living on *Pinus ponderosa*, and secondary attraction was suspected in 1948 in the case of male *Ips pini* colonising *Pinus banksiana*. It was however only from 1960 on that intensive research was carried out in several countries in North America and Europe.

Attacks on trees by scolytids usually take place in two stages. Primary attraction is exerted by the host plant. In the first stage, pioneers of one or other sex arrive in small numbers. Secondary attraction is a consequence of settlement by the pioneers and takes the form of massive and rapid colonisation which enables the beetles to overwhelm the tree's defences. This pattern was discovered in the course of research on scolytids living on conifers; it would appear to apply, with minor variations, to those scolytids of broadleaf trees that have been studied.

2.1. Primary attraction

Not all trees are attractive to scolytids. The beetles are only attracted to those suffering physiological deficiency resulting from wounds, drought, fire, or that have been uprooted by wind, or newly-felled. Attacks on healthy trees are infrequent, but may occur during outbreaks, as beetles move on to healthy trees when all the more vulnerable trees have already been invaded. Vagaries of climate give rise to such outbreaks: the outbreak of *Ips sexdentatus* in the Landes (SW France) in 1944-45 followed years of exceptional drought.

Conifers have a network of resin canals in which oleoresin circulates under pressure. Oleoresin (or simply resin) is a mixture of substances, which varies from one tree to another, formed essentially of hydrocarbons belonging to the terpene group and called turpentine. These hydrocarbons contain a solution of a solid residue called resinic acid or colophane which can be obtained by distillation. The three main groups of terpenes are monoterpenes with the basic formula C_5H_8, which are the most volatile and abundant, diterpenes $(C_5H_8)_2$ and sesquiterpenes $(C_5H_8)_6$. The exudation pressure of oleoresins is usually lower than normal in infested trees. Pressure is related to the tree's water content, and consequently to the osmotic pressure of cells. Attacks by scolytids start when pressure is 7-8 atmospheres, 10 atmospheres being the normal, and trees do not die until it falls to 4 atmospheres. Trees infested by *Ips typographus* when osmotic pressure is 6 to 8 atmospheres are subsequently invaded by *Pityogenes chalcographus*, and then, after a long interval, *Dryocoetus* and *Hylurgops* settle when the osmotic pressure of cells may be as low as 1.6 atmospheres.

Detection of suitable trees by beetles on the wing depends on olfactory stimuli. Substances that attract scolytids to conifers are complex mixtures of compounds present in resin. They include terpenes and their alcohol derivatives produced by oxidation. Five diterpenes are found in maritime pine, α-pinene, β-pinene, myrcene, limonene and Δ-3-carene. Stone pine (*Pinus pinea*) contains limonene almost exclusively (figure 15.5).

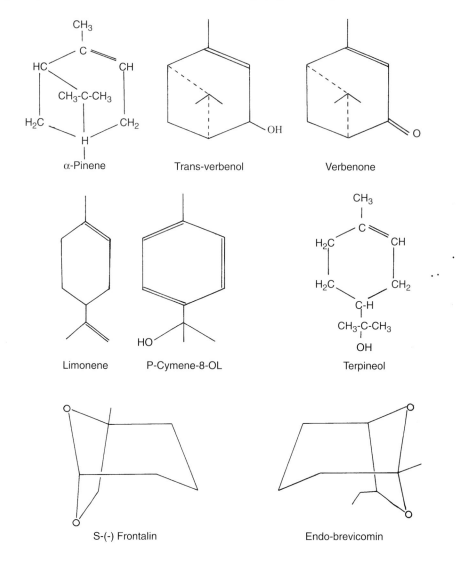

Figure 15.5. *Formulae of some terpene compounds of conifers and of constituents of scolytid aggregation pheromones derived from them.*

Different species of scolytid react to attractant mixtures which are often specific. Mixtures are often more attractive than pure substances, and substances other than terpenes may be responsible for primary attraction. *Trypodendron lineatum*, which lives on conifers, is drawn to ethanol, as is *T. domesticum* which lives mostly on oak. A mixture of α-pinene and ethanol has even greater attraction for *T. lineatum* but repels *T. domesticum*, which is understandable since the latter species lives only on broadleaf trees which are devoid of terpenes. Addition of the aggregation pheromone 3-hydroxy-3-methylbutane-2-ane (or HMB) does not increase the attractant power of the ethanol / α-pinene mixture (Nijholt & Schönherr, 1976; table 15.1). The α-terpineol of Scots pine is attractive to *Tomicus piniperda*. Other substances which have no effect on this beetle in isolation reinforce the effect of α-terpineol when they are added in well-defined quantities. These substances are chiefly *trans*-carveol and *cis*-carveol (Oksanen *et al.*, 1968). The α-pinene / ethanol mixture is attractive to various species such as *Tomicus piniperda*, *Trypodendron lineatum*, *Hylurgops palliatus* and the predatory clerid *Thanasimus formicarius* (Schrœder & Lindelöw, 1989).

The attraction of some scolytids, as well as their associated species, either to α-pinene or to ethanol or to a mixture of the two has been demonstrated. *Tomicus piniperda* and its predators *Thanasimus formicarius* and *Rhizophagus ferrugineus* are attracted to α-pinene; *Hylurgops palliatus* and *Trypodendron lineatum* are attracted to ethanol. The α-pinene / ethanol mixture attracts large numbers of *Hylurgops palliatus*, *Rhizophagus ferrugineus*, *Epuraea* spp. and *Glischrochilus* spp., and catches increase with levels of ethanol. *Tomicus piniperda* on the other hand is less strongly attracted to the α-pinene / ethanol mixture than it is to α-pinene alone. This may be related to the beetle's biology, since it prefers to colonise fallen trees which emit large quantities of terpene but little ethanol (Schrœder, 1988).

Bait	*Trypodendron domesticum*	*Trypodendron lineatum*
Water (control substance)	3	72
α-pinene	7	153
Ethanol	88	414
Ethanol + α-pinene	12	735
Ethanol + α-pinene + HMB	11	718

Table 15.1.   *Numbers of insects caught in summer in a Canadian forest according to bait in traps (Nijholt & Schönherr, 1976).*

Attractant substances have been found that act on xylophagous insects other than scolytids. The terpenes of Scots pine repel the weevil *Pissodes notatus*. The substances that provoke the feeding bites of this species are flavones present in phloem (Blanc & Blanc, 1975). Monoterpenes are not attractive to *Hylobius abietis*, but sesquiterpenes and terpenic alcohols are. Females of the cerambycid *Hylotrupes bajulus* are attracted by α-pinene and by β-pinene, but much less readily to Δ-3-carene. These compounds trigger oviposition (Becker, 1949). Two other cerambycids that live on conifers, *Ergates faber* and *Spondylis buprestoides*, are also attracted to terpenes, whereas *Cerambyx cerdo*, which lives mainly on oak, responds to ethanol and ethyl acetate (Döhring, 1955).

If scolytids are oriented chiefly by volatile substances, responses to visual stimuli such as shape or colour also play a part in colonising trees. Coloured traps have shown that yellow holds little attraction, and that most beetles land on brown, black or red traps. The attractiveness of design also varies. Compact shapes are preferred to divided figures; vertical stripes attract more beetles than horizontal bands. Attractive colours correspond to those of bark.

Visual attraction seems to be efficient at short range, and olfactory attraction at greater distances. *Trypodendron domesticum* are not induced to land by one stimulus alone; smell and suitable shape must be combined for the insects to land (Schönherr, 1977).

Variations in scolytid behaviour depend on their physiological state and on environmental conditions. *Trypodendron lineatum* only responds to odours emitted by its potential host tree after it has spent 30 to 90 minutes on the wing; after this time has elapsed its phototactic reactions are reversed. In *Dendroctonus pseudotsugae*, as in other scolytids, flight entails heavy loss of lipid reserves; individuals with the greatest reserves are those that most tend to disperse by flight (Atkins, 1969).

Tomicus piniperda hibernates in the ground at the foot of trees. Activity is resumed when temperatures reach + 3 °C, and flight may carry them up to 2 km. Orientation towards reproduction sites is largely controlled by microclimate. As soon as they have overwintered *Tomicus* settle in places in the tree tops where the temperature rapidly rises to 15 or 16 °C, at which point they can take off. After a certain amount of time on the wing the beetles cease to react to light, and odour emitted by trees becomes the main stimulus. Since α-pinene is repellent and α-terpineol is attractive, it is the relative concentrations of these two substances that determine whether or not a tree is attractive.

Water content of wood also affects the choice made by scolytid beetles. *Trypodendron lineatum* prefers to settle in wood with humidity levels of 63 to 144%. When humidity falls below 53%, as happens in wood stripped of bark and exposed to air, adults leave their galleries and go to settle on a tree better suited to their larvae. Tree trunk diameter and appearance are also factors that may affect a tree's attractiveness to scolytids. Percentage of invaded trees increases with their diameter (Rolling & Kearby, 1977).

2.2. Secondary attraction and aggregation pheromones

Aggregation pheromones cause dispersed individuals of the same species to congregate in a given place in the vicinity of the individuals that emitted them. Aggregation pheromones are only known in any detail in the Scolytidae and in a few Curculionidae. Studies of aggregation pheromones, their chemical composition, method of action and synthesis have been undertaken mainly with a view to developing methods of biological control based on mass trapping of insect species that are major pests of forests (cf. chapter 6, 7.2). Numerous papers on this subject have been published since attention was first focused on it (Bakke, 1973; Carle, 1974; Silverstein, 1970; Vité, 1978; Wood, 1970 etc.). More than two hundred papers had appeared by 1976. Pioneering individuals that produce the pheromone are the males in polygamous species of *Ips, Pityokteines, Orthotomicus, Pityogenes*, etc. In monogamous species belonging to the genera *Dendroctonus, Trypodendron* and *Tomicus* it is the females that arrive first and produce the pheromone. There are a few exceptions to this rule: *Gnathotrichus sulcatus* is monogamous, but attacks are initiated by males. In *Leperesinus fraxini* (which lives on broadleaf trees) females are the pioneers, but the aggregation pheromone is produced by the males. In some cases each sex produces a different pheromone at a different stage of invasion of the tree. In the initial stage of invasion by *Dendroctonus frontalis*, males produce verbenone which reduces the number of males that are attracted. In *Dendroctonus brevicomis*, males produce both verbenone and frontalin at the same stage of the invasion, the latter compound being specially attractive to females. Males of *Dendroctonus frontalis, D. ponderosae and D. rufipennis* all produce an anti-aggregation substance once they have joined the female in her gallery.

All the complex aspects of the relationships between scolytids and their host trees may be seen as the result of long coevolution. Production of aggregation pheromones is an adaptation to exploitation of the relatively scarce and widely-dispersed specialised environment that is a sick or moribund tree, one that would be hard to find by random searching. The aggregation pheromone, which is synthesised from constituents of the host tree, is a neat way of facilitating the beetles' search for suitable breeding habitats and overcoming the tree's resistance as rapidly as possible. Resistance takes the form of abundant resin flux that glues and kills some of the insects. Mass colonisation is made possible by the aggregation pheromones. Scolytids have evolved two kinds of strategy for overcoming the trees they attack. The first consists in an assault by a large number of beetles, each one of which bores its own gallery system. This is the strategy used by most species. The second strategy is that used by *Dendroctonus micans* and a few other species belonging to this genus. Adults are not attracted to suitable trees by an aggregation pheromone; instead, when a female happens on a suitable tree, her offspring, consisting of a great many gregarious larvae, manage to overcome the tree's defences by a collective assault on the sub-cortical layer.

Sexual maturity is a prerequisite to triggering secondary attraction. Adults must spend a certain minimum period on the wing before they can mature sexually. Intake of food plays a determining role in pheromone production in many species, as does food quality. In an environment devoid of terpenes *Tomicus destruens* produces few pheromones (Carle, 1974).

2.2.1. Nature of pheromones and production sites

With the single exception known to date of *Scolytus multistriatus*, aggregation pheromones are associated with the hindgut and the Malpighian tubes, and are liberated with the insect's frass. It is probable that these pheromones are largely produced in parts of the body other than the digestive tract, the latter's role being merely to concentrate and excrete them. The Malpighian tubes are one possible point of production. In the terminal part of the gut of *Trypodendron lineatum* major histological differences were found between attractive females producing pheromones and non-attractive females. In the attractive females there are highly developed and active secretory cells which are inactive in non-attractive females (Pitman *et al.*, 1965; Schneider & Rudinsky, 1969).

One characteristic of aggregation pheromones is that they are not secretions of the beetles themselves, but substances of plant origin, most often resin terpenes that are modified as they pass through the digestive tract. Micro-organisms present in the digestive tract almost certainly play an important part in this transformation (Brand *et al.*, 1975; Chararas *et al.*, 1979).

Much research has been done on American species of *Dendroctonus* and *Ips*, which are serious pests, and on various European species such as *Ips typographus* (Bakke, 1970; Rudinsky *et al.*, 1977), *Ips acuminatus* (Bakke, 1967), *Ips sexdentatus* (Vité *et al.*, 1974), *Ips duplicatus* (Bakke, 1975) *Pityokteines curvidens* (Harring *et al.*, 1975), *Leperesinus fraxini* (Schönherr, 1970), *Trypodendron domesticum* (Francke, 1973) and *Scolytus multistriatus* (Gore, 1975; Lanier *et al.*, 1976). The aggregation pheromone may have one constituent or several depending on species (table 15.2). In species of *Dendroctonus*, six different substances have been isolated and synthesised: *cis*-verbenol, *trans*-verbenol, verbenone, frontalin, exobrevicomin and endobrevicomin. Each species has its own characteristic arsenal of these compounds. *D. brevicomis*, for example, has five constituents (it lacks *cis*-verbenol), while *D. frontalis* has only four (it lacks *cis*-verbenol and exobrevicomin) (table 15.3).

These pheromones are produced by insects from monoterpenes according to metabolic processes that are as yet poorly understood. *Bacillus cereus*, which is able to convert α-pinene into verbenol, has been found in the gut of various *Ips* and *Dendroctonus* species (Brand *et al.*, 1975). In *Dendroctonus brevicomis* there is a linear relationship between the length of exposure of beetles to α-pinene and quantity of *trans*-verbenol present in the hindgut. Production of *trans*-verbenol ceases when α-pinene is no longer provided. Of the metabolic ways of oxidising terpenes into alcohols and ketones

Species	Host tree	Sympatric with	Pheromone constituents			
			Ipsdienol	Ipsenol	*cis*-verbenol	MB
1. *Ips sexdentatus*	Pine, spruce	2, 3, 4, 7	×	–	–	–
2. *Ips duplicatus*	Spruce	1, 6, 7	(R) – (–)	–	–	–
3. *Pityokteines curvidens*	Fir	1, 4, 6, 7	–	(S) – (–)	–	–
4. *Pityokteines vorontsovi*	Fir	1, 3, 6, 7	(R) – (–)	(S) – (–)	–	–
5. *Ips cembrae*	Larch	1, 3, 4	×	×	–	(3,3,1)
6. *Ips acuminatus*	Pine	1, 2, 3, 4, 7	×	×	×	–
7. *Ips typographus*	Spruce	1, 2, 3, 4, 5, 6	×	–	(S)	(2,3,2)

Table 15.2. *Pheromone composition in a few species of* Ips *and* Pityokteines *of the European fauna. x designates a pheromone constituent whose optical character is indifferent. In column MB the symbol (3,3,1) designates 3-methyl-3-buten-1-ol and symbol (2,3,2) designates 2-methyl-3-buten-2-ol (after Vité, 1978).*

Species and sex	Pheromone	Sex(es) attracted	Sex(es) repelled
D. frontalis (F)	Frontalin and *trans*-verbenol	M and F	-
D. frontalis (M)	Verbenone	-	M
	endo-brevicomin	-	M and F
D. brevicomis (F)	(+) brevicomin	M	-
D. brevicomis (M)	(-) Frontalin	F	-
	Verbenone	-	M and F
D. adjunctus (F)	Frontalin	M	-
D. adjunctus (M)	Brevicomin	F	M
D. ponderosae (F)	*trans*-verbenol	M and F	-
D. ponderosae (M)	Brevicomin	-	M and F
D. pseudotsugae (F)	Frontalin + seudenol	M and F	-
D. pseudotsugae (M)	MCH	M and F	M and F

Table 15.3. *Main constituents of aggregation pheromone and effects on both sexes in six species of* Dendroctonus. *MCH is 3-methyl-3-cyclo-hexene-1-ol (after Vité, 1978).*

having the properties of pheromones one may mention oxidation of α-pinene to *trans*-verbenol, to *cis*-verbenol and then to myrtenol, and oxidation of myrcene to ipsdienol and to ipsenol (figure 15.6). The optical configuration of the molecule is important; only the isomer S-*cis* of verbenol is active on *Ips calligraphus* and *Ips typographus* (Vité *et al.*, 1976).

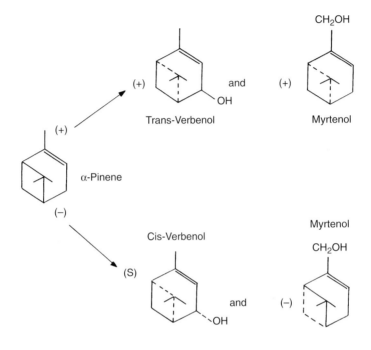

Figure 15.6. *Conversion by oxidation of a resin monoterpene, α-pinene, to two components of aggregation pheromones: verbenol and myrtenol (Vité & Francke, 1976).*

Pheromone constituents have been identified in the faeces of American species of *Ips*; they are *cis*-verbenol, ipsdienol, ipsenol and *trans*-verbenol. These compounds are not active in isolation, but only in the form of a mixture. In the hindgut of male *Ips typographus* at the stage when they begin to bore their galleries there are various compounds including *cis*-verbenol, ipsdienol and ipsenol, which are the components of the pheromone (figure 15.7).

The scolytid *Xylocleptes bispinosus* is a polygamous species which lives in the woody climber *Clematis*. Males produce an aggregation pheromone whose main constituent is ipsenol, a substance which was previously only known to occur in scolytids that feed on conifers (Klimetzek *et al.*, 1989b).

Geographical variations in composition of the aggregation pheromone and in insect responses have been shown to exist in the American species *Ips*

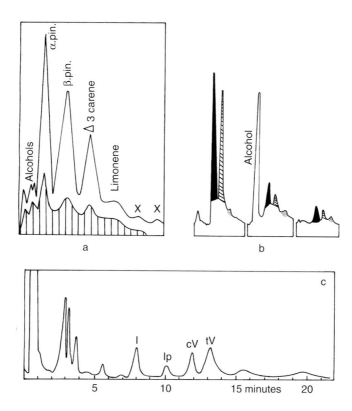

Figure 15.7. *a. Main terpene constituents of Scots pine wood. Peaks correspond to alcohols, α-pinene, β-pinene, Δ-3-carene, limonene and two unidentified substances indicated by x. The area of vertical hatching represents deteriorated wood that is no longer attractive to scolytids. b. Diagrams showing variations in abundance of terpenes in three Scots pine trees. Left: vigorous tree; middle: less vigorous, highly attractive tree with a high alcohol peak; right: very deteriorated tree abandoned by scolytids (Pesson & Chararas, 1969). c. Chromatograph of volatile substances present in the hind gut of* Ips typographus *after it has begun to bore galleries. I: ipsenol; Ip: ipsdienol; cV:* cis-verbenol; *tV:* trans-verbenol *(Bakke, 1973).**

pini and in species with which it is frequently associated, which include *Ips integer* and various predatory clerids and trogositids on which the pheromone acts as a kairomone. Geographical variations in pheromone composition have also been found in *Dendroctonus frontalis*. This observation is of great practical importance, as traps used to catch scolytids will have to have different baits according to region (Miller *et al.*, 1997).

Genetic variations within a widely-distributed species are known to occur in various xylophagous insects such as the weevil *Pissodes strobi*. A genetic variability gradient has been found in this species, and it is linked to variations in the choice of host tree. Populations of the Pacific coast of North America are basically monophagous on *Picea sitchensis*; those of the Rocky Mountains live on several species of *Picea*, and those of the eastern part of the continent colonise various *Picea* and *Pinus* species (Phillips & Lanier, 1985). Another case has been observed in *Dendroctonus frontalis* and *D. brevicomis*, whose isoenzymes are very different in localities in Arizona and Texas from those found in Virginia and Louisiana. These genetic variations may be linked to differences in behaviour and in likelihood of outbreaks (Namkoong *et al.*, 1979). These facts must be taken into account in the context of pest control, for they may be linked to geographical differences in the composition of aggregation pheromones.

Aggregation pheromone synthesis is controlled by hormones, particularly by the juvenile hormone and the cerebral hormone (Hughes & Renwick, 1977).

The pheromone composition of the various species sometimes varies subtly within a single genus. It may for instance take the form of the substitution of one enantiomer (or optical isomer) by another. Table 15.2 which relates to various European species of *Ips* and *Pityokteines*, and table 15.3 which relates to a few American species of *Dendroctonus* illustrate this phenomenon. One tempting explanation for this has been put forward: since many of these species are sympatric, differences in the chemical composition of the pheromones may avoid colonisation of the same tree by individuals belonging to different species, and thus limit competition (Vité, 1978).

2.2.2. How do pheromones act?

Sensory cells on which pheromones act have been studied in several species of scolytid (Dickens, 1979, 1981; Dickens & Payne, 1978). These studies have shown that various types of sensory cell in the antennal club react to constituents of the pheromone. In *Ips confusus* there are five kinds of sensory cell. Experiments have shown that it is the *sensilla trichodea* that perceive the pheromone (Wood, 1970). There is a threshold above which beetles are attracted; in *Dendroctonus brevicomis* 1 μg of brevicomin suffices to attract males (Silverstein, 1970). Wind speed, temperature and light also have an effect. Chemical stimuli are only really efficient in localising the source from which they are emitted over short distances. This explains the part played by visual stimuli such as the size and silhouette of trees (Pitman & Vité, 1970).

Olfactory sensory cell responses in the antennae of *Dendroctonus micans* and its predator *Rhizophagus grandis* have been recorded. Both species have several types of cell, and each type of cell is specialised in the detection of one compound. In *D. micans* there are seven types of cell, two of which detect, respectively, exobrevicomin and (+) - ipsdienol. In *R. grandis* the clearest responses are obtained with (+) - ipsdienol and (-) - verbenone, both of which are produced by *D. micans* larvae. These data suggest that exo-brevicomin is one element of *D. micans'* pheromone (Tommeras *et al.*, 1984). In the curculionid *Hylobius abietis*, sensory cells that react to attractant substances are also situated on the antennal club, and are of the same types as those of scolytids (figure 15.8).

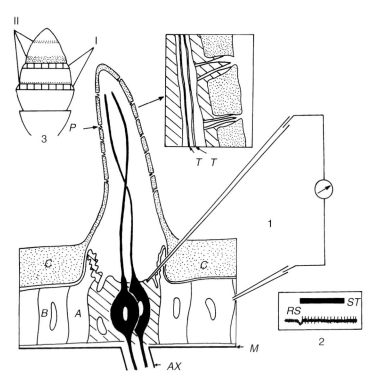

Figure 15.8. *Schematic diagram 1 shows a* sensilla trichodea. *C: cuticle; P: pore; M: basal membrane. Framed, detail of part of the* sensilla *with two pores and microtubules T. Two micro-electrodes able to receive waves from the sensory cells when these are excited by olfactory substances are shown. 2 is a schematic representation of recordings obtained. ST is the stimulus and RS the cell's response. 3 is a schematic diagram of the antennal club of the weevil* Hylobius abietis. *Areas with vertical hatching bear* sensilla basiconica *(I) and the stippled areas bear* sensilla trichodea *(II) (Mustaparta, 1975).*

Some cells react to frontalin, some to terpenes or to other substances. It thus seems reasonable to suppose that these sensilla are *Hylobius abietis'* pheromone receptors. The pheromone has, unfortunately, not yet been isolated (Mustaparta, 1975). These data show some of the complexity of the sensory equipment of xylophagous insects and explain why their behaviour varies according to the composition of olfactory effluvia they receive.

2.2.3. Processes of invasion of trees

The processes are complex and vary from one scolytid species to another. They may be described by means of a few examples.

In *Dendroctonus ponderosae* there is an initial stage during which female pioneering individuals attack a tree to which they have been attracted by its α-pinene. Then the settled females emit *trans*-verbenol, a pheromone that, in association with α-pinene, attracts numerous individuals of both sexes. When production of *trans*-verbenol by females busy boring their galleries increases, the number of males that arrive also rises. Once the tree's resistance is overcome and sap flux ceases the attack ceases too, and colonisation is complete. Thus the signal that tells *Dendroctonus* that another host must be sought is simply the cessation of resin exudation.

Events are more complex with *Dendroctonus brevicomis*. Initial attack seems to be triggered by the presence of Δ-3-carene which attracts females. These pioneer females release small quantities of frontalin which is highly attractive when combined with the tree's terpenes. It is however mostly females that are attracted. Another constituent, exobrevicomin, is produced by females that have fed, and the new mixture attracts more males than females. Males that settle emit substantial quantities of frontalin and verbenone; the latter substance has an inhibiting effect on the attractant power of the other substances, and colonisation gradually comes to an end.

The colonisation mechanism of *Ips paraconfusus* is also very complex (Birch, 1984). Potential competitors such as *Dendroctonus brevicomis* and *Ips pini* are repelled by the pheromone of *Ips paraconfusus*, and the pheromones of *Ips pini* and *Dendroctonus brevicomis* are equally repellent to *Ips paraconfusus*, and prevent it from settling on trees already colonised by these two species (figure 15.9).

Various mechanisms are available to avoid overpopulation on infested trees. One of these works through the effect of high concentrations of pheromones when beetles that have already settled are numerous. Beyond a certain threshold pheromones have a repellent effect and induce new arrivals to settle on neighbouring trees instead (Hughes & Pitman, 1970). In *Trypodendron lineatum* the mechanism that prevents overpopulation appears to be release by males of a pheromone that prevents other individuals of the same species from responding to aggregation pheromones (Nijholt, 1970).

Larval mortality, attributable to competition for food, begins at low population densities in scolytids, and rises with increase in density. Competition is due either to insufficient available quantities of phloem or to uneven distribution of larval galleries. *Scolytus scolytus* galleries in elm tend to be concentrated on the lower parts of trunks. Competition appears to be common in scolytids due to the efficiency of aggregation pheromones which draw large populations to suitable trees (Beaver, 1977).

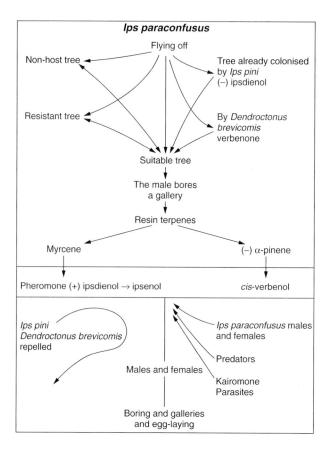

Figure 15.9. *Schematic diagram of different stages of colonisation of a tree by* Ips paraconfusus.

Intraspecific competition in *Ips typographus* and its consequences have been studied by Anderbrant *et al.* (1985). Rising density in laboratory-reared populations causes a decrease in number of progeny per female; these progeny have lower body weights and lipid reserves; males reared from high density populations produce fewer pheromones; descendants of *Ips* so reared are only half as numerous as those of *Ips* reared in low density populations (figure 15.10). All these data show that population density of *Ips typographus* may play a major part in population dynamics.

Colonisation of a tree, the weakening of its defence mechanisms, and completed occupation by scolytids are facilitated by pathogenic fungal spores of the genus *Ceratocystis*, which are carried by the beetles, and induce an acquired defence reaction and hypersensitive reaction in the tree (cf. chapter 3.1.8). Structures involved in transporting the pathogenic fungi are often localised in depressions of the prothoracic tegument, and are associated with glandular cells whose role is still unknown (Levieux & Cassier, 1994).

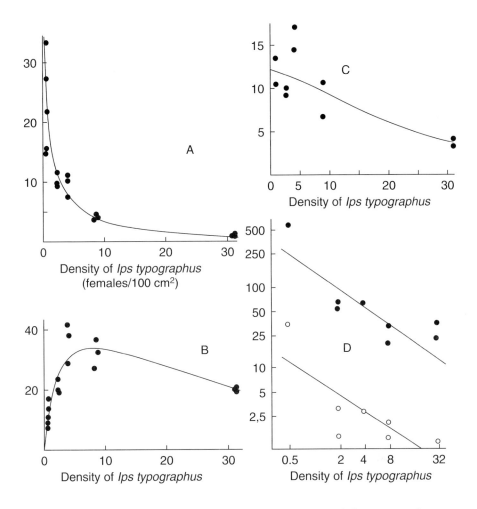

Figure 15.10. *Effects of density on population dynamics in* Ips typographus.
A: number of offspring per female at different densities.
B: number of beetles produced per unit of surface aera of bark.
C: number of offspring per female as a function of density in the
previous generation. D: quantity of pheromones present in the
hindgut of males as a function of density. Black dots: 2-methyl-3-
buten-2-ol; circles: 2-phenylethanol (Anderbrandt et al., 1985).

Outbreaks of scolytids are usually caused by two phenomena: presence of an already dense population of beetles, and lowered resistance of trees which may be due to a variety of causes. Berryman (1982) built a model explaining variations in abundance of scolytids taking these two observations into account. The model supposes a "minimum threshold of abundance" necessary for scolytids to successfully invade a tree. Low density populations of scolytids may overcome resistance in enfeebled trees and then settle and breed. In the case of trees with high resistance, assault by a greater number of beetles is necessary (figure 15.11). In an area in which trees are resistant, epidemics may occur in two different situations: (a) where scolytid populations increase by taking advantage of suitable reproduction sites such as trees uprooted by wind; (b) where tree resistance is lowered, for instance by drought. The validity of this model was confirmed by studies on invasions of spruce by *Ips typographus* in Norway (Worrell, 1983).

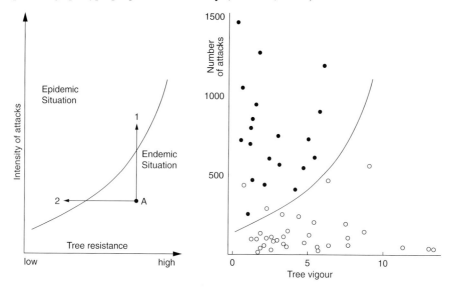

Figure 15.11. *Left, Berryman's model. The curve corresponds to the minimum threshold of abundance. From point A onwards an endemic situation may become epidemic either by increase in scolytid abundance (arrow 1) or by lowered resistance of trees (arrow 2). Right, relationship between vigour of spruce trees in Norway and intensity of attacks by Ips typographus. Tree vigour is evaluated by the ratio (in %) between the area of the last growth ring and the total area of wood. The curve represents the threshold that must be reached if an attack is to be successful. Black dots: wilting trees; circles: surviving trees (Mulock & Christiansen, 1986).*

The physical characteristics of trees, particularly their spacing and diameter, affect the population dynamics of scolytids, as has been shown by research carried out on lodgepole pine (*Pinus contorta*) and *Dendroctonus ponderosae* (figure 15.12). (a) Phloem is usually thicker in trees of a large diameter than in small trees. (b) Trees in dense stands have thinner phloem than those growing in open situations. (c) *Dendroctonus* production has a quasi-linear relationship with the phloem thickness of invaded trees. (d) The greater the diameter of a tree the greater the ratio of insects that emerge to those that invaded the tree. (e) Of trees that have survived an earlier year's infestation, *Dendroctonus* usually kill those with the greatest diameter. (f) Large diameter trees are those that die in greatest numbers during an infestation.

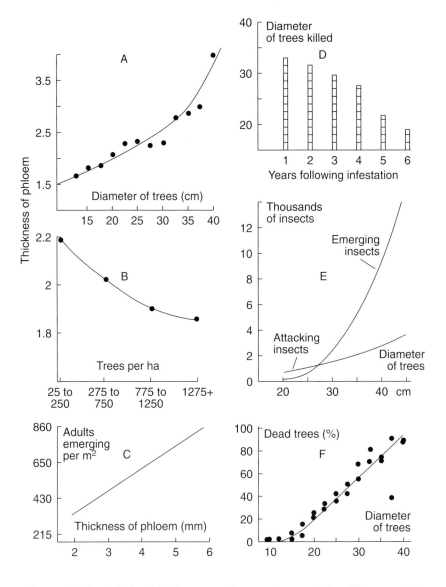

Figure 15.12. *Relationships between different characteristics of* Pinus contorta *and attacks by* Dendroctonus ponderosae. *A: relationship between trunk diameter at breast height (or DBH) and secondary phloem thickness. B: relationship between number of trees per hectare and average thickness of secondary phloem. C: relationship between number of adults emerging per square metre of bark and phloem thickness. D: average diameter (DBH) of trees killed according to year of infestation; E: relationship between diameter (DBH) and number of* Dendroctonus *that attack or emerge from a tree. F: relationship between percentage of trees killed and trunk diameter (DBH) (Safranyik, 1970).*

In the case of *Ips calligraphus*, which lives in the secondary phloem of *Pinus elliotii*, it was found that the beetle's demographic characteristics vary according to the thickness of secondary phloem. In trees in which it is less than 1 mm thick reproduction rates are low, larval growth is slower, and the number of adults produced by each female is smaller than in trees with secondary phloem 3 mm thick. Sex ratio is unbalanced in favour of females (one male to two females) in beetles reared in thin phloem. The reason for this change in sex ratio is not clearly understood. The great evolutionary plasticity of its biological characteristics enables *Ips calligraphus* to colonise a great variety of trees (Haack *et al.*, 1987).

2.2.4. Stridulation and its role

As soon as they land on a tree, scolytids of the same sex as the pioneers begin to bore their galleries under the bark. Individuals of the opposite sex search the galleries for pioneer individuals that have not yet paired. Most scolytids are able to stridulate. The sound-producing organ includes a rasp-shaped part called "pars stridens" and a "plectrum" formed of spines or tubercles which rub against the pars stridens. The stridulatory organ is usually possessed only by the opposite sex to the one that first bores the gallery, but it exists in both sexes in monogamous species of *Dendroctonus*. When stridulations are emitted by a female attracted to a mating chamber excavated by a male they condition the male to accept the female. A male will refuse entry to a female experimentally deprived of the ability to stridulate. Similarly an *Ips pini* female that lands on a tree occupied by males will wander on the bark and attempt to enter a gallery when it finds one. The male is normally stationed at the entrance to the gallery, which it blocks with its elytral declivity; the female pushes the male and stridulates vigorously until the male moves forward into the nuptial chamber. A male will usually accept two to three females (Swaby & Rudinsky, 1976). Similar behaviour is known in *Dendroctonus pseudotsugae* (figure 15.13).

2.2.5. Effects of pheromones on other species

Pheromones emitted by one species of scolytid may carry a message to other scolytid species, and especially to various parasites and predators. Pheromones that work on individuals of a species other than the one emitting them are known as kairomones (cf. chapter 6.7.2). Synthetic *Dendroctonus pseudotsugae* pheromones attract more than thirty species of scolytid, particularly those belonging to the genera *Hylastes* and *Pseudohylesinus*. Predators such as beetles of the families Cleridae and Colydiidae as well as parasitic flies such as *Medetera* are also attracted to scolytid pheromones. This explains the almost simultaneous arrival of scolytids and of their enemies on trees, and the often large numbers of predators caught in pheromone

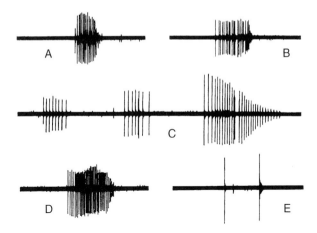

Figure 15.13. *Oscillograms obtained by recording various types of stridulation emitted by* Dendroctonus pseudotsugae. *A: attraction stridulation emitted by a male when it approaches a gallery occupied by a female. B: courting stridulation emitted just before mating. C: rivalry stridulation emitted on encountering another male. D: alarm stridulation emitted when the beetle is handled. E: single stridulation emitted by a female guarding the entrance of the gallery in which it has laid its eggs.*

traps. The weevil *Hylobius abietis* is sensitive to conifer terpenes as well as to derivatives of terpenes such as verbenone, which serves as a pheromone among scolytids. *Hylobius* does not, on the other hand, react to the cotton boll weevil pheromone which is very different from those of conifer scolytids such as *Ips*. These observations show that pheromone specificity is linked to the nature of the host plant and not to the taxonomic position of the insect.

2.2.6. Arrival sequence of insects in the wake of scolytids

This sequence has been described in the case of *Pinus ponderosa* infested by *Dendroctonus brevicomis* (Stephen & Dahlsten, 1976). More than one hundred species of insect are associated with this beetle. The main predators arrive during the mass colonisation stage or shortly after. The clerid *Enoclerus lecontei*, the trogositid *Temnochila chlorodia* and the colydiid *Aulonium longum*, which predate both larvae and adults, are among the first to appear, together with the fly *Medetera aldrichii* which predates larvae. Specialised predators arrive later when the suitable stage in the beetle's life cycle is present; these include in particular *Roptrocerus xylophagorum* and *Dinotiscus burkei*. Increased diversity is seen from the time of massive arrival of the scolytid until their offspring fly away. This is due to colonisation by a host of detritivorous, fungivorous and xylophagous species and parasites, which may be very numerous (figure 15.14).

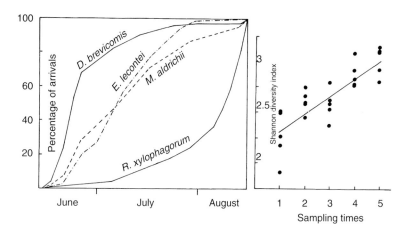

Figure 15.14. *Left, arrival sequence of the scolytid* Dendroctonus brevicomis *and of three of its enemies,* Medetera aldrichii, Roptrocerus xylophagorum *and* Enoclerus lecontei, *on* Pinus ponderosa. *In ordinate are cumulative percentages of arrivals. Right, Shannon diversity index for the sub-cortical fauna during different stages of colonisation until the emergence of* Dendroctonus *(Stephen & Dahlsten, 1976).*

3. BIOLOGY OF SOME SCOLYTID SPECIES

This section presents data on the biology of some of the commoner, more harmful or more remarkable species of the European fauna (and of the North American fauna in the case of *Dendroctonus*).

3.1. Elm bark beetles

Since 1970 we have witnessed the spectacular spread of a disease that kills elm trees, caused by the pathogenic fungus *Ceratocystis ulmi* which is carried by two species of scolytid beetle; these are *Scolytus scolytus*, the large elm bark beetle, which measures 5 to 6 mm, and *Scolytus multistriatus*, the small elm bark beetle, which measures 2.5 to 3.5 mm. The disease was named "Dutch elm disease" because it was first identified in Holland, but it is of Asian origin, and has been introduced in the United States. The disease first appeared in Europe in 1918 (Pinon, 1976; Gibbs *et al.*, 1977). Infested trees bud late; leaves on some branches suddenly turn yellow and fall. A cross-section of a branch shows irregular brown stains on one or more growth rings; these spots result from invasion of the parenchyma and the vessels by the mycelium, gums, vesicles and spores, which disrupts the flow of water to leaves and makes them wilt. The tree dies after a period of time that varies according to the virulence of the fungus (Sinclair & Campana, 1978).

Ceratocystis ulmi is transmitted to healthy trees by the adult beetles which swarm in two generations in May and late August. They make feeding bites on the bark of young branches, usually near the bases. Females bore galleries under the bark; larval galleries, as in all species of *Scolytus*, are arranged in regular rows on either side of the female gallery. Larvae pupate in the bark or in the outer layers of wood, and the adult emergence holes riddle the surface of branches. Adults swarm from May to October of the following year under favourable temperature conditions, carrying *Ceratocystis* spores to other trees. The fungus may be transmitted to other trees through fusion of roots that may unite several elm trees. *Ceratocystis ulmi* is an ascomycete that is the imperfect form of *Graphium ulmi*.

The *Ceratocystis ulmi* life cycle can be summarised as follows. Starting from one or another type of spore (conidiospore or ascospore) the fungus multiplies and proliferates by forming a mycelium that invades the wood. The mycelium is in a vegetative state which has been named *Graphium ulmi* (which accounts for the term "elm graphiosis" which is sometimes given to Dutch elm disease in French). *Graphium* spores give rise to a new asexual generation of the fungus. When cells belonging to two sexual types that differ in their nuclear structure meet, they fuse to produce a stage whose cells, which have two nuclei, form first a perithecium and then ascospores. This sexual stage is that of an Ascomycete which has been named *Ceratocystis ulmi* (= *Ophiostoma ulmi*).

The two species of elm bark beetle are accompanied by a whole suite of predators and parasites that seem unable to check growth of the beetles' numbers. The predators include Coleoptera that are habitual inhabitants of scolytid galleries such as *Corticeus*, *Thanasimus* and *Rhizophagus*. Parasitoids belong to the Braconidae and Chalcidoidea. The braconid *Dendrosoter protuberans* is the main parasitoid of the small elm bark beetle and has been used in attempts at biological control. *Coeloides scolyticida* is a major parasitoid of the large elm bark beetle. The fly *Dolichopoda nitida* parasitises both species.

The aggregation pheromone of *Scolytus multistriatus*, which lives on broadleaf trees, differs from those of species that live on conifers. The pheromone is a mixture of three substances (figure 15.15). Two of them are produced by virgin females in different parts of the metathorax and the abdomen as well as in an accessory gland in the tip of the abdomen. The third substance is present in elm wood, where its levels rise significantly in trees invaded by *Ceratocystis ulmi*. A mixture of the three substances obtained synthetically is known by the name multilure; its power of attraction appeared great enough for it to be tried in biological control. Other insects are drawn to multilure, particularly various predators and parasitoids of *S. multistriatus*. Systemic fungicides can successfully combat *Ceratocystis ulmi*; the two main ones are benomyl and thiabendazole (Stipes, 1973).

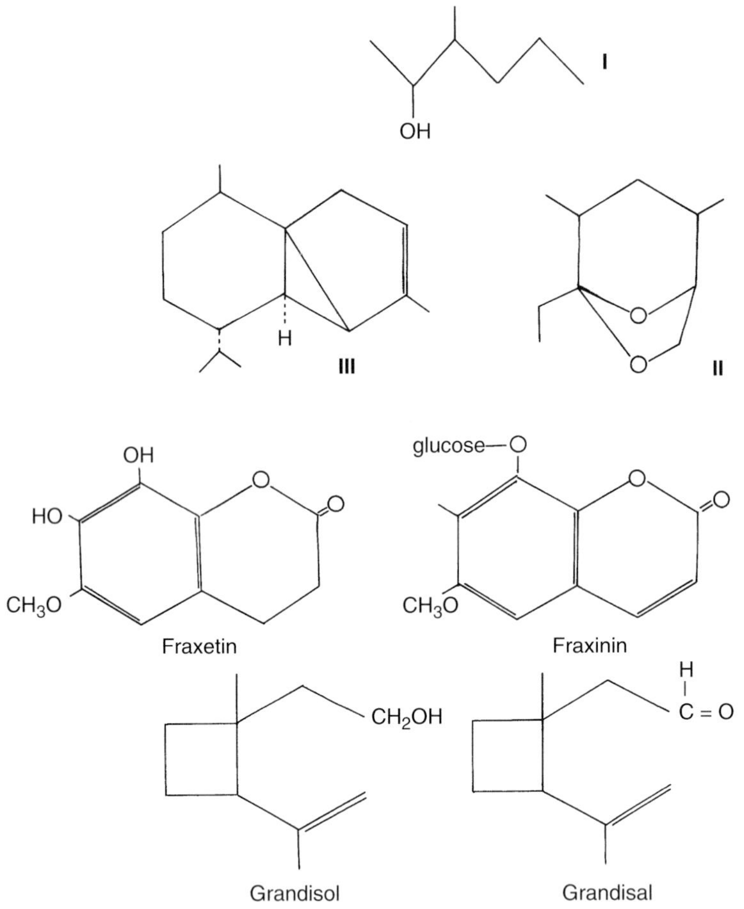

Figure 15.15. *Formulae of three molecules which together constitute the Scolytus multistriatus pheromone. I: (-)-4-methyl-3-heptanol; II: multistriatin; III: (-)-α-cubebene (Lanier et al., 1976). Formulae of fraxetin and fraxinin, substances characteristic of ash and attractive to Leperesinus fraxini. Formulae of grandisol and grandisal, constituents of the Pissodes weevil pheromone.*

Selection of resistant varieties or species of elm is certainly the best way of avoiding the total extinction of elm in Europe. Various Asian species have good levels of resistance, whereas all the American species are sensitive, particularly *Ulmus americana*, which has been greatly affected by the disease. Interspecific hybrids are the most promising. The resistance mechanism appears to be linked to anatomical properties of the wood which impede propagation of the fungus.

3.2. *Pteleobius vittatus* and the genus *Leperesinus*

The scolytid beetle *Pteleobius vittatus* lives on several species of elm on which it may be very common. It is a small species, 2 mm long, with the body covered in characteristically-shaped scales of three different colours. Its galleries are of the transverse type, with larval galleries perpendicular to and on either side of the main gallery (figure 15.4). *Pteleobius vittatus*, which only attacks wilting or moribund elms, deserves mention because its aggregation pheromone has an unusual composition. It consists of a mixture of the following three molecules: vinylcarbinol [or 2-methyl-3-buten-2-ol] which was already known to be a constituent of the aggregation pheromone of some scolytids that live on conifers, *cis*-pityol [or *cis*-2 (1-hydroxy-1-methylethyl)-5-methyltetrahydrofurane] which is unique to *Pteleobius* as a pheromone constituent, and *cis*-vittatol [or *cis*-3-hydroxy-2,2,6-trimethyl-tetrahydropyrane] which is also unique to *Pteleobius* and was previously unknown (Klimetzek *et al.*, 1989a). The mixture of these three molecules also attracts a specialised predator, the ostomatid beetle *Nemosoma elongatum*.

Leperesinus fraxini and *L. orni* are two species associated with the ashes *Fraxinus excelsior* and *Fraxinus oxyphylla*. They may occupy healthy trees as well as wilting ones. The adult bores galleries that cause the formation of galls known as "ash roses" in average sized branches. It seems certain that the attractant substances in these beetles are two compounds characteristic of ash – fraxetin and fraxinin (figure 15.15).

3.3. *Tomicus piniperda*, the large pine shoot beetle

Three species belonging to the genus *Tomicus* (= *Blastophagus*) occur in France, where they live on pines. *T. piniperda* is common and has a vast distribution range; *T. minor* is scarcer but is just as widely distributed; *T. destruens* has a more southerly range. *Tomicus piniperda* very often becomes a primary pest which may infest perfectly healthy trees. Its main host is Scots pine; *T. piniperda* invaded pine plantations in central France at the same time as did *Ips acuminatus* and *Ips sexdentatus*. Adults emerging from hibernation swarm in two and sometimes three waves spread over time as soon as temperatures reach 10 to 22 degrees centigrade. They settle on trunks with bark at least 4 mm thick. Eggs are laid in two or three stages corresponding with swarming and thus two or three sibling generations are produced. Larvae live two to three months and pupate in June, and adults emerge in June to July. These adults bore into the axis of young shoots on which they feed until they reach sexual maturity, excavating the pith which they eat. This causes desiccation of the buds and the tips of shoots which break off. *T. piniperda* is a monogamous species; its galleries under bark are of characteristic design (figure 15.16).

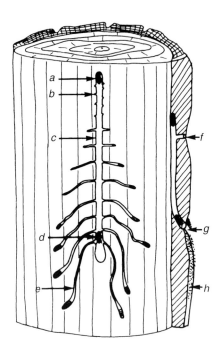

Figure 15.16. Tomicus piniperda *family gallery system. a: female boring the egg-laying gallery; b: incubation notch; c: egg-laying gallery; d: male cleaning the egg-laying gallery; e: larval gallery; f: ventilation shaft; g: entrance; h: sawdust and resin flux.*

The large pine shoot beetle's predator and parasitoid complex has been studied in Britain. It resembles that of many other scolytids and includes the following species:

Parasitoid Hymenoptera
– Pteromalidae   :   *Rhopalicus tutela*; *Dinotus calcaratus*
– Braconidae   :   *Coeloides abdominalis*; *Dendrosoter middendorffi*

Predatory Diptera
– Xylophagidae   :   *Xylophagus ater*
– Dolichopodidae :   *Medetera* sp.
– Lonchaeidae   :   *Lonchaea laticornis*; *Palloptera usta*

Raphidioptera   :   *Raphidia notata*

Coleoptera
– Staphylinidae   :   *Baptolinus affinis*; *Leptusa angusta*, *Quedius* sp.
– Colydiidae   :   *Aulonium ruficorne*
– Tenebrionidae   :   *Corticeus fraxini*
– Cleridae   :   *Thanasimus formicarius*
– Rhizophagidae   :   *Rhizophagus* sp.

Tomicus minor is localised in coastal regions of Mediterranean Europe where it made an appearance following outbreaks of the scale insect *Matsucoccus* on maritime pine.

3.4. *Ips typographus*, the engraver beetle

The engraver beetle is a major pest of spruce in the whole of Eurasia, and may occasionally infest pine, fir and larch. Outbreaks of this beetle have been recorded since the eighteenth century. Losses sustained reached 4 million cubic metres of wood in Germany in the years 1857-1862 and 30 million in 1899-1948. In Sweden losses were 2 million cubic metres in 1976-1979 and in Norway 5 million cubic metres in the period 1970-1982. Temperature is a major factor in determining the beetle's life cycle, which explains why there is only one generation a year in Finland and two or three generations in central Europe. In France two generations are produced in areas below 800 m, and one generation above 1,100 m. After hibernating in litter at the foot of trees, the adults lay their eggs in April. The larvae give a first generation of adults in July to August; these swarm in turn and their offspring form a second generation that hibernate as larvae, pupae or adults. Above 1,100 metres there is only one period of reproduction, in June.

The engraver is a polygamous species; its galleries resemble those of *Ips acuminatus*. In normal conditions the beetle breeds on recently fallen or very enfeebled trees. Populations expand when the species is able to exploit numbers of trees uprooted by wind or trees weakened or stricken in other ways, when it can become a serious primary pest.

Massive trapping using aggregation pheromones has been tried in order to reduce numbers of *Ips typographus* and of *Trypodendron lineatum*. Synthetic pheromones, marketed under the names "Linoprax®" and "Pheroprax®" were used, one trial giving the following results:

– Traps baited with Linoprax®: 24,349 insects caught, of which 78.5% *T. lineatum*, 19.9% *I. typographus* and 1.6% other insects.

– Traps baited with Pheroprax®: 22,815 insects caught, of which 92.8% *I. typographus*, 5.8% *T. lineatum* and 1.4% other insects.

– Unbaited control traps: 9,429 insects, of which 52.5% *I. typographus*, 38.4% *T. lineatum* and 9.1% various other insects. These results demonstrate the effectiveness and selectivity of pheromone traps (Babuder *et al.*, 1996).

An outbreak that occurred in recent years in Norway and Sweden was mainly triggered by drought during the years 1974-1976 and by storms during the 1970s that uprooted many trees (Bakke, 1983). Outbreaks were facilitated by the structure of spruce forests which consisted of old trees particularly vulnerable to attack. Forest hygiene measures were taken on extensive areas in order to stem these infestations: removal of infested trees, setting networks of attractant traps designed to capture *Ips* on a massive scale, and elimination of excessively old trees. Detailed study of the invasion made it possible to calculate a "risk index" and to identify those parts of the forest most threatened by future outbreaks (Worrell, 1983). Observations relating to intensity of infestation and to tree vigour confirmed the validity of Berryman's model (cf. figure 15.11).

3.5. *Ips sexdentatus*

Ips sexdentatus is a common species on pine. It mainly colonises trees with a diameter between 20 and 40 cm. The species is polygamous; the male bores an entrance gallery which ends in a mating chamber from which the egg-laying galleries of 2 to 5 females radiate (figure 15.17). In France this beetle produces two generations a year in the North and three generations in the Mediterranean region and the Landes. Attacks are usually made on enfeebled trees but during outbreaks healthy trees are also colonised. Infested trees are recognisable by their yellowing foliage, the beetles' entrance holes and the sawdust that comes out of them. In the Orleans area infestations by *Ips sexdentatus* were facilitated by prior defoliation due to outbreaks of the sawfly *Diprion pini* (Lieutier *et al.*, 1984).

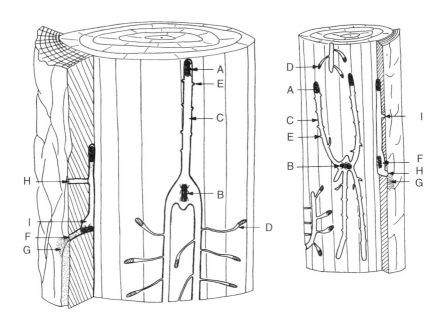

Figure 15.17. *Larval galleries of two scolytids. Left:* Ips sexdentatus; *right:* Ips acuminatus. A: female boring the egg-laying gallery; B: male; C: egg-laying gallery; D: larval gallery; E: incubation notch; F: entrance; G: sawdust; H: ventilation shaft; I: mating chamber.

3.6. *Ips acuminatus*

This scolytid, which used to be confined to the great natural conifer forests of eastern France and the mountains, has spread to central France. The beetle is about 3 mm long, is polygamous and has two generations a year. It is usually a secondary pest which speeds the death of trees that are physiologically unbalanced. Its gallery system includes a central vestibule which serves as mating chamber and 8 to 10 cm longitudinal galleries leading off it (figure 15.17).

The distribution of some of the scolytids of Scots pine has been studied in detail in Norway by Bakke (1968). A general outline showing the main tendencies in distribution of six species as a function of the part of the tree on which they settle, measured from the base of the trunk, and consequently according to the thickness of bark, can be drawn up. *Ips acuminatus* and *Tomicus minor* occupy more or less the same level, and the other four species are distributed according to characteristic patterns (figure 15.18).

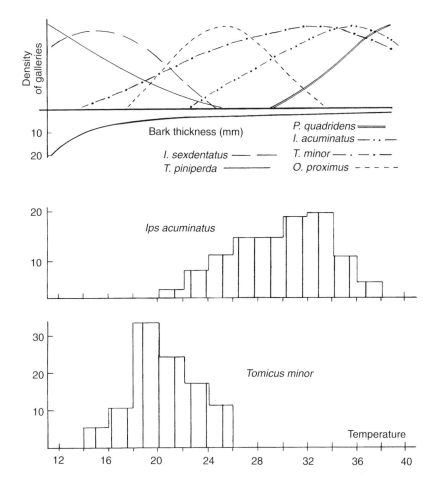

Figure 15.18. *Above: schematic representation of the distribution and abundance of six species of scolytid on a Scots pine trunk according to distance from the base of trunk and bark thickness.*
Below: distribution of entrance holes of Ips acuminatus *and* Tomicus minor *as a function of temperature to which pine trunks are subjected (Bakke, 1968).*

T. minor however occupies the lower surfaces of trunks and does not settle on the upper parts that other species favour. This distribution may be explained by thermal preferences. The preferred temperature of *T. minor* lies between 14 and 26 °C, with an average of 20.5 °C; *Ips acuminatus* has a wider range of tolerance, between 20 and 38 °C with an average of 29.9 °C, which explains why it is found on the upper parts of trunks.

3.7. *Hylastes ater*

Species of the genus *Hylastes* often settle in the roots and bases of trunks of dead or dying conifers, and even under the bark of fallen trees lying on damp soil. *Hylastes* are attracted by humidity, which explains their localisation on trees. *Hylastes ater* and *Hylastes brunneus* mainly feed on pines, and *Hylastes cunicularius* on spruces (Scott & King, 1974).

3.8. *Pityokteines curvidens*, the fir engraver beetle

This is a common species in all mountain ranges as well as in lowlands, and mainly feeds on fir. The male bores an entrance gallery ending in a mating chamber. Two to four maternal galleries bored by females lead off from the chamber, forming a stellate pattern whose rays, which reach about 4 cm in length, lie perpendicular to the grain of the wood (figure 15.19). *P. curvidens* is able to attack trees that are still quite green, which makes it a pest.

The three sympatric European species, *Pityokteines curvidens*, *P. spinidens* and *P. vorontzovi* may occur on the same tree. While *P. vorontzovi*, the smallest species, colonises the crown and branches, the largest species, *P. spinidens*, occupies the trunk. *P. curvidens* may also colonise the trunk, but it often does so alone or in association with *Cryphalus piceae*. The *P. curvidens* aggregation pheromone is S-(-)-ipsenol, which does not attract the other two species. *P. vorontzovi* males produce a pheromone which is a mixture of ipsenol and R-(-)-ipsdienol. This explains the spatial separation of *P. spinidens* whose aggregation pheromone remains unknown. Experiments carried out on *Pityokteines* confirm that the juvenile hormone plays a part in biosynthesis of the pheromone (Harring, 1978).

3.9. *Trypodendron lineatum*

This species, which ranges over the whole of Eurasia and North America, feeds on fir and spruce and even larch and Scots pine. *T. lineatum* lives exclusively on dying or fallen trees, and causes great losses to the economic value of wood, which becomes unusable when it is riddled with deep galleries which are moreover stained blue by the *Ceratocystis* fungus which the beetle carries.

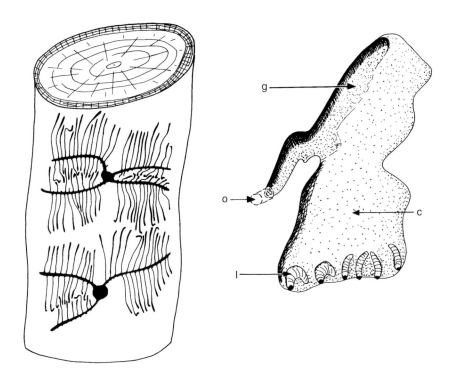

Figure 15.19. *Left:* Pityokteines curvidens *galleries in a fir branch.*
Right: Dendroctonus micans *galleries. o: entrance hole;*
c: communal larval chamber; l: larvae attacking wood collectively;
g: maternal gallery.

Trypodendron lineatum is a monogamous and mycetophagous species, one of the "ambrosia beetles". Swarming takes place as early as March when temperatures hardly exceed 12 °C. The female settles first and begins to bore a gallery; mating takes place outside. The 2 mm diameter galleries reach a length of 9 cm; they may ramify into two or three branches and penetrate the wood at right angles to the surface.

3.10. *Dendroctonus micans*

The genus *Dendroctonus* is represented in North and central America by a number of species whose biology is unusual and differs from those of most other scolytids. The only European species, *D. micans*, is sometimes known as the giant spruce beetle. It ranges over the whole of Siberia and westward as far as France. *D. micans* prefers to feed on various species of spruce: common spruce (*Picea excelsa*), oriental spruce (*Picea orientalis*) and Sitka spruce (*Picea sitchensis*) and very occasionally attacks fir, larch and Scots pine. Infested trees are usually old ones; trees between 55 and 70 years old are particularly vulnerable. The beetle's choice seems to depend on finding rough bark, which makes penetration easier, and on the presence of certain chemical stimuli. Young trees have smooth bark and are less attractive. Massive infestations usually take place near felled areas which create large clearings and a microclimate that suits the beetle. Trees on the edge of clearings are exposed to direct sunlight and higher temperatures than those situated inside dense stands of trees. Mixed populations of spruce and other trees appear to be an effective way of limiting populations of *D. micans*; infestations are always less intense in mixed forests than in pure populations.

D. micans is a primary pest able to occupy apparently healthy trees, especially during outbreaks. Infestation is facilitated by anything that causes physiological unbalance in the tree: drought, poor soils, wounds incurred during forestry work or "ring disease" caused by the fungus *Fomes annosus*. Mating takes place in larval galleries before the adults leave the tree in which they grew. Males usually remain on the spot; fertilised females disperse and colonise new places. Sex ratio is about 8 to 10 in favour of females (Carle, 1975; Carle *et al.*, 1979; Grégoire *et al.*, 1981; Grégoire *et al.*, 1985).

Dispersal of females takes place in four different ways. (a) Since this species' galleries may penetrate roots, it may spread from one tree to another via the root system. Spruce trees that have made contact at root level gradually contaminate each other; this is what Russian authors have named dendroctonosis. (b) When ambient temperature exceeds 21 °C, beetles are able to take to the wing and colonise more or less distant trees. (c) Temperatures between 13 and 21 °C lie below the flight threshold; in these circumstances females migrate to new trees on foot. (d) Other, as yet intact parts of the infested tree may be colonised; in this case beetles leave their galleries, wander for a while on the bark, and bore new galleries elsewhere in the tree.

The female alone bores the entrance orifice and a sub-cortical gallery which may reach a length of 20 cm (figure 15.19). The gallery ends in an irregular pouch-shaped egg-laying chamber; on Sitka spruce the female lays 150 to 200 eggs, but little over 50 on European spruce which is less suitable for *D. micans*. Infested trees are easily recognisable as they become covered with cones of yellowish to reddish brown resin. Larval behaviour is probably what distinguishes *D. micans* most obviously from other scolytids. Larvae exploit the walls of the larval chamber collectively by gnawing the secondary phloem together. Larval growth lasts one to three years depending on temperature. The gregarious behaviour of the larvae is stimulated by a pheromone consisting of a mixture of *trans*-verbenol, *cis*-verbenol and myrtenol. This feature distinguishes *D. micans* from other scolytids, in which aggregation pheromones are produced by adults. The pheromone is found in the excrement of those larvae at the front line of attack; the excrement is reabsorbed by larvae left behind the front line. In short, in *Dendroctonus micans* the pheromone is produced by larvae in which it induces congregation into family groups, whereas in other scolytids pheromones are produced by adults, enabling them to congregate in large numbers on the same tree.

Among insects that frequent *Dendroctonus micans* galleries is a 4 to 5 mm long beetle, *Rhizophagus grandis*, that appears to be a host-specific predator of both the larvae and adults. Trials at using this beetle in biological control of *D. micans* were made as early as 1968. Results in developing techniques to rear the beetle on a large scale and to release it appear promising.

3.11. American species of the genus *Dendroctonus*

The genus *Dendroctonus* includes 15 species, only two of which live in the Old World; all the others are native to North and central America where they cause great damage to conifer forests.

These beetles are responsible for the loss of some 150,000 cubic metres of wood every year. *Dendroctonus* are distinguished from other scolytids by their large size and by their behaviour. They infest either healthy trees at least 20 cm in diameter that have suffered from drought or other causes of weakness or, during outbreaks, healthy vigorous trees. The American species have four larval instars, whereas other scolytids have five.

Larval gallery structure and larval behaviour vary considerably from one species to another, as may be appreciated from a few examples (figure 15.20). In *Dendroctonus brevicomis* the maternal gallery is slightly sinuate and largely transverse; a single egg is laid in each of several relatively large chambers ranged alternately on either side of the gallery. This is reminiscent of some other scolytids such as *Tomicus piniperda*. In *Dendroctonus ponderosae* the maternal gallery is rectilinear and lies in the axis of the trunk. Eggs are deposited in individual chambers grouped in clusters of up to eight on either side of the gallery. Larval galleries are expanded as the larvae near maturity, and may unite. In *Dendroctonus valens* the maternal gallery is short and broad, often even circular; eggs are laid in this single gallery in heaps mixed with decaying sawdust. The larvae live collectively, as in the European species *D. micans*. In *Dendroctonus obesus* the maternal gallery is rectilinear and eggs are deposited in irregular heaps. First instar larvae bore individual galleries and in the second instar often gather together; they become independent once more in the third instar and congregate again in the fourth and final instar. In *Dendroctonus murrayanae* the maternal gallery is short and longitudinal. Eggs are laid on the sides in irregular clusters mixed with decaying sawdust. The larvae collectively excavate a large communal transverse chamber until they reach the second instar; this chamber is then extended at right angles and becomes longitudinal, reaching up to 20 cm in length. These few examples show lifestyles ranging from a solitary to an embryonic social one, a rare phenomenon among Coleoptera.

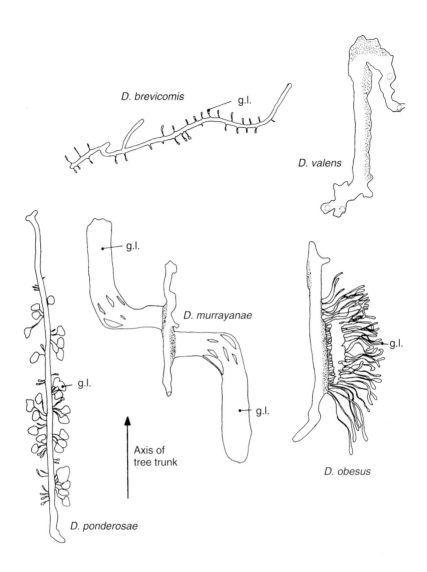

Figure 15.20. *Gallery systems of a few species of* Dendroctonus. *Maternal galleries are blank and laying areas are stippled. gl: larval galleries (Wood, 1963).*

3.12. The genus *Xyleborus*

In the genus *Xyleborus* only the females are winged and males are rare. They are xylomycetophagous, or ambrosia beetles, which bore deep galleries ending in egg-laying corridors out of which the larvae bore their own galleries. The fungal mycelium grows to form a carpet that the larvae consume at the same time as they eat wood. During the entire life cycle the female cleans the gallery and disposes of frass or else accumulates it to be used as a culture medium for the fungus. Loss of the female leads to the death of her progeny. *Xyleborus monographus* and *X. saxaseni* are two polyphagous species that live on various broadleaf trees and are distributed in the whole northern hemisphere.

4. BIOLOGY OF SOME PLATYPODIDS AND CURCULIONIDS

A number of species belonging to the families Platypodidae and Curculionidae have a biology which to some extent resembles that of scolytids.

4.1. *Platypus cylindrus*

This fairly widely distributed species lives mainly on oak, and occasionally in chestnut. Its deep galleries, lined with the mycelium of one of the ambrosia fungi, cause a change in the wood known as "black spot" (figure 15.21). *P. cylindrus* lives on dying trees and sometimes, as in Morocco and in Portugal, on healthy cork oaks which it may kill in a single year. It also infests felled and still fresh tree trunks, especially those in well-lit sites. Scolytid beetles whose biology is similar to that of *Platypus*, such as *Xyloterus domesticus*, *Xyleborus monographus* and *Xyleborus dryographus*, also cause black spot. The *Platypus cylindrus* female alone bores galleries, while the male expels the characteristic very fine pale sawdust that betrays the species' presence in trees. *Platypus oxyurus* is a much rarer species which only survives in a few localities in the Pyrénées, in Corsica, Italia and Greece. It is a relict species associated with fir. Nearly all other platypodids are tropical insects. They may be fearsome enemies of trees of tropical forests in Africa and South America (Cachan, 1957).

4.2. Xylophagous curculionids

This vast family, which includes over 60,000 described species, contains few xylophagous species. Those there are may be primary pests able to attack healthy trees, such as *Pissodes notatus* and *Hylobius abietis*. *Magdalis* species live on various species of dead or dying trees in which they bore subcortical galleries which mark the sapwood in the same way as do those of scolytids. The tribe Cossonini includes about thirty species in France that almost all live in more or less damp and rotten wood in which they mate and lay over several generations without leaving the substrate (figure 15.22).

4.2.1. Pissodes notatus

The genus *Pissodes* includes black or brown weevils covered in scales that form patterns of yellow or white spots. It is distributed in Europe and

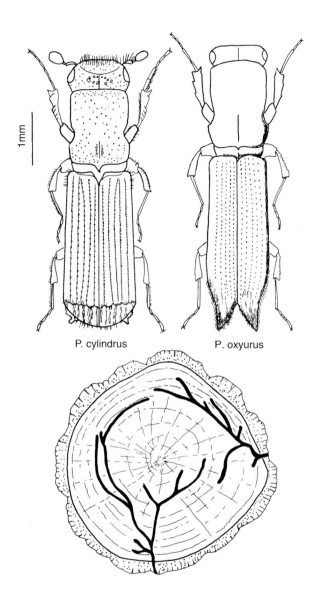

Figure 15.21. *Appearance of the two European species of* Platypus, *and larval galleries of* P. cylindrus *in a cross-section of an oak branch.*

North America. The commonest species in Europe is *P. notatus*, which lives on pines, and settles on weakened trees, either at the collars of saplings or on the trunks of older trees. In some circumstances *P. notatus* may become a primary pest. Its larvae cause the tree to weaken and speed its death by enabling other xylophagous species to settle (cf. figure 11.3).

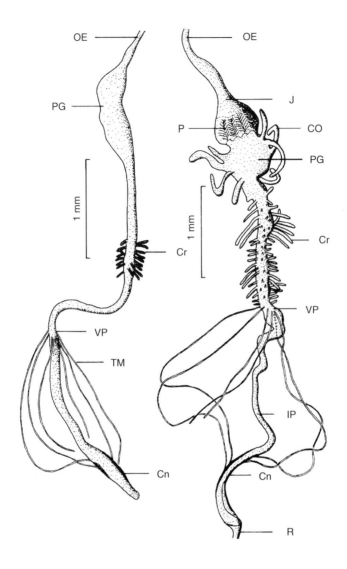

Figure 15.22. *Digestive tract of* Rhyncolus porcatus, *a xylophagous curculionid of the tribe Cossonini. Left: larva. Right: adult. Cn: cryptonephridism area; CO: caeca; Cr: crypts of the midgut; J: crop; IP: hindgut; OE: oesophagus; P: proventriculus; PG: gastric pouch; R: rectum; TM: Malpighian tubes; VP: pyloric valvule.*

The life cycle varies according to climatic conditions. In Provence there are two adult emergence periods, the first in May to July, the second in September to December. Young adults fly to pine trees and settle on the current year's shoots where they gnaw the bark and phloem. This feeding and maturation stage lasts until they mate. The female perforates the bark and lays one to three eggs in the cavity thus formed. Each female has one or more egg-laying cycles that last on average 40 days interrupted by periods of rest; about 200 eggs are laid in each cycle. Adults live up to 397 days in the Mediterranean region. Activity ceases twice a year, during aestivation and hibernation. Larval growth takes place under bark in four stages. Beetles produced from eggs laid in May and June emerge before winter, and those produced from eggs laid in summer only emerge in April of the following year. A list of *Pissodes* parasites and predators has been compiled in the South of France (Alauzet, 1982). The family Braconidae is the one most strongly represented since 70% of the entomophagous species belong to the family, whose members may play a part in the regulation of *Pissodes* populations. The pteromalid *Rhopalicus tutela* also plays a considerable role (table 15.4).

Species	Relative importance	Stage of *P. notatus* predated
Parasitic mites		
Pediculoides ventricosus	+	Pupae, adults
Unidentified gamasids	+	Adults
Parasitoid Hymenoptera		
Braconids		
Coeloides abdominalis	+++	Larvae
Coeloides sordidator	+++	Larvae
Habrobracon palpebrator	++	Larvae
Spathius rubidus	+	Larvae
Dendrosoter middendorffi	+	Larvae
Eubazus semirugosus	++++	Larvae
Ichneumonid		
Dolichomitus terebrans	++	Larvae
Pteromalid		
Rhopalicus tutela	+++	Larvae
Predatory Coleoptera		
Clerid		
Thanasimus formicarius	+	Adults

Table 15.4.   *Predators, parasites and parasitoids found in* Pissodes notatus *galleries (Alauzet, 1982).*

Fir has been in gradual decline in central Europe since the middle of the nineteenth century, and the process has accelerated during the past thirty years. It appears to be due to abiotic factors, but also to pests such as the weevil *Pissodes piceae* which is associated with bark beetles such as

Pityokteines curvidens and *Cryphalus piceae* (Starzyk, 1996). More than twenty species of Coleoptera have been found associated with *Pissodes piceae* in Polish fir plantations.

Aggregation pheromones have been isolated in several species of *Pissodes* such as *P. strobi*, *P. approximatus* and *P. nemorensis*. The pheromones are very different from those of scolytids, and are formed of a mixture of two molecules, grandisol and its corresponding aldehyde, grandisal, which have been isolated from matter in the hindgut of males (cf. figure 15.15). Grandisol is one of the four components of the pheromone of the male cotton boll weevil, *Anthonomus grandis* (Booth *et al.*, 1983; Phillips *et al.*, 1984).

4.2.2. Hylobius abietis

The genus *Hylobius* includes about fifty species distributed in the northern hemisphere; most of them are associated with conifers. *Hylobius abietis*, the large pine weevil, feeds on all conifers and particularly on maritime pine in South West France, Douglas-fir in central France, and spruce and Scots pine in South East France. Outbreaks occur mainly following clear felling and replanting of pine or Douglas-fir, which provides the beetle with ideal conditions in which to lay its eggs and for its larvae to grow under the bark of stumps and the undersides of logs that have not been stripped of their bark. Feeding bites are made by adults before they become sexually mature, on the bark of saplings. These bites are usually harmless to older trees but may kill saplings one to two years old. *Hylobius abietis* is an example of a species that has spread as a result of particular forestry practices.

Hylobius abietis measures 7 to 15 mm and is recognisable by its blackish-brown colour and its covering of sparse yellow pubescence which is denser on parts of the elytra where it forms irregular patterns. It is distinguished from *Pissodes notatus* by the position of the antennae on the rostrum. The life cycle is fairly complex; it lasts one year in France but may take up to four or five years in Scandinavia where average temperatures are lower (figure 15.23). Adults live up to four years and breed every year.

5. COMMENSALS, PREDATORS AND PARASITES OF BARK BEETLES

Many published works describe the bacteria, fungi and animals associated with scolytid beetles as predators, parasites or parasitoids, competitors, or merely as commensals; these works deal mainly with Europe and North America. Of those that list comprehensive numbers of taxonomic groups mention may be made of Dahlsten (1970), Whitney (1982), Mills (1983), Mœck & Safranyik (1984) and Gara *et al.* (1995). Many lists have also been

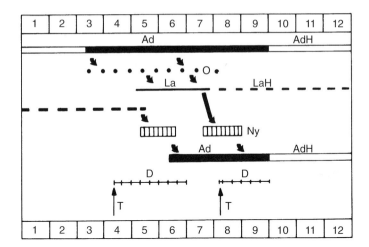

Figure 15.23. *Average life cycle of* Hylobius abietis *in France. Ad: active adults;*
AdH: adults hibernating in litter or in pupal chambers; O: eggs;
La: larvae under tree stump bark; LaH: larvae hibernating in pupal
chambers; Ny: pupae; D: damage period; T: period suitable for
treatment with insecticide.

compiled of species associated with a single species of scolytid or with a
taxonomic group of associated species (Beaver, 1966; Dahlsten & Stephen,
1974; Deyrup & Gara, 1978; Thatcher *et al.*, 1981; Langor, 1991; Weslien,
1992; Dajoz, 1993; etc.).

Although a large number of these organisms has been identified, precise
data on the biology of many of them are still lacking. Complex trophic networks develop in larval galleries, as species exploit not only the scolytids but
also other insects, as well as bacteria, fungi, dung and corpses, decayed
wood, phloem and sapwood (figure 15.24). More than one hundred species
have been found in the galleries of *Ips sexdentatus* and *Ips typographus* in
Scots pine (table 15.5). In North America at least 175 species of arthropod
have been recorded as associated with *Dendroctonus frontalis* (Thatcher *et
al.*, 1981). This list contains the following elements:

– Predators. Coleoptera: 75 species including 9 histerids, 6 trogositids,
7 clerids and 7 colydiids. Diptera: 11 species. Hymenoptera: 17 species of
ant. Heteroptera: 6 species. Mites: 32 species.

– Parasitoids. Hymenoptera: 34 species.

The abundance of species associated with scolytids is often high, and
both numbers of species and numbers of individuals rise rapidly with the age
of infestation by the scolytid (table 15.6). In North America an average of
525.5 predatory insects has been observed under one square metre of pine

Coleoptera: 50 species	**Diptera**: 12 species
Scolytidae:	Anthomyiidae: one unidentified species
Ips sexdentatus	Cecidomyiidae
Ips typographus	*Medetera pinicola*
Cerambycidae	*Medetera signaticornis*
Rhagium inquisitor	*Medetera nitida*
Cucujidae	Lonchaeidae
Silvanus unidentatus	*Lonchaea zetterstedti*
Laemophloeus duplicatus	Stratiomyidae
Nitidulidae	*Pachygaster minutissima*
Epuraea limbata	Sciaridae: one unidentified species
Histeridae	**Nematoda**
Paromalus flavicornis	*Parasitorhabditis* sp.
Hetaerius ferrugineus	*Fuchsia* sp.
Platysoma oblongum	*Diplogasteroides* sp.
Platysoma compressum	*Cryptaphelenchus* sp.
Rhizophagidae	*Bunonema* sp.
Rhizophagus depressus	*Macrolaimus* sp.
Tenebrionidae	*Panagrolaimus* sp.
Corticeus pini	**Hemiptera**: 8 species
Corticeus fraxini	Anthocoridae:
Corticeus unicolor	*Scoloposcelis pulchella*
Corticeus linearis	**Acari**: 8 species
Hymenoptera	Uropodidae
Braconidae	Tarsonemidae
Dendrosoter sp.	**Raphidioptera**: 1 unidentified species
Coeloides bostrychorum	
Torymidae	
Roptrocerus xylophagorum	

Table 15.5. *Main species present in galleries of* Ips sexdentatus *and* Ips typographus *in Scots pine in Fontainebleau forest (Lieutier, 1974).*

bark infested by *Dendroctonus frontalis* (Thatcher *et al.*, 1981). Abundance of these predators was as follows:

Heteroptera		Cleridae	
Anthocoridae		*Thanasimus dubius*	37.8
Lyctocoris elongatus	7.2	Tenebrionidae	
Scoloposcelis mississipensis	47.0	*Corticeus* spp	289.9
Coleoptera		Colydiidae	
Histeridae		*Aulonium ferrugineum*	45.8
Platysoma cylindrica	6.9	*Aulonium tuberculatum*	17.0
Platysoma parallelum	5.7	**Diptera**	
Plegaderus spp.	24.4	Dolichopodidae	
		Medetera bistriata	43.9

The following beetles were observed concentrated on a Jeffrey pine tree that had been broken by wind and had an abundant sap flow from the wound over an area of three square decimetres: 21 examples of the scolytid *Ips pini*, and 50 colydiids (*Lasconotus simplex* and *Aulonium longum*); 14 tenebrionids

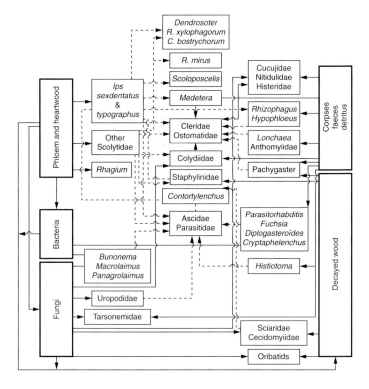

Figure 15.24. *Main elements of the fauna present in galleries of* Ips sexdentatus *and* Ips typographus *in Scots pine in Fontainebleau forest, and trophic relationships. Solid lines show trophic relationships of xylophages and detritivores; dashed lines show trophic relationships of predators and parasites (Lieutier, 1974).*

Insects	Week "0"	Week "8"
Ips typographus	3,072 ± 461	543 ± 229
Thanasimus, Quedius, Nudobius,	0	23 ± 11
Medetera	0	166 ± 52
Rhizophagus, Plegaderus, Chernes	0	8 ± 8
Small staphylinids	7 ± 8	198 ± 112
Epuraea	13 ± 21	46 ± 24
Lonchaea	0	64 ± 29
Parasitoides of *Ips*	6 ± 9	445 ± 130
Rhagium	0	19 ± 10
Dryocoetes	0	32 ± 19
Pityogenes	1 ± 3	174 ± 265

Table 15.6. *Number of emerging insects per square metre of bark on spruce tree trunks left on the ground in a forest during initial infestation by* Ips typographus *(indicated by week "0") and eight weeks later (indicated by week "8"). Average values of 6 experiments and confidence interval 95% (Weslien, 1992).*

(*Corticeus substriatus*) and 10 trogositids (3 *Temnochila chlorodia* and 7 *Tenebrioides corticalis*).

The insects that colonise scolytid galleries in Europe and in North America often belong to the same genera. Numerical abundance of Coleoptera living in scolytid galleries varies from 0.055 to 2.26 individuals per square decimetre in those parts of Europe where it has been determined; it varies from 2.068 to 6.56 individuals per square decimetre in North America (table 15.7). Abundance and species diversity seem to increase with lower latitudes, which may be accounted for by higher temperatures, more generations of scolytids per year, and their presence throughout the year or almost so.

	Tomicus piniperda	*Ips acuminatus*	*Ips typographus*	*Ips pini*	*Ips calligraphus*	*Dendroctonus frontalis*
Corticeus	0.0001	0.26	–	2.48	0.192	2.899
Histerids	0.002	0.34	0.007	0.14	0.392	0.370
Staphylinids	0.033	0.43	0.825	0.06	0.020	-
Colydiids	-	-	-	2.48	0.224	0.628
Clerids	0.001	0.25	1.145	0.19	0.240	0.378
Nitidulids	0.0005	0.18	1.23	-	-	-
Rhizophagids	0.016	0.28	0.053	0.08	-	-
Trogositids	0.0001	-	-	0.44	0.036	-
Total Coleoptera	0.055	2.26	2.20	6.56	2.068	4.275

Table 15.7. *Numbers of individuals per dm² of bark for eight groups of Coleoptera associated with various species of scolytid in Europe (*Tomicus piniperda, Ips typographus*) and in North America (the other four species). (After various authors in: Dajoz, 1993).*

Different ecological categories may be defined among organisms associated with scolytids: (a) those responsible for diseases, such as bacteria and pathogenic fungi; (b) parasitic species that are protozoa, mites and nematodes; (c) predatory species such as various mites and many Diptera and Coleoptera; (d) parasitoid species, which belong to the Diptera and Hymenoptera; (e) species which compete with the scolytids by exploiting the same habitat, such as other scolytids or cerambycid larvae such as *Monochamus*; (f) symbiotic species such as mycangia fungi; (g) commensals, which live in the same habitat but do not affect the scolytids (phoretic mites, cecidomyiids).

5.1. Bacteria, fungi and protozoans

Long lists of the bacteria and fungi (more than 100 species) found in scolytid galleries have been published (Whitney, 1982). The pathogenic function of *Beauveria bassiana* has been established in the case of *Ips typographus*. Many symbiotic fungi are transmitted by way of mycangia. Diseases caused by bacteria, fungi, protozoans and perhaps also viruses may be factors regulating populations of *Dendroctonus frontalis* (Moore, 1971, 1972). The part played by bacteria is still little understood; some are pathogenic; others metabolise precursors of pheromones in the beetle's digestive tract. Of the protozoa, the Microsporidia are the greatest enemies of scolytids. These sporozoan micro-organisms are widely distributed but little is known of their impact on scolytids; infection is transmitted through the alimentary tract when contaminated elements are ingested.

5.2. Nematodes (Nematoda)

Commensal or parasitic nematode worms are numerous (ten different species are sometimes found in the same scolytid gallery), but the parts they play are still little known (Ruhm, 1956; Laumond & Ritter, 1971). There are parasitic species mainly localised in the fatty tissues, and commensal species which station themselves in the galleries or under the adults' elytra. *Parasitorhabditis obtusa* is found in the galleries of *Ips sexdentatus* and *Ips typographus*, as are *Fuchsia buetschlii, Diplogasteroides halleri* and a *Cryptaphelenchus* sp. *Parasitorhabditis obtusa* may infest up to 40% of individuals of *Ips cembrae* with levels of 20 to 55 larvae in each insect's digestive alimentary tract; its larvae grow in sawdust in the galleries. When the host beetles swarm, the worms' larvae are carried by the beetles to new habitats which they colonise. Nematodes are known to lower fertility in *Dendroctonus frontalis* and in *D. pseudotsugae* (Thong & Webster, 1975). One species of *Cryptaphelenchus* is met with in large numbers in the galleries of *Dendroctonus simplex* and appears to be mycetophagous (Langor, 1991). Another nematode worm, *Steinernema carpocapsae*, parasitises *Scolytus scolytus*. Attempts have been made at rearing it on a large scale but production has proved costly.

5.3. Mites (Acarina)

Many predatory, parasitic, phoretic or commensal mites, mostly belonging to the families Ascidae, Uropodidae and Parasitidae, are found in scolytid galleries. Mites are not usually host-specific where scolytid species are concerned, rather they prefer certain habitats to others (Lindquist, 1969). The same mites are found with *Ips typographus, Hylurgops palliatus* and *Dryocoetes autographus* which all cohabit in the same spruce tree trunks. The great majority of phoretic mites are little specialised and live with scolytids as well as with other insects associated with them. Tenebrionids of the genus *Corticeus* and some clerids may carry more mites than the scolytids whose galleries they frequent. Diets of mites vary: ascids are often predators of the eggs of scolytids or of other mites; uropodids appear to be mostly mycetophagous. Levels of predation by mites are variable. The tarsomenid *Iponemus gaebleri* parasitises the eggs of *Ips typographus*. This mite is carried under its host's elytra, thus enabling it to colonise other trees. The female lays 40 to 80 eggs; larvae do not feed, but live on their reserves; the life cycle lasts about two weeks. *Iponemus* may destroy between 10 and 90% of scolytid eggs according to circumstances.

Some Acarina have mutually beneficial relationships with scolytids. Such is the case of the phoretic species *Dendrolaelaps neodisetus* which is beneficial to *Dendroctonus frontalis* as it predates parasitic nematodes (Kinn, 1980).

5.4. True flies (Diptera)

The main Diptera that predate scolytids belong to the families Lonchaeidae, Asilidae, Xylophagidae and especially the Dolichopodidae. Dolichopodids of the genus *Medetera* occur in Europe as well as in North America (Hopping, 1947; Schmid, 1971; Lieutier, 1974). Larvae of *Medetera excellens*, which predate the larvae and pupae of *Ips cembrae*, overwinter and pupate in their host's galleries. Some species of *Medetera* would appear to be parasites rather than predators; such is the case of various species that attack *Dryocoetes autographus, Tomicus piniperda, Ips typographus* and *Pityogenes chalcographus* in Finland. Adults mate on tree trunks infested by the beetles and eggs are laid at the entrance to or inside the beetle's mating chambers, or in crevices in bark. Females are guided in their choice of egg-laying sites by olfactory compounds released by the tree or by the scolytids; larvae are similarly guided after they hatch (Camors & Payne, 1973).

Many flies of the genus *Lonchaea* (family Lonchaeidae) are associated with scolytids but their biology is poorly known. Some species are predators of species such as *Dendroctonus micans* or *Tomicus piniperda*; others, such as *Lonchaea zetterstedti,* appear to feed on detritus. The small whitish larvae of sciarids live as saprophages or mycetophages in the galleries; adults fly on sunny days. Anthomyiids have larger larvae reaching 2 cm that are saprophages. *Pachygaster minutissima* seems to be both saprophagous and necrophagous, ridding the galleries of various dead insects. Larvae of *Xylophagus* spp. predate both scolytids and cerambycids in their galleries. The family Phoridae includes the only flies that parasitise scolytids. *Megaselia aletiae* parasitises adult *Scolytus ventralis* which it hunts on the wing; parasitised female beetles do not lay eggs. Other species of *Megaselia* are parasites of *Dendroctonus*.

The arrival of Diptera on pines infested by *Ips* takes place in a distinct sequence. Sciarids arrive as soon as a tree has been felled, before it has been occupied by *Ips sexdentatus*, which suggests olfactory attraction, possibly to the terpenic substances released by the tree. *Medetera* arrive as soon as *Ips* have settled; anthomyiids arrive almost simultaneously, whereas *Pachygaster minutissima* turns up later when the beetles are in the pupal or young imaginal stages.

Diptera are often the most numerous insects after Coleoptera in the subcortical habitat. In *Ips sexdentatus* galleries studied in Fontainebleau forest, Diptera, with 309 individuals per square metre, amounted to 21.8% of all insects. They were distributed among different families (in percentage of total number of individuals) as follows: Sciaridae: 37.7%; Lonchaeidae: 24.4%; Dolichopodidae (genus *Medetera*): 6.7%; Anthomyiidae: 1.6%; Stratiomyidae: 20.8%; others: 8.8%.

5.5. Wasps (Hymenoptera)

More than one hundred species of Hymenoptera that parasitise scolytids have already been found. They belong mainly to the families Pteromalidae and Braconidae. *Roptrocerus xylophagorum* is a pteromalid known to parasitise at least 25 species of scolytid. The female lays its eggs in larval galleries by penetrating the entrance hole or through a crevice in the bark. The larva may attach itself to a host larva as an ectoparasite or live free in the gallery. Another pteromalid, *Rhopalicus tutela*, predates many scolytids in large parts of Europe; levels of parasitism vary between 6 and 84% according to circumstances.

In the family Braconidae, *Coeloides bostrychorum* is widely distributed in Europe where it parasitises *Ips, Pityokteines* and *Orthotomicus*. Female *Coeloides brunneri* locates scolytid larvae under bark by detecting body heat, the heat-sensitive organ apparently being the antenna (Richerson & Borden, 1972). *Dendrosoter middendorffi* is a species frequently found in the galleries of *Polygraphus* and *Ips*. This wasp pierces thin bark to deposit its eggs on the host's larvae, and its larvae are ectoparasitic. *Dendrosoter protuberans* is a common European parasite of *Scolytus multistriatus*. It was introduced to the United States in an attempt to control the scolytid *Hylurgopinus rufipes* which is a vector of Dutch elm disease in America, but results proved unsatisfactory (Gardiner, 1976).

Of the ichneumonids, *Dolichomitus terebrans* parasitises about twenty species belonging to several different insect orders, and particularly the scolytid *Dendroctonus micans*.

5.6. Beetles (Coleoptera)

The beetles that live as commensals or predators of scolytids belong to many different families. Coleoptera are the most abundant insects, amounting to up to 80% of the total. Their diet is in many cases poorly known. Predators may attack other prey such as mites or nematodes; they may also have mixed diets and feed on fungi, debris of the tree's tissues and even the faeces of other insects.

5.6.1. Clerids

Clerids are voracious predators which have been used in attempts at biological control. Their larvae are often of a pinkish-white; their bodies are covered with erect hairs and the last abdominal segment bears two sclerified urogomphs of characteristic shape. Adults are brightly coloured, often in patterns of red, blue, green or yellow. The most widely distributed clerid in France is *Thanasimus formicarius*, a polyphagous species which devours the larvae and adults of many species of xylophagous Coleoptera. Laboratory-reared beetles have shown that female *Thanasimus formicarius* lays an average of 162 eggs, over a period of 66 to 123 days. In this period one pair of *Thanasimus* would consume between 66 and 132 *Ips typographus* (Weslein & Regnander, 1992). Forty three clerid larvae, including 31 *T. formicarius*, were found under 12 dm^2 of bark of a pine infested with scolytids. This diet is the rule in many other members of the family. The American species *Enoclerus sphegens* eats an average of one adult *Dendroctonus* a day; it accounts for the destruction of 1% of adults of the scolytid, while its larvae cause the loss of 5 to 11% of the daughter generation (Schmid, 1970). Protecting tree trunks infested by *Dendroctonus brevicomis* by placing them in screen cages prevents arrival of the clerid *Enoclerus lecontei*. The density of *E. lecontei* reaches 200 larvae per square metre of bark in trees to which it has access, and 577 adult *Dendroctonus* per square metre emerge from such trunks. In trunks to which *E. lecontei* is denied access 1,622, or three times as many *Dendroctonus* emerge per square metre (Person, 1940). Among other clerids that predate scolytids in France mention may be made of several species of *Thanasimus* and of *Opilo mollis*, a polyphagous carnivore whose larvae also prey on the weevil *Pissodes notatus*, on anobiids and on cerambycids.

Clerids that predate scolytids are more numerous on conifers than on broadleaf trees. They produce one generation a year and emergence is synchronised with that of their host in spring (Kleine & Rudinsky, 1964). They are attracted either by the scent of a tree's oleoresins or by scolytid pheromones (Pitman & Vité, 1971). This explains why considerable numbers of clerids are often found in pheromone traps designed to catch scolytids. The reaction of predators to scolytid pheromones was first discovered when many *Enoclerus lecontei* were taken in traps containing the three constituent substances of the *Ips paraconfusus* pheromone. Another species, *Thanasimus dubius*, is attracted to the *Dendroctonus frontalis* pheromone. In Europe *Thanasimus formicarius* and *T. rufipes* are both drawn to the *Ips typographus* pheromone (Bakke & Kvamme, 1978).

5.6.2. Trogositids (= Ostomatids)

Beetles of this family are closely related to clerids and the larvae of the two families are very similar in appearance. Both larvae and adults prey on various scolytids and other insects that live under bark. Some, like *Temnochila*, are attracted by scolytid pheromones. *Temnochila coerulea* occurs in the South of France, and a larger number of species live in North America. It appears that females of one of these, *Temnochila chloridia*, must feed on adult scolytids before they are able to reproduce. *Nemosoma elongatum* is a predator of several species of Scolytidae both on conifers and on broadleaf trees. The species is attracted in large numbers to pheromone traps. According to Dippel (1996) *N. elongatum* is effective at controlling *Pityogenes chalcographus*.

5.6.3. Rhizophagids

Representatives of the genus *Rhizophagus* are readily recognised by their clubbed antennae and by the elytra which leave the tip of the abdomen exposed. Many species live on exuvia and various other debris, thus playing the part of cleaners in scolytid galleries in a similar way to some fly larvae. *Rhizophagus grandis* is a specific predator of all stages of *Dendroctonus micans*, and is used in biological control in several western European countries. Its high rate of activity even at low temperatures, its rapid rate of reproduction (three generations a year), and the ease with which it locates prey (by olfactory detection of pheromones) make it a valuable auxiliary that appears to be able to control *Dendroctonus micans*. Another species, *Rhizophagus dispar*, is found in the galleries of *Hylurgops palliatus* (figure 15.25).

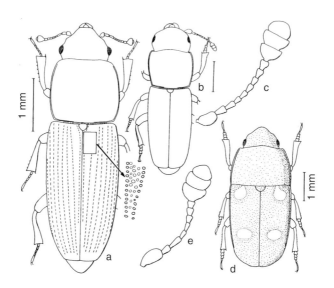

Figure 15.25. *Predatory beetles that live under bark. a:* Rhizophagus grandis; *b:* Pityophagus laevior, *belonging to the family Nitidulidae, and antenna, in c; d:* Glischrochilus quadripunctatus, *which belongs to the same family, and antenna, in e.*

5.6.4. Colydiids

Several species of the family Colydiidae are more or less specialised predators of small arthropods that live in wood, and particularly of scolytids. The genus *Aulonium* is represented by two species in Europe. *Aulonium trisulcum* hunts *Scolytus multistriatus* and *Scolytus scolytus* in elm or in ash, while *Aulonium ruficorne* lives in pines where its prey are *Orthotomicus erosus*, *Pityogenes calcaratus* and *Tomicus piniperda*. This colydiid attacks all stages of the life cycle of *Orthotomicus erosus* and *Pityogenes calcaratus*, but both its larvae and adults may also chew bark, and wood fragments have been found in their digestive tracts. In Israel, combined predation by larval and adult *Aulonium ruficorne* reduces populations of scolytids living on *Pinus brutia* and *Pinus halepensis* to 90% of their original value in spring and autumn and by 30 to 50% in summer (Podoler *et al.*, 1990).

The genus *Lasconotus*, which is very rare in Europe where it is represented only by a single species, includes about twenty North American species whose biology is still poorly known although many of them are thought to be predators. *Lasconotus referendarius* and *L. pusillus* live in *Ips* galleries in several species of pine. When reared in captivity, this species may complete its life cycle by feeding solely on *Ips* eggs, larvae and pupae; it is probable that in the wild *L. pusillus* is partly predatory (Rohlfs & Hyche, 1981, 1984). The arrival of *L. pusillus* is synchronised with peak arrival of *Ips*, whereas *L. referendarius* turns up on average two weeks later. Other colydiids present with *Lasconotus* in *Ips* galleries are *Pycnomerus terebrans*, *Aulonium tuberculatum*, *Aulonium ferrugineum* and *Colydium nigripenne*. One species, *Lasconotus intricatus*, is attracted to traps baited either with *Picea mariana* logs infested by the scolytid *Polygraphus rufipennis* or with the latter species' male pheromone, 3-methyl-3-butene-1-ol. In some cases the abundance of these small colydiids may be far greater than that of the scolytids.

5.6.5. Tenebrionids

This family is represented under bark invaded by scolytids by species belonging to several genera. Species of the genus *Corticeus* have more or less elongate cylindrical bodies and measure 3 to 6 mm according to species. They are polyphagous predators, although some appear to be dependent on scolytids. In Europe *C. pini* attacks *Ips sexdentatus* and *C. linearis* preys on *Pityogenes bidentatus*. In America several species of *Corticeus* may abound in *Dendroctonus* galleries. They are occasional predators which may also feed on scolytid faeces or corpses, or on fungi such as *Ceratocystis*. The two most thoroughly studied American species are *Corticeus glaber* and *C. parallelus* (Goyer & Smith, 1981; Smith & Goyer, 1980).

5.6.6. Histerids

This easily recognisable family of beetles with hard, smooth, shiny black bodies includes several species that live under bark. Those belonging to the genera *Paromalus* and *Platysoma* are sometimes very common in trees invaded by scolytids. Four hundred histerids belonging to five species were counted in two laricio pine trunks. In *Platysoma* and *Hololepta* the body is flattened dorso-ventrally, this being an adaptation to the sub-cortical habitat. Histerids are all predators, but the part they play in this habitat is little known.

5.6.7. Cucujids

Two species of *Silvanus*, which live on a variety of prey including scolytids, are quite often found under the bark of both conifer and broadleaf trees. *Laemophloeus* species are remarkable for their flattened bodies. Their diet is fairly eclectic and still poorly known. *Laemophlaeus juniperi* hunts *Phloesinus thuyae* in juniper, and *Laemophlaeus ater* feeds on larval exuvia and other debris. *Placonotus testaceus* and *Cryptolestes ferrugineus* live mainly in pine at the expense of *Ips* and *Tomicus*.

5.6.8. Nitidulids

In this family, species of the genera *Glischrochilus* and *Pityophagus* have been recorded as preying on scolytids. *Epuraea angustula* lives in the galleries of *Trypodendron* spp., either as a saprophage or as a predator. Many other *Epuraea* species live on sap flowing from wounds on trees. Several species live in dying trees infested by scolytids and are attracted to the same substances that draw the scolytids. This is the case of *Epuraea bickhardti* and *E. boreella* which are attracted to a mixture of ethanol and α-pinene. Attraction to α-pinene alone is weak (table 15.8). In laboratory conditions

	Release speed in mg/hour		Number of *E. bickhardti* attracked
	α-pinene	Ethanol	
Control trap	-	-	6[a]
α-pinene alone	10	-	46[ab]
α-pinene + ethanol	10	1.5	111[b]
α-pinene + ethanol	10	13	182[bc]
α-pinene + ethanol	10	128	837[cd]
α-pinene + ethanol	10	2,116	2,155[d]

Table 15.8. Numbers of Epuraea bickhardti *caught in traps baited with α-pinene alone, a mixture of α-pinene + ethanol, and in unbaited traps. Figures followed by the same letter are not significantly different at threshold P = 0.05 (Schrœder, 1993).*

E. bickhardti can be reared on a diet of scolytid eggs alone, which suggests that the species is predatory (Schrœder, 1993).

5.6.9. Carabids

Species of this family that live under bark are few. *Tachyta nana* is a small, very agile and entirely black species 2 to 3 mm long that feeds on a variety of prey as well as on the faeces of xylophagous species. In the genus *Dromius*, *D. quadrinotatus*, hunts young larvae of the weevil *Pissodes notatus* under pine bark, and *D. schneideri* hunts *Dendroctonus micans*.

5.6.10. Staphylinids

This enormous family almost certainly contains many species that prey on scolytids and other xylophagous insects, but the biology of individual species is poorly known. *Placusa complanata* preys on *Tomicus minor, Ips sexdentatus* and *Pityokteines curvidens*. An other species, *Phloeopora testacea*, is common under pine bark.

Members of many other families of Coleoptera live under bark and in scolytid galleries; their roles are little known. Cerambycid larvae such as those of *Monochamus* are competitors which exploit the same habitat and kill scolytid larvae by burrowing at will through the phloem. Curculionids of the genus *Cossonus* that often cohabit with bark beetles have very similar lifestyles and are also probably potential competitors. Many other sub-cortical beetles settle in wood when it is already decaying and scolytids have departed.

5.7. True bugs (Heteroptera)

Nine families of Heteroptera include bugs that predate or are otherwise associated with scolytids. Anthocorids are the commonest. *Xylocoris cursitans* frequents conifers and broadleaf trees and hunts various arthropods at the larval and adult stages, particularly scolytids, Collembola and thrips. Adults are brown to black, 2 to 2.5 mm long; the larvae, in common with many other anthocorids, are reddish, making them readily recognisable. *Dufouriella ater* also lives under bark and predates scolytids on pines, fir, oaks, beech and various fruit trees.

REFERENCES

ALAUZET C., (1982). Biocénose de *Pissodes notatus* F. ravageur des pins maritimes en forêt de Bouconne (Haute-Garonne: France). *Nouv. Rev. Ent.*, **12**: 81-89.

ANDERBRANT O., SCHLYTER F., BIRGERSSON G., (1985). Intraspecific competition affecting parents and offspring in the bark beetle *Ips typographus*. *Oikos*, **45**: 89-98.

ATKINS M. D., (1969). Lipid loss with flight in the Douglas-fir beetle. *Canad. Ent.*, **101**: 164-165.

BABUDER G., POHLEVEN F., BRELIH S., (1996). Selectivity of synthetic aggregation pheromones Linoprax and Pheroprax in the control of the bark beetles (Coleoptera, Scolytidae) in a timber storage yard. *J. Appl. Ent.*, **120**: 131-136.

BAKKE A., (1970). Evidence of a population aggregating pheromone in *Ips typographus* (Coleoptera Scolytidae). *Contr. Boyce Thompson Inst.*, **24**: 309-310.

BAKKE A., (1973). Bark beetle pheromones and their potential use in forestry. *OEPP/EPPO*, Bull no. **9**: 5-15.

BAKKE A., (1975). Aggregation pheromone in the bark beetle *Ips duplicatus* (Sahlberg). *Norw. J. Ent.*, **22**: 67-69.

BAKKE A., (1983). Host tree and bark beetle interaction during a mass attack of *Ips typographus* in Norway. *Zeit. ang. Ent.*, **96**: 118-125.

BAKKE A., KVAMME T., (1978). Kairomone response by the predators *Thanasimus formicarius* and *Thanasimus rufipes* to the synthetic pheromone of *Ips typographus*. *Norw. J. Ent.*, **25**: 41-43.

BALACHOWSKY A.-S., (1949). *Coléoptères Scolytidae*. Faune de France, vol. 50. Lechevalier, Paris.

BEAVER R. A., (1966). Notes on the fauna associated with elm bark beetles in Wytham wood, Berks. I. Coleoptera. *Ent. Mon. Mag.*, **102**: 163-170.

BEAVER R. A., (1977). Intraspecific competition among bark beetle larvae (Col. Scolytidae). *J. Anim. Ecol.*, **46**: 455-467.

BECKER G., (1949). Beitrage zur Ökologie der Hausbockkäferlarven. *Zeit. ang. Ent.*, **31**: 134-174.

BERRYMAN A. A., (1982). Biological control, thresholds, and pest outbreaks. *Envir. Entomol.*, **11**: 544-549.

BIRCH M. C., Aggregation in bark beetles. *In*: W.-J. BELL, R.-T. CARDE (1984). *Chemical ecology of insects*. Chapman & Hall, London, 331-353.

BLANC A., BLANC M., (1975). Étude du rôle des substances extraites du pin maritime sur l'orientation et la prise de nourriture chez *Pissodes notatus* F. (Coleoptera, Curculionidae). *Ann. Zool. Écol. Anim.*, **7**: 525-533.

Booth D. C., Phillips T. W., Claesson A. *et al.*, (1983). Aggregation pheromone components of two species of *Pissodes* weevils (Coleoptera: Curculionidae): isolation, identification and field activity. *J. Chem. Ecol.*, **9**: 1-12.

Bowers W. W., Borden J. H., (1992). Attraction of *Lasconotus intricatus* Kraus (Coleoptera: Colydiidae) to the aggregation pheromone of the four-eyed spruce bark beetle, *Polygraphus rufipennis* (Kirby) (Coleoptera: Scolytidae). *Canad. Ent.*, **124**: 1-5.

Brand J. M., Bracke J. W., Markovetz A. J. *et al.*, (1975). Production of verbenol pheromone by a bacterium isolated from bark beetles. *Nature*, **254**: 136-137.

Cachan P., (1957). Les Scolytoidea mycétophages des forêts de basse Côte-d'Ivoire. Problèmes écologiques et biologiques. *Rev. Path. Vég. Ent. Agr. Fr.*, **36**: 1-126.

Camors F. B., Payne J. L., (1973). Sequence of arrival of entomophagous insects to trees infested with southern pine beetle. *Environ. Ent.*, **2**: 267-270.

Carle P., (1974). Les phéromones chez les Scolytidae des Conifères. *Ann. Zool. Écol. Anim.*, **6**: 131-147.

Carle P., (1974). Mise en évidence d'une attraction secondaire d'origine sexuelle chez *Blastophagus destruens* Woll. (Col., Scol.). *Ann. Zool. Écol. Anim.*, **6**: 539-550.

Carle P., (1975). *Dendroctonus micans* Kug. (Col. Scolytidae), l'hylésine géant ou dendroctone de l'épicéa. *Rev. For. Fr.*, **27**: 115-128.

Carle P., Granet A.-M., Perrot J.-P., (1979). Contribution à l'étude de la dispersion et de l'agressivité chez *Dendroctonus micans* Kug. (Col. Scolytidae) en France. *Bull. Soc. Ent. Suisse*, **52**: 185-196.

Chararas C., (1962). *Scolytides des conifères*. Lechevalier, Paris.

Dahlsten D. L., Parasites, predators and associated organisms reared from western pine beetle infested bark samples. *In:* R. W. Stark, D. L. Dahlsten (1970). *Studies on the population dynamics of the western pine beetle, Dendroctonus brevicomis* Le Conte (Coleoptera: Scolytidae). *Univ. of Calif.*, Berkeley, California, 75-79.

Dahlsten D. L., Relationships between bark beetles and their natural enemies. *In:* J. B. Mitton, K. B. Sturgeon (1982). *Bark beetles in North American conifers. Univ. of Texas Press*, Austin, 140-182.

Dahlsten D. L., Stephen F. M., (1974). Natural enemies and insect associates of the mountain pine-beetle, *Dendroctonus ponderosae* (Coleoptera: Scolytidae), in sugar pine. *Canad. Ent.*, **106**: 1,211-1,217.

Dajoz R., (1993). Écologie et biogéographie des peuplements de Coléoptères associés à douze espèces de Scolytidae dans le sud de la Californie. *Bull. Soc. Ent. Fr.*, **98**: 409-424.

DEYRUP M. A., GARA R. I., (1978). Insects associated with Scolytidae (Coleoptera) in Western Washington. *Pan. Pacific Entomol.*, **54**: 270-282.

DICKENS J. C., (1979). Electrophysiological investigations of olfaction in bark beetles. *Bull. Soc. Ent. Suisse*, **52**: 203-216.

DICKENS J. C., (1981). Behavioural and electrophysiological responses of the bark beetle, *Ips typographus*, to potential pheromone components. *Physiol. Entomol.*, **6**: 251-261.

DICKENS J. C., PAYNE T. L., (1978). Structure and function of the sensilla on the antennal club of the southern pine beetle, *Dendroctonus frontalis* (Zimmerman) (Coleoptera: Scolytidae). *Int. J. Insect. Morphol. Embryol.*, **7**: 251-265.

DIPPEL C., (1996). Investigations on the life history of *Nemosoma elongatum* L. (Col. Ostomatidae), a bark beetle predator. *J. Appl. Ent.*, **120**: 391-395.

DÖHRING E., (1955). Zur Biologie des grossen Eichenbockkäfers (*Cerambyx cerdo*) unter besonderer Berucksichtigung der Populationsbewegung im Areal. *Zeit. ang. Ent.*, **3**: 251-273.

FRANCKE W., (1973). Nachweis und Identifizierung von Aggregations-substanzen in dem Ambrosiakäfer *Xyloterus domesticus* L. (Coleoptera Scolytidae). *Diss. Fachbereich Chemie, Univ. Hamburg.*

GARA R. I., WERNER R. A., WHITMORE M. C., HOLSTEN E. H., (1995). Arthropod associates of the spruce beetle *Dendroctonus rufipennis* (Kirby) (Col. Scolytidae) in spruce stands of south-central and interior Alaska. *J. Appl. Ent.*, **119**: 585-590.

GARDINER L. M., (1976). Tests of an introduced parasite against the native elm bark beetle. *Env. Canada, For. Serv. Bi-Mo. Res. Notes*, **32**: 11.

GIBBS J. N., BURDEKIN D. A., BRASIER C. M., (1977). Dutch elm disease. *Forestry Commission, Forest Record*, **115**. HMSO, London.

GORE W. E., (1975). The aggregative pheromone of the European elm bark beetle (*Scolytus multistriatus*): isolation, identification, synthesis and biological activity. Ph. D. dissertation, *State University of New York*, Syracuse.

GOYER R. A., SMITH M. T., (1981). The feeding potential of *Corticeus glaber* and *Corticeus parallelus* (Coleoptera: Tenebrionidae), facultative predators of the southern pine beetle, *Dendroctonus frontalis* (Coleoptera: Scolytidae). *Canad. Ent.*, **113**: 807-811.

GRÉGOIRE J.-C., BRAEKMAN J.-C., TONDEUR A., (1982). Chemical communication between the larvae of *Dendroctonus micans* Kug. (Coleoptera, Scolytidae). *Les Médiateurs chimiques*, éditions de l'INRA, 253-257.

GRÉGOIRE J.-C., MERLIN J., PASTEELS J.-M. *et al.*, (1985). Biocontrol of *Dendroctonus micans* by *Rhizophagus grandis* Gyll. (Col., Rhizophagidae) in the Massif Central (France). *Zeit. ang. Ent.*, **99**: 182-190.

HAACK R. A., WILKINSON R. C., FOLTZ J. L., (1987). Plasticity in life-history traits of the bark beetle *Ips calligraphus* as influenced by phlœm thickness. *Œcologia*, **72**: 32-38.

HARRING C., (1978). Aggregation pheromones of the European fir engraver beetles *Pityokteines curvidens*, *P. spinidens and P. vorontzovi* and the role of juvenile hormone in pheromone biosynthesis. *Zeit. ang. Ent.*, **85**: 281-317.

HARRING C.-M., VITÉ J.-P., HUGHES P.-R., (1975). Ipsenol, der Populationslockstoff des Krummzähnigen Tannenborkenkäfers. *Naturwis*, **62**: 488.

HOPPING G. R., (1947). Notes on the seasonal development of *Medetera aldrichii* Wheeler (Diptera, Dolichopodidae) as a predator of the Douglas-fir bark beetle, *Dendroctonus pseudotsugae* Hopk. *Canad. Ent.*, **79**: 150-153.

HUGHES P. R., PITMAN G. B., (1970). A method for observing and recording the flight behavior of tethered bark beetles in response to chemical messengers. *Contr. Boyce Thompson Inst. Plant Res.*, **24**: 329-336.

HUGHES P. R., RENWICK A. A., (1977). Neural and hormonal control of pheromone biosynthesis in the bark beetle *Ips paraconfusus*. *Physiol Entomol.*, **2**: 117-123.

HUGHES P. R., RENWICK A. A., (1977). Hormonal and host factor stimulating pheromone synthesis in female western pine beetle, *Dendroctonus brevicomis*. *Physiol. Entomol.*, **2**: 289-292.

KINN D. N., (1980). Mutualisms between *Dendrolaelaps neodisetus* and *Dendroctonus frontalis*. *Envir. Ent.*, **9**: 756-758.

KIRKENDALL L. R., (1983). The evolution of mating system in bark and ambrosia beetles (Coleoptera: Scolytidae and Platypodidae). *Zool. J. Lin. Soc.*, **77**: 293-352.

KLIMETZEK D., BARTELS J., FRANCKE W., (1989a). Das Pheromon-System des Bunten Ulmenbastkäfers *Pteleobius vittatus* (F.) (Col. Scolytidae). *J. Appl. Ent.*, **107**: 518-523.

KLIMETZEK D., KÖHLER J., KROHN S., FRANCKE W., (1989b). Das Pheromon-System des Waldreben-Borkenkäfers, *Xylocleptes bispinosus* Dufts. (Col., Scolytidae). *J. Appl. Ent.*, **107**: 304-309.

LANGOR D. W., (1991). Arthropods and nematodes co-occuring with the eastern larch beetle, *Dendroctonus simplex* (Col.: Scolytidae), in Newfoundland. *Entomophaga*, **36**: 303-313.

LANIER G. N., SILVERSTEIN R. M., PEACOCK J. W., Attractant pheromone of the European elm bark beetle (*Scolytus multistriatus*): isolation, identification, synthesis and utilization studies. *In*: J. F. ANDERSEN, H. KAYA (1976). *Perspectives in forest entomology*. Academic Press, 149-175.

LAUMOND C., RITTER M., (1971). Les Nématodes parasites des insectes xylophages. La lutte biologique en forêt, no. spécial. *Ann. Zool. Écol. Anim.*, 195-205.

LEVIEUX J., CASSIER P., (1994). La vection de champignons pathogènes des résineux par les xylophages forestiers européens. *L'Année Biologique*, **63**: 19-37.

LIEUTIER F., (1974). *Recherches sur la zoocénose des galeries de Coléoptères Scolytides.* Thesis, Université Paris, 176 pp.

LIEUTIER F., GÉRI C., GOUSSARD F., ROUSSEAU G., (1984). Problèmes entomologiques actuels du pin sylvestre en région centre. *La forêt privée*, no. **155**: 25-36.

LINDQUIST E. E., (1969). Review of holarctic tarsonemid mites (Acarina, Prostigmata) parasiting eggs of pine bark beetles. *Mem. Soc. Ent. Canada*, **60**: 1-111.

MILLER D. R., GIBSON K. E., RAFFA K. F. *et al.*, (1997). Geographic variation in response of pine engraver, *Ips pini*, and associated species to pheromone, lanierone. *J. Chem. Ecol.*, **23**: 2,013-2,031.

MILLS N. J., (1983). The natural enemies of scolytids infesting conifer bark in Europe in relation to the biological control of *Dendroctonus* spp. in Canada. *Biocontrol News and Info.*, **4**: 305-328.

MŒCK H.-A., SAFRANYIK L., (1984). *Assessment of predator and parasitoid control of bark beetles.* Information report BC-X-248. Pacific Forest Research Centre. Canadian Forestry Service.

MOORE G. E., (1971). Mortality factors caused by pathogenic bacteria and fungi of the southern pine beetle in North Carolina. *J. Invert. Pathol.*, **17**: 28-37.

MOORE G. E., (1972). Southern pine beetle mortality in North Carolina caused by parasites and predators. *Env. Ent.*, **1**: 58-65.

MORIMOTO K., (1962). Comparative morphology and phylogeny of the superfamily Curculionoidea of Japan. *J. Fac. Agric. Kyushu Univ.*, **11**: 331-373.

MULOCK P., CHRISTIANSEN E., (1986). The threshold of successful attack by *Ips typographus* on *Picea abies*: a field experiment. *Forest Ecol. Manag.*, **14**: 125-132.

MUSTAPARTA H., (1975). Response of the pine weevil, *Hylobius abietis* L. (Col. Curculionidae) to bark beetle pheromone. *J. Comp. Physiol.*, **88**: 395-398.

NAMKOONG G., ROBERDS J. H., NUNNALLY L. B., THOMAS H. A., (1979). Isozyme variations in populations of southern pine beetles. *Forest Sci.*, **25**: 197-203.

NIJHOLT W. W., (1970). The effect of mating and the presence of the male ambrosia beetle *Trypodendron lineatum* on secondary attraction. *Canad. Ent.*, **102**: 894-897.

NIJHOLT W. W. SCHONHERR J., (1976). Chemical response behavior of Scolytids in West Germany and Western Canada. *Bi-Monthly Research Notes*, **32**: 31-32.

OKSANEN H., KANGAS E., PERTUNNEN V., (1968). The chemical composition of the breeding material of *Blastophagus piniperda* L. (Col. Scolytidae) and its significance in the olfactory orientation of this species. *Ann. Ent. Fenn.*, **34**: 1-13.

PERSON H. L., (1940). The clerid *Thanasimus lecontei* (Wolc.) as a factor in the control of the western pine beetle. *J. Forestry*, **38**: 390-396.

PHILLIPS T. W., LANIER G. N., (1985). Genetic divergence among populations of the white pine weevil, *Pisodes strobi* (Coleoptera: Curculionidae). *Ann. Ent. Soc. Amer.*, **78**: 744-750.

PHILLIPS T. W., WEST J. R., FOLTZ L. *et al.*, (1984). Aggregation pheromone of the deodar weevil, *Pissodes nemorensis* (Coleoptera: Curculionidae): isolation and activity of grandisol and grandisal. *J. Chem. Ecol.*, **10**: 1,417-1,423.

PINON J., (1976). Faits nouveaux sur le dépérissement de l'orme. *Rev. For. Fr.*, **28**: 323-342.

PITMAN G. B., KLIEF R. A., VITÉ J. P., (1965). Studies on the pheromone of *Ips confusus* Lec. II. Further observations on the site of production. *Contrib. Boyce Thompson Inst.*, **23**: 13-17.

PITMAN G. B., VITÉ J.-P., (1970). Field response of *Dendroctonus pseudotsugae* (Col., Scol.) to synthetic frontalin. *Ann. Ent. Soc. Amer.*, **63** : 661-664.

PITMAN G. B., VITÉ J.-P., (1971). Predator prey response to western pine beetle attractants. *J. Econ. Ent.*, **64**: 402-404.

PODOLER H., MENDEL Z., LIVNE H., (1990). Studies on the biology of a bark beetle predator, *Aulonium ruficorne* (Coleoptera: Colydiidae). *Envir. Ent.*, **19**: 1,010-1,016.

RICHERSON J. V., BORDEN J. H., (1972). Host finding by heat perception in *Coeloides brunneri* (Hymenoptera: Braconidae). *Canad. Ent.*, **104**: 1,877-1,881.

ROHLFS W.-M., HYCHE L.-L., (1981). Colydiidae associated with *Ips* in southern pines: relative abundance and time of arrival of adults at pines under attack by *Ips* spp. *J. Econ. Ent.*, **74**: 458-460.

ROLLING M. P., KEARBY W. H., (1977). Influence of tree diameter aspect on the behaviour of Scolytids infesting black oaks. *Canad. Ent.*, **109**: 1,235-1,238.

RUHM W., (1956). *Die Nematoden der Ipiden*. G. Fischer, Iéna.

SAFRANYIK L., (1970). Host characteristics, brood density and size of mountain pine beetles emerging from lodgepole pine. *Canadian Forestry Service, Bi-Monthly Research Notes*, **26**: 35-36.

SCHMID J.-M., (1970). *Enoclerus sphegus* (Coleoptera, Cleridae) a predator of *Dendroctonus ponderosae* in the Black Hills. *Canad. Ent.*, **102**: 969-977.

SCHMID J.-M., (1971). *Medetera aldrichii* (Diptera, Dolichopodidae) in the Black Hills. II. Biology and densities of the immature stages. *Canad. Ent.*, **103**: 848-853.

SCHNEIDER I., RUDINSKY J.-A., (1969). The site of pheromone production in *Trypodendron lineatum* (Col., Scol.): bio-essay and histological studies of the hindgut. *Canad. Ent.*, **101**: 1 181.

SCHÖNHERR J., (1970). Evidence of an aggregating pheromone in the ash-bark beetle *Leperesinus fraxini* (Coleoptera, Scolytidae). *Contr. Boyce Thompson Inst.*, **24**: 305-307.

SCHÖNHERR J.-S., (1977). Importance of visual stimuli in the host selection of bark beetles (*Dendroctonus ponderosae* and *Ips montanus*). *Comportement des insectes et milieu trophique*, 187-193. Colloques internationaux du CNRS no. 265.

SCHRŒDER L.-M., (1988). Host recognition in *Tomicus piniperda* (Coleoptera: Scolytidae) and other bark beetles attacking Scots pine. *Dissertation, University of Uppsala*, Sweden.

SCHRŒDER L.-M., (1988). Atttraction of the bark beetle *Tomicus piniperda* and some other bark-and wood-living beetles to the host volatiles α-pinene and ethanol. *Entomol. Exp. Appl.*, **46** : 203-210.

SCHRŒDER L.-M., (1993). Attraction of *Epuraea bickhardti* St Claire Deville and *E. boreella* (Zetterstedt) (Coleoptera, Nitidulidae) to ethanol and α-pinene. *Entomol. Fenn.*, **31**: 133-135.

SCHRŒDER L.-M., LINDELÖW A., (1989). Attraction of Scolytids and associated beetles by different absolute amounts and proportions of α-pinene and ethanol. *J. Chem. Ecol.*, **15**: 807-817.

SCOTT T. M., KING C. J., (1974). The large pine weevil and black pine beetles. *Forestry Commission Leaflet* no. **58**, 1-11.

SILVERSTEIN R. M., Attractant pheromones of Coleoptera. *In*: M. BEROZA (1970). *Chemical controlling of insect behavior*. Academic Press, London, 21-40.

SINCLAIR W. A., CAMPANA R. J., (1978). Dutch elm disease. Perspectives after 60 years. *Agriculture*, **8**: 1-52.

SMITH M. T., GOYER R. A., (1980). Relative abundance and seasonal occurrence of *Corticeus glaber* and *Corticeus parallelus* (Coleoptera: Tenebrionidae), associates of the southern pine beetle, *Dendroctonus frontalis* (Coleoptera: Scolytidae). *Canad. Ent.*, **112**: 515-519.

STARZYK J.-R., (1996). Bionomics, ecology and economic importance of the fir weevil, *Pissodes piceae* (Ill.) (Col., Curculionidae) in mountain forests. *J. Appl. Ent.*, **120**: 65-75.

STEPHEN J. A., DAHLSTEN D. L., (1976). The temporal and spatial arrival pattern of *Dendroctonus brevicomis* (Coleoptera Scolytidae), in Ponderosa pine. *Canad. Ent.*, **108**: 271-282.

STEPHEN J. A., DAHLSTEN D. L., (1976). The arrival sequence of the Arthropod complex following attack by *Dendroctonus brevicomis* (Coleoptera Scolytidae) in Ponderosa pine. *Canad. Ent.*, **108**: 283-304.

STIPES R. J., (1973). Control of Dutch elm disease in artificially inoculated American elms with soil-injected benomyl, captan, and thiabendazole. *Phytopathology*, **63**: 735-738.

SWABY J. A., RUDINSKY J. A., (1976). Acoustic and olfactory behaviour of *Ips pini* (Say) (Coleoptera: Scolytidae) during host invasion and colonisation. *Zeit. ang. Ent.*, **81**: 421-432.

THATCHER R. C., SEARCY J. L., COSTER J. E., HERTEL G. D., (1981). *The southern pine beetle. U S Department of Agriculture, Technical bulletin* **1,631**, 266 pp.

THONG C. H. S., WEBSTER J. M., (1975). Effects of the bark beetle nematode *Contortylenchus reversus*, on gallery construction, fecundity and egg viability of the Douglas-fir beetle *Dendroctonus pseudotsugae* (Coleoptera: Scolytidae). *J. Invert. Pathol.*, **26**: 235-238.

TOMMERAS B.-A., MUSTAPARTA H., GRÉGOIRE J.-C., (1984). Electrophysiological recording from olfactory receptor cells in *Dendroctonus micans* and *Rhizophagus grandis*. *In:* Biological control of bark beetles (*Dendroctonus micans*). *Commission of the European Communities*, Brussels, 98-106.

VITÉ J.-P., (1978). Insektenhormone im Waldschutz: Erreichtes und Erreichbares. *Biologie in unsere Zeit*, **8**: 112-119.

VITÉ J.-P., BAKKE A., HUGHES P. R., (1974). Ein Populationslockstoff des zwölzahnigen Kiefernborkenkäfers *Ips sexdentatus*. *Naturwiss.*, **61**: 365-366.

VITÉ J.-P., FRANCKE W., (1976). The aggregation pheromones of bark beetles: progress and problems. *Naturwiss.*, **63**: 550-555.

VITÉ J.-P., KLMITZEK D., LOSKANT G. *et al.*, (1976). Chirality of insect pheromones: response interruption by inactive antipodes. *Naturwiss.*, **63**: 582-583.

WESLIEN J., (1992). The arthropod complex associated with *Ips typographus* (L.) (Coleoptera, Scolytidae): species composition, phenology, and impact on bark beetle productivity. *Entomol. Fennica*, **3**: 205-213.

WESLIEN J., REGNANDER J., (1992). The influence of natural enemies on brood production in *Ips typographus* (Col. Scolytidae) with special reference to egg-laying and predation by *Thanasimus formicarius* (Col.: Cleridae). *Entomophaga*, **37**: 333-342.

WHITNEY H. S., Relationships between bark beetles and symbiotic organisms. *In:* J. B. MITTON, K. B. STURGEON (1982). *Bark beetles in North American Conifers.* University of Texas Press, Austin, 183-211.

WOOD D. L., Pheromone control of bark beetles. *In:* D. L. WOOD, R. M. SILVERSTEIN, M. NAKAJIMY (1970). *Control of insect behavior by natural products.* Academic Press, New York, 301-316.

WOOD S. L., (1982). The bark and ambrosia beetles of North and Central America (Coleoptera, Scolytidae), a taxonomic monograph. *Great Basin Naturalist Memoirs,* **6**: 1-1,359 (general account pp. 1-54).

WOOD S. L., (1987). A catalog of Scolytidae and Platypodidae (Coleoptera). Part 1. Bibliography. *Great Basin Naturalist Memoirs,* **11**: 1-685.

WORRELL R., (1983). Damage by the spruce bark beetle in South Norway 1970-80: A survey, and factors affecting its occurrence. *Norsk Inst. Skogforsk,* **38** (6): 1-34.

16

Saproxylic insects

All animals that live in living or dead wood are called saproxylic organisms, even though their biology may vary greatly. The relatively recently coined term saproxylic includes several categories of species:

– Xylophagous species are those that feed on living or dead wood. Species that live on rotting wood are often called saproxylophagous. True xylophagous species may be phloem-eaters that often also attack part of the bark. This is the case in most scolytids, in various buprestids and cerambycids and in a few Lepidoptera. Xylophagous insects may also be phloem- and xylem-eaters that attack living or dead wood; cerambycids and buprestids come into this category. Insects that eat only xylem are usually secondary destroyers such as *Sirex*, ants, termites and some cerambycids.

The main xylophagous insects in temperate regions belong to four orders: Coleoptera, Diptera, Hymenoptera and Lepidoptera. The order Coleoptera is the richest in xylophagous species, some families being entirely or largely composed of species dependent on wood, at least in their larval state. In the tropics, the Isoptera, or termites, play a considerable part; termites are usually of very minor importance in temperate regions.

– Many insects that live in wood are mycetophagous, or more precisely xylomycetophagous, feeding on mycelium filaments that invade wood, or are detritivorous, feeding for example on the corpses and dung of other insects.

– The third category of saproxylic insects is composed of a vast group of predators, detritivores and commensals that accompany xylophagous species in their galleries.

Some insects are found only on freshly dead wood, others in decomposed wood, and yet others in an intermediate stage. Dead wood also serves as shelter and a temporary refuge during extreme climatic conditions. Various gastropods, isopods, myriapods and lumbricid worms migrate to dead wood in winter. The carabid beetle *Pterostichus oblongopunctatus*, which lives in litter, hibernates in a chamber excavated in wood under the bark and also spends the summer there (Penney, 1967). Insects use wood as a place in which to nest; these include ants of the genus *Camponotus*, termites, carpenter bees (*Xylocopa*) and even the honey bee. Some aquatic insects may use submerged logs in the same ways.

1. DIVERSITY OF SAPROXYLIC COLEOPTERA

Many families of xylophagous beetles are able to exploit wood directly or indirectly, and there are many inventories, both in regard to the European fauna (Saalas, 1917; Palm, 1951, 1959; Dajoz, 1965, 1990; Mamaev, 1977; Muona & Viramo, 1986; Biström & Väisänen, 1988; Martin, 1989; Schmitt, 1992, etc.) and to that of North America (Deyrup, 1975, 1976, etc.). The main families, listed in decreasing order of importance, are as follows.

1.1. Cerambycids

The family Cerambycidae includes more than 20,000 species. It is the fifth largest family of Coleoptera, and includes about 250 species in France. These beetles are commonly called longhorns because of their often long antennae, especially in males. The antennae of male *Acanthocinus aedilis* are five times the length of the body. The taxonomy and biology of cerambycids have been much studied (Duffy, 1953; Linsley, 1959, 1961; Villiers, 1978, etc.). Members of the family vary greatly in size, from a length of 2 mm in the North American species *Cyrtinus pygmaeus* to 250 mm in the South American *Titanus giganteus*, one of the largest of all beetles. In the French fauna *Cerambyx cerdo* reaches 55 mm. With few exceptions, the larvae live in wood. In northern Germany 26% of species are associated with broadleaf trees, 50% with conifers, 9% live in either, 13% (genera such as *Agapanthia* and *Phytoecia*) live in herbaceous plants, and 2% (species of *Dorcadion*) live in the soil. Larval size is commensurate with that of adults. The larva of *Ergates faber*, which feeds on the decaying wood of old pine tree stumps in France, measures up to 8 cm, and the larva of the South American species *Macrodontia cervicornis* reaches 20 cm.

1.1.1. Adult morphology

Cerambycids mostly have elongate and almost cylindrical bodies (figure 16.1). Sexual dimorphism is pronounced, especially in the antennae which are longer in males than in females. Dimorphism is weak in species of the subfamily Prioninae and is most extreme in the subfamily Lamiinae. Antennal growth is of the allometric type in most species. Where X is body length and Y antennal length the equation is

$$\log Y = a \log X + b$$

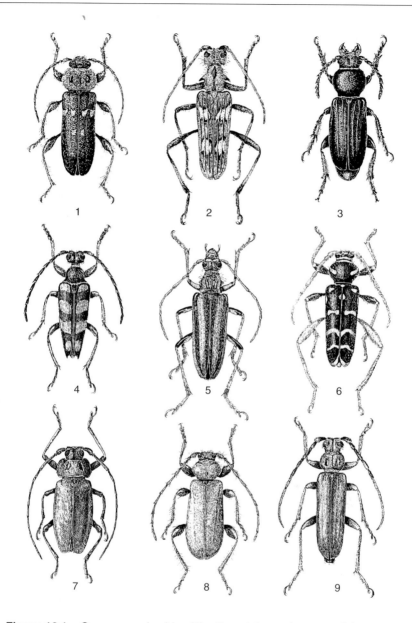

Figure 16.1. *Some cerambycids of the French fauna (not to scale).*
1: Hylotrupes bajulus *10 to 20 mm; 2:* Rhagium bifasciatum *15 to 18 mm; 3:* Spondylis buprestoides *15 to 24 mm; 4:* Leptura aurulenta *13 to 18 mm; 5:* Oxymirus cursor *18 to 24 mm; 6:* Plagionotus arcuatus *10 to 18 mm; 7:* Ergates faber *26 to 48 mm; 8:* Pyrrhidium sanguineum *9 to 11 mm; 9:* Arhopalus rusticus *15 to 25 mm (Villiers, 1978).*

a being the allometric coefficient which varies from 0.6 to 2.7 according to species. Less pronounced sexual dimorphism is seen in the mandibles and the forelegs of some species. In the American species *Dendrobias mandibu-laris* there are moreover two distinct forms of male differing in the size of their mandibles (figure 16.2).

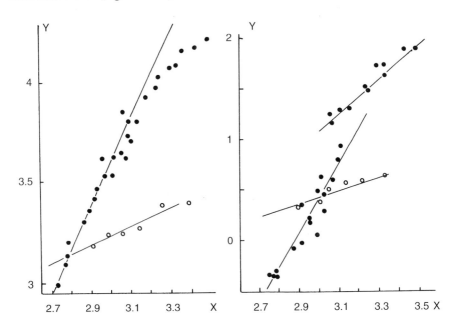

Figure 16.2. *Left: allometric relationship in each sex between antennal length Y and body length X in the cerambycid* Dendrobias mandibularis. *Females have shorter antennae than males. Right: relationship between mandible size and body size. There are two forms of male which differ in the relative size of their mandibles. Black dots: males; circles: females (Dajoz, 1995).*

Cerambycids vary in colour. Diurnal species are usually brightly col-oured, and nocturnal species are dull, often brownish. The black and yellow patterns of species of the tribe Clytini show that they mimic various species of wasp. American species of the genus *Moneilema* mimic tenebrionids of the genus *Eleodes* both in appearance and by their behaviour; other ceramby-cids imitate cantharids (soldier beetles) and lycids; this resemblance to species that are effectively protected by defensive secretions is a fine example of Batesian mimicry.

The conformation of the head differs from one subfamily to another (figure 16.3). In genera such as *Callidium* or *Clytus* the head lies in a hori-zontal plane and the antennae are inserted close to the eyes. These beetles deposit their eggs on the surface of bark or insert them into crevices with their ovipositor. The head is inclined to the vertical plane in genera such as *Monochamus* that use their mandibles to bore a hole through the bark to reach the cambium; these species have no exertile ovipositor. Finally genera such as *Acanthocinus* use both their mandibles and ovipositor to bore a well

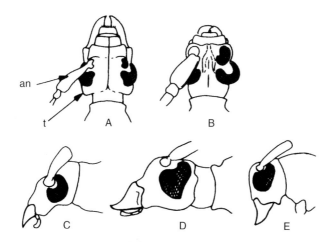

Figure 16.3. *Head of adult cerambycids in dorsal and lateral views. A and D: a species of Lepturinae of the genus* Leptura. *B and C: a species of Cerambycinae of the genus* Cerambyx. *E: a species of Lamiinae of the genus* Oberea. *an: base of antenna; t: tempus.*

in which the egg is deposited. The ovipositor in these species is exertile and the antennae are inserted towards the middle of the frons so they do not get in the way when the beetle is boring. The well bored for egg-laying may reach a depth of 15 mm in some species, the funnel-shaped upper part being 4 mm and the cylindrical lower part 11 mm long (figure 16.4).

Six subfamilies of cerambycid occur in the European fauna. Adult Asemiinae and Cerambycinae are difficult to distinguish at subfamily level but their larvae are quite distinct. Subfamilies may be identified by means of the following key to adults.

1. All five segments of tarsi distinct. Antennae short, not extending beyond base of pronotum. Pronotum without lateral border . . . .*Spondylinae*
   Fourth tarsal segment more or less concealed between the third and fifth segments. Antennae long, extending beyond base of pronotum. 2

2. Head vertically inclined to longitudinal axis of body, the frons at right angles to the vertex. Inner surface of pro-tibia grooved. . . . .*Lamiinae*
   Pro-tibia not grooved. 3

3. Lateral margin of pronotum raised, forming a more or less acute border. Front coxae transverse. . . . .*Prioninae*
   Lateral margin of pronotum not raised, not acute. Front coxae not transverse 4

4. Mesonotal stridulatory plate without a smooth median line. Sides of head not strongly retracted behind eyes, the temples indistinct.
   Front coxae globular . . . .*Cerambycinae*
   Mesonotal stridulatory plate divided by a smooth mid-longitudinal line 5

5. Head retracted behind the eyes, temples distinct, and neck separated from head by a deep furrow. Front coxae conical. . . . .*Lepturinae*
   Head not retracted behind eyes, without distinct temples.
   Front coxae globular . . . .*Asemiinae*

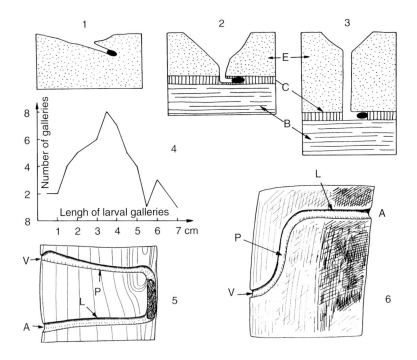

Figure 16.4. *Different forms of oviposition by cerambycids. 1: egg inserted in a crevice in bark; 2: egg deposited at the bottom of a well excavated with the mandibles; 3: egg deposited in a well bored with both mandibles and ovipositor; 4: variations in gallery length in* Monochamus sutor; *5 and 6:* Monochamus sutor *larval galleries bored in a large trunk (5) and in a trunk 14 cm in diameter (6); B: wood; C: cambium; E: bark; A: entrance orifice; V: emergence orifice; L: larval gallery prior to hibernation; P: pupal chamber (Tragardh, 1930).*

1.1.2. Larval morphology

Cerambycid larvae have a characteristic appearance (figure 16.5). The tegument is soft and depigmented and legs are reduced or lacking; the body is swollen and the abdominal segments bear ambulatory ampullae that give the larva a purchase on gallery walls as it moves through them. The head is either transverse or longer than broad. The alimentary tract is remarkably long, up to three times the body length. In *Rhagium bifasciatum* there is a short oesophagus leading to a gastric pouch into which mycetomes full of symbiotic fungi, known as "yeast crypts", open. These crypts do not exist in the adult. The male is devoid of symbiotes, and those of the female are included in accessory glands of the genital apparatus. The midgut is prolonged into a cylindrical portion and then into an area covered in diverticula that are regeneration crypts; it ends at a pyloric valvule at the point at which the

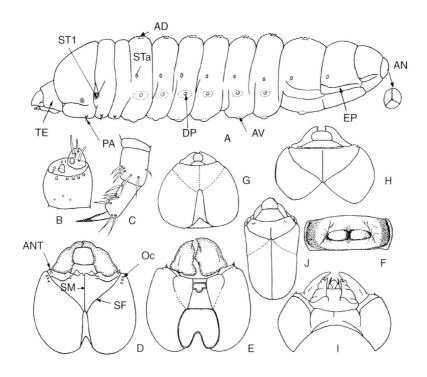

Figure 16.5. *Cerambycid larval morphology.* A: Ergates faber *larva in lateral view. B: tip of antenna. C: greatly reduced mesothoracic leg. D and E: head in dorsal and ventral views (labium and maxillae removed). F: first abdominal segment in dorsal view. G: Cerambycinae head. H and I: Lepturinae head. J: Lamiinae head. ST1: prothoracic stigma; STa: first abdominal stigma; AD and AV: dorsal and ventral ambulatory ampullae; PA: legs; DP: pleural discs; EP: epipleuron; AN: anus; SM: medio-frontal suture; SF: frontal suture; ANT: antenna; Oc: ocelli.*

Malpighian tubes open. The midgut represents about 55% of the length of the alimentary tract and the hindgut 40%. A similar morphology is seen in other cerambycids. According to Semenova & Danilevsky (1977) the alimentary tract in cerambycids that live in relatively dry wood poor in nutrients is long and the hindgut very large; in species that live in non-decomposed wood richer in water and absorbable substances the alimentary tract is short and the hindgut less long (figure 16.6). Cerambycid larvae that live in phloem are compressed like those of buprestids; those that bore their galleries in xylem have cylindrical bodies.

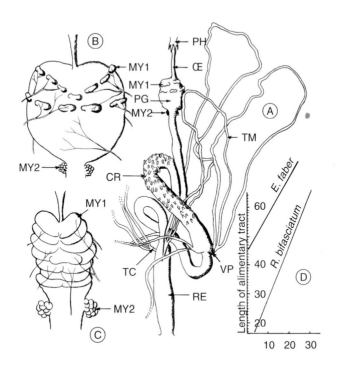

Figure 16.6. *A: Larval alimentary tract of* Rhagium bifasciatum. *PH: pharynx; Œ: oesophagus; MY1 and MY2: mycetomes; PG: gastric pouch; CR: regeneration cell crypts of the midgut; VP: pyloric valvule; TC: common duct of Malpighian tubes; RE: rectum; TM: Malpighian tubes; B and C: detail of the gastric pouch in* Rhagium bifasciatum *and in* R. inquisitor. *D: length of alimentary tract as a function of body length (in mm) in* Rhagium bifasciatum *and in* Ergates faber.

Cerambycid larvae can be identified to subfamily level using the following key:

1. Head longer than broad. One occipital foramen. Legs lacking or vestigial . . . .*Lamiinae*
   Head not longer than broad, usually transverse. Legs usually present albeit reduced 2

2. Occipital foramen divided in two 3
   Occipital foramen entire 4

3. Front margin of frons toothed or carinate, forming a ridge over the labrum. Legs fully formed. Tips of mandibles more or less acute . . . .*Prioninae*
   Front margin of frons not toothed or carinate. Legs rudimentary or lacking. Tips of mandibles blunt. . . . .*Cerambycinae*

4. Head strongly dilated with deeply emarginate base. Body compressed. Last abdominal segment either with a small terminal spine or, more often, simple . . . . .*Lepturinae*
   Head less dilated, the base less deeply emarginate. Body cylindrical. Last abdominal segment bearing two short urogomphs. . . . .*Asemiinae & Spondylinae*

Cerambycid larval galleries have characteristic layouts and are sited differently according to species.

– Some galleries are bored in trees to which the bark still adheres. These galleries may be superficial and lie entirely in the bark or just under the bark. *Rhagium inquisitor* and *Acanthocinus aedilis* larvae spend their entire lives in thick pine bark which they only leave in order to excavate their pupal chambers beneath the bark. Others, such as various species of *Clytus* and *Callidium*, bore galleries which begin by being sub-cortical and later penetrate the outer layers of the sapwood in which pupation takes place. *Cerambyx*, *Ergates* and *Monochamus* bore deep galleries that begin just beneath the bark but soon reach the sapwood and even the heartwood.

– Other galleries are bored in trees devoid of bark. This is the case in species that live in decayed wood or that bore in timber, such as *Hylotrupes bajulus*, *Arhopalus* spp. or *Clytus pilosus*. *Hylotrupes bajulus*, the house longhorn, may mate and lay eggs within its larval galleries, which accounts for the difficulty in spotting the damage it does to building timber.

All cerambycid galleries are filled with the larva's faeces which form compacted masses of sawdust. The length of galleries is characteristic for each species. In some cases gallery layout depends on the diameter of the tree trunk (figure 16.4).

1.1.3. Predators and parasites of cerambycids

Many of the beetles that predate scolytids also predate cerambycids. This is the case of trogositids and clerids. To this list must be added the larvae of various species of elaterid. Among Diptera that predate cerambycids are the larvae of asilids, or robber flies. The part played by parasitoid Hymenoptera is still poorly known; they are mainly ichneumonids and braconids. *Rhyssa* species are remarkable for their large size and long ovipositors capable of piercing thick bark to reach quite deeply imbedded larvae. *Rhyssa persuasoria* parasitises *Spondylis buprestoides*, among other species (figure 16.7). Another wasp, *Scleroderma domestica*, attacks various xylophagous beetles such as anobiids, lyctids and bostrychids, but particularly *Hylotrupes bajulus*. Its use in biological control of this species has been considered.

1.1.4. Some biological and morphological characters of cerambycids

– Adults of many species belonging to genera such as *Leptura* and *Clytus* are diurnal and are found on flowers to which they come for nectar. Others are drawn to fruits or to sap flowing from trees. Several diurnal cerambycids such as *Rosalia alpina* are never found on flowers, but wander on old trees. Species such as *Cerambyx cerdo* are crepuscular or nocturnal. Adult life span is usually short; many emerge in summer; others undergo pupal ecdysis at the onset of winter and remain in their pupal chambers until spring. This is

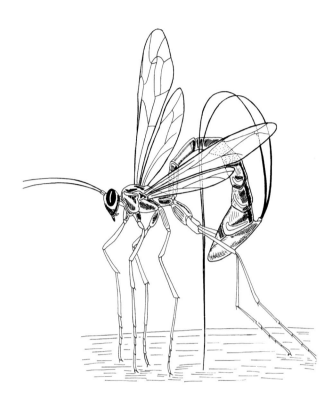

Figure 16.7. *A female of the ichneumonid* Thalassa lunator *inserting its oviposi-tor in wood in order to lay eggs.* Rhyssa *are of similar appearance.*

the case in *Rhagium* species which construct a circular chamber lined with compressed wood fibres under bark (figure 16.8).

– Cerambycid larvae live much longer than adults. Larval life may last up to six years in *Hylotrupes bajulus*; it lasts two years in *Ergates faber* and three in *Cerambyx cerdo*. The larvae of most species live one year. Variations in the length of larval life, which are due to environmental condi-tions, are considerable; they are determined particularly by humidity and the nutritional value of wood.

– Primary pests are few among the Cerambycidae. An exception are *Saperda* species which feed on poplars. Most members of the family are secondary destroyers which exploit either sick or recently felled trees (*Callidium, Clytus, Acanthocinus*), or else wood that is long dead and more or less decayed (*Ergates faber, Leptura aurulenta, Oxymirus cursor, Rosalia alpina*). In the latter habitat several generations may breed in the same tree trunk. In the tropics the proportion of cerambycid species that feed on living trees is higher, which may be due to competition with termites for dead wood. Cases of sudden changes of food plant are also known, and suggest

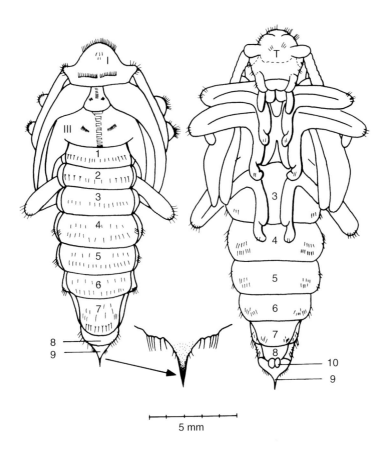

5 mm

Figure 16.8. Rhagium bifasciatum *pupa in dorsal (left) and ventral (right) views. The spine at the tip of the abdomen is characteristic of the genus* Rhagium. *I, III: thoracic segments; 1 to 10: abdominal segments; T: head.*

that sympatric speciation occurs in the Cerambycidae. *Tetropium gabrieli* lives on larch, but females can be induced to lay their eggs on pine and spruce, on both of which the larva grows normally. In some cases host plant recognition is achieved by means of substances released by the plant. *Hylotrupes bajulus* is attracted to α-pinene; *Ergates faber* and *Spondylis buprestoides* are also attracted to terpenes; *Cerambyx cerdo* is drawn to ethanol and to ethyl acetate.

– There are polyphagous species such as *Ergates faber* that live in oak, chestnut, ash, elm, willow, horse chestnut, cherry, pine etc. or *Rhagium bifasciatum* which is just as eclectic. Monophagous species are few, and many cerambycids feed on a small number of tree species. *Cerambyx cerdo* feeds mainly on oak, and *Monochamus galloprovincialis* lives in a few species of pine. Cerambycids that live on decayed wood are more polyphagous than those that feed on recently dead trees. The most polyphagous of all species is probably *Stromatium barbatum* which in India has been recorded as feeding on 331 different species of tree, both broadleaf and coniferous.

— The mechanism of digestion of wood has been described in chapter 14. Species that use cellulose have long digestive tracts, slow intestinal transit which may last for hours, and a poor coefficient of digestive use. A relationship exists between the average size of larvae, the duration of larval growth, and the part of the tree they exploit (table 16.1). In *Hoplocerambyx spinicornis* the larva ingests 584 mg (dry weight) of wood per gram of body weight per day, 40.4% of which is assimilated. Substances used are cellulose, hemicelluloses, starch and a few sugars. The larva also acquires nitrogen through micro-organisms in its alimentary tract. Savely (1939) calculated that the surface area of larval galleries of *Callidium antennatum* varies from 2.26 to 6.07 cm^2, and dry weight of larvae varies from 0.015 to 0.042 g. The quantity of wood consumed by a larva in the course of its life varies from 1.26 to 3.37 g. All woody parts of the tree may be exploited, from small branches to the trunk and even the roots. Most cerambycid larvae live just under the bark, and some species penetrate deep into the wood, or in bark when it is sufficiently thick.

Parts of tree exploited	Number of species	Average size (mm)	Average duration of growth (years)
Bark	2	19	2.3
Phloem	11	14	1.1
Phloem + sapwood	24	17	1.6
Phloem + sapwood + heartwood	6	20	2.2
Sapwood + heartwood	3	27	3.1
Heartwood	20	30	3.6

Table 16.1. *Average size and duration of larval growth period in cerambycids exploiting different parts of a tree.*

— Although many cerambycids do not feed as adults, some species must obtain some nourishment to enable them to produce eggs. This is the case in those species that frequent flowers in search of pollen or nectar. Umbelliferous flowers are particularly sought after by this family.

— The number of larval moults varies. *Monochamus sutor* has ten larval instars and *Prionus coriarius* fourteen; *Arhopalus syriacus* may go through eight to fifteen instars (Carle, 1973). In this species growth is rapid in the first five instars which feed on the secondary phloem of pines, and then slows when the larva penetrates the deeper layers of the wood; in the last instars leading up to pupation growth is almost nil. Oxygen levels in dead wood may be low, and those of carbon dioxide high. *Orthosoma brunneum* larvae tolerate oxygen concentration as low as 2% and carbon dioxide levels of up to 25% (Paim & Becker, 1963).

– Variations in body size in xylophagous insects, and cerambycids in particular, are greater than in other insects. The ratio of maximum to minimum body length of a species varies from 1.4 to 2.1 in xylophagous Coleoptera and from only 1.1 to 1.4 in other Coleoptera (Andersen & Nielssen, 1983). This great plasticity appears to be due to variability of food quality.

– Xylophagous beetles such as cerambycids and lucanids feed on xylem, which is poor in nitrogen compounds (0.0005% on average). Their life style, however, shelters them from predators and the long larval life makes it possible for them to accumulate sufficient nitrogen compounds to build a large adult exoskeleton. Fertility is relatively low, which again is compensated for by the secure life style of the larvae, and is desirable since overpopulation would lead to shortage of absorbable nitrogen. In contrast, beetles such as chrysomelids are leaf-eaters with relatively short life spans which do not allow them to accumulate many nitrogen compounds. Chrysomelid exoskeletons are much less sclerotised and they invest more energy in egg production, which is necessary because their larvae are exposed to numerous predators (Rees, 1986).

– Some cerambycids such as females of the American genus *Oncideres* improve the nutritional value of the wood eaten by their larvae by using their mandibles to make incisions in small diameter branches that block sap flow. Before the leaves fall in autumn, nutritive elements produced by the leaves accumulate above the incised area and are available as food for the larvae. This behaviour only occurs in species that live on broadleaf trees and is absent in those that live on conifers. (Forcella, 1982).

1.1.5. Some French cerambycids

– Subfamily Spondylinae. *Spondylis buprestoides* is recognisable by its antennae, which are unusually short for a cerambycid. The species lives on pine and fir and occurs almost everywhere in France.

– Subfamily Asemiinae. They have relatively short antennae that do not reach the length of the body. *Asemum striatum* lives in conifer forests in mountain regions in the whole northern hemisphere. *Arhopalus rusticus* also colonises various conifers and sometimes timber. *Tetropium castaneum* occurs in the whole of Europe on conifers; the related species *Tetropium gabrieli* appears to be restricted to larch.

– Subfamily Prioninae. These large longhorn beetles are often nocturnal and dully coloured. *Ergates faber* lives in decaying conifers in the South of France. *Prionus coriarius* and *Aegosoma scabricorne* live mainly in old beech and other broadleaf trees. Both species are disappearing in France, probably due to loss of their habitats. *Tragosoma depsarium* lives on conifers particularly on mountain pine in the Pyrénées. This species is heading for extinction in France but is still common in Scandinavia and even more so in the great forests of the American West.

– Subfamily Cerambycinae. The house longhorn *Hylotrupes bajulus* is a European species that has spread around the world through the timber trade. It attacks standing dead trees as well as more or less decayed wood and building timber. *Cerambyx cerdo* reaches a length of 55 mm and occurs in most of France; the adult is crepuscular and nocturnal. *Pyrrhidium sanguineum*, which measures about 1 cm and lives mainly in oak, owes its name to its blood-red colour. *Poecilium alni* is a 4 to 7 mm long species which lives in small oak branches and is still very common in France. *Clytus* and related genera are often patterned in black and yellow and mimic wasps; they are found both on flowers and on dead trees. *Rosalia alpina* is one of the most spectacular longhorns and is found around old beech trees, the more or less decayed trunks of which its larvae feed on. It reaches a length of 40 mm and is a powdery blue with velvety black spots of variable extent on the thorax and elytra. The genus *Rhagium* includes four species in France; they are robust species with relatively short antennae and spines on the sides of the prothorax.

– Subfamily Lepturinae. *Oxymirus cursor* can be found up to 2,000 m in mountains and lives on pine. *Corymbia rubra* is remarkable for its sexual dimorphism: the female is a uniform red and the male has a black prothorax and ochre-yellow elytra. Many *Leptura* and related genera are diurnal and feed on flowers as adults.

– Subfamily Lamiinae. *Morimus asper* is a dull black, wingless, polyphagous species up to 35 mm long that moves around on old dead tree trunks during the day. It appears to be vanishing in France. *Monochamus sutor* is the commonest species of its genus. *Acanthocinus aedilis* is a widely distributed species which lives on conifers; it is remarkable for its very long antennae.

Cerambycid beetles, together with representatives of other families, include saproxylic species that are good indicators of woodland biodiversity, and inventories of such species have begun to be made in some protected areas (Parent, 1996).

1.2. Buprestids

This family includes about 15,000 species that are mainly distributed in the warm parts of the globe and are easily recognised by their often bright metallic colours which has given rise to the name "jewel beetles".

1.2.1. Larval morphology

Larvae belonging to this family can be recognised by the strong dilation of the prothorax which partly conceals the head. This appearance has earned them the name "hammer larvae" in some countries (figure 16.18). Adult buprestids are diurnal, sun-loving beetles; some, such as many small species of *Anthaxia* and *Agrilus*, settle on flowers, with a special predilection for yellow blooms; others, such as *Chrysobothris*, go in search of old tree trunks on which to lay their eggs. With few exceptions buprestid larvae are xylophagous. Their enzymatic arsenal includes a cellulase, an amylase and a pectinase. Those of a few species however live underground, like *Julodis onopordi*, while others, such as *Trachys*, live in herbaceous plants. The heat-loving nature of adults is also found in larvae which settle under bark on the upper parts of trunks exposed to the sun. Larval galleries are bored either within the bark, between the bark and the sapwood, or deep inside the wood. Galleries are flattened, wide and winding and are filled with a mixture of sawdust and faeces. The duration of larval growth varies, lasting one or two years in larger species. The alimentary tract of buprestid larvae is of characteristic structure, with a crop at the level of the foregut, a paired caecum at the limits of the fore- and midgut, and a hindgut that is shorter than that of cerambycids.

Buprestids may become primary destroyers, attacking healthy, living trees, but are more often secondary destroyers which live in dead trees. Host specificity varies. *Chrysobothris affinis* feeds on many different broadleaf species; *Coroebus undatus* is confined to oak, as are *Anthaxia hungarica* and *Eurythyrea quercus*. *Buprestis* and *Chalcophora* species live on conifers.

1.2.2. Biology of a few species

– *Agrilus biguttatus* lives in oak. Pupation takes place at the end of a long gallery, and adults emerge through the bark. Tree mortality may be high in heavily infested areas. In France, following the hot summers of 1945, 1947 and 1949 damage was reported in the Vosges (Alsace), the Loire valley, the Paris area and Normandy. The larvae cause blocking of vessels by tylosis, which impedes water circulation. Oaks are particularly vulnerable since 75% of water is carried in the outer layers of the wood. Tumours appear, characterised by new formation of ligneous tissue outside the secondary phloem overrun by larval galleries. This is caused by auxinic substances that have not yet been isolated and which are produced by the larvae. These newly-formed tissues only appear two or more years after the galleries are bored, and therefore do not serve as food for the larvae. This situation is quite different from that of gall-forming insects, which induce formation of plant tissue destined as food for the larvae (Jacquiot, 1976).

– Outbreaks of *Agrilus viridis* sometimes occur on beech, especially following hot dry years (Bovey, 1953). This species' galleries are mainly found under those parts of the bark exposed to the sun at the southern edges of woods; it is therefore difficult to make a distinction between damage done by the beetle and that caused by drought and insolation. *Agrilus viridis* lives mainly in beech, but also in birch, willow, oak, hornbeam and maple. The green or bronze adult is 6 to 10 mm long; the larva, which grows to 13 mm, bores winding, flattened galleries elliptical in cross-section through secondary phloem and sapwood. Where bark is thin the presence of galleries is betrayed by slight swellings on the surface. The life cycle lasts two years. In many other small species belonging to the genera *Agrilus* and *Anthaxia*, adults are to be found on flowers.

– Species of the genus *Chrysobothris* have shallow circular depressions on the elytra. *C. solieri* lives in the trunks and branches of pines, while *C. affinis* is a polyphagous species that lives mainly in oak and beech. The genus *Buprestis* includes fairly large (up to 20 mm) species that live in conifers. *Melanophila acuminata* is common in scorched pine trees following forest fires, a habitat shared by another buprestid, *Phaenops cyanea*. *Lampra festiva* feeds on juniper and cypress, and *Lampra rutilans* on lime.

– *Coroebus florentinus* (= *Coroebus bifasciatus*) feeds on oaks. The female lays eggs in June-July in the bark of young healthy branches of various species of oak, often near the tips of the current year's growth of twigs. The larva goes to the cambium layer and bores a gallery that may reach a length of 1.5 m, towards the base of the branch. When it is fully grown the larva bores an almost complete ring under the bark. This transverse gallery cuts the flow of sap and leads to desiccation and death of the whole distal part of the branch. Affected oak trees are recognisable by the branches that wilt above the level of the pupal chambers. Larvae live between 20 months and four years. Deterioration on cork oak and evergreen oak may be considerable, particularly on young trees under cover of older ones. Adult beetles live on the tops of trees in which the larvae grew; they are 12 to 18 mm long, brassy green with three curved transverse blue stripes on the elytra. Many other wood-boring beetles are able to colonise the dead branches in the wake of *Coroebus florentinus*. In the past few years in the South of France this species has caused increasingly frequent damage which is seen in the form of defoliated branches and the abnormal colour of foliage. Another member of this genus, *Coroebus undatus*, also lives on oaks and is common in cork oak (figure 16.9).

– *Chalcophora mariana*, the pine borer, is the largest buprestid of the French fauna; it reaches a length of 30 mm and is greenish-bronze with a coppery sheen. The genus *Chalcophora* is distributed in the whole northern hemisphere. The larvae grow up to 7 cm long and live in old pine trees, especially in the more southern regions. They are most frequently found in decaying tree stumps, often in the company of cerambycid larvae such as those of *Ergates faber*. Adults emerge in summer, when they fly about in broad daylight and settle on old tree trunks.

1.3. Anobiids

This family includes about a thousand species, few of which are larger than 5 or 6 mm. Most species are xylophagous and live in dead, even very

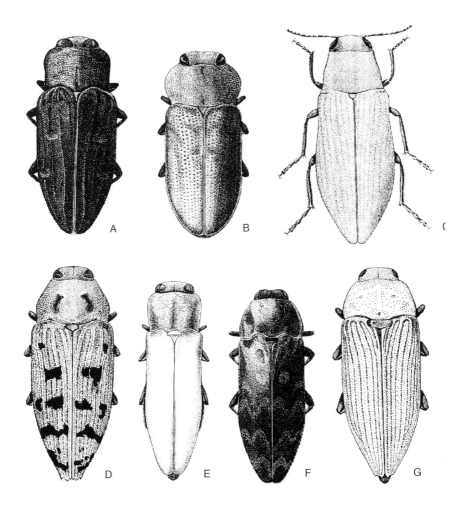

Figure 16.9. *A few buprestids of the French fauna. A:* Chrysobothris affinis; *B:* Anthaxia nigritula; *C:* Kisanthobia ariasi; *D:* Lampra festiva; *E:* Agrilus sulcicollis; *F:* Coroebus undatus; *G:* Eurythyrea austriaca.

dry wood that they may sometimes reduce to dust. These species are danger-ous destroyers of woodwork, but this family in the wild seems to cause little damage. The commonest and most familiar genera are *Anobium* and *Xestobium*; their larvae are typically grub-like in appearance, whitish, hunched, and with a dilated thorax and short slender legs (figure 16.10).

Adult anobiids are recognisable by their more or less cylindrical bodies, the hood-shaped prothorax that often covers the head in dorsal view, and their antennae, the last three segments of which are longer than the preceding ones (figure 16.11). *Ernobius* species are close to *Anobium,* but their elytra

Figure 16.10. *Larva of the anobiid species* Xyletobius walsinghami. *Note the hunched shape of the body and fully formed but small legs.*

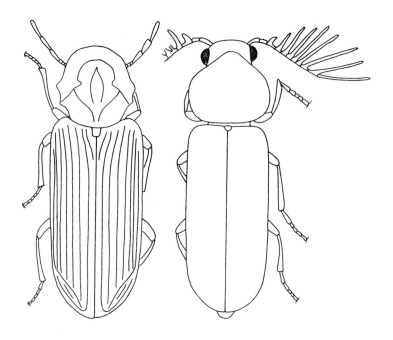

Figure 16.11. *Two anobiids. Left,* Anobium emarginatum; *right, male* Ptilinus pectinicornis.

are not striate, and most of them live in conifer cones. *Ernobius mollis* bores the wood of conifers. *Ptilinus pectinicornis,* which feeds on dead wood, is recognisable by the male's strongly pectinate antennae. *Anobium* species, and *Anobium punctatum* in particular, are known as furniture beetles or woodworms. *A. punctatum* has spread throughout the world in woodwork. Various other beetles predate it, including the clerids *Corynetes coeruleus, Opilo mollis* and *Opilo domesticus.* Another common species is *Xestobium rufovillosum,* called the death watch beetle on account of the habit of both sexes of rapping their heads against their gallery walls, producing a "sinister" ticking sound; this sound may well be a mating call. *Xestobium rufovillosum* is apterous, and has been spread in building timber and furniture.

Many anobiids possess a mycetome which does not appear to play a part in digesting wood but only in supplying necessary vitamins. Larval life-span of a species such as *Xestobium rufovillosum* varies according to the condition of the wood on which it feeds. It lasts up to 50 months in intact wood but may be reduced to 16 months in wood invaded by various fungi (Fischer, 1941).

A number of other saproxylic beetles are associated, like anobiids, with dry wood. The most striking example is that of the dermestid *Ctesias serra* which lives under detached dry bark where it survives on a diet of spider webs and the corpses of insects caught in the webs.

1.4. Bostrychids

This is a small family of some 300 species of mostly xylophagous species. The larvae live in dead wood in which they bore deep galleries. Their bodies are recurved like those of scarabaeid larvae and have strongly developed thoracic segments (figure 16.12). Adults often have almost cylindrical bodies with the head covered dorsally by the prothorax, and the tip of the elytra is often truncate and bears protruberances in the same manner as scolytids of the genus *Ips.* These structures are considered to be organs (called annitive organs) that facilitate purchase on gallery walls. This structural similarity may also explain a long-standing confusion of names, scolytids being termed "bostryches" in much of the older literature. In English the family is sometimes called "false powder post beetles" alluding to their similarity to lyctids. Adults are active by day or by night according to species. Most bostrychids are one to two cm long. The giant of the group is *Dinapate wrightii* which measures 5 cm and lives in the endemic Californian palm tree *Washingtonia filifera.* The reputed rarity of this beetle induced museums in the nineteenth century to pay several hundred dollars per specimen.

Bostrychus capucinus is recognisable by its black forebody and bright red elytra; it lives in old tree stumps, especially those of oak. *Lichenophanes varius* lives mainly in old beech trees. *Sinoxylon sexdentatum* is highly polyphagous and often lives on fallen oak branches on the ground. The same is

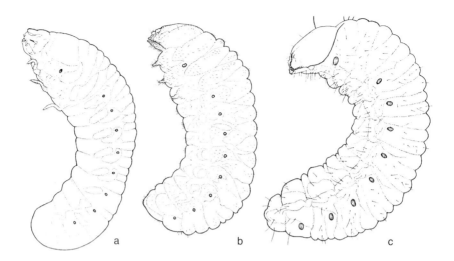

Figure 16.12. *Larvae of xylophagous beetles.* A: Bostrychus capucinus
(Bostrychidae); B: Platyrrhinus resinosus *(Anthribidae);*
C: Hylobius abietis *(Curculionidae).*

true of *Sinoxylon perforans*, which occurs mainly in the South of France and is common in cork oak and evergreen oak on which it exploits dead or dying branches. Among the latter species' predators are the histerids *Teretrius picipes* and *T. parasitica*, the malachiid *Axinotarsus pulicarius*, and especially the clerids *Tillus unifasciatus* and *Opilo mollis*.

1.5. Curculionids

The Curculionidae, or weevils, includes some 60,000 described species which mostly live on herbaceous plants or on tree leaves. Xylophagous species are few, but of some economic importance. Some, such as *Pissodes notatus* (cf. chapter 11.1.2) and *Hylobius abietis* (cf. figure 11.3) are primary pests of conifers. *Magdalis* species live on various diseased or dead trees in which they bore sub-cortical galleries that mark the sapwood in the same way as those of scolytids. *Hylobius abietis* is a pest of pine (cf. chapter 15.4). The curculionid tribe Cossonini in France includes ten genera and about thirty species which are nearly all found in more or less humid and decayed wood in which they may mate and lay without leaving the substrate for several generations. *Rhyncolus elongatus* lives on pine trees that have fallen and lain on the ground for a long time and that were previously exploited by scolytids (cf. figure 15.22). Cossonini occur in the whole northern hemisphere, and are particularly abundant in dead coniferous trees in the American West.

1.6. Lyctids

This family, the powder post beetles, only has about sixty species world-wide. The larvae have a recurved abdomen, the front part of the body dilated, short legs, and the stigmata of the eighth abdominal segment larger than the others (figure 16.13). Adults are elongate, somewhat cylindrical, with the head visible in dorsal view; the eleven-segmented antennae have a two-segmented club. They are under 7 mm long and a uniform brown or black.

The most widely distributed species is *Lyctus brunneus*, which mainly lives on building timber. Before laying eggs the female tests the wood with bites to ascertain its starch content. Wood must contain at least 3% starch for the larvae to grow. Eggs are deposited in the lumen of a vessel. The larvae can only exploit the starch in feeding. Parasites are braconids, pteromalids and bethylids; predators are other beetles, the commonest of which is the clerid *Tarsostenus univittatus*.

1.7. Lymexylonids

This family is represented in France by three species, the commonest of which is *Hylecoetus dermestoides* which is easily recognised at the larval stage by its hump-backed prothorax which projects above the head and by the long process at the tip of the abdomen (figure 16.15). Adults have a soft tegument, a cylindrical body, and measure about 15 mm. The palpi are elaborately structured, especially in the male (figure 16.14).

Hylecoetus dermestoides larvae live in broadleaf trees such as beech as well as in conifers such as fir. Infested wood is often damp, viscous and fetid owing to the fungus with which the beetle lives symbiotically (cf. chapter 14.5 and figure 14.7). Larval galleries have a characteristic layout; they penetrate the bark deep into the wood. The larva clears the gallery of sawdust by moving backwards and pushing the sawdust with the appendage at the tip of its abdomen, leaving characteristic rings of compacted sawdust around the gallery entrance on the underside of the bark. The life cycle lasts one year.

1.8. Scarabaeoidea

This vast superfamily is composed of several families, many representatives of which live in dead wood which is often in an advanced state of decay.

1.8.1. Lucanids

This family is represented in France by a few large species. The larvae are typical grubs and differ from scarabaeid larvae, such as those of the

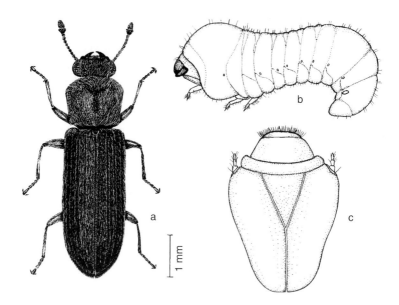

Figure 16.13. Lyctus brunneus. *a: adult. b: larva. c: head of larva in dorsal view.*

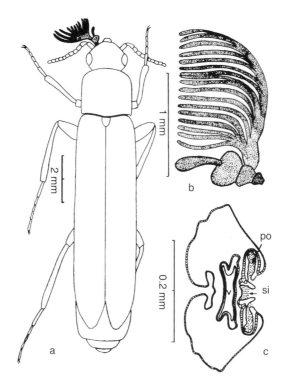

Figure 16.14. *Adult* Hylecoetus dermestoides. *a: dorsal view; b: male maxillary palp; c: cross-section of the female ovipositor; po: pouches containing fungal spores; si: median furrow; v: vagina.*

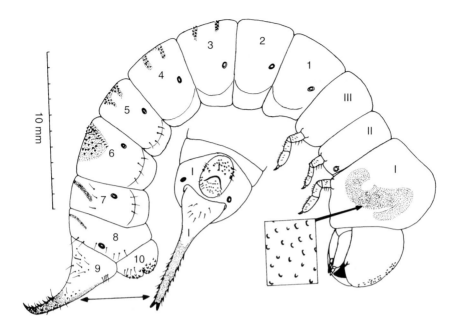

Figure 16.15. Hylecoetus dermestoides *larva in lateral view and tip of abdomen in dorsal view. I, II, III: thoracic segments; 1 to 10: abdominal segments.*

cockchafer which live in the soil, in a number of characters considered to be adaptations to life in wood: the body is relatively shorter, pubescence sparser, the antennae and legs shorter and the mandibles are shorter as a result of reduction of the distal part. Three species of lucanid are quite common in dead wood. Male stag beetles (*Lucanus cervus*) reach a length of 85 mm and have hypertrophied mandibles, while females are under 60 mm and have normal sized mandibles. Adults fly in the evening in early summer. They prefer old oak trees but also settle in many other trees. The larvae live four to five years. Pupation takes place either in the tree or in the ground in a cocoon made of wood fibres or soil particles (figure 16.16). Another species, *Dorcus parallelipipedus*, reaches a maximum size of 35 mm and its sexual dimorphism is minimal; its larvae live four years. Adults emerge in late summer and often live more than a year; they lay their eggs in May. This species wanders at night on old trees in which its larvae live. *Sinodendron cylindricum* is a scarcer species that lives mainly in beech, especially in mountain areas.

1.8.2. Dynastids and cetonids

Dynastids play a major role in the tropics where they are represented by large species such as *Oryctes*, *Augosoma* and *Dynastes*. *Oryctes nasicornis* is

Figure 16.16. *Metamorphosis of the stag beetle. Male, female, larva and pupa in its opened pupal cocoon (Blanchard, 1868).*

the commonest species in France. Its larvae live in dead wood, but may also live underground on young tree roots, which makes this species a pest in nurseries. *Callicnemis latreillei* is a dynastid that lives on driftwood on the coast of South West France (Caussanel & Dajoz, 1967). Cetonids (chafers) spend their larval lives in very decayed wood and in rot-holes in trees. The most characteristic species in Europe are *Osmoderma eremita*, *Potosia cuprea*, *Gnorimus nobilis*, and *Gnorimus octopunctatus* (cf. chapter 18).

Some families of Scarabaeoidea play a greater role in the tropics than in temperate regions where they are fewer. This is the case of melolonthids and rutelids. Passalids are a tropical family that has almost no representatives in temperate regions. They have highly developed social behaviour and live in very damp wood. One of the few temperate species is *Popillius disjunctus* which is common in the eastern United States. All stages of the passalid life cycle are found in the galleries bored by the adults. Coprophagy, which is a form of true rumination, has been shown in *Popillius disjunctus*, which dies if it is prevented from re-ingesting its own faeces (Mason & Odum, 1969).

All members of the superfamily Scarabaeoidea that live in wood and have been studied have a fermentation chamber which is a dilatation of the hindgut that is permanently almost completely filled with wood fragments and symbiotic bacteria (figure 16.17). In *Dorcus* this chamber may occupy most

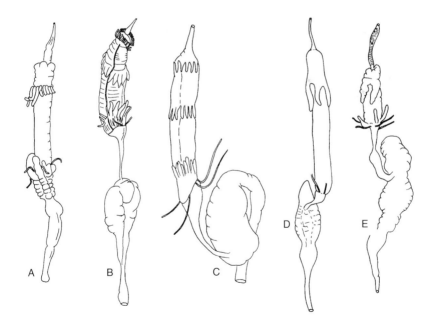

Figure 16.17. *Alimentary tracts of a few xylophagous insects having a rectal fermentation chamber. Coleoptera. A:* Sinodendron cylindricum *(lucanid); B:* Oryctes nasicornis *(dynastid); C:* Potosia cuprea *(cetonid). Diptera, Tipulidae. D:* Tipula maxima; *E:* Ctenophora flavicornis.

of the larva's abdominal cavity; adults do not possess it. The bacteria present assist in digesting wood (cf. chapter 14.5).

1.9. Other families of xylophagous beetles

Many other families of Coleoptera include xylophagous species that may sometimes play major roles in the process of wood decay. One of the most curious is the elmid *Lara avara* which lives in rivers in Oregon on fragments of wood that fall into the water. This obligate xylophage has a larval life of four to six years, which is commensurate with the poor nutritional quality of its food. The larva absorbs substances produced by decomposition of wood by bacteria; it does not possess cellulase or a symbiotic intestinal flora (Steedman & Anderson, 1985). *Lara avara* combines some of the features that enables a herbivore that feeds on matter of poor nutritional quality to obtain sufficient nitrogen to complete its life cycle. These include its large size for an elmid (the larva measures 10 mm) and long larval life. Similarly the xylophagous chironomid fly *Xylotopus par* is one of the largest to be found in temperate lotic environments, measuring 1.5 to 2 cm. This species assimilates wood efficiently thanks to its intestinal bacteria, a long digestive tract in which food remains for seven hours, and a midgut having an alkaline pH which releases wood proteins.

In Oregon as much as 40 kg of dead wood per square metre may be found in rivers that flow through the great conifer forests. In Europe the water-logged wood habitat has virtually disappeared as a result of river management, however the beetle *Rhizophagus aeneus* (family Rhizophagidae) can still be found under the thick bark of submerged trunks or those lying in very damp places. *Nacerda melanura* (family Oedemeridae) and *Pselactus spadix* as well as species of the genus *Mesites* (family Curculionidae) live in driftwood carried by rivers and washed up on beaches and thus impregnated with salt. Many weevils of the tribe Cossonini live in decayed wood floating far out to sea like rafts; this may explain the presence of the many species belonging to this subfamily on some islands, on which they speciated following the arrival of a few immigrants.

1.9.1. Eucnemids

Adult beetles of this family, the false click beetles, resemble click beetles (Elateridae), but they are unable to jump by "clicking" and their bodies are more cylindrical. The larvae, which live in dead wood, resemble those of buprestids in having a dilated prothorax and flattened body (figure 16.18). Many eucnemids are rare; the commonest species is *Melasis buprestoides* which lives mainly in beech in which the larvae bore flattened galleries at right angles to the grain of the wood.

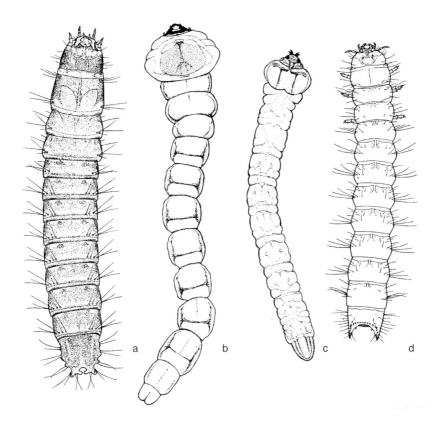

Figure 16.18. *Saproxylic Coleoptera larval types. A:* Denticollis linearis *(Elateridae); B:* Chalcophora mariana *(Buprestidae); C:* Melasis buprestoides *(Eucnemidae); D:* Pytho depressus *(Pythidae).*

1.9.2. Melandryids

These beetles, sometimes known as "false darkling beetles", live mainly in fungi that grow on wood. Others such as *Melandrya caraboides* and *Serropalpus barbatus* live in wood.

1.9.3. Alleculids

There are many saproxylic species in this family, including *Prionychus ater* and *Pseudocistela ceramboides*; they mostly live in cavities in trees (cf. chapter 18).

1.9.4. Mordellids

Adult mordellids of the genera *Tomoxia* and *Mordella* are easily recognised by their laterally compressed black bodies ending in a long point extending beyond the tips of the elytra. Mordellids move by jumping in an apparently clumsy way, which is why they are also known as "tumbling flower beetles". Their larvae live in more or less decayed wood.

1.9.5. Pyrochroids

The family Pyrochroidae, or cardinal beetles, has only a single genus in France, *Pyrochroa*, which owes both its Latin and English names to the bright red colour of the more familiar species, the head being black in one species (figure 16.19). The larvae are flattened and live under damp detached

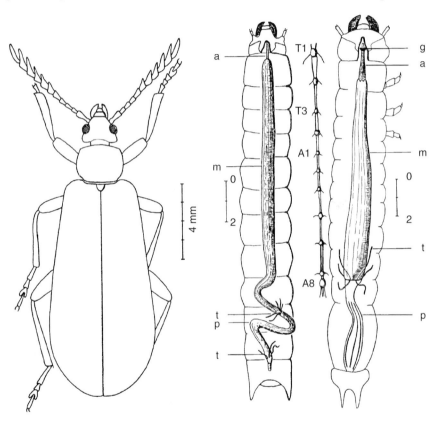

Figure 16.19. *Left: Adult* Pyrochroa coccinea. *Centre:* Pytho depressus *larva and structure of the alimentary tract. Right:* Pyrochroa coccinea *larva, alimentary tract and nervous system. a: foregut; m: midgut; p: hindgut; t: Malpighian tubes; g: cerebral ganglia; T_1, T_3, A_1, A_8: ganglia of the nervous system.*

bark of dead trees where they are sometimes abundant. Their diet is partly lignivorous and detritivorous and partly carnivorous. They are readily identifiable by their large size (25 to 30 mm), their flattened bodies with large heads and the two spikes at the tip of the last abdominal segment.

Unlike larvae that are entirely or almost entirely xylophagous, *Pyrochroa* larvae have a short alimentary tract that is scarcely longer than the body.

1.9.6. Pythids

The family Pythidae is represented principally by *Pytho depressus*, a metallic blue beetle that lives under the bark of various conifers in mountain areas. Its larvae resemble those of *Pyrochroa*, but they appear to be ligni- vorous. Notwithstanding this the alimentary tract is short and almost rectilinear and shows no adaptation to the specialised diet. The closely related family Oedemeridae contains species that live in decayed wood, such as those of the genus *Xanthochroa*.

1.9.7. Tenebrionids

The enormous family Tenebrionidae (darkling beetles) forms a group of ground-dwelling beetles that is well represented in arid regions but also contains a few species that live in decayed dead wood. *Melasia culinaris* is fairly widespread in dead oak and beech wood; *Helops coeruleus* is found mainly in the South of France in a wide range of trees. *Tenebrio obscurus*, a close relative of the meal-worm (*Tenebrio molitor*) lives in very decayed wood.

1.9.8. Anthribids

This small family is close to the Curculionidae. Most species are xylo- phagous. Adults are found on old trees and, more rarely, on flowers of trees such as hawthorn. The larvae bore winding galleries in dead wood invaded by fungi. *Platyrrhinus resinosus* (figure 16.12) lives in beech wood attacked by the fungus *Ustulina vulgaris* and in ash wood attacked by another fungus, *Daldinia concentrica*. Chestnut wood invaded by *Sphaeria stigma* is inhab- ited by *Choragus sheppardi*. Anthribids are not uncommon in bundles of kindling wood. Species of the genus *Anthribus* are unusual in that they are predators whose larvae live under the shields of lecanine scale insects. *Anthribus nebulosus*, which is fairly common in the whole of France, lives by predating *Eulecanium coryli* and *E. corni*, and also attacks *Physokermes piceae* and *P. hemicryphus*.

This list of beetle families ought also to include brenthids, micromalthids, cupedids etc., but detailed knowledge of the diets and enzyme equipment of these beetles is almost non-existent.

1.10. Non-xylophagous saproxylic beetles

Besides xylophagous beetles, the main families of which have just been listed, dead wood harbours species belonging to many other families and with very varied diets. To list these families would entail mentioning almost every family of Coleoptera.

1.10.1. Smaller families

Many of the smaller families have members associated with scolytids (cf. chapter 15) and others that cohabit with xylophagous insects. *Colydium elongatum*, belonging to the family Colydiidae, has a cylindrical body that enables it to crawl into xylophagous insect galleries. The commonest member of this family is *Bitoma crenata*, a black and red species that feeds on many micro-arthropods under bark. Several species belonging to the family Cucujidae that live under bark have flattened bodies and are probably detritivores or predators. *Uleiota planata*, several species of *Laemophloeus*, *Dendrophagus crenatus* and *Prostomis mandibularis* are the most noteworthy (figure 16.20). The genus *Cucujus* contains species that are on the

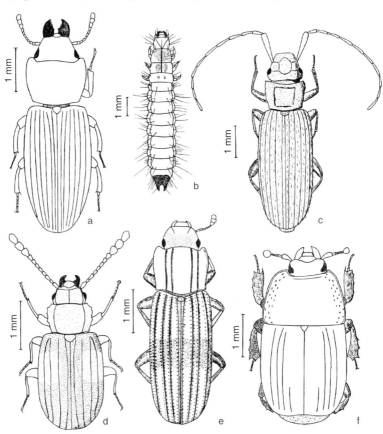

Figure 16.20. *A few Coleoptera that live under bark. a and b:* Tenebroides mauritanicus *(Trogositidae), adult and larva; c:* Uleiota planata *(Cucujidae); d:* Laemophloeus monilis *(Cucujidae); e:* Bitoma crenata *(Colydiidae); f:* Platysoma compressum *(Histeridae).*

way to extinction in Europe but are still abundant in North America; they are red beetles with bodies up to 20 mm long. Their larvae are predatory, live under bark and resemble those of *Pyrochroa*. This striking morphological convergence between species belonging to different families is due to adaptation to life in the very special sub-cortical habitat. Many species of the family Staphylinidae are found under bark and in dead wood, even after scolytids have left it. Their diets are mostly poorly known; some are predators and others detritivores.

1.10.2. Elaterids

Beetles of the family Elateridae play a major role in the dead wood habitat. They are elongate species mostly between 10 and 20 mm long and sometimes larger. They are known as click beetles because they are able to jump with a spring-like click when on their backs. Larvae are of two types. The first, known as wire-worms, are cylindrical, brown and smooth and with the ninth abdominal segment furnished with spikes of different shapes according to genus. This type of larva characterises the genera *Ampedus* and *Melanotus* which are common in dead wood. The other type of larva has a more flattened body, is lighter in colour, and the ninth abdominal segment is emarginate and furnished with numerous denticles. This type of larva is seen in genera of the French fauna such as *Adelocera*, *Lacon* and *Athous* (figure 16.21).

Elaterids whose larvae live in dead wood are at least partly predatory. There are five larval instars spread over two to four years according to species and depending on environmental conditions. Young larvae are saprophagous and eat ligneous debris, moulds and dead insects; older larvae attack the larvae of xylophagous insects such as cerambycids and buprestids. *Adelocera punctata* is a common species that mainly lives in oak wood; it is almost black and 13 to 20 mm long. *Ludius ferrugineus* is a large uniformly brown species that lives mainly in rot-holes in trees (cf. chapter 18). Most species of the genus *Ampedus* are recognisable by their red or yellow elytra sometimes adorned with black spots. Adults of many elaterids do not leave the old trees in which they lived as larvae. Some, such as *Cardiophorus gramineus*, are to be found feeding on pollen on the flowers of various trees on warm days; others, such as *Stenagostus villosus*, are nocturnal and wander at night on the trunks of old trees. Several species of saproxylic Elateridae have become rare and threatened. An inventory of endangered species has been drawn up for Denmark (Martin, 1989) and is probably valid for France (Dajoz, 1962). Among those species that are threatened or may already have vanished are *Lacon lepidopterus*, *Limoniscus violaceus*, *Ludius ferrugineus* (= *Elater ferrugineus*) and *Ampedus quadrisignatus*.

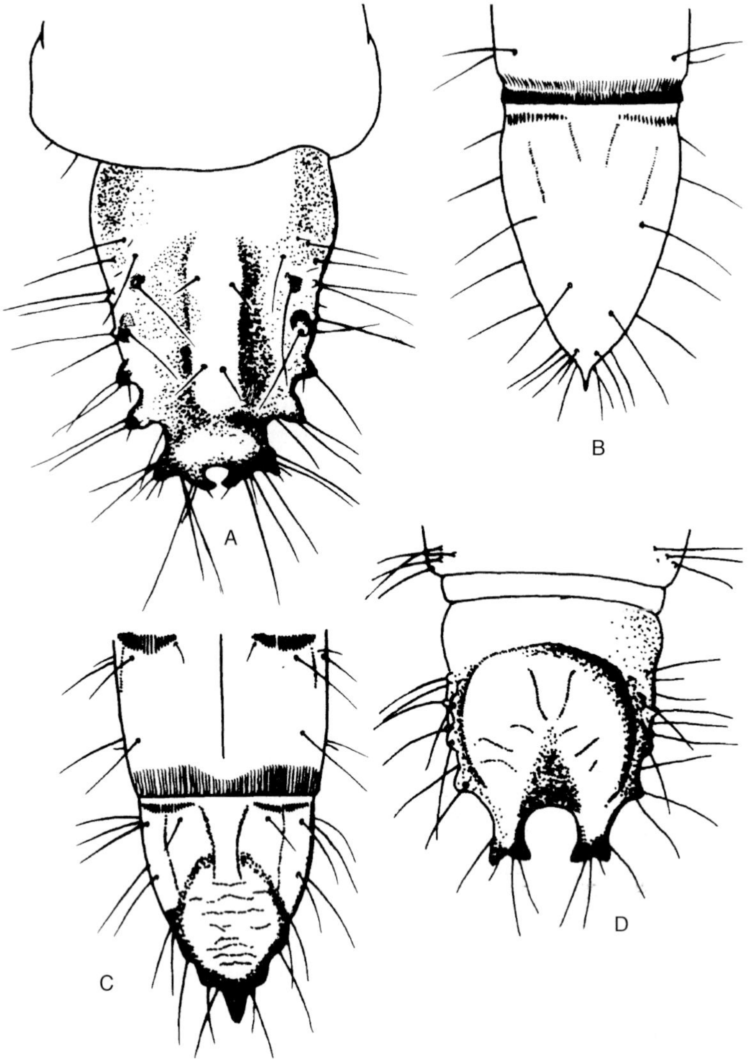

Figure 16.21. *Apex of abdomen of a few elaterid larvae. A:* Lacon fasciatus*; B:* Ampedus *sp.; C:* Melanotus rufipes*; D:* Athous hirtus.

2. DIVERSITY OF SAPROXYLIC FLIES (DIPTERA)

The order Diptera contains many species whose larvae live in wood. These species belong to 45 different families (Krivosheina & Mamaev, 1967; Teskey, 1976). Larvae able to attack the wood of living trees are few but include, for example, tipulids of the genus *Tipulodina*. All other saproxylic Diptera probably only colonise dead wood that is already well invaded by other insects and by fungi. These flies are often the dominant invertebrate group in small-sized woody debris (e.g. twigs) of oak and beech in England, and of these, tipulids such as *Tipula flavolineata*, which was found in 39% of branches, are the most abundant (Swift *et al.*, 1976).

Dipterous larvae vary greatly, but are readily divided into three large groups or sub-orders: Nematocera, Brachycera and Cyclorrhapha. Nematocera larvae have a well-developed head with prominent mouthparts. In some tipulids however the head may be partly retracted into the prothorax; the head is small and weakly sclerotised with rudimentary or no mouthparts in cecidomyiids, and is lacking in hyperoscelids. Larvae are devoid of eyes and antennae are rudimentary, and the body is usually cylindrical and elongate. Brachycera larvae exhibit various stages of head reduction and retraction into the prothorax. All Cyclorrhapha larvae have a characteristically shaped and reduced head capsule.

2.1. Nematocera

2.1.1. Tipulids

This family of flies, the crane flies, is the one that plays the greatest role in the dead wood habitat. Larvae of species that live underground have well-developed anal lobes, but these are progressively reduced in species that live in wood (figure 16.22). Tipulid larvae are of the metapneustic type, which means that only the last pair of abdominal stigmata is functional; the head is well-developed but sunken in the prothorax; legs are lacking as in all dipterous larvae; the tegument is soft and white and most often glabrous or nearly so. The larvae of *Ctenophora ornata*, which live in wood invaded by white rot fungi, require wood with a water content equal to 240% of dry weight for optimal growth. *Gnophomyia* larvae live in fermenting sap flows found on bark before the onset of decay. Saproxylic tipulid larvae have various predators including the larvae of the elaterid beetles *Melanotus rufipes* and *Denticollis linearis* and the fly *Xylophagus ater*. Adult tipulids are mostly typically long-legged, slow-flying crane flies that frequent damp undergrowth in woodland.

2.1.2. Cecidomyiids

Cecidomyids are best known as gall midges (cf chapter 12.4), but some species lay their eggs in the heart of freshly cut tree trunks and branches. Willows harbour a dozen species of the genus *Rhabdophaga*. Several species live under the decayed bark of dead trees. The larvae, which measure less than 5 mm, are often yellow, orange or reddish; the head is small, with rudimentary mouthparts. In addition to the divisions between the thoracic and

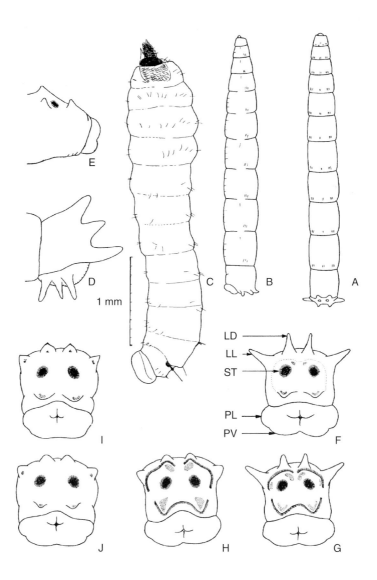

Figure 16.22. *Morphology of tipulid larvae. A and B:* Tipula paludosa, *a ground-dwelling species, in dorsal and lateral views. C:* Tanyptera atrata, *a xylophagous species, lateral view. D and E: anal segment of* Tipula luteipennis, *a ground-dwelling species, and* Tanyptera atrata, *a xylophagous species, in lateral views; F to J: hind views of the anal segment of* Dictenidia bimaculata, Ctenophora pectinicornis, Ctenophora ornata, Tipula flavolineata *and* Tipula atrata. *LD: dorsal lobe; LL: lateral lobe; ST: stigma; PL: lateral anal papilla; PV: ventral anal papilla.*

the abdominal segments the body usually clearly shows other subdivisions. The respiratory system is of the peripneustic type, which means that all abdominal stigmata are functional. The most characteristic feature of these larvae is the I- or T- shaped "sternal spatula" on the ventral surface of the prothorax of older larvae. The adults are small midges whose antennomeres bear verticils of variously shaped appendages.

The most remarkable of the cecidomyiids of dead wood are *Miastor* species. The female *Miastor metraloas* lays eggs that are remarkable for their large size and small number (four to five). The larvae that hatch from these very soon produce between five and forty eggs that produce other larvae just like their mothers. Several of these "paedogenic" generations succeed each other from autumn until spring under the bark of dead trees. This type of asexual reproduction ceases when climatic conditions improve in spring and adults emerge. Paedogenesis continues indefinitely in a tree stump subjected to constant temperature and high humidity and shaded from light.

Species living in dead wood have been recorded in a dozen other families of Nematocera, including culicids, bibionids, mycetophilids, sciarids, chironomids and ceratopogonids.

2.2. Brachycera

2.2.1. Asilids

This vast family includes species whose predatory adults are known as robber flies. They are large flies, often with very hairy bodies (figure 16.23);

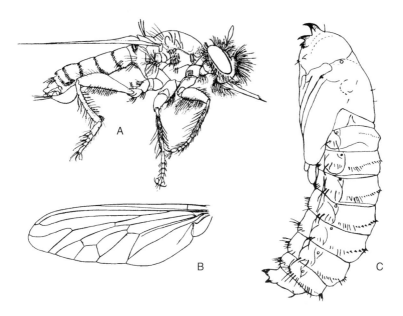

Figure 16.23. *Morphology of asilids. A: male* Laphria fulva; *note the proboscis. B:* Laphria ignea *wing. C:* Laphria gilva *pupa.*

the mouthparts form a short, hard, heavily sclerotised proboscis that enables these flies to pierce the very tough teguments of such insects as buprestids. The asilid larvae that live in dead wood belong to the subfamily Laphriinae (larvae of other subfamilies live underground). Laphriinae larvae show morphological convergence with cerambycid beetle larvae; they are white, more or less cylindrical, with a black head partly sunk into the prothorax and prolonged posteriorly by a dorsal metacephalic plate (figure 16.24). The labrum is triangular and very elongate, as are the mandibles, while the maxillae are broad and toothed, and are used to bore galleries; the antennae are reduced.

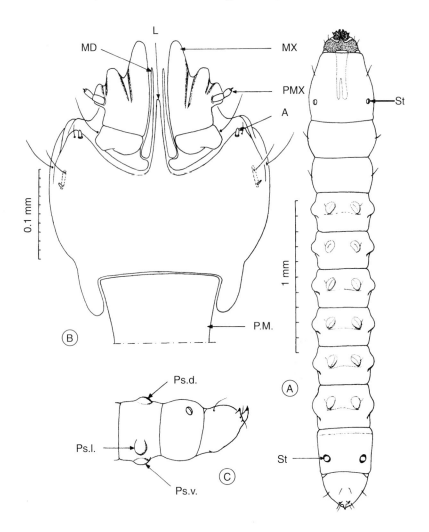

Figure 16.24. *An asilid larva of the genus* Laphria. *A: dorsal view of whole larva. B: head in dorsal view. C: tip of abdomen in lateral view. St: stigmata; MD: mandible; MX: maxilla; PMX: maxillary palp; A: antenna; PM: metacephalic plate; L: labrum; Ps.d., Ps.l., Ps.v.: dorsal, lateral and ventral pseudopods.*

There are four pairs of turgescent pseudopods on each of the first six abdominal segments and the last segment bears apical hooks. The only functional stigmata are those of the prothorax and the seventh abdominal segment. *Laphria* larvae are found in the decayed wood of various trees. They bore galleries or use those of xylophagous Coleoptera. Although asilid larvae are regarded as predators, particularly of cerambycid and buprestid larvae, many species that live in dead wood may have partly or wholly phytophagous diets (Melin, 1923; Musso, 1978).

2.2.2. Xylophagids

Larvae of *Xylophagus* species live under detached bark and in decaying wood. They are carnivorous, preying on other insects, and their life cycle lasts two years. These larvae are easily recognised by their black, elongate, conical and strongly sclerotised heads; one or two of the thoracic segments bear black plates. The last abdominal segment ends in a sclerotised plate furnished with two short processes (figure 16.25). The larvae reach a length of 2 cm.

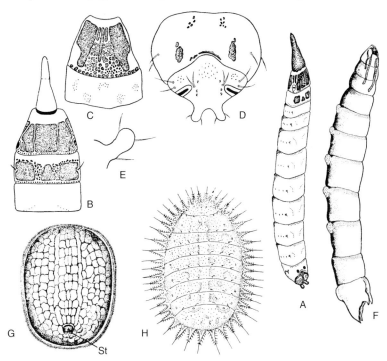

Figure 16.25. *Larvae of saproxylic Diptera.* Xylophagus ater *(Xylophagidae)*
A: view of entire larva; B: front part of body in dorsal view;
C: pro- and mesothorax; D: tip of abdomen in dorsal view;
E: appendages of last abdominal segment. Dolichopus *sp.*
(Dolichopodidae) F: entire larva in lateral view. Microdon *sp.*
(Syrphidae) G: dorsal view with stigmata in St. Callomyia *sp.*
(Platypezidae) H: dorsal view.

2.2.3. Xylomyids

This family includes the genus *Solva* whose larvae are variously considered to be predators, xylophagous or detritivorous.

2.2.4. Pantophthalmids

Pantophthalmids live in South American forests. They are very large flies with bodies up to 50 mm long and wingspans of 95 mm. Their eggs, which resemble small wheat grains, are laid in batches under bark or in an insect gallery and hatch into cylindrical larvae whose bodies bear two rows of stout spines on the hind segments. These larvae bore galleries that may fill with sap, forcing the larvae to live as amphibians. Special organs at the hind end of the body then act as gills. Wood fibres detached by the gnawing action of the maxillae are ejected outside the galleries by a shield formed by the last abdominal segment.

2.2.5. Dolichopodids

This family includes species which live in dead wood as predators, especially of scolytid larvae. The last abdominal segment of these cylindrical fly larvae is obliquely truncate and bears four protuberances.

2.3. Cyclorrhapha

2.3.1. Syrphids

The family Syrphidae, also known as hoverflies, are mainly represented in dead wood by members of the subfamily Milesiinae, which are either saprophagous or xylophagous. They are found under bark or in galleries bored in the wood. *Myiatropa*, *Xylota* and *Criorhina* species live in rot-holes and in decaying tree trunks. *Fernandinea* exploits sap flowing from wounds. Syrphid larvae vary in shape. Those of the subfamily Milesiinae have more or less truncate front ends and the tip of the abdomen extended into a respiratory siphon; the body is covered with numerous, often stellate, clusters of hairs. The larva of *Microdon* is hemispherical, with a reticulate tegument devoid of any trace of segmentation and two stigmata borne on a short stalk. This aberrant morphology explains how it came to be described as a mollusc (figure 16.25). It lives in dead trees invaded by ants, perhaps as a commensal of the latter. Adult syrphids are sun-loving flies and are often brightly coloured, either in patterns of black and yellow or with a metallic sheen. They are active fliers and many feed on flowers. According to Kula (1997) the syrphids of central European spruce forests are unusual in that woods subjected to industrial atmospheric pollution (especially sulphur dioxide) are richer in species than unpolluted or little-polluted woodland.

2.3.2. Agromyzids

These flies are represented in wood by species of the genus *Dendromyza* such as *D. cambii* and *D. tremula*, which feed in the cambium of trees such as poplars. The larval gallery may reach a length of one metre. These galleries open the wood to penetration by micro-organisms that exploit fermenting sap.

2.3.3. Lonchaeids

Lonchaeid flies include the genus *Lonchaea* whose larvae live under the bark of many dead or dying trees. Some species predate scolytids, while others are saprophagous.

2.3.4. Aquatic xylophagous Diptera

The greatest variety of aquatic xylophagous insects is certainly to be found among the Diptera, ten families of which are represented in North American rivers (Dudley & Andersen, 1982); the main ones are the chironomids, mycetophilids and sciarids. The discovery of chironomid larvae that live in submerged wood is a recent one, but xylophagous species do seem to be fairly numerous in this family (Anderson *et al.*, 1984; Cranston, 1982). *Stenochironomus* and related genera form a group of 65 species whose larvae are highly modified by their life style; they have flattened heads, dilated thoraxes and elongate movable abdomens, making them look rather like buprestid larvae. Among other chironomids that mine submerged wood are representatives of the subfamily Orthocladiinae such as *Chaetocladius*, *Orthocladius*, *Symphiosiocladius* and *Limnophyes*.

The large size of members of the family Tipulidae makes them the best known group of aquatic xylophagous flies; a whole range of genera that live in wood has been described. *Gnophomyia* lives in fermenting sap under bark before the wood begins to decay; *Ctenophora* and *Epiphragma* penetrate relatively hard wood; *Lipsothrix* only bores in partially submerged wood. Aquatic xylophagous Diptera show one or more of the following evolutionary adaptations: long larval life, which compensates a slow growth rate; symbiotic intestinal flora; ability to seek a more nutritious medium than dead wood during their final larval instars. Xylophagous species have also been found in other insect orders that have aquatic larvae, including the Trichoptera, Plecoptera, Ephemeroptera and helodid Coleoptera (Anderson, 1982; Anderson *et al.*, 1984; Dudley & Anderson, 1982; Wisseman & Anderson, 1987; O'Connor, 1991).

3.   SAPROXYLIC HYMENOPTERA

The vast order Hymenoptera, which is at least as rich in species as the Coleoptera, contains few species that live in wood.

3.1. Siricid wasps

3.1.1. Biology of siricids

The biology of siricids has been the subject of much research (Wolf 1968, 1969). The subfamily Xyphydriinae contains the genus *Xyphydria* whose larvae live in the wood of a great many trees: alder, poplar, willow, oak, birch, elm and pine. Adults are characterised by a long neck, and, as in all wasps of the suborder Symphyta, the thorax is fused to the abdomen. The subfamily Siricinae is composed of larger wasps. The commonest species are *Sirex noctilio*, *Sirex juvencus*, *Urocerus gigas* and *Xeris spectrum*. Siricids are only rarely abundant or harmful in areas in which they are indigenous. They may attack healthy trees when their populations grow large, but normally they are secondary pests that settle on trees weakened by other species or that have suffered from drought. Pine trees planted too densely may be attacked, as is building timber and wood stocked in sawmills. In Europe and the Mediterranean region the host trees are various species of pine and fir, as well as cedar, spruce and larch. Preferences vary according to species. *Sirex noctilio* prefers pines and chooses standing trees, while *Urocerus gigas* looks for firs and fallen trees.

Egg-laying takes a different form in each species of siricid. *Urocerus gigas* females drill a single relatively long shaft in which they deposit several eggs one on top of the other with the symbiotic fungus massed between them. The average length of the ovipositor is 17.5 mm and drilling lasts 11.4 minutes; shafts are 9.7 mm deep and an average of 4.4 eggs are laid per shaft. In *Xeris spectrum* the female bores a rather deep central shaft leading to up to 5 galleries; the ovipositor has an average length of 22.3 mm; drillings lasts 46.1 minutes; the average length of holes is 12.7 mm and 4.3 eggs are deposited in each gallery. This species does not have symbiotic fungi, but lays in trees already invaded by other siricids. In *Sirex noctilio* there are up to four perforations leading out of a common well; the symbiotic fungus is deposited in a special gallery without eggs. The average length of the ovipositor is 12.4 mm, drilling lasts 9.4 minutes, the average length of galleries is 4.2 mm and the average number of eggs per gallery is 1.6 (figure 16.26).

The duration of siricid life cycles varies, as low temperatures trigger diapause. *Urocerus gigas* completes its cycle in one year in Spain and in four years in Belgium. Life cycles are shorter in trees of a small diameter and last two or more years in those of a large diameter. A typical three year cycle with ten larval instars was described in Belgium.

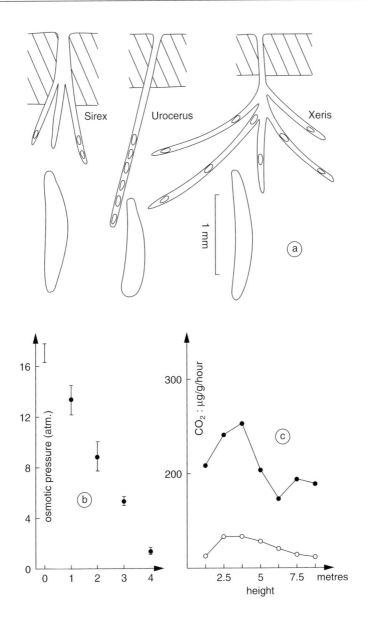

Figure 16.26. *Siricid biology. a: oviposition and egg shapes in* Sirex noctilio, Urocerus gigas *and* Xeris spectrum. *b: oviposition in* Sirex noctilio *as a function of the osmotic pressure of secondary phloem sap. Figures 1, 2, 3 and 4 correspond to the number of perforations leading from a common shaft. c: tissue respiration under* Pinus radiata *bark at different heights in control trees (circles) and in trees infested seven days before (black dots) (Madden, 1974, 1977; Spradberry, 1977).*

The main parasites of siricids are ichneumonids such as *Rhyssa persuasoria*, *R. amoena* and *Megarrhyssa emarginata*, as well as two ibaliids, *Ibalia leucospoides* and *I. rufipes*. Levels of parasitism may reach 40% in some cases.

3.1.2. Biology of Sirex noctilio

The accidental introduction of this species in New Zealand and then in neighbouring regions resulted in its becoming a serious pest in the *Pinus radiata* plantations that cover large expanses there. This led to much research on this wasp. Females lay in trees in which the osmotic pressure of the secondary phloem sap is reduced. When osmotic pressure is above 12 atmospheres the female makes a single perforation per oviposition; when osmotic pressure falls, the number of perforations per well rises up to four (figure 16.26). Females are attracted to trees suffering from physiological deficiencies by volatile substances in the cambium. These may be terpenic substances similar to those that attract scolytids.

Once an attack on a *Pinus radiata* tree by *Sirex* has begun, physiological changes in the tree take place that open up new parts of the tree to invasion by the wasp. Transpiration increases, as does respiration of phloem; cell permeability rises and so do levels of evaporation of monoterpenes and water through the bark. All these characteristics in the tree induce *Sirex* to come and lay. Moreover, high concentrations of resin and polyphenols accrue around the wasps' oviposition holes, and the affected tissues necrose, which causes formation of a new layer of wood outside the affected area.

As they lay, female *Sirex noctilio* inject spores of the symbiotic fungus *Amylostereum* as well as mucus produced by accessory glands of the genital apparatus, a feature that is not found in other European siricids. The mucus has toxic properties and is responsible for the changes undergone by the tree, these changes being a reduction of radial growth of branches, a lowering of starch reserves in needles, and a change in colour of the foliage which becomes senescent and falls prematurely. Moreover the mucus impedes water circulation in the conductive tissues. This mucus is a polysaccharide which appears to be essential for growth of the wasp's symbiotic fungus (Coutts, 1969).

Chemical control of *Sirex noctilio* has proved impossible because the wasp lives in the wood as a larva and because its adult life is short. Twenty one species of parasitoids were introduced in New Zealand and then in Australia, five of which became established, the most widespread being the ichneumonids *Ibalia leucospoides*, *Megarrhyssa nortoni* and *Rhyssa persuasoria* (Taylor, 1976). Hundreds of millions of nematodes were spread in infested Australian forests with encouraging results; these are the two species *Deladenus siricidicola* and *D. wilsoni*, which effectively stop ovarian growth in the wasp.

3.2. Ants

A few species of ant build their nests in dead wood; none seems to invade living trees. Excavated wood is ejected in the form of more or less fine sawdust. The most remarkable of these ant genera is *Crematogaster*. The species *C. scutellaris* is common in cork oak woods in Provence, Corsica, Sardinia and Morocco. When they are disturbed, these ants are able to raise their abdomens almost above their heads by means of a special articulation (figure 16.27). *Crematogaster scutellaris* measures 4 to 5 mm, and has a

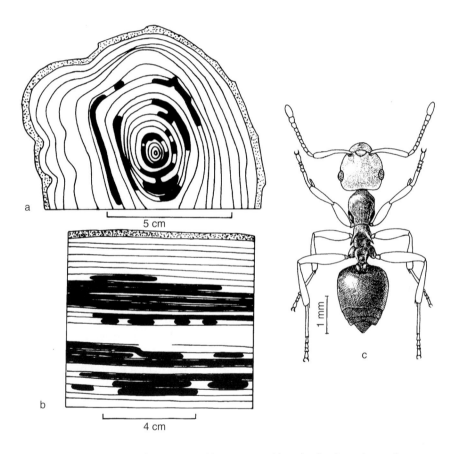

Figure 16.27. *Arboreal ants. a and b: cross and longitudinal sections of a spruce trunk showing nest galleries of* Camponotus herculeanus. *c:* Crematogaster scutellaris *worker.*

light red head and prothorax with the rest of the body steel blue. This ant builds its nest mainly in and under oak bark, and the workers are very aggressive. The cavities they excavate are closed with a sort of papier maché which the workers make from wood pulp with the help of their cephalic glands. In the case of cork oak the ants prefer trees with thick bark; the ant becomes a pest when it damages cork by riddling it with holes. *C. scutellaris* is an omnivorous species that exploits scale insects and aphids for honey-dew. Commensals are common in its nests, particularly the cricket *Myrmecophila acervorum*.

Camponotus herculeanus, which is black with reddish-brown spots, reaches a length of 15 mm and is the largest European ant. It lives in forests except in the Mediterranean region. The nest is excavated in wood by the workers; it may occupy a space several metres long inside dead trees. Fairly regular walls separate numerous galleries. The ant's diet includes various insects and detritus, but also wood which is digested by symbiotic fungi and bacteria in its gut. *Camponotus vagans* is a completely black, closely related species which is common in the Mediterranean region where it also nests in old trees. *Camponotus* species never settle in wood invaded by the mycelium of fungi, but only in healthy, more or less decayed wood.

4. XYLOPHAGOUS MOTHS

Xylophagous species are only known to occur in four families of Lepidoptera: Hepialidae, Cossidae, Tortricidae and Sesiidae. All bore in living wood, often beginning with phloem and then moving to the sapwood and finally the heartwood. Penetration by micro-organisms in the wake of the larvae contributes to the deterioration of trees.

4.1. Cossids

The two main species of this family are *Cossus cossus*, the goat moth, and *Zeuzera pyrina*, the leopard moth, which feed on many trees including several forest species. The adult goat moth is nocturnal, reaches a wingspan of 80 mm, and lives 10 to 15 days. Larvae grow over a period of two to three years and have 11 to 14 instars depending on climatic conditions. They reach a length of 10 cm, have a black head, the dorsal surface of the body is pink-ish or brownish and the ventral surface is greenish-yellow. Their presence is betrayed by the piles of reddish, fetid droppings that are ejected from the galleries and fall to the base of the tree. These caterpillars tolerate a tem-perature of – 20 °C during hibernation. They live in physiologically deficient trees as well as in dead ones. Leopard moth caterpillars reach 60 mm and have a bright yellow body with black spots and shiny black head and pro-thorax.

These two moths do not possess digestive enzymes able to attack cellu-lose and can only use starch, pectin and inulin, and cellobiose and soluble sugars such as saccharose. The many micro-organisms (bacteria and yeasts) in the gut convert cellulose to utilisable cellobiose. *Cossus* possesses two fungi, *Mucor spinosus* and *Fusarium rubiginosum*, that attack cellulose *in vitro* to free cellobiose, glucose and glycuronic acid.

4.2. Tortricids

A few xylophagous species occur in this family. *Laspeyresia pactolana* lives on fir, spruce and pine, and may in some cases do damage in young spruce plantations (Baurant, 1968). Three other related species, *L. duplicana*, *L. coniferana* and *L. cosmophorana* have been observed with it on young spruce (Postner, 1957). Larval galleries of these four species are similar. Each female lays about forty eggs which are deposited at the base of verticils on the proximal ends of branches. All verticils may be infested but those nearer the ground are preferred. The whitish or pale red caterpillars grow to 11 mm; they bore short, winding, irregular galleries filled with faeces and silk threads under bark. Their presence is seen as a slight swelling of the bark and the presence of piles of brownish-black faeces. Areas damaged by caterpillars are characterised by the brown discoloration and soft consistency of underlying layers of wood, which prevents formation of scar tissue. This may be due to infestation by the fungus *Nectria cucurbitula* whose mycelium penetrates by way of the larval galleries. Adults are olive brown moths with a wingspan of 15 mm. The torticid moths *Rhyacionia buoliana* and *Petrova resinella* which are mentioned in chapter 12 (devoted to gall insects) and in chapter 13 (devoted to cone insects) may also be regarded as xylophagous species.

4.3. Sesiids

Sesiids, or clearwing moths, are wasp mimics. They have transparent wings, a long abdomen ringed with white or yellow and black, and are fast fliers. Their larvae bore galleries in wood. The hornet clearwing, *Sesia apiformis*, resembles a hornet, has a wingspan of 40 mm, and lives in poplar, ash, lime and birch; its larvae live two years. The clear underwing moth, *Paranthrene tabaniformis*, feeds on poplar and is occasionally a pest.

One xylophagous moth belonging to the small family Hepialidae, is known in North America. It is *Sthenopis quadriguttatus*, the ghost moth, whose larvae bore galleries in the roots of aspen, poplar and willow and may cause damage in poplar plantations. The large moth flies in the evening.

5. TERMITES

Termites, which make up the insect order Isoptera, play a major part in the tropics (Collins, 1989). France only has three species (Clément, 1977).

They are not strictly forest insects; rather they are insects able to attack cellulose in different forms such as wood or paper.

The genus *Reticulitermes* belonging to the family Rhinotermitidae contains two French species. *R. santonensis*, which was originally localised around La Rochelle, is gradually colonising neighbouring areas. This termite is resistant to cold and drought and lives out of doors as well as in buildings in Charente-Maritime, which is the most heavily affected area. The other species is *R. lucifugus*, which is divided into four geographically isolated subspecies, the limits of range of each being apparently determined by climate. *R. lucifugus banyulensis* occupies Roussillon, and *R. lucifugus grassei* South West France; *R. lucifugus corsicus* is confined to Corsica, and *R. lucifugus lucifugus* lives in Provence and Italy. This species is found in towns as well as in rural areas. Wooden buildings are infested, and few species of tree are spared except perhaps false acacia (*Robinia*). Colonies of the termites have been reported as far as Bourges and Paris. *Kalotermes flavicollis* (family Kalotermitidae) is a less synanthropic, Mediterranean species which colonises tree stumps, dead trees and old vine stocks but only rarely buildings. Damage to timber has however been reported in Venice. In France this species is distributed mainly in Languedoc. Combating termites, and particularly *R. lucifugus santonensis*, is difficult; wood invaded by the fungus *Lenzites trabea* is attractive to the termites and may be used as a trap when it is impregnated with insecticide. Wood invaded by *Ganoderma applanatum* is toxic to some termites, a fact that might be used to control these species.

Termites are more numerous and more diverse in North America. *Hodotermes* are large species that mainly colonise damp wood. Their soldiers have large heads and powerful mandibles up to 20 mm long. These species are common in dead trees in the western United States. The family Kalotermitidae contains species that are more often found in dry wood and particularly in building timber. Members of the family Rhinotermitidae live underground and may also cause much damage to buildings.

Lower termites (belonging to all families except the Termitidae) live symbiotically with hypermastigine flagellates that phagocytise and digest wood fragments ingested by the termites. Higher termites of the family Termitidae only possess symbiotic bacteria and are devoid of flagellates. *Microcerotermes edentatus* is an example of one of these higher termites (figure 16.28). Its hindgut has two rectal pouches, the second of which is voluminous and swarming with bacteria. *Microcerotermes* utilise the cellulose and hemicelluloses of healthy wood as well as those of wood attacked by fungi, and are also able to break down lignin.

Termites that eat wood containing 62% cellulose, 6% pentosans (a group of hemicelluloses) and 26% lignin produce faeces that contain 10% cellulose, 4% pentosans and 65% lignin, which shows that there is a large attack on polysaccharides and little or no attack on lignin. According to some authors however, one species of *Reticulitermes* may digest 80% of lignin in wood.

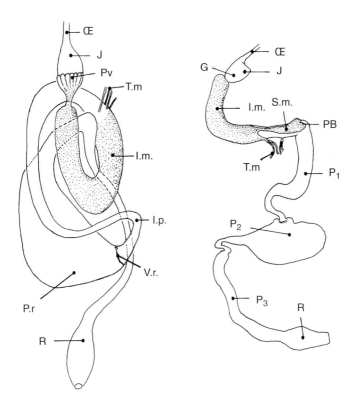

Figure 16.28. *Alimentary tract of two termites. Left: a lower termite of the genus Eutermes. Right: the higher termite Microcerotermes edentatus. Midgut stippled. Œ: oesophagus; J: crop; Pv: proventriculus; Tm: Malpighian tubules; Im: midgut; Ip: hindgut; Vr: rectal valve; R: rectum; G: gizzard; Sm: mixed segment; PB: bacteria pouch; P_1, P_2, P_3: three segments of hindgut; P_2: pouch; P_3: colon; Pr: rectal pouch.*

REFERENCES

ANDERSEN J., NIELSSEN A.-C., (1983). Intrapopulation size of free-living and tree-boring Coleoptera. *Canad. Ent.*, **115**: 1,453-1,464.

ANDERSON N. H., (1982). A survey of aquatic insects associated with wood debris in New Zealand streams. *Mauri Ora*, **10**: 21-33.

ANDERSON N. H., STEEDMAN R. J., DUDLEY T., (1984). Patterns of exploitation by stream invertebrates of wood debris (xylophagy). *Verh. Internat. Verein Limnol.*, **22**: 1,847-1,852.

BAURANT R., (1968). Note sur *Laspeyresia pactolana* Zell.: la pyrale des verticilles de l'épicéa. *Bull. Rech. Agron. Gembloux*, **3**: 217 -225.

BISTRÖM O., VÄISÄNEN R., (1988). Ancient-forest invertebrates of the Pyhän-Häkki national park in Central Finland. *Acta. Zool. Fenn.*, **185**:1-69.

BLANCHARD E., (1868). *Mœurs et métamorphoses des insectes.* Baillière, Paris.

BOVEY P., (1953). Le bupreste vert (*Agrilus viridis* L.) ravageur du hêtre en Suisse. *Bull. Soc. Ent. Suisse*, 152-153.

CARLE P., (1973). Le dépérissement du pin mésogéen en Provence. Rôle des insectes dans les modifications d'équilibre biologique des forêts envahies par *Matsucoccus feytaudi* Duc. (Coccoidea, Margarididae). Thesis, Université de Bordeaux.

CAUSSANEL C., DAJOZ R., (1967). Morphologie et biologie d'un Coléoptère des plages sableuses: *Callicnemis latreillei* Cast. (Scarabaeidae, Dynastinae). *Cah. Nat.*, **23**: 25-37.

CLÉMENT J.-L.-A., (1977). Écologie des *Reticulitermes* français (Isoptères). Position systématique des populations. *Bul. Soc. Zool. Fr.*, **102**: 169-185.

COLLINS N. M., (1989). *Termites. In: Ecosystems of the world. Tropical rain forest ecosystems,***14B**: 455-471.

COUTTS M. P., (1969). The mechanism of pathogenicity of *Sirex noctilio* on *Pinus radiata*. I. Effects of the symbiotic fungus *Amylostereum* sp. (Thelephoaceae). *Austr. J. Bot. Sci.*, **22**: 915-924. II. Effects of *S. noctilio* mucus. *Austr.. J. Bot. Sci.*, **22**: 1,153-1,161.

CRANSTON P. S., (1982). The metamorphosis of *Symphiosiocladius lignicola* (Kieffer) n. gen; n. comb., a wood mining Chironomidae (Diptera). *Entomol. Scand.*, **13**: 419-429.

DAJOZ R., (1962). Les espèces françaises du genre *Ampedus*. *Rev. Fr. Ent.*, **20**: 5-26.

DAJOZ R., (1965). Catalogue des Coléoptères de la forêt de la Massane. *Vie et Milieu*, supplément, 209 pp.

DAJOZ R., (1990). Diptères et Coléoptères du pin à crochets dans les Pyrénées-Orientales. Étude biogéographique et écologique. *L'Entomologiste*, **46**: 253-274.

DAJOZ R., (1993). Croissance des antennes, sex ratio, variations de taille et dimorphisme sexuel chez les Coléoptères Cerambycidae. *Cahiers des Naturalistes*, **49**: 15-38.

DEYRUP M. A., (1975). The insect community of dead and dying Douglas-fir. I. The Hymenoptera. *Ecosyst. Anal. Stud. Bull.*, **6**: 1-104.

DEYRUP M. A., (1976). The insect community of dead and dying Douglas-fir : Diptera, Coleoptera and Neuroptera. Ph. D, University of Washington, 540 pp.

DUDLEY T., ANDERSON N. H., (1982). A survey of invertebrates associated with wood debris in aquatic habitats. *Melanderia,* **39**: 2-21.

DUFFY E. A. J., (1953). *A monograph of the immature stages of British and imported timber beetles (Cerambycidae).* British Museum, London.

FISHER R. C., (1941). Studies on the biology of the death watch beetle *Xestobium rufovillosum* De Geer. IV. The effect of type and extent of fungal decay on the rate of development of the insect. *Ann. Appl. Biol.,* **28**: 244-260.

FORCELLA F., (1982). Why twig-girdling beetles girdle twigs. *Naturw.,* **69**: 398-399.

HICKIN N. E., (1963). *The insect factor in wood decay.* Hutchinson, London.

JACQUIOT C., (1976). Tumors caused by *Agrilus biguttatus* Fab. attacks on the stems of oak trees. *Marcellia,* **39**: 61-67.

JOLY R., (1975). *Les insectes ennemis des pins,* 2 volumes. École Nationale du génie rural et des eaux et forêts, Nancy.

KRIVOSHEINA N.-P., MAMAEV B.-M., (1967). *Identification keys to larvae of Diptera that live in wood* (in Russian). Moscow, Nauka Books.

KULA E., (1997). Hoverflies (Dipt.: Syrphidae) of spruce forest in different health condition. *Entomophaga,* **42**: 133-138.

LINSLEY E. G., (1959). Ecology of Cerambycidae. *Ann. Rev. Ent.,* **4**: 99-138.

LINSLEY E. G., (1961). *The Cerambycidae of North America.* Part I. Introduction. University of California Press, Berkeley.

MADDEN J. L., (1977). Physiological reactions of *Pinus radiata* to attack by woodwasp, *Sirex noctilio* F. (Hym. Siricidae). *Bull. Ent. Res.,* **67**: 405-425.

MAMAEV B.-M., (1977). Biology of xylophagous insects (in Russian). *Itogi Nauki i Tekniki, ser Entomologia,* **3**: 1-213.

MARTIN O., (1989). Click beetles (Coleoptera, Elateridae) from old deciduous forests in Denmark. *Ent. Meddr.,* **57**: 1-107.

MASON W. H, ODUM E. P.,. The effect of coprophagy in retention and bioelimination of radionucleides of detritus feeding animals. *In:* D. J. NELSON, F. C. EVANS (1969). *Proc. Second Nat. Symp. Radioecology,* 721-724.

MELIN D., (1923). Contribution to the knowledge of the biology, metamorphosis and distribution of the Swedish Asilids. *Zool. Bidrag Uppsala,* **8**: 1-317.

MUONA J., VIRAMO J., (1986). The Coleoptera of the Koillismaa area (Ks), North east Finland. *Oulanka Reports,* **6**: 3-50.

MUSSO J.-J., (1978). Recherches sur le développement, la nutrition et l'écologie des Asilidae (Diptères, Brachycera). Thèse, Université Aix-Marseille.

O'CONNOR N., (1991). The effects of habitat complexity on the macroinvertebrates colonising wood substrates in a lowland stream. *Œcologia,* **85**: 504-512.

Paim U., Becker W.-E., (1963). Seasonal oxygen and carbon dioxide content of decaying wood as a component of *Orthosoma brunneum* (Forster) (Coleoptera: Cerambycidae). *Canad. J. For. Res.*, **41**: 1,133-1,147.

Palm T., (1951). Die Holz und Rindenkäfer der nordschwedischen Laubbaüme. *Medd. fran. Stat. Skogforsk*, **40**: 242 pp.

Palm T., (1959). Die Holz und Rindenkäfer der süd und mittelschwedischen Laubbaüme. *Opus. Ent.*, suppl. **16**, 374 pp.

Parent E., (1996). *Étude sur les Cerambycidae saproxyliques dans le Parc Naturel Régional du Haut Jura.* (Unpaginated booklet)

Penney M. M., (1967). Studies on the ecology of *Feronia oblongopunctata* (F.) (Coleoptera: Carabidae). *Trans. Soc. Br. Entomol.*, **17**: 129-139.

Postner M., (1957). Beitrag zur Kenntnis der Rindenwickler *Laspeyresia duplicana* Ztt., *L. coniferana* Rtz., und *L. cosmophorana* Tr. (Lep. Tortricidae). *Zeit. ang. Ent.*, **41**: 312-319.

Rees C., (1986). Skeletal economy in certain herbivorous beetles as an adaptation to a poor dietary supply of nitrogen. *Ecol. Ent.*, **11**: 221-228.

Saalas U., (1917-1923). Die Fichtenkäfer Finlands. *Ann. Acad. Sc. Helsinki*, **8**, 546 pp. and **23**, 746 pp.

Savely H. E., (1939). Ecological relations of certain animals in dead pines and oak logs. *Ecol. Monog.*, **9**: 327-385.

Schmitt M., (1992). Buchen-Totholz als Lebensraum fur xylobionte Käfer : Untersuchungen im Naturwaldreservat "Waldhaus" und zwei Vergleichs-flachen im Wirtschaftswald. *Waldhygiene*, **19**: 1-192.

Semenova L.-M., Danilevsky M.-L., (1977). Structure of digestive system in larvae of longhorn beetles. *Zool. Zhurn.*, **56**: 1,168-1,174.

Spradberry J.-P., (1977). The oviposition biology of siricid woodwasps in Europe. *Ecol. Ent.*, **2**: 225-230.

Steedman R. J, Anderson N. H., (1985). Life history and ecological role of the xylophagous aquatic beetle, *Lara avara* LeConte (Dryopoidea: Elmidae). *Freshw. Biol.*, **15**: 535-546.

Swift M. J, Healey I. N., Hibberd J. *et al.*, (1976). The decomposition of branch-wood in the canopy and floor of a mixed deciduous woodland. *Œcologia*, **26**: 139-149.

Teskey H. J., (1976). Diptera larvae associated with trees in North America. *Memoirs. Entomol. Soc. Canad.*, **100**: 1-53.

Tragardh I., (1930). Some aspects in the biology of longhorn beetles. *Bull. Ent. Res.*, **21**: 1-8.

Villiers A., (1978). *Faune des Coléoptères de France. I. Cerambycidae.* Éditions Lechevalier, Paris.

Vuattoux R., (1968). Le peuplement du palmier rônier (*Borasssus aethiopium*) d'une savane de Côte-d'Ivoire. *Ann. Univ. Abidjan*, sér. E, **1**: 1-138.

WISSEMAN R. W., ANDERSON N. H., (1987). The life history of *Cryptochia pilosa* (Trichoptera: Limnephilidae) in an Oregon coast range watershed. *Proc. 5 th. Int. Symp. Trichoptera*, 243-246.

WOLF F., (1968). Éthologie des Siricidae. *Bull. Ann. Soc. R. Belg.*, **104**: 427-457.

WOLF F., (1969). Les Siricidae en Belgique. Leurs mœurs et leur importance en sylviculture. *Bull. Soc. Roy. For. Belgique*, 1-39.

WOLF F., (1969). Sur les ennemis des Hyménoptères Siricides. *Bull. Ann. Soc. R. Ent. Belg.*, **105**: 202-224.

17

Wood decay and insect successions

In "*The pattern of animal communities*" Elton (1966) points out that "Dying and dead wood provides one of the two or three greatest resources for animal species in a natural forest... if fallen timber and slightly decayed trees are removed the whole system is gravely impoverished of perhaps more than a fifth of its fauna".

Wood amounts to a large proportion of a forest's biomass, closely followed by foliage. But while leaves are renewed every year and decay rapidly, within a few months at most, wood accumulates and is much less easily degradable owing to its chemical composition. Decomposition takes several years, sometimes even several decades. In the great conifer forests of the north-western United States it takes 480 to 530 years for 90% of a Douglas-fir trunk 80 cm in diameter to decompose. This demonstrates the great stability of the saproxylic environment. Where trees grow on poor soils and under harsh climatic conditions such as drought they grow very slowly and to a very great age. Wood formed in this way is hard and survives for several millennia without being attacked by fungi or insects. This is what has occurred in the bristle cone pine (*Pinus aristata*) forest of the Sierra Nevada in California, where dead tree trunks several thousand years old with very hard wood and very thin growth rings lie strewn on the ground without any visible trace of insects or fungi.

The presence of various forms of woody debris is an important element in maintaining biodiversty in forests. Many animals depend on dead wood, not only insects but also many other invertebrates and vertebrates; the latter have been studied in detail in the conifer forests of western North America. Dead wood in these ancient forests provides indispensable habitats to several dozen species of vertebrate (Franklin *et al.*, 1981; Maser & Trappe, 1984; Thomas, 1988). Decaying tree trunks provide suitable environments for germinating seeds and the growth of saplings; thus they act as nurseries and contribute to natural regeneration of forest. Standing dead trees, known as snags, provide an essential habitat for many living creatures, including 63 species of vertebrate, the latter dominated by birds that nest in tree cavities and which are all insectivorous. Each species has rather narrow requirements in such things as height from the ground, diameter of the trunk and state of decomposition of the wood. Woodpeckers are a group of birds whose abundance and numbers of species present in a forest depend on forest management methods and on an abundance of dead wood (Angelstram & Mikusinski, 1994; Virkkala *et al.*, 1994). The physico-chemical characteristics of soil under fallen trees may be altered. Ausmus (1977) points out that in broadleaf forest the quantity of organic matter in soil is increased fourfold, that of the adenosine 5'-triphosphate or ATP (which is related to microbial biomass) is multiplied eight-fold, root biomass is doubled, and calcium levels are increased five-fold.

It is estimated that more than 20% of beetle species depend on dead wood in Sweden, and that 41% of threatened moss species also depend on it. Fungi, especially basidiomycetes belonging to the three families Polyporaceae, Corticiaceae and Hymenochaetaceae, play an important part in wood decay and many of them are suffering from the way forests are being managed. Sixty six of the two hundred species of Polyporaceae known in Sweden are threatened (Bader *et al.*, 1995; Gilbertson, 1984). These fungi, like some insects such as beetles of the family Elateridae, are good indicators of forest continuity and of the permanent presence of sufficient quantities of dead wood in all stages of decomposition over a very long period.

In Europe it is estimated that 1,500 species of fungus and 1,300 species of Coleoptera are dependent on dead wood. In France many basidiomycetes such as *Dryodon coralloides* and *Stereum insignitum* are now found only in the non-exploited reserves of Fontainebleau forest (Jacquiot, 1978). Myxomycetes thrive on dead wood, which is why Fontainebleau forest still has 76 species, or one quarter of the world's flora of this group (Heim, 1955).

The fauna associated with a tree changes with time according to the stage of decomposition of the wood. A progression leads from species of invertebrate more or less dependent on one species of tree to invertebrates dependent on dead wood, so that at the end of the progression the stage of decomposition becomes more important than the taxonomic status of the tree.

Early study of successions goes back to the nineteenth and early twentieth centuries. Among pioneering works are those of Adams (1915), Blackman & Stage (1924), Graham (1925), Régnier (1925) and Savely (1939). Many studies have been conducted in temperate regions; a few relate to the tropics or sub-tropics (Eidmann, 1943; Becker, 1955; Vuattoux, 1968; Moron, 1985; Torres, 1994 etc.). Successions are very diverse in nature, for they depend on many different factors such as tree species and climatic conditions.

1.   EXAMPLES OF SUCCESSIONS IN TEMPERATE REGIONS

Successions of insects, fungi and other organisms that live in dead trees in temperate regions, and resulting changes in the physical and chemical properties of wood have been much studied (Howden & Vogt, 1951; Schimitschek, 1952; Mamaev, 1961; Dajoz, 1967, 1990; Ausmus, 1977; Swift, 1977; Mazur, 1979; Heliövaara & Väisänen, 1984; etc.).

1.1. Successions in pine and oak

Insect successions in pine and in oak have been described by Savely (1939) in Pennsylvania, and by Mamaev (1960) and Mamaev and Sokolov (1960) in Russia. Four stages can be defined in oak in central Russia: the cerambycid stage, the lucanid stage, the ant stage and the earthworm stage (figure 17.1). At stage 1 the cerambycid *Leptura sexguttata* and the melandryid *Hypulus quercinus* dominate and decompose 13.5% of the wood. At stage 2 the lucanid *Ceruchus chrysomelinus*, Diptera larvae (mainly belonging to the family Lycoridae) and elaterids of the genera *Melanotus* and *Ampedus* take over. *Ceruchus chrysomelinus* decomposes 50% of the wood at this stage, with a single larva consuming 0.45 cubic centimetres of wood per day; 13 to 15% of remaining wood debris come from this beetle's faeces. At stage 3 the ants *Lasius niger, L. fuliginosus* and *Myrmica rubra* move into the wood together with other species. At stage 4 the remaining wood undergoes a process of decomposition contributed to by diplopods and earthworms; the part played by insects becomes negligible.

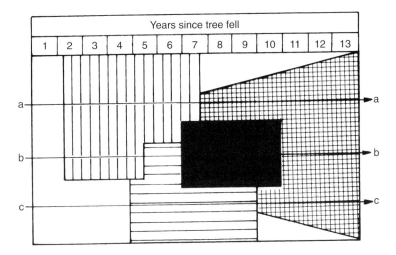

Figure 17.1. *Schematic representation of possible invertebrate decomposers in oak wood. Arrows show the direction of successions. b: complete succession; a and c: incomplete successions. Blank areas: decomposition by bacteria and fungi alone; vertical hatching: cerambycid stage; horizontal hatching: lucanid stage; black: ant stage; crossed hatching: earthworm stage (Mamaev, 1960).*

An evaluation of the role of invertebrates in pine and oak wood decay was made in the Ukraine (Mamaev, 1961). Control trunks invaded by macrofauna (mainly insects and earthworms) were left on the ground, and other trunks were protected from the macrofauna, without which bacteria and fungi alone play a part in decay. A pine trunk deprived of macrofauna decays in 12 years and an oak trunk in 20. With macrofauna present decay is completed in 7 and 12 years respectively.

Under the conditions of this particular experiment invasion of wood occurs in three stages: a stage during which cerambycids act, a second stage in which lucanids and ants take over, and a third stage in which earthworms that have come up from the soil finish the process (figure 17.2).

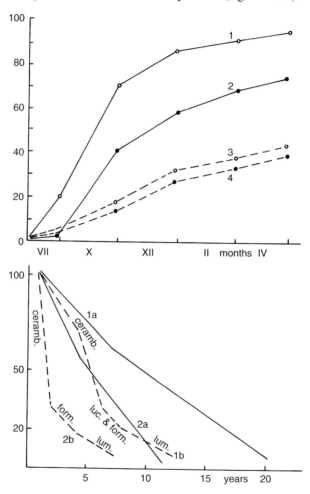

Figure 17.2. *Decay of dead plant matter by invertebrates. Above, rate of decay of oak leaves (curves 1 and 3) and beech leaves (curves 2 and 4), with (solid lines) and without (dashed lines) the help of soil invertebrates (Edward & Heath, 1963). Below: decay of oak wood aged 100 years (curves 1a and 1 b) and a 50 year old pine tree (curves 2a and 2b) with (dashed lines) and without (solid lines) the help of invertebrates (Mamaev in Ghilarov, 1967).*

1.2. Successions in beech

The faunal successions in dead beech trees are very similar to each other over a large part of Europe. A study conducted in a southern European beech wood shows that there are several stages, each of which is characterised by a particular fauna.

– First stage. The first stage of decay lasts one to four years, and may be labelled the buprestid and cerambycid stage. The first beetles to settle under bark are buprestids such as *Chrysobothris affinis, Agrilus viridis, Dicerca berolensis* and a very few scolytids such as *Taphrorychus bicolor* and *Dryocoetinus villosus* (this paucity is characteristic of broadleaf trees, which harbour fewer scolytids than do conifers in temperate regions), and the cerambycid *Rhagium bifasciatum*. These beetles' predators turn up at the same time; these are clerids such as *Pseudoclerops mutillarius*, cucujids such as various *Laemophloeus*, and colydiids such as *Bitoma crenata*. The few Coleoptera that penetrate the wood at this stage are cerambycids such as *Rhagium* and *Xylotrechus*, and eucnemids.

– Second stage. This is a cerambycid and anobiid stage. It lasts on average from three to seven years after the tree's death. The sub-cortical habitat is occupied by the few buprestids and scolytids left over from the first stage as well as by cerambycid larvae belonging to the genera *Rhagium, Leptura* and *Grammoptera*. Predators become more numerous, including cucujids of the genus *Laemophloeus*, trogositids (*Nemosoma elongatum* and *Tenebrioides fuscus*) and clerids (*Opilo mollis* and *Tillus elongatus*). It is within the wood that fauna becomes more abundant, particularly in anobiids of the genera *Anobium* and *Xestobium* and *Ptilinus pectinicornis*, and in cerambycids such as species of *Rhagium, Leptura* and *Strangalia*.

– Third stage. This lasts from six to ten years and is characterised by cerambycids and especially by lucanids and tenebrionids. *Pyrochroa* larvae often abound under bark as do the larvae of elaterids of the genera *Ampedus, Melanotus* and *Athous*. The wood is penetrated by the larvae of large cerambycids such as *Morimus asper* and *Aegosoma scabricorne*, and tenebrionids such as *Melasis culinaris* and *Helops coeruleus*, and lucanids are also common (figure 17.3).

When, after about ten years, these waves of insects have passed, what is left of the wood consists of a friable mass which Silvestri (1913) named the saproxylic complex. These complexes contain invertebrates, including Coleoptera, although the latter are no longer the dominant group. In the case of the beech wood the invertebrates are worms (lumbricids and enchytraeids), woodlice, snails, various mites and, among the insects, Diptera larvae and Coleoptera, including numerous representatives of soil fauna such as pselaphids and ptiliids.

A quantitative study of the fauna and of its role revealed a number of characteristics:

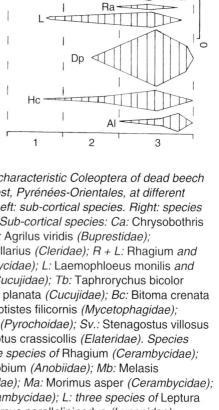

Figure 17.3. *Abundance of the most characteristic Coleoptera of dead beech wood in la Massane forest, Pyrénées-Orientales, at different stages of wood decay. Left: sub-cortical species. Right: species penetrating dead wood. Sub-cortical species: Ca:* Chrysobothris affinis *(Buprestidae); Av:* Agrilus viridis *(Buprestidae); Pm:* Pseudoclerops mutillarius *(Cleridae); R + L:* Rhagium *and* Leptura *larvae (Cerambycidae); L:* Laemophloeus monilis *and* Placonotus testaceus *(Cucujidae); Tb:* Taphrorychus bicolor *(Scolytidae); Up:* Uleiota planata *(Cucujidae); Bc:* Bitoma crenata *(Colydiidae); Pf:* Parabaptistes filicornis *(Mycetophagidae); Pc:* Pyrochroa coccinea *(Pyrochoidae); Sv.:* Stenagostus villosus *(Elateridae); Mc:* Melanotus crassicollis *(Elateridae). Species within the wood: R: three species of* Rhagium *(Cerambycidae); An: three species of* Anobium *(Anobiidae); Mb:* Melasis buprestoides *(Eucnemidae); Ma:* Morimus asper *(Cerambycidae); Ra:* Rosalia alpina *(Cerambycidae); L: three species of* Leptura *(Cerambycidae); Dp:* Dorcus parallelipipedus *(Lucanidae); Hc:* Helops coeruleus *(Tenebrionidae); Al: three species of* Alleculidae.*

Scale, above: time in years; scale, below: the three stages of decay.

– Progressive increase in biomass of insects from stage 1 to stage 3. The biomass of xylophagous Coleoptera (in grams of fresh weight per 100 kg of dry wood) is 17 g at stage 1, is 18 g at stage 2, and reaches 395 g at stage 3.

– Increase in relative numbers of predators per number of xylophagous animals.

– Increase in species richness and in the diversity index.

– As wood deteriorates, species associated with a particular species of tree become fewer and the fauna generalises to include a large number of species associated with a range of broadleaf trees or even with conifers.

An evaluation of biomass per hectare of saproxylic Coleoptera gave values of 6.34 kg for the fauna of beech and 1.62 kg for that of oak, or a total biomass of Coleoptera of 7.8 kg. This amounts to between 90 and 95% of the total biomass of the invertebrates of dead wood.

One method of ascertaining the role of insects consists in evaluating the quantity of wood ingested each day, the duration of larval life, and the abundance of the different species. In the cases of the buprestid *Chrysobothris femorata* and the cerambycid *Callidium antennatum*, each larva consumes 80 g of secondary phloem to produce 1 g of tissue (Savely, 1939). Given that a fully grown *Callidium* larva weighs on average 0.1 g, one cubic foot of phloem can feed 22 larvae, which is equivalent to one larva per 42 square centimetres. Mamaev (1960) has shown that a larva of the lucanid *Aesalus* weighing 0.09 g consumes 0.14 cubic centimetres of wood in one day, and a *Lucanus* larva weighing about 1 g consumes 22.5 cubic centimetres of wood. Among the cetonids, in the course of its lifetime a *Potosia lugubris* larva consumes a quantity of wood equal to 142 times its fully grown body weight (Pawlowsky, 1961). In 100 days a larva of the cerambycid *Ergates faber* weighing 9 g consumed about 850 g of pine wood the water content of which was 40%, corresponding to 510 grams of dry weight.

This body of data lets us assume that xylophagous species in general have an average daily consumption equal to 1/20 of their body weight. Throughout the three stages, xylophagous Coleoptera use about 38% of the original mass of wood. Wood that is not degraded by Coleoptera is degraded by other invertebrates and by fungi, which play a major role. The basidiomycete *Coriolus versicolor* may cause beech wood to lose 99% of its weight by means of enzymatic attack followed by respiratory oxidation of hydrolysed products.

1.3. Successions in fir

The fauna of fir was studied in Rotwald primary forest in Austria by Schimitschek (1952) who determined four stages defined by their fauna and by the appearance of phloem and of the cambial area which gives an idea of the state of decay of the wood:

– Stage 1. Immediately following the death of the tree, wood is little altered and the cambium has not yet changed colour. The sub-cortical and cambial areas of tree trunks exposed to sunlight are invaded by scolytids such as as *Ips typographus* and *Ips amitinus*, cerambycids of the genus *Tetropium* and numerous predators such as beetles belonging to the families Cleridae and Staphylinidae, and flies of the genus *Xylophagus* and *Lonchaea*. The bark of trunks in shaded spots harbours scolytids of the genera *Dryocoetes, Hylurgops* and *Polygraphus*, the curculionid *Pissodes piceae*, cerambycids of the genus *Tetropium* and various clerid and staphylinid predators.

Deeper layers of the wood are penetrated by larvae of cerambycids (such as *Rhagium*), lucanids (*Ceruchus chrysomelinus* and *Sinodendron cylindricum*) and elaterids (such as *Melanotus*) as well as by those of *Hylecoetus dermestoides*.

– Stage 2. Wood is in a more advanced state of decay; the secondary phloem is brown or spotted with brown; fungi invade. Various insects are still found under bark: scolytids such as *Hylurgops* and *Dryocoetes*, cerambycids such as *Rhagium, Pyrochroa*, and various carabid, staphylinid and clerid predators, and also Diptera such as syrphids and the genus *Xylophagus*. The fauna within the wood is the same as that of the first stage but begins to be enriched by soil fauna and contains more elaterids, lucanids and staphylinids.

– Stage 3. Bark is by now largely detached; lignicolous fungi such as *Fomes fomentosus* and *Ganoderma applanatum* form fruiting bodies. Wood is attacked by red rot and has a high water content. The bark shelters larvae of *Rhagium* and *Pyrochroa*, elaterids of the genera *Melanotus* and *Ampedus*, and various staphylinids and Diptera. Many small arthropods become common; ants of the genus *Myrmica* build nests; the fauna of the deeper layers of wood becomes poorer in cerambycids, but lucanids and elaterids become more abundant.

– Stage 4. Wood is now in an advanced state of decomposition; the bark falls off. Insects of dead wood are by now only represented under remaining bark by a few *Pyrochroa* and elaterid larvae and by staphylinids and Diptera. Lucanids and elaterids persist inside the wood, and there is a massive invasion of soil fauna: Collembola, Psocoptera, various Diptera, Annelida and Diplopoda.

1.4. Conifers of boreal forests

Heliövaara and Väisänen (1984) identify four stages in the decay of dead wood in the boreal conifer forest in Scandinavia. Decomposition is slow and may take over a hundred years. The duration of the different stages increases from one stage to the next, which shows that the rate of decomposition slows progressively (figure 17.4). The four stages are as follows:

– Stage A. This stage, which is of short duration, mainly involves species that feed under bark, such as scolytids and cerambycids and their parasites, predators and commensals. Many of the species of this stage are innocuous in primary forest but become pests in managed forest.

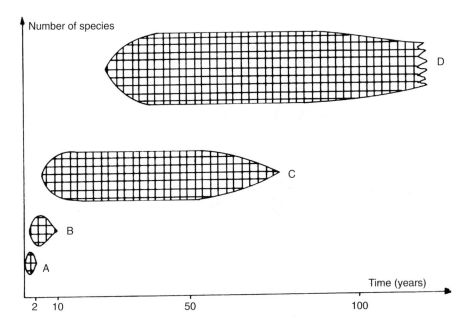

Figure 17.4. *The four stages of dead wood decay in Scandinavian boreal forest. Schematic representation after Heliövaara & Väisänen (1984).*

– Stage B. This stage also lasts only a short time. It involves sub-cortical species and those that feed in the outer layers of wood and also species that feed on fungi. It is at this stage that bark begins to peel away.

– Stage C. The majority of species involved in this long stage (several decades) live within the wood. Most of the species threatened with extinction belong to this stage, at least in Scandinavian boreal forest.

– Stage D. This stage is by far the longest one. Many of the species involved shelter under the trunk and rise from the soil. The trunk begins to fragment. Fauna associated with this stage seems to be little affected by forestry management methods, as may be the soil fauna itself.

Swift (1977) defines three stages in the successions of fauna, micro-organisms and fungi in dead wood. The colonisation stage is characterised by the death of living cells in cambium and phloem, and is dominated by the arrival of scolytids which open the way to invasion by fungi and bacteria. The most prominent fungi are *Fomes annosus* and *Armillaria mellea*. The decaying stage is characterised by dominance of fungi that cause white rot and brown rot and by invasion by new xylophagous insects, mainly cerambycids and anobiids, and the presence of the last scolytids and siricids. The terminal stage sees invasion by fungi and by various representatives of soil fauna such as Collembola, mites and spiders. These species are accompanied by predatory insects such as staphylinid beetles and Diptera.

1.5. Successions in Douglas-fir

In the conifer forests of the north-western United States trees are gigantic and levels of wood biomass are very high. In a wood consisting mainly of *Pseudotsuga menziesii, Thuya plicata* and *Tsuga heterophylla* with an average age of 450 years aerial biomass lies between 492 and 976 t ha⁻¹, and that of large-sized woody debris on the ground between 143 and 215 t ha⁻¹. Authors who have studied successions in Douglas-fir in these forests have defined five stages, and at the same time monitored variations in the characteristics of dead wood fallen to the ground (figure 17.5). During wood decay water content rises and then stabilises; nitrogen levels rise and wood density decreases.

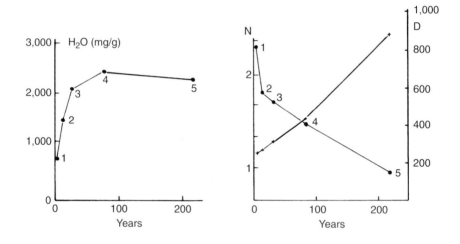

Figure 17.5. *Changes in some characteristics of dead Douglas-fir wood fallen to the ground through different succession stages numbered 1 to 5, and as a function of the duration of decomposition. Left: water content (in mg/g). Right: nitrogen content N in mg/g (crosses) and density D (black dots) in mg/cm³ (Franklin et al., 1981).*

In the case of a standing dead tree the following stages can be distinguished (Cline *et al.*, 1980):

– Stage 1. The tree has been dead less than six years. Xylophagous Coleoptera settle under the bark and open the way to lignivorous fungi; needles and twigs fall; wood and cambium remain hard and do not discolour; a few small pieces of bark may peel off.

– Stage 2. The tree has been dead between 7 and 18 years. Bark and sapwood are intensely mined by insects which penetrate the wood, causing further deterioration; small branches fall; the wood turns brown and becomes less hard, opening the way to birds which excavate nests; the upper part of the tree begins to collapse as it is attacked by insects and fungi.

– Stage 3. The tree has been dead between 19 and 50 years. The sapwood is partly decayed; insects mine the heartwood; pieces of bark and wood fall; a large proportion of the upper part of the tree collapses leaving only large limbs.

– Stage 4. The tree has been dead 51 to 125 years. Sapwood is entirely decayed and the heartwood riddled right through; no bark remains; woody debris accumulated at the foot of the tree begins to be colonised by vegetation.

– Stage 5. The tree has been dead more than 125 years. Only standing trees with a diameter of at least 50 cm ever reach this stage; bark has long fallen; remaining dead wood is increasingly invaded by vegetation; the height of standing trunks is greatly reduced.

In the case of fallen tree trunks there are five analogous stages of decay with roughly similar characteristics. Some 300 species of insect are associated with dead Douglas-fir trunks, including xylophagous beetles belonging to the families Scolytidae, Cerambycidae and Lucanidae, ants of the genus *Camponotus* and tipulid flies (Deyrup, 1975, 1976, 1981; Maser & Trappe, 1984).

Penetration of the wood by scolytid beetles opens the way to fungi, bacteria and various invertebrates such as mites and nematodes. Attacks by scolytids may be intense: 12,000 galleries and 24,000 individuals of the scolytid *Trypodendron lineatum* were counted on a tree with a bark surface of 40 m^2. Trophic networks are very complex. Eighteen months after the fall of Douglas-fir trees, abandoned *Trypodendron lineatum* galleries still contained, per 1 cm length, 10^{10} bacteria, 250 metres of fungal hyphae, 5.10^4 amoebae, 3.10^4 ciliates, 80 nematodes feeding on bacteria and 23 mycetophagous nematodes (Carpenter *et al.*, 1988).

Douglas-fir has a rich fauna. It includes 223 species of beetle including 52 cerambycids, 22 buprestids and 50 scolytids. Three species of Scolytidae are abundant and settle very soon after the death of a tree; these are *Scolytus unispinosus*, *Pseudohylesinus nebulosus* and *Dendroctonus pseudotsugae*. The number of species associated with them is high: 33 for *P. nebulosus* and 43 for *D. pseudotsugae*. The predators of *S. unispinosus* are mostly clerids such as *Enoclerus lecontei* and the colydiid *Lasconotus subcostulatus*. The larvae of the latter species feed on fungi during their first two instars and later attack the scolytid's larvae. Among the buprestids, *Buprestis aurulenta* is an abundant species in all Douglas-fir populations. The curculionid *Pissodes fasciatus* has larvae that live under the bark of recently dead trees. The cerambycid *Anoplodera crassipes* prefers partially decayed and shaded wood. Diptera are few in the early stages of decay due to the structure of their mouthparts which makes them virtually unable to gnaw wood, and due to the absence of fungi on which their larvae feed. There are large numbers of parasitic wasps in Douglas-fir populations, but only four species of ant.

1.6. Successions of Diptera in beech

Few successions of Diptera in dead wood have been described. Brauns (1954) defined three stages in the Diptera fauna of stumps of felled beech trees.

– In the three years following felling, bark frequently detaches. Fungi such as *Polyporus* and *Trametes* as well as mosses grow on the stumps, and they are invaded by molluscs and worms. In drier situation it is ants that invade the stumps. Syrphid flies of the genus *Microdon* and muscids of the genus *Fannia* are also encountered. In damp situations the Diptera larvae belong to the families Xylophagidae, Coenomyiidae, Asilidae and Therevidae. Polyporous fungi harbour cecidomyiids and tipulids of the genus *Limonia*.

– After nine or ten years the stumps are covered with moss and sometimes phanerogams. These shelter tipulids, bibionids, rhagionids and psychodids.

– Eleven to twelve years after felling the stumps are very damp. Terricolous Diptera larvae risen from the soil predominate.

2. EXAMPLES OF SUCCESSIONS IN THE TROPICS

Few studies of invertebrates in this context, and particularly of insects have been made in the tropics. Eidmann (1943) described stages in the decay of broadleaf trees in West Africa. He showed the extent of groups that are absent or poorly represented in temperate regions, particularly termites and passalid and platypodid beetles.

2.1. Decay of pines in Guatemala

Becker (1955) has described attacks on pines at high altitude in Guatemala.

– The first stage involves healthy trees invaded by primary destroyers such as scolytids of the genus *Dendroctonus* and Lepidoptera of the family Sesiidae. Cecidomyiid fly larvae appear under bark; they are followed by various scolytids and platypodids such as *Gnathotrichus sulcatus, Xyleborus scopularum* and *Platypus quadridens*. Curculionids of the genus *Cossonus* also arrive, followed finally by cerambycids of the genera *Tetropium* and *Acanthocinus*.

– In the second stage, wood is still hard but bark becomes detached. The cambial area is occupied by scolytids of the genera *Ips* and *Dendroctonus* accompanied by their predators and commensals: tenebrionids of the genus *Corticeus*, colydiids of the genus *Lasconotus*, various histerids, clerids, lycids, staphylinids and pseudoscorpions. Cerambycids such as *Callipogon barbatum* and *Astyochus tenebrosus* live in the wood, as well as several weevils of the tribe Cossonini. In the warmest regions termites belonging to several genera are also present.

– At the third stage, wood is attacked by fungi. Termites, cerambycids, buprestids and curculionids are present; they are joined by feeders on decayed wood and on mycelium as well as by predators such as elaterids, eucnemids, tenebrionids, staphylinids, carabids and histerids. Ants, scorpions, spiders and earwigs are also encountered.

2.2. Decay of dead trees in Mexico

Moron (1985) and Moron *et al.* (1988) have addressed the quantitative aspects of wood decay of various trees in areas of Mexico with a subtropical climate. These authors write of three stages, the last of which corresponds to the saproxylic complexes. The main groups of macro-Coleoptera are the Cerambycidae, Passalidae, Melolonthidae and Tenebrionidae. Each taxonomic group has its own preferences. Cerambycids of the subfamily Prioninae are mainly present in the early stage of decay (stage 1), whereas melolonthids are mainly present during stage 3 in wood that is already in an advanced state of decay. Passalids and tenebrionids are present at the intermediate stage (stage 2). Large Scarabaeoidea belonging to the families Lucanidae and Passalidae play a major part. These species do not occur in temperate regions. In a Mexican pine wood cerambycids amount to 36% of macro-Coleoptera, passalids 33%, and rutelids 31%. In a mixed broadleaf forest rutelids account for 49% of the larger Coleoptera, passalids 42% and cerambycids 9%. Melolonthids of the subfamily Rutelinae include endemic American genera such as *Plusiotis* and *Chrysina*. The larvae of these species live two years during which each individual consumes between 1.3 and 2.6 kg of wood. The average monthly quantity of pine wood eaten by *Plusiotis* is estimated at 5 kg/ha, and that of *Liquidambar* wood consumed by *Chrysina* at 2.5 kg/ha. These figures amount to 19 to 31% of wood consumed each month by all the macro-Coleoptera of these forests.

2.3. Wood decay in Puerto Rico

Characteristics of decaying *Cyrilla racemiflora* wood and the fauna that occupies it have been studied in a tropical forest in Puerto Rico (Torres, 1994). A group of 138 species of invertebrate was identified in dead wood, which is a small number. Termites dominate; they are followed by ants, and only then by Coleoptera. Parasitic Hymenoptera are absent, which may be due to the scarcity of scolytids, and siricids are also absent. There are only 3 species of cerambycid and 3 species of buprestid. These features of the fauna of a tropical area contrast with those of temperate regions, but they may be an exception due to insularity, particularly considering the paucity of the fauna. Other studies are needed to confirm the general features.

2.4. Fauna of borassus palm in the Ivory Coast

This fauna was described in detail by Vuattoux (1968). The average biomass of arthropods that inhabit these palm trees is around 2 to 3 kg per hectare, which is much lower than that of the herbaceous stratum which is estimated at 15 kg per hectare. The average biomass per dead borassus tree, excluding termites, is 275 grams.

2.4.1. Fauna of standing dead borassus palms

Attacks on these palm trees begins at the top of the stem. The pith of the upper part of the stem decays over a length of 50 cm to 1 m, leaving only a tube of fibrous bark. The first insects to arrive are Diptera; they are followed by large Coleoptera: curculionids such as *Rhynchophorus phoenicis*, and especially dynastids of the genus *Oryctes* (rhinoceros beetles). Other, less important species such as cetonids, cerambycids and lignivorous caterpillars also live in this habitat. The distribution of the fauna of standing dead borassus palms is as follows:

- xylophagous and saprophagous species: 67.5% of species and 75% of biomass
- predatory arthropods: 31% of species and 5% of biomass
- vertebrates: 1.5% of species and 20% of biomass.

Finally the entire upper part of the stem is hollowed out while the lower part is packed with humus recycled by termites. This latter habitat has a special fauna including the elaterid *Agrypnus caliginosus*.

2.4.2. Fauna of fallen dead borassus palms

Stems lying on the ground are colonised by insects attracted by fermenting sap such as nitidulid beetles and fruit flies (*Drosophila*); they are followed by curculionids and various Diptera. At the next stage fungi grow on the bark where they harbour mycetophagous beetle larvae. Ants and termites abound as well as larvae of *Oryctes* and elaterids. Numbers of individuals and biomass of this population later decrease. The stem is hollowed out and shelters various invertebrates, most of which belong to typical savannah species; vertebrates may also find a home in hollow stems.

*

The foregoing glimpse of a few of the successions that have been described and of their causes shows that the term "dead wood" is misleading, for this habitat is the centre of intense activity that is vital to the functioning of forest by recycling organic matter in the form of humus and mineral elements that are essential to forest regeneration. A great diversity of processes and species contribute to wood decay; however a few general rules emerge.

– Wood usually decays more rapidly in the tropics than in temperate regions, largely owing to the action of termites, which are present all over the tropics.

– In all cases simultaneous action by bacteria, fungi and invertebrates, including insects, is necessary. In the rare instances where insects act alone wood decay is very slow.

– There is a contrast, at least in temperate regions, between litter formed of dead leaves, which is decomposed mainly by Diptera larvae, and wood, which is mainly decomposed by Coleoptera (cf. chapter 19).

– The first species to arrive are usually those associated with one or a few closely related tree species; these insects are small, short-lived species. Species that arrive later at wood that is more or less decayed are not dependent on a host species but on a microhabitat having particular physical and chemical properties; these species are larger, have longer life cycles, and may live for several generations in the same tree.

– There are well determined successions of species governed by the nature of the tree and environmental conditions. The usual pattern is one of increasing species-richness until the time when wood is converted to a saproxylic complex containing a fauna largely originating in the soil.

3. SUCCESSIONS AND TREE AGE

Another type of succession involves changes in the fauna of woodland composed of trees that are all about the same age, and that age simultaneously. The composition of the insect fauna in such woods, involving both leaf-eaters and saproxylic species, is seen to change as the trees age (Starzyk & Witkowski, 1981; Gutowski, 1995). A study of the succession of cerambycids according to the age of woodland plots was made by Starzyk (1977), who showed that some species are associated through preference with a particular age group of trees. The most detailed study is that by Gutowski (1995) relating to the cerambycids and buprestids of pine in Bialowieza in Poland. This author distinguishes four stages in the growth of Scots pine populations: stage 1, or "cultivation" stage involving trees aged 1 to 4 years; stage 2, or "young" stage; stage 3, or "tall tree" stage; stage 4, or "mature" stage. There are major differences between the two families of Coleoptera, especially in species-richness at different stages of tree growth:

Stage	Number of species	
	Cerambycidae	Buprestidae
1	17	9
2	12	5
3	25	6
4	29	9
Total number of species	49	16

The buprestids are most numerous in species at the "cultivation" stage and the cerambycids at the "mature" stage of woodland. This may be explained by the greater exposure to light in the first stage which makes it more suitable to the sun-loving buprestids. The "young" stage has the smallest number of species of both families, because decaying trunks necessary to larval growth are still few and the bark of small branches is too thin. The "mature" stage is the richest in rare species such as *Monochamus urussovi* which is a very rare member of the European fauna.

In the family Cerambycidae males predominate in young woodland and females in old woodland. Seventy males and 27 females were counted in a young plot, and 87 males and 140 females in mature forest (Starzyk & Witkowski, 1986). Variations in sex ratio probably depend on the diameter of trunks in which the larvae live, and particularly on the thickness of bark. It appears that bark thickness affects the microclimate of the sub-cortical habitat and leads to different mortality rates in male and female larvae.

REFERENCES

ADAMS C. C., (1915). An ecological study of prairie and forest invertebrates. *Bull. Illinois State Lab.,* **11**: 31-279.

ANGELSTAM P., MIKUSINSKI G., (1994). Woodpecker assemblages in natural and managed boreal and hemiboreal forest – a review. *Ann. Zool. Fenn.,* **31**: 157-172.

AUSMUS B. S., (1977). Regulation of wood decomposition rates by arthropod and annelid populations. *Ecol. Bull.,* **25**: 180-192.

BADER P. *et al.,* (1995). Wood-inhabiting fungi and substratum decline in selectively logged boreal spruce forests. *Biol. Cons.,* **72**: 355-362.

BECKER G., (1955). Grundzüge der Insektensuccession in *Pinus* Arten der Gebirge von Guatemala. *Zeit. ang. Ent.,* **37**: 1-28.

BLACKMAN M. W., STAGE H. M., (1924). On the succession of insects living in the bark and wood of dead, dying and decaying hickory. *N.-Y. State Coll. Forestry*, Tech. Publ. **24**: 1-269.

BRAUNS A., (1954). Die Sukzession der Dipterenlarven bei der Stockhumifizierung. *Zeit. Morph. Okol. Tiere,* **43**: 313-320.

CARLE P., (1971). SLes phénomènes présidant aux successions d'insectes dans le dépérissement du pin maritime du Var. *Ann. Zool. Écol. Anim.,* no. spécial, la lutte biologique en forêt, 177-192.

CARPENTER S. E., HARMON M. E., INGHAM E. R. *et al.,* (1988). Early patterns of heterotroph activity in conifer logs. *Proc. Roy. Soc. Edinburgh,* **94B**: 33-43.

DAJOZ R., (1967). Écologie et biologie des Coléoptères xylophages de la hêtraie. *Vie et Milieu,* **17**, série C: 523-763.

DAJOZ R., Les insectes xylophages et leur rôle dans la dégradation du bois mort. *In*: P. PESSON (1974). *Écologie forestière.* Gauthier-Villars, Paris, 257-307.

DAJOZ R., (1990). Coléoptères et Diptères du pin à crochets dans les Pyrénées-Orientales. Étude biogéographique et écologique. *L'Entomologiste,* **46**: 253-270.

DELGADO-CASTILLO L., DELOYA C., (1990). Observaciones sobre los macro-coleopteros saproxilofilos de los bosques tropicales subcaducifolios de Acahuizotla, Guerrero, Mexico. *Fol. Ent. Mexicana,* **80**: 281-282.

DEYRUP M. A., (1975). *The insect community of dead and dying Douglas-fir. 1. The Hymenoptera.* Bulletin no. 6. Coniferous forest biome, Univ. Washington.

DEYRUP M. A., (1976). The insect community of dead and dying Douglas-fir: Diptera, Coleoptera and Neuroptera. Ph. D., Univ. Washington.

DEYRUP M. A., (1981). Deadwood decomposers. *Nat. Hist.,* **90**: 84-91.

EDWARD C. A., HEATH G. W., The role of soil animals in breakdown of leaf material. *In:* J. DŒKSEN, J. VAN DER DRIFT (1963). *Soil Organisms.* North Holland Publishing Company, 76-84.

EIDMANN H., (1943). Successionen westafrikanischer Holzinsekten. *Mitt. Akad. Dtsch. Forstwiss.,* **1**: 241-271.

ELTON C. S., (1966). *The pattern of animal communities.* Methuen, 1966 (dying and dead wood pp. 279-305).

FAGER O. S., (1964). The community of invertebrates in decaying oak wood. *J. Anim. Ecol.,* **37**: 121-142.

FRANKLIN J. F. et al., (1981). *Ecological characteristics of old growth Douglas-fir forests.* USDA forest service, General Tech. Rep. PNW-118, 1-48.

GHILAROV M. S., (1967). Abundance, biomass and vertical distribution of soil animals in different zones. *Secondary productivity of terrestrial ecosystems,* **2**: 611-629.

GILBERTSON R. L., Relationships between insects and wood rotting Basidiomycetes. *In:* Q. WHEELER, M. BLACKWELL (1984). *Fungus-insect relationships. Perspectives in ecology and evolution.* Columbia Univ. Press, New York, 130-165.

GRAHAM S. A., (1925). The felled tree trunk as an ecological unit. *Ecology,* **6**: 397-411.

GUTOWSKI J.-M., (1995). Changes in communities of longhorn and buprestid beetles (Coleoptera: Cerambycidae, Buprestidae) accompanying the secondary succession of the pine forests of Puszcza Bialowieska. *Frag. Faun.,* **38**: 389-409.

HEIM R., (1955). Inventaire et raretés mycologiques de la forêt de Fontainebleau. *Trav. Nat. Forêt Fontainebleau*, **12**: 45-48.

HELIÖVAARA K., VÄISÄNEN R., (1984). Effects of modern forestry on northwestern European forest invertebrates: a synthesis. *Acta For. Fenn.*, **189**: 1-32.

HOWDEN H. F., VOGT G.-B., (1951). Insect communities of standing dead pine (*Pinus virginiana* Mill.). *Ann. Ent. Soc. Amer.*, **44**: 581-595.

JACQUIOT C., (1978). *Écologie des champignons forestiers*. Gauthier-Villars, Paris.

KÄÄRIK A.-A.,. Decomposition of wood. *In*: C.-H. DICKINSON, G.-J.-F. PUGH (1974). *Biology of plant litter decomposition*. Academic Press, London, 129-174.

MAMAEV B.-M., (1961). Activity of larger invertebrates as one of the main factors of natural destruction of wood (in Russian). *Pedobiologia*, **1**: 38-52.

MASER C., TRAPPE J.-M., (1984). *The seen and unseen world of the fallen tree*. U S Dept Agric., General technical report PNW-164: 1-56.

MAZUR S., (1979). Beetle succession in feeding sites of the pine shoot beetle (*Tomicus piniperda* L.) (Coleoptera Scolytidae) in one-species and mixed stands. *Memorab. Zool.*, **30**: 63-87.

MORON M.-A., (1985). Observaciones sobre la biologia de dos especies de Rutelinos saproxilofagos en la Sierra de Hidalgo, Mexico (Coleoptera: Melolonthidae: Rutelinae). *Folia Ent. Mexic.*, **64**: 41-53.

MORON M.-A., (1985). Los insectos degradadores, un factor poco estudiado en los bosques de Mexico. *Folia Ent. Mexic.*, **65**: 131-137.

MORON M.-A., VALENZUELA J., TERRON R.-A., (1988). La macro-coleopterofauna saproxilofila del Soconusco, Chiapas, Mexico. *Folia. Ent. Mexic.*, **74**: 145-158.

PAWLOWSKY J., (1961). Lamellicorns cariophags in forest biocenosis in Poland. *Ekol. Polska*, sér. A, **9**: 355-437.

RÉGNIER R., (1925). Du rôle des insectes dans la désorganisation d'un arbre. La faune entomologique des peupliers. Thesis Université de Paris.

SAVELY H. E., (1939). Ecological relations of certain animals in dead pine and oak logs. *Ecol. Monog.*, **9**: 321-385.

SCHIMITSCHEK E., (1952-1953). Forstentomologische Studien im Urwald Rotwald. *Zeit. ang. Ent.*, **34**: 178-215 ; **34**: 513-542 ; **35**: 1-54.

SIMANDL J., (1993). The spatial pattern, diversity and niche partitioning in xylophagous beetles (Coleoptera) associated with *Frangula alnus* Mill. *Acta Œcol.*, **14**: 161-171.

STARZYK J.-R., WITKOWSKI Z., (1981). Changes of the parameters describing the cambio- and xylophagous insect communities during the secondary succession of the oak-hornbeam associations in the Niepolomice forest near Krakow. *Zeit. ang. Ent.*, **91**: 525-533.

STARZYK J.-R., WITKOWSKI Z., (1986). Dependence of the sex ratio of cerambycid beetles (Col. Cerambycidae) on the size of their host trees. *J. Appl. Ent.*, **101**: 140-146.

SWIFT M. J., (1977). The ecology of wood decomposition. *Sci. Prog.*, **64**: 175-199.

THOMAS D. W., (1988). The distribution of bats in different ages of Douglas-fir forests. *J. Wildl. Manag.*, **52**: 619-626.

TORRES J. A., (1994). Wood decomposition of *Cyrilla racemiflora* in a tropical montane forest. *Biotropica*, **26**: 124-140.

VUATTOUX R., (1968). Le peuplement entomologique du palmier rônier (*Borassus aethiopum*) d'une savane de Côte-d'Ivoire. *Ann. Univ. Abidjan.*, sér. E: Écologie, **1**: 1-138.

18

Two special habitats: tree holes and fungi

Forests have two habitats that are closely bound up with dead wood: cavities in trees, and fungi. These two habitats, which survive more and more precariously in Europe, are of great importance for they contain species with very unusual biology.

1. TREE HOLES

Tree holes constitute a habitat that is both very unevenly distributed and ephemeral, although cavities in willows and chestnut may last more than a hundred years.

1.1. Water-filled holes or dendrotelms

Dendrotelms are formed in trees with hard, compact wood either in the axial part of the trunk or at a fork in the trunk. They are a common habitat in the tropics but much less so in temperate regions. The wood of some trees such as beech, lime, horse chestnut and oak is conducive to the formation of such cavities; that of elm and conifers is less so.

Dendrotelms harbour a special fauna that may be divided into three categories: species that are dependent on this habitat, those that favour the habitat but not exclusively, and those that occur in it accidentally. Study of these species goes back a long way (Keilin, 1927; Mayer, 1938). The list of European species belonging to the first two categories is a short one (Kitching, 1971); it includes only about a dozen species all but one of which are Diptera, as follows:

– Species found exclusively in dendrotelms:

Diptera, Culicidae: *Aedes geniculatus; Anopheles plumbeus; Orthopodomyia pulchripalpis*

Diptera, Chironomidae: *Metriocnemus martinii*

Diptera, Ceratopogonidae: *Dasyhelea lignicola, Dasyhelea dufouri*

Diptera, Syrphidae: *Myatropa florea*

Coleoptera, Helodidae: *Prionocyphon serricornis*

– Species not found exclusively in dendrotelms:

Diptera, Tipulidae: *Ctenophora pectinicornis*

Diptera, Psychodidae: *Pericoma fagocavatica; Pericoma* sp.

Diptera, Ceratopogonidae: *Bezzia* sp.

Diptera, Anisopodidae: *Mycetobia* sp.; *Anisopus fenestralis*

Diptera, Anthomyiidae: *Phaonia mirabilis*

This list should be expanded to include dolichopodid Diptera of the genus *Systenus* that live in cavities in broadleaf trees as well as in those in conifers (Vaillant, 1978).

The greater part of the fauna of dendrotelms of cork oak in Morocco consists of Diptera larvae of the family Culicidae. The dominant species is *Aedes echinus*, three generations of which may breed each year in the same cavity. Culicids are predated by other Diptera of the families Chironomidae and Culicidae. *Culex pipiens* moves into a cavity when it begins to dry out; as dendrotelms continue to dry they are colonised by soil fauna species in search of sufficient humidity to survive the summer (Metge & Belakoul, 1984; Villemant & Fraval, 1991).

The dendrotelm habitat is characterised by the nature of substances dissolved in the water, such as humic acids and potassium salts. Dendrotelm water is strongly alkaline; only insects that are physiologically adapted can live in them, which explains why the habitat is so poor in species.

The number of species present in dendrotelms is a function of the stability of climate and particularly of temperature. A comparison of dendrotelms in England and in Australia illustrates this. In England, climate is variable and winters are cold; in Australia, climate is much more stable. There are only two trophic levels in the fauna of dendrotelms in England whereas there are four in Australia. All but one dendrotelm species in England are Diptera, whereas in Australia this fauna includes a frog and several mites (figure 18.1).

1.2. Dry holes filled with wood mould

The wood mould contained in cavities in trees results from wood decayed by various fungi which cause brown rot, such as *Polyporus sulfureus*, and insects play an additional part. The physicochemical properties of such wood mould are those of very degraded wood and resemble those of humus, and particularly those of mull. Levels of organic nitrogen are low in healthy wood; they rise in decaying wood and become very high in this wood mould.

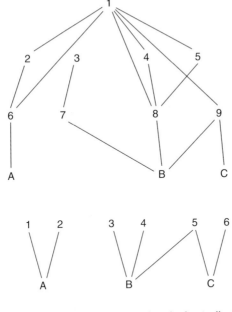

Figure 18.1. *Trophic networks in dendrotelms in Australia (4 trophic levels) and in England (2 trophic levels). A: large-sized organic debris; B: small-sized organic debris; C: organic matter in suspension. Australian fauna. 1: frog* Lechriodus fletcheri*; 2: chironomid* Anatopynia pennipes*; 3: ceratopogonid* Culicoides angularis*; 4: culicid* Aedes sp.*; 5: acarian* Arrhenurus sp.*; 6: helodid beetle* Prionocyphon niger*; 7: acarian* Clodgia sp.*; 8: culicid* Aedes sp.*; 9: ostracod crustaceans. English fauna: 1: syrphid* Myatropa flavea*; 2: helodid* Prionocyphon sp.*; 3: chironomid* Metriocnemus sp.*; 4: ceratopogonid* Dasyhelea sp.*; 5: culicid* Aedes sp.*; 6: culicid* Anopheles sp. (Kitching, 1981 in Pimm, 1982).*

Simultaneously the C/N ratio, which is very high in healthy wood, gradually falls (table 18.1).

	Organic nitrogen (%)	Ammoniacal nitrogen (%)	Nitric nitrogen (%)	Carbon (%)	Organic matter (%)	Ash (%)	C/N
Healthy wood	0.24	28	12	41.2	97.9	2.1	171.7
Dry rot	0.31	0	32	48	98.6	1.4	154.8
Damp rot	0.47	0	56	48.9	97.8	2.2	104
Wood mould	0.77 - 0.99	16 - 20	64 - 88	40.5 - 43.8	87 - 94.8	1.3 - 5.2	44.2 - 51.3
Soil	0.33	-	-	0.65	-	-	2.2

Table 18.1. *Chemical composition of healthy wood, decaying wood and wood mould of chestnut compared with the chemical composition of soil (Kelner, 1967).*

The microclimate in tree holes is akin to that in wood. The thermal regime depends on the position of the cavity in relation to the ground, on the amount of sunshine it receives, on the quantity of wood mould and the latter's humidity. Variations in outside temperature are greatly attenuated in cavities owing to high levels of humidity. Water reserves that may evaporate in smaller or greater quantities act as a thermal control. High levels of colloids in the mould determine its high water retention which in turn acts as a thermal regulator. The fauna of rot-holes is highly hygrophilous in nature.

1.2.1. Tree hole fauna

– In willow and chestnut tree wood mould the best represented families belong to the order Coleoptera followed a long way behind by the Diptera. (Kelner, 1967). Abundance of the different families is as follows:

Alleculidae: 51% and six species including *Pseudocistela ceramboides, Prionychus ater* and *Allecula morio.*

Elateridae: 32% and ten species including *Ludius ferrugineus, Cardiophorus gramineus* and various *Ampedus.*

Scarabaeoidea: 4% and the single species *Gnorimus octopunctatus.*

Cucujidae: 3% and the single species *Prostomis mandibularis.*

Tenebrionidae: 1% and two species, *Helops coeruleus* and *Tenebrio opacus.*

Others: 2% including a few carabids, pselaphids and silphids.

The best represented family of Diptera is the Phoridae with four species of the genus *Aphiochaeta*; other families present are the Tipulidae, Therevidae and Empididae.

Where chestnut tree wood mould is very damp and rich in fragments of decaying wood the two dominant species are *Gnorimus octopunctatus* and *Prostomis mandibularis.* As soon as the wood mould contains a high proportion of fine particles but remains rich in organic matter and has a C/N ratio between 40 and 50 the dominant species are *Ludius ferrugineus* and *Pseudocistela ceramboides.*

– The fauna of beech tree cavities is very similar. In the southern French beech wood of la Massane the dominant species belong to the families Cetoniidae, Alleculidae and Elateridae; the fauna found in this forest is listed in table 18.2. Together with the larger species listed (at least one centimetre long) were thirteen species of micro-Coleoptera belonging to eight families: Scydmaenidae, Pselaphidae, Ptiliidae, Clambidae, Cryptophagidae, Histeridae, Erotylidae and Endomychidae.

Rot-holes in beech trees in Fontainebleau forest harbour a fauna that appears rather richer. It includes elaterids (*Limoniscus violaceus, Procraerus tibialis, Megapenthes lugens, Ischnodes sanguinicollis, Ampedus* sp., *Melanotus* sp.), cerambycids (*Rhamnusium bicolor, Aegosoma scabricorne, Necydalis ulmi*), curculionids (*Rhyncolus truncorum and R. lignarius*),

Species	Frequency % (ten samples)
Constants (constancy > 50 %)	
Alleculidae *Prionychus ater*	14
Allecula morio	23
Cetoniidae *Osmoderma eremita*	7.4
Cetonia aurata	4.9
Elateridae *Ampedus rufipennis*	2.8
Ampedus corsicus	10.2
Ludius ferrugineus	8.2
Melanotus tenebrosus	8.2
Occasionals (constancy 25 to 50 %)	
Alleculidae *Prionychus fairmairei*	2
Lucanidae *Dorcus parallelipipedus*	1.7
Accidentals (constancy < 25 %)	
Cetoniidae *Gnorimus nobilis*	
Alleculidae *Hymenalia rufipes*	21.3
Others (13 species of small Coleoptera)	

Table 18.2.   *Main species of Coleoptera found in the wood mould of ten beech tree rot-holes.*

cetoniids including *Osmoderma eremita*, and the lucanid *Dorcus paralleli-pipedus*.

– Tree holes form in broadleaf trees where they harbour a characteristic fauna in all temperate regions. According to Park & Auerbach (1954) the fauna of dry rot-holes in the United States is very varied. The Arthropoda is the most richly represented group, in the following proportions:

Taxonomic group	Biomass (in g per kg of wood mould)	Frequency (in %)	Constancy (in %)
Acari	1.515	61.20	100
Collembola	0.674	26.96	100
Coleoptera	0.080	3.20	100
Diptera	0.079	3.16	88
Other insects	0.080	3.20	5 to 83
Other arthropods	0.049	1.96	5 to 44

In the southern United States rot-holes in trees such as plane and oaks contain larvae of various large scarabaeoid beetles such as *Dynastes granti* and *Dynastes tityus*. The type of tree hole that harbours *Osmodera eremicola* in the south-eastern United States is the same as those in which *Dynastes tityus* lives, but the two species are almost allopatric, their ranges only narrowly overlapping, which suggests interspecific competition between the larvae of these two species (Glaser, 1976).

1.2.2. *Biology of a few species*

– Elateridae. The biology of elaterids that live in tree holes in oak and beech has been studied, in particular in Fontainebleau forest (Iablokoff, 1943). Some species such as *Limoniscus violaceus* and *Ischnodes sanguinicollis* live in tree holes at ground level. Others such as *Megapenthes lugens, Procraerus tibialis* and *Ludius ferrugineus* live in wood mould that harbours scarabaeoid larvae of the genera *Osmoderma, Cetonia* and *Serica*. Some of these cavities may be over one metre in depth. Their inner walls are lined with very dry wood decayed by red rot inhabited by *Ampedus aurilegulus*. The bottom of the cavity is filled to an average depth of 50 cm with wood mould inhabited by cetoniid larvae and their predator, the elaterid *Ampedus megerlei*.

The polyphagous larva of *Ludius ferrugineus* (= *Elater ferrugineus*) mainly attacks the larvae of cetoniids such as *Osmoderma eremita* and *Cetonia aurata* and alleculids such as *Pseudocistela ceramboides*, and lives in the wood mould of various broadleaf trees. Its life cycle may last up to four or five years and includes seven or eight larval instars (figure 18.2). Pupation takes place out of the mould in fissures of decayed wood in a cocoon similar to that of *Osmoderma eremita*. Adults emerge in June to August; they rarely leave the rot-holes in which they grew, and are active at dusk.

– Other families and orders. The alleculid *Pseudocistela ceramboides* grows in four years and undergoes six larval instars. The larvae somewhat resemble those of elaterids; they consume fine particles of mould consisting mainly of the digestive residues of other xylophagous species. Pupation takes place in a cocoon. Adults are diurnal and are often found on the flowers of various forest trees. *Prostomis mandibularis* belongs to the family Cucujidae and has a flattened body. Its life cycle lasts two years and includes three larval instars. Both larvae and adults are encountered in old decayed trunks in fissures of very damp wood that has turned a chocolate brown colour. This species' diet is unknown. *Enoicyla pusilla*, belonging to the family Limnophilidae, is one of the few caddis flies (order Trichoptera) whose larvae are adapted to a terrestrial life, and is a typical component of the fauna of some forests. Its narrow hygrothermic requirements explain why it only occurs locally in widely scattered localities. Both larvae and adults live in mosses and dead leaves; fully grown larvae often take refuge, when they can, in rot-holes in trees in order to pupate; they prefer damp holes in chestnut, beech and oak trees (Kelner, 1960).

– Cetoniidae. A few members of the family Cetoniidae are more or less closely associated with rot-holes. Like other beetles of this habitat they are

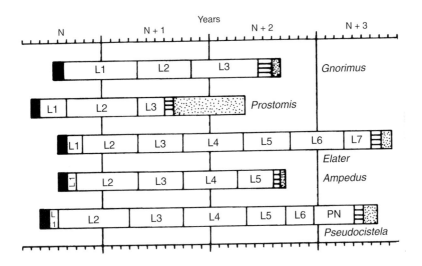

Figure 18.2. *Life cycles of five beetles of tree holes:* Gnorimus octopunctatus, Prostomis mandibularis, Ludius (= Elater) ferrugineus, Ampedus *sp.,and* Pseudocistela ceramboides. *L1 to L7: larval instars; eggs in black; pupae in horizontal hatching; adults in stippling (Kelner, 1967).*

classed as micro-cavernicolous species. The list of species of the French fauna includes *Osmoderma eremita, Liocola lugubris, Cetonischema aeruginosa, Gnorimus nobilis, Eupotosia koenigi, Eupotosia affinis, Potosia opaca* and *Potosia fieberi.* Several of these species have become rare and are threatened with extinction in France (Luce, 1995). Each species occupies a characteristic ecological niche, which would appear to avoid competition. In the forest of Fontainebleau *Gnorimus nobilis* and *Potosia fieberi* occupy small holes situated high in the tree's crown; *Gnorimus nobilis* and *Potosia fieberi* live in decayed wood or wood mould under the bark of a dead part of the tree; *Cetonischema aeruginosa* chooses medium-sized cavities on the upper part of the trunk; *Osmoderma eremita* and *Liocola lugubris* are found in large holes in the middle part of the trunk; finally, *Gnorimus variabilis* colonises holes at the foot of the tree. Each species has a preference for holes of a certain volume; *Osmoderma eremita* is the species that requires the largest ones (figure 18.3). The three species *L. lugubris, O. eremita* and *P. fieberi* have long life-spans, are of large size, have a high adult survival rate and low fertility, but are iteroparous (meaning they lay a new brood of eggs every year of their adult lives), and they show a strong tendency to intraspecific competition, all of which makes them typical of species using the K type demographic strategy.

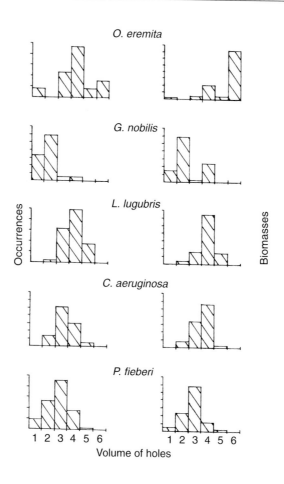

Figure 18.3. *Distribution of occurrence and biomass of five micro-cavernicolous cetoniids in Fontainebleau forest according to the volume of tree holes (Luce, 1995).*

2. FUNGI AND THEIR ASSOCIATED FAUNA

Many forest fungi are symbiotically associated with trees by means of their mycorrhizae; other species grow on dead wood. The abundance of fungi and of mycetophagous insects makes these two groups important elements in forest biodiversity. Much has been published on the subject of mycophagy in insects (Benick, 1952; Borden & McClaren, 1972; Dajoz, 1967, 1981, 1996; Graves, 1960; Klimaszewski & Peck, 1986; Lawrence, 1973, 1977; Okland, 1994; Pielou & Matthewman, 1966; Pielou & Verma, 1968; etc.).

2.1. Diet and biology of mycetophagous insects

Mycetophagous insects belong mainly to two orders, Coleoptera and Diptera. Many of these insects live in the fruiting bodies, called carpophores or sporophores, of the higher fungi, or macromycetes, which are the basidiomycetes and ascomycetes, and are obligate mycetophagous insects. Other insects feed on lower or micromycete fungi, or on myxomycetes. Those that feed on fungal hyphae in dead wood are only partially mycetophagous.

The chemical composition of cell walls in fungi is different from that of vascular plant cells. They are devoid of cellulose, lignin and pectin, but possess chitin and polyholosides that differ from cellulose. Fungi also contain urea, which is a source of nitrogen for insects, as well as other nitrogen compounds. Levels of nitrogen in basidiomycetes reach 0.72% to 1.13% in the fruiting bodies and 3% in spores, levels that are far higher than those found in wood (Merrill & Cowling, 1966). Fungi are a rich source of food for those insects that possess the appropriate enzymes. A study of the enzymes of eleven species of mycetophagous insect showed that they are able to digest chitin as well as compounds related to but different from cellulose, such as β-(1,3) glucane and α-(1,4) glucane, which are present in fungi (Martin, 1979; Martin *et al.*, 1981).

Forests harbour hundreds of species of basidiomycetes of the Agaricales group whose fruiting bodies are usually ephemeral and decay within a few days or a few weeks, and dozens of species of Polyporaceae and Corticiaceae which often grow on dead wood, are tough and long-lasting, some even being perennial. Growth of a carpophore takes place in several stages. Different species of insect occupy them at different stages of growth. Those that arrive early come in search of spores, while later arrivals feed on other parts of the fungus.

Attacks on fungi may begin very early. The erotylid beetle *Cypherotylus californicus* devours the whole fruiting body of *Polyporus adustus* before it has even had time to produce spores (Graves, 1965). Many mycetophagous insects are polyphagous; this may be explained by the unpredictable and ephemeral nature of the presence of carpophores. Some species however are monophagous or oligophagous. This is confirmed by the greater frequency of oligophagia in species that feed on long-lasting Polyporaceae than in those that live on ephemeral Agaricales. One example of oligophagia is provided by *Mycetoma suturale*, a very rare European melandryid whose larvae and adults live exclusively in the sporophores of *Ischnoderma benzoinum*, which grows on various pines, and on *Ischnoderma resinosum* which grows on beech (Burakowski, 1995).

Speciation in various beetles of the family Ciidae (= Cisidae), almost all of which live on Polyporaceae, has been studied by Lawrence (1973) in the case of North American species. In these beetles, choice of the host fungus is largely determined by the structural complexity of carpophores. Three different categories, monomitic, dimitic and trimitic fungi, can be distinguished according to the arrangement of the filaments of the mycelium. Monomitic species are usually not very tough and their sporophores are formed by a single category of hyphae whose cell walls are thin. Species with a dimitic structure such as *Piptoporus betulinus* are formed of two types of hyphae that build a substrate having a woody consistency. Trimitic species such as *Ganoderma applanatum* or *Coriolus versicolor* possess a body of intermixed hyphae forming a sporophore that has the consistency of cork or leather. Consistency is often a major factor when a species of ciid chooses to colonise a fungus. Two groups of ciid have been identified in an English wood (Paviour-Smith, 1960). The "*Polyporus betulinus - Cis bidentatus*" group includes five species of ciid that colonise nine species of fungus with monomitic or dimitic structures. The "*Polystictus versicolor - Octotemnus glabriculus*" group contains five species of ciid that inhabit five species of fungus with a trimitic structure (table 18.3). This classification was confirmed by Lawrence (1973) for the whole of the North American fauna.

Group	Ciidae species	Fungus species	Hyphae structure
Polyporus betulinus– Cis bidentatus	*Cis nitidus, C. bidentatus, C. fagi, C. bilamellatus, Ennearthron cornutum*	*Pleurotus sapidus, Polyporus squamosus, P. sulfureus, P. betulinus, P. dryadeus, P. adustus, Fomes ulmarius, Ganoderma applanatum, Irpex* sp.	Monomitic with only hyphae producing hymenium, or dimitic, with in addition rarely ramified hyphae with thick walls.
Polyporus versicolor– Octotemnus glabriculus	*Cis boleti, C. hispidus, Rhopalodontus fronticornis, Ennearthron affine, Octotemnus glabriculus*	*Polystictus versicolor, P. hirsutus, Trametes gibbosa, Daedalea* sp., *Lenzites betulina*	Trimitic with hyphae producing hymenium and ramified hyphae uniting other hyphae.

Table 18.3. *Specificity of different species of Ciidae towards different species of polyporous fungi in an English wood (Paviour-Smith, 1960).*

Choice of fungus varies greatly according to species. In North America *Cis acritis* is strictly confined to the monomitic polypore *Hischioporus abietinus*, whereas *Cis americanus* has been found in 24 different species of polypore. Features other than the structure of the fungus may however determine ciid specificity. According to Thunes (1994) ciid distribution may be regulated by interspecific competition, as is shown by the almost vicarious distributions, from one forest to another, of species such as *Cis alter* and *Cis jacquemarti* (figure 18.4).

Several dozen species of Coleoptera or Diptera may sometimes be found in a single carpophore. Very different species live on the same parts of a fungus. Ciids that live in tough species of Polyporaceae compete with

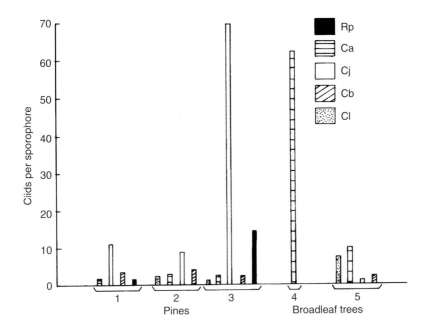

Figure 18.4. *Numbers of individuals of ciid per sporophore of* Fomes fomentarius *and of* Piptoporus betulinus *in five Norwegian forests. Sites 1 to 3: pine woods; sites 4 and 5: broadleaf woods. Pine wood species are almost all vicariants of those of broadleaf woods. Cl:* Cis lineatocribratus; *Ca:* C. alter; *Cj:* C. jacquemarti; *Cb:* C. bidentatus; *Rp:* Rhopalodontus perforatus *(Thunes, 1994).*

tenebrionids belonging to the tribes Bolitophagini and Diaperini. Intra- and interspecific competition are fuelled by high population density. Graves (1960) has recorded the occurrence of 1,000 insects in a single carpophore. The nutritive quality of a fungus also plays a part, as is shown by the ciid beetle *Hadraule blaisdelli* which chooses one out of several species even when the carpophore has already been reduced to dust (Klopfenstein, 1972). Preferences of *Gyrophaena* and related genera (Staphylinidae) have been described in detail by Ashe (1984). These beetles feed by grazing the spore-covered gills of a great many species of fungus.

In the Coleoptera, mycophagia is accompanied by great variability in the structure of mouthparts according to the fungus eaten and to its physical characteristics (Lawrence, 1989).

Aposematic coloration, both in larvae which live on the surface of sporophores and in adults, and which makes the insects highly visible, is common in mycetophagous beetles; it is a signal to potential predators that these beetles are inedible (Leschen, 1994). Larvae are often white with black spots, and patterns of red and black are frequent in adults.

2.2. The main mycetophagous insects

There are no complete inventories of mycetophagous insects, but many partial ones have been compiled. In the Holarctic region 1,160 species of Coleoptera that live in fungi have been counted, including 202 true mycetophages (Benick, 1952). In Norway 260 species of mycetophilid Diptera live in an unexploited forest (Okland, 1996). About sixty species of Coleoptera are found in Norway in *Fomes fomentarius* and *Piptoporus* (= *Polyporus*) *betulinus* (Thunes, 1994). In Canada, Pielou and Verma (1968) found 198 species of insect and 59 species of mite in *Piptoporus betulinus*. Benick (1952) records the presence of 246 species of Coleoptera in *Polyporus squamosus* (table 18.4).

Species	Family	Number of Coleoptera
Fistulina hepatica	Fistulinaceae	22
Fomes fomentarius	Polyporaceae	85
Piptoporus betulinus	Polyporaceae	102
Polyporus giganteus	Polyporaceae	47
Polyporus squamosus	Polyporaceae	246
Polyporus versicolor	Polyporaceae	32
Pholiota squarrosa	Cortinariaceae	18
Gymnopilus spectabilis	Cortinariaceae	31
Armillaria mellea	Tricholomataceae	90
Volvariella bombycina	Volvariaceae	36
Coprinus micaceus	Coprinaceae	17

Table 18.4. *Numbers of species of Coleoptera found in a few species of basidiomycete causing wood decay in Europe (after data provided by Benick, 1952).*

As a general rule species that live in the Agaricales or other short-lived fungi have short life cycles that are completed in a few weeks or less. Growth is much slower in the insects that live on Polyporaceae. The bracket fungus *Fomes fomentarius* is a fungus whose carpophores may live nine years and last for several years after they have died. Ciids may survive several years in the same carpophore, in which they produce several generations. The same is true of the tenebrionid *Bolitophagus reticulatus* which also lives in *Fomes fomentarius*. Agaricales are mainly occupied by Diptera (of which the family Mycetophilidae dominate). *Gyrophaena* and other staphylinids of the subfamily Aleocharinae that live in ephemeral fungi lay eggs that hatch in a matter of hours and have larvae that grow to maturity in three to eleven days (Ashe, 1987), whereas in the tenebrionid *Bolitotherus cornutus*, which lives in long-lasting fungi, eggs hatch on average in 16 days and the larvae live 72 days (Liles, 1956).

2.2.1. Coleoptera

Many families of Coleoptera have members found in fungi. Some are accidental, such as dung beetles of the genus *Geotrupes*, which are attracted by the ammoniacal scent that emanates from some decomposing fungi. The main families of Coleoptera that include mycetophagous beetles are as follows:

– Staphylinidae. Members of this family are often dominant in numbers of individuals and species in many fungi. The two most common genera are *Gyrophaena* and *Atheta*.

– Nitidulidae. Various species of *Epuraea* and *Cychramus* are constant inhabitants of fungi.

– Cucujidae. A few species of *Laemophloeus* are mycetophagous and live in fungi.

– Cryptophagidae. Many species of this family live in fungi, particularly ascomycetes and micromycetes.

– Erotylidae. Most species in this family are mycetophagous. Species of *Triplax* are particularly abundant in *Pleurotus*, which is their preferred habitat.

– Lathridiidae. Many species, particularly in the genera *Enicmus* and *Corticaria*, are mycetophagous. Some are specialised feeders on myxomycetes.

– Endomychidae. This family includes many mycetophagous beetles. *Lycoperdina* species live in puff balls (*Lycoperdon*) whose spores they eat.

– Mycetophagidae. Almost all the species of this family are mycetophagous. Most of them choose less tough fungi in which they may abound.

– Ciidae. Members of this family (known in all the older literature as Cisidae), nearly all exploit Polyporaceae.

– Tenebrionidae. Many species of this family are mycetophagous and live in a wide variety of fungi. The two most characteristic tribes are the Bolitophagini and Diaperini. *Diaperis boleti* is a common species in *Polyporus sulfureus*.

– Melandryidae. A few species such as *Mycetoma suturale* and *Tetratoma fungorum* are mycetophagous.

2.2.2. Diptera

The main family of flies that feed on fungi is the Mycetophilidae. Many drosophilids also live in fungi. Nearly all mycetophilids live in fungi that grow on the ground or on dead wood, although a few species have been found outside fungi. Okland (1996) studied 17 localities in a vast area of south-eastern Norway dominated by spruce. These localities were chosen in primary forest that is not or has only been recently and partially exploited.

He obtained 260 species of Mycetophilidae, which shows how rich this group is in species. Removal of fallen tree trunks reduces the quantity of dead wood, and thus of fungi and number of mycetophilid species. The abundance of dead wood and lignicolous fungi also regulates the species diversity of other mycetophagous insects. Erotylid beetles of the genus *Triplax*, which are particularly associated with fungi of the genus *Pleurotus*, include nine species in the French fauna. Forests in which these fungi are still abundant harbour four or five species of *Triplax*, but most forests now have none or only one common species, and beetles of this genus are now becoming rare (Dajoz, 1985).

Faunal successions do not appear to have been described in the case of mycetophagous Diptera, which almost all belong to the family Mycetophilidae. These have been studied in great detail in Norwegian forests by Okland (1994, 1995, 1996).

2.3. Faunal successions

Communities of arthropods associated with a particular species of fungus change as the sporocarp ripens and decays. This leads to successions in which the greatest numbers of species are to be found in the later stages of decay of the carpophore. Two examples may illustrate the form of these successions.

2.3.1. *Fauna of* Pleurotus ostreatus

The oyster mushroom, which breaks down wood, has a vast distribution in the whole northern hemisphere. It occurs in Europe as well as in North America together with other species of the same genus. The vast distributional range of many fungi contrasts with the much more restricted ranges of insects that live on them. Not one of the beetles that live on *P. ostreatus* in Europe is found in America. Erotylids of the genus *Triplax* and mycetophagids of the genus *Mycetophagus* occur on both continents, but are represented by different species. This discrepancy in ranges may either be explained by the fact that insects speciate more rapidly than fungi, or by the fact that thanks to their spores fungi have more efficient dispersal methods than insects. Four stages can be distinguished in the changes in *Pleurotus* carpophores and in their fauna in a beech wood in the Pyrénées-Orientales:

– Stage 1. The first stage includes 29 species of Coleoptera, of which four constant and three occasional species belong to the family Staphylinidae and include three species of *Gyrophaena* which alone amount to 49.7% of individuals. This fauna settles soon after the carpophores appear, between the fifth and seventh day. It consists of species that are less than 3 mm long.

– Stage 2. This stage is clearly defined between the 15th and 25th days of the carpophore's existence. *Gyrophaena* become rare and are replaced by *Atheta*, a genus which includes six constant out of a total of seven species, and two occasional out of three species, and amount to 31.2% of all individuals. Diversity increases, with a total of 54 species.

– Stage 3. The third stage is defined between the 40th and 50th days and has 40 species. Staphylinids by now amount to only 3.2% of the fauna. Erotylids of the genus *Triplax* and mycetophagids of the genus *Mycetophagus* account for 65.4% of individuals. Various corticolous species penetrate the carpophore.

– Stage 4. It is at this final stage, in which the fungus desiccates and begins to fall apart, that the fauna is poorest. Only 13 species remain; these are mainly erotylids of the genera *Triplax* and *Dacne*, mycetophagids of the genera *Mycetophagus* and *Litargus* and a very few staphylinids of the genus *Bolitobius*. Sometimes carpophores rot instead of drying up. In these cases they are colonised by particular species such as staphylinids of the genus *Proteinus* and even by *Geotrupes* dung beetles.

2.3.2. Fauna of Polyporellus squamosus

Ninety eight species of Coleoptera were collected from this fungus in Silesia (Klimaszewski & Peck, 1986). As in *Pleurotus ostreatus,* the fauna is dominated both in numbers of individuals and species by staphylinids of the genera *Gyrophaena* and *Atheta* (table 18.5). Four stages can be seen:

– Stage 1. The first stage involves young carpophores before spores have matured. The fauna is dominated by two families, the Staphylinidae (represented mainly by *Gyrophaena*, which may reach a density of 185 individuals per dm^2 of hymenium) and Scaphidiidae, with species of the genus *Scaphisoma*. Dominant families are represented by 33 species.

Family	Stage 1	Stage 2	Stage 3	Stage 4
Leiodidae	0.04 (1)	3.1 (3)	-	0.08 (2)
Scaphidiidae	1.7 (2)	0.04 (1)	-	0.4 (2)
Staphylinidae	39.4 (25)	9.3 (17)	3.0 (21)	13.8 (20)
Mycetophagidae	-	0.2 (1)	-	0.9 (1)
Ciidae	-	0.2 (2)	-	3.7 (5)
Erotylidae	0.04 (1)	0.7 (1)	-	-
Nitidulidae	0.2 (3)	0.1 (1)	-	0.1 (2)
Lathridiidae	-	0.1 (1)	-	1.1 (10)

Table 18.5. *Main families of Coleoptera on* Polyporellus squamosus *in percentage of total number of individuals (first figure) and in number of species (second figure in brackets) (Klimaszewski & Peck, 1986).*

– Stage 2. This stage extends from the initial maturation of spores until the beginning of decay. Three families predominate: the Staphylinidae, dominated by species of *Atheta*, Leiodidae and a few Ciidae. The number of species belonging to dominant families is somewhat lower than at the first stage, being only 25.

– Stage 3. At this stage the sporophores are decomposing. The number of families present falls, but there is a slight increase in the number of species. Staphylinids of the genus *Gyrophaena* have become very scarce (only 0.4 individual per dm^2 of hymenium), but other genera such as *Oxytelus, Omalium* and *Xantholinus* appear.

– Stage 4. The last stage is that in which the carpophores are dried out or entirely decayed. It is characterised by a greater number of both families and species. Staphylinids of the genus *Atheta* reach maximum abundance as do *Omalium*. Among families that now appear for the first time the best represented are the Cryptophagidae, Mycetophagidae and Lathridiidae.

REFERENCES

ASHE J. S., Major features of the evolution of relationships between gyrophaenine staphylinid beetles (Coleoptera: Staphylinidae: Aleocharinae) and fresh mushroom. *In:* Q. D. WHEELER, M. BLACKWELL (1984). *Fungus-Insect Relationships: Perspectives in Ecology and Evolution.* Columbia Univ. Press, New York, 227-255.

ASHE J. S., (1986). Subsocial behavior among gyrophaenine staphylinids (Coleoptera: Staphylinidae: Aleocharinae). *Sociobiology*, **12**: 315-320.

ASHE J. S., (1987). Egg chamber production, egg protection and clutch size among fungivorous beetles of the genus *Eumicrota* (Coleoptera: Staphylinidae) and their evolutionary implications. *Zool. J. Linn. Soc.*, **90**: 255-273.

BENICK L., (1952). Pilzkäfer und Käferpilze. *Acta Zool. Fenn.*, **70**: 1-250.

BORDEN J. H., McCLAREN M., (1972). *Cryptoporus volvatus* (Peck) Shear (Agaricales, Polyporaceae) in southwestern British Columbia: life-history, development and arthropod infestation. *Syesis*, **5**: 67-72.

BURAKOWSKI B., Biology and life-history of *Mycetoma suturale* (Panzer) (Coleoptera: Melandryidae) with a redescription of the adult. *In*: J. PAKALUK, S.-A. SLIPINSKI (1995). *Biology, Phylogeny and Classification of Coleoptera.* Institut Zoology, Warszawa, 491-502.

DAJOZ R., (1967). Écologie et biologie des Coléoptères xylophages de la hêtraie. *Vie et Milieu*, **17**, sér. C: 523-763.

Dajoz R., (1981). Note sur les Coléoptères d'un champignon Ascomycète de Tunisie. *L'Entomologiste*, **37**: 203-211.

Dajoz R., (1985). Répartition géographique et abondance des espèces du genre *Triplax* Herbst (Coléoptères, Erotylidae). *L'Entomologiste*, **41**: 133-145.

Dajoz R., (1996). Inventaire et biologie des Coléoptères du champignon Basidiomycète Polyporaceae *Coriolopsis gallica* en Arizona. *Bull. Soc. Ent. Fr.*, **101**: 241-250.

Glaser J. D., (1976). The biology of *Dynastes tityus* (Linn.) in Maryland (Coleoptera: Scarabaeidae). *Coleop. Bull.*, **30**: 133-138.

Graves R. C., (1960). Ecological observations on the insects and other inhabitants of woody shelf fungi (Basidiomycetes: Polyporaceae) in the Chicago area. *Ann. Ent. Soc. Amer.*, **53**: 61-78.

Graves R. C., (1965). Observations on the ecology, behavior and life cycle of the fungus-feeding beetle, *Cypherotylus californicus*, with a description of the pupa (Coleoptera: Erotylidae). *Coleop. Bull.*, **19**: 117-122.

Iablokoff A.-Kh., (1943). Éthologie de quelques Élaterides du massif de Fontainebleau. *Mém. Mus.*, **18**: 81-160.

Kelner S., (1967). Étude écologique du peuplement entomologique des terreaux d'arbres creux. *Ann. Sci. Nat.*, **9**: 1-228.

Kitching R. L., (1971). An ecological study of water filled tree holes and their position in the woodland ecosystem. *J. Anim. Ecol.*, **40**: 281-302.

Klimaszewski J., Peck S. B., (1986). Succession and phenology of beetle faunas (Coleoptera) in the fungus *Polyporellus squamosus* (Huds: Fr.) Karst. (Polyporaceae) in Silesia, Poland. *Canad. J. Zool.*, **65**: 542-550.

Klopfenstein P.-C., (1972). The ecology, behavior and life cycle of the mycetophilous beetle *Hadraule blaisdelli* (Casey) (Insecta: Coleoptera: Ciidae). Quoted by Lawrence 1973.

Lawrence J. F., (1973). Host preference in Ciid beetles (Coleoptera: Ciidae) inhabiting the fruiting bodies of Basidiomycetes in North America. *Bull. Museum Comp. Zool.*, **145**: 163-212.

Lawrence J. F., (1977). Coleoptera associated with *Hypoxylon* species (Ascomycetes: Xylariaceae) on oak. *Coleop. Bull.*, **31**: 309-312.

Lawrence J. F., Mycophagy in the Coleoptera: feeding strategies and morphological adaptations. *In*: N. Wilding, N. M. Collins, P. M. Hammond, J. F. Webber (1989). *Insect-Fungus interactions*. Academic Press, London, 2-23.

Leschen R. A. B., (1994). Ecological and behavioral correlates among mycophagous Coleoptera. *Folia Entomol. Mex.*, **92**: 9-19.

Liles M. P., (1956). A study of the life history of the forked fungus beetle, *Bolitotherus cornutus* (Panzer) (Coleoptera: Tenebrionidae). *Ohio J. Sci.*, **56**: 329-337.

Luce J.-M., (1995). Écologie des cétoines (Insecta: Coleoptera) micro-cavernicoles de la forêt de Fontainebleau. Niches écologiques, relations interspécifiques et conditions de conservation des populations.Thesis, Muséum National d'Histoire Naturelle, Paris.

Martin M. M., (1979). Biochemical implications of insect mycophagy. *Biol. Rev. Cambridge Philos. Soc.*, **54**: 1-21.

Martin M. M., Kukor J. J., Martin J. S., O'Toole T. E., Johnson M. W., (1981). Digestive enzymes of fungus-feeding beetles. *Physiol. Zool.*, **54**: 137-145.

Mayer K., (1938). Zur Kenntniss der Buchenhohlenfauna. *Archiv. fur Hydrobiol.*, **33**: 388-400.

Merrill W., Cowling E. B., (1966). Role of nitrogen in wood deterio-ration: amount and distribution of nitrogen in fungi. *Physiopathology*, **56**: 1,083-1,090.

Metge G., Belakoul N., (1989). Colonisation d'un nouvel habitat par *Culex pipiens* (Diptera, Culicidae): le creux d'arbre des subéraies en pays Zaër, Maroc. *Ann. Limnol.*, **25**: 73-80.

Nilsson T., (1997). Survival and habitat preferences of adult *Bolitophagus reticulatus*. *Ecol. Ent.*, **22**: 82-89.

Okland B., (1994). Mycetophilidae (Diptera), an insect group vulnerable to forestry practices? A comparison of clearcut, managed and semi-natural spruce forests in southern Norway. *Biodiversity and Conservation*, **3**: 68-85.

Okland B., (1995). Diversity patterns of two insect groups within spruce forests of southern Norway. Thesis, Agricultural University of Norway.

Okland B., (1996). Unlogged forests: important sites for preserving the diversity of Mycetophilids (Diptera: Sciaroidea). *Biol. Cons.*, **76**: 297-310.

Okland B., Bakke A., Hagvar S., Kvamme T., (1996). What factors influence the diversity of saproxylic beetles? A multiscaled study from a spruce forest in southern Norway. *Biodiv. and Conservation*, **5**: 75-100.

Park O., Auerbach S., (1954). Further study of the tree-hole complex with emphasis on quantitative aspect of the fauna. *Ecology*, **35**: 208-222.

Park O., Auerbach S., Corley G., (1950). The tree hole habitat with emphasis on the Pselaphid beetle fauna. *Bull. Chicago Acad. Sci.*, **9**: 19-41.

Paviour-Smith K., (1960). The fruiting-bodies of macrofungi as habitat for beetles of the family Ciidae (Coleoptera). *Oikos*, **11**: 1-71.

Pielou D.-P., Matthewman W. G., (1966). The fauna of *Fomes fomen-tarius* (Linnaeus ex Fries) Kickx growing on dead birch in Gatineau Park, Québec. *Canad. Ent.*, **98**: 1,308-1,312.

Pielou D.-P., Verma A.-N., (1968). The arthropod fauna associated with the birch bracket fungus *Polyporus betulinus* in eastern Canada. *Canad. Ent.*, **100**: 1,179-1,199.

Pimm S. L., (1982). *Food webs.* Chapman and Hall, London.

Thunes K.-H., (1994). The coleopteran fauna of *Piptoporus betulinus* and *Fomes fomentarius* (Aphyllophorales: Polyporaceae) in western Norway. *Ent. Fenn.,* **5**: 157-168.

Vaillant F., (1978). Les *Systenus* et leur habitat dendrotelme. *Bull. Soc. Ent. Fr.,* **83**:73-85.

Villemant C., Fraval A., (1991). *La faune du chêne-liège.* Éditions Actes, Rabat.

Virkkala R., Rajasärkkä A., Väisänen R.-A. *et al.,* (1994). Conservation value of nature reserves: do hole-nesting birds prefer protected forests in southern Finland? *Ann. Zool. Fenn.,* **31**: 173-186.

19

Insects of forest soils

Soil is where organic matter from fallen litter returns to a mineral state, which shows how important soil is in biochemical cycles and in the life of the forest. Conversion of litter into humus and minerals is mainly done by micro-organisms (bacteria and fungi), while animals prepare the substrate by fragmenting and mixing it by means of the physical and chemical changes they make it undergo, and also by direct action which favours micro-organisms. Let us take, for example, a pine needle 60 mm long, 1 mm wide and 0.5 mm thick. This needle has a surface area of 180 mm^2; after it has been fragmented into 60 pieces by an earthworm its surface area is 240 mm^2; following attacks by mites that reduce it to cubes measuring 10 µm there are 30,000 fragments with a total surface area of 1,800,000 mm^2, or 1.8 m^2. This considerable increase in surface area greatly aids later action by micro-organisms. It would seem that the whole of the forest floor litter passes at least once through the alimentary tracts of soil animals (Nef, 1957).

Forest soil fauna contains many groups of invertebrates which vary greatly in nature and in numbers (Bornebusch, 1930; Van der Drift, 1951; Birch & Clark, 1953; Wallwork, 1976; Nef, 1957; Bachelier, 1978). Figure 19.1 gives an idea of the range of body sizes that exist among the main groups of soil arthropods in Europe. The figures in table 19.1 show average values of the main groups of soil organisms; it is size that matters rather than absolute values. There is a difference between the two main types of humus, mull and mor. Mull, or sweet humus, is characteristic of broadleaf forests. It is formed in soil that is rich in earthworms (Lumbricidae and Enchytraeidae); decay of plant debris is rapid, and horizon A is thin. Mor, or raw humus, is that found in conifer forests; it forms where earthworms are few. Litter decay is slow, which leads to the accumulation of thick layers of dead leaves. The difference between mull and mor fauna is shown by data relating to the main groups of invertebrates obtained in Belgian woods (Nef, 1957):

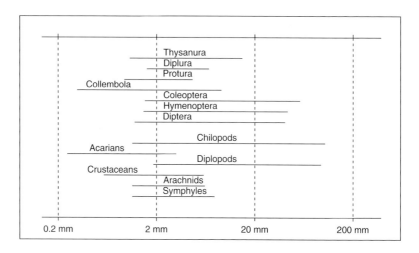

Figure 19.1. *Variations in size in the main groups of arthropods of soil fauna in Europe. Species over 2 mm long are traditionally classed as "macro-arthropods" and those under 2 mm as "micro-arthropods".*

Taxonomic group	Number per m^2	Biomass (in grams) per m^2	Metabolism (in kcal per year) per m^2
Bacteria	10^{15}	1,000	-
Fungi	-	400	-
Protozoans	5.10^8	38	-
Nematodes	10^7	12	113
Lumbricids	10^3	120	180
Enchytraeids	10^5	12	355
Molluscs	50	10	62
Myriapods	500	12.5	96
Isopods	500	5	38
Opiliones	40	0.4	5
Parasitic Acarina	5.10^3	1.0	64
Oribatid Acarina	2.10^5	2.0	30
Araneida	600	6.0	34
Coleoptera	100	1.0	8
Diptera	200	1.0	6
Collembola	5.10^4	5.0	153

Table 19.1. *Average values of abundance, biomass and metabolism of the main groups of soil organisms.*

Mull		Mor	
Biomass	1,000 kg ha^{-1}	Biomass	300 kg ha^{-1}
Lumbricids	70%	Arthropods	50%
Arthropods	20%	Enchytraeids	30%
Nematodes	10%	Nematodes	20%

Decomposition of forest floor litter is five times slower when fauna has been eliminated with naphthalene, which acts as a repellent (Ghilarov, 1970). Edward & Heath (1963) have shown the effect of macro-fauna invertebrates on the decomposition of oak and beech litter (cf. figure 17.2). Soil fauna is richer in woodland and meadows than in cultivated land, which is a result of cultivation practices. In cultivated soils there are on average 1,000 pterygote insects (larvae and adults) per square metre, whereas there are 3,000 per square metre in woodland and 4,500 in meadows. Litter accumulated under conifers is less suitable for large, non-burrowing arthropods such as woodlice or carabid, staphylinid, scarabaeid and elaterid beetles; these groups are more abundant in broadleaf forests.

1. SOIL INSECTS OF A SPRUCE FOREST

The soil fauna of a spruce forest at Solling in Germany has been studied in detail (Thiede, 1976, 1977). Samples were made using photoeclectors and emergence traps placed on the ground over a three year period. Pterygote insects accounted for 65% of the 305,000 arthropods collected, apterygotes 20% and arachnids and chilopods 2%. This represents an average density of 2,700 to 4,300 individuals per square metre and per year for the pterygotes. Diptera dominate, amounting to 60 to 90% of the total of samples. The two main families are the Cecidomyiidae, with 1,000 to 2,600 individuals per square metre, and the Sciaridae, with 1,300.

Production of pterygote insects varies from 5.2 to 14.3 kg ha^{-1} year^{-1} (dry weight), which represents 29 to 79.10^3 kcal ha^{-1} year^{-1}. Coleoptera and Diptera Nematocera account for the greater part of this production. The biomass of insects that emerged from the soil was 5 kg ha^{-1} year^{-1} (dry weight) in 1973, with the different groups distributed, in percentages of biomass, as follows:

Diptera Nematocera:

Sciaridae: *Ctenosciara*: 2 species and 45%
Tipulidae: *Tipula*: 1 species and 5%
Other Nematocera: 7.6%
Diptera Brachycera: 3.5%

Hymenoptera: 3.2%

Coleoptera:

Curculionidae with rhizophagous larvae: 12%
Elateridae: *Athous subfuscus*: 7%
Scolytidae: 5.5%
Other Coleoptera: 3.4%

Scolytidae in this list are merely accidental occurrences of beetles that had hibernated on the ground. The approximate distribution of species for the total of 870 that were studied is as follows:

Diptera: 290 species including 55 mycetophilids, 45 sciarids, at least 30 cecidomyiids, 26 empidids and 20 phorids.

Hymenoptera: 310 species including 95 ichneumonids, 60 to 80 chalcids, 70 proctotrupoids and 30 braconids.

Coleoptera: 80 species including 70 staphylinids.

Others: 90 species (including apterygotes).

This list demonstrates the dominance of Diptera in the soil fauna, whereas Coleoptera dominate in dead wood. Soil is also a temporary growth site or a transitory refuge for many members of the arboreal fauna. Finally, many soil insects such as ants and carabid beetles are predators that often play a major part in regulating populations of arboreal species.

2. COLLEMBOLA

Collembola, or springtails, are the most abundant apterygote insects; they by far surpass pterygote insects in numbers of individuals. In a spruce plantation their numbers vary from 13,600 to 100,000 individuals per square metre; 58,000 springtails were counted per square metre in a beech wood. According to Nef (1957) an average abundance of 200,000 Collembola per square metre may be assumed, which amounts to a biomass of 2 grams. Collembola are rarely predatory; they feed on dead wood, various plant debris, fungal hyphae, pollen, unicellular algae and even on humus. They carry out a major part of the physical fragmentation of litter. Fragments found in faeces have an average size of 30 x 150 µm in the case of *Tomocerus flavescens* and 60 µm in *Folsomia*.

There are three fairly distinct types of morphology in the Collembola (Gisin, 1943). Surface-dwelling forms such as *Bourletiella hortensis* and *Entomobrya muscorum* live on herbaceous plants or on the surface of litter. These are large species, fully pigmented and have well-developed eyes and appendages (legs, antennae and furca). Species that live deep underground such as *Neelus minutus* and *Cyphoderus albinus* are less than one millimetre long, elongate, depigmented, with eyes reduced or lacking and with short appendages. Forms that live in litter such as *Sminthurus elegans* and

Lepidocyrtus lanuginosus also colonise mosses and lichens; they are inter-mediate in build between the two more extreme types.

The irregular distribution of Collembola in soil is a function of the prevailing microclimatic conditions. In the vicinity of trees in a Scots pine plantation, Collembola are twenty times more abundant than at a distance from the trees. The number of species present varies from 9 to 11 near trees and is only 3 to 4 away from trees (Stebayeva, 1975).

3. DIPTERA

Phytophagous or saprophagous Diptera larvae that live in the soil consume large quantities of litter that they mix into the mineral part of the soil in the same way as earthworms. In many cases Collembola and Diptera are the main groups responsible for breaking down litter and opening the way to smaller invertebrates such as enchytraeids. Mor soils are the richest in Diptera larvae. An average of 250 to 1,000 per square metre are found in these soils, which corresponds to a biomass of 1 to 7 grams (cf. chapter 4.5). In Belgium the biomass of soil Diptera varied from 1,641 to 5,752 g ha^{-1} in a hornbeam wood, and from 2,453 to 6,636 g ha^{-1} in a hazel wood (Krizelj, 1969).

Two families play a major part in litter decomposition, the Tipulidae (Perel et *al.*, 1971) and Bibionidae (Karpachevsky et *al.*, 1968). Their larvae favour and accelerate litter decay. Decomposition is twice as rapid where tipulids are present. The faeces of bibionids have higher levels of nitrogen, organic carbon, hemicelluloses and water than the litter.

Larvae of *Bibio marci* are found in autumn in masses of 40 to 1,400 individuals at the limit of forest soil and litter, where they remain active throughout the winter. The young larva feeds on matter in the humus and from the second instar on dead leaves. A comparative study of the litter of chestnut, oak and grasses shows that the C/N ratio is lower in the faeces of these larvae than in their food. They thus act mechanically (by fragmenting litter) and chemically, bringing the organic components of litter closer to the optimum value C/N = 10 that characterises balanced soil humus. This is comparable to what happens in rot-holes in trees (cf. chapter 18). Adult *Bibio* are large black hairy flies with a slow clumsy flight, and are known as "March flies" because they are sometimes very abundant in spring, especially on grasslands.

Other Diptera of forest soils include the mycetophilids, which mainly feed on fungi. Sciarids sometimes swarm in great numbers. Chironomids and cecidomyiids are sometimes also abundant. Adult sciarids are midges that are common in woodland; their larvae are cylindrical with oval, partially retractile heads. *Sciara militaris* lives in heavily-shaded areas, especially in beech and hornbeam woods. Their larvae eat dead leaves, leaving only the veins; they may congregate in lines of several thousand individuals, forming a greyish ribbon two to four metres long and 2 centimetres wide, hence the name "army worms" by which they are sometimes known.

4. COLEOPTERA

Beetles play an important part in soil ecology. In contrast with most other insect orders this one contains many predators. The main ones belong to the families Carabidae and Staphylinidae. Elaterids may also abound; forest soils also harbour many small Coleoptera belonging to many different families such as the Pselaphidae, Scydmaenidae, Ptiliidae, etc.

4.1. Carabidae

The ecology and biology of Carabidae, or ground beetles, have been much studied (Thiele, 1977). The fauna of woodland soils differs from that of open environments such as meadows or cultivated land, few species occurring in both environments (table 19.2). A single genus of carabid will include forest species and species that live in open environments. Thus *Pterostichus vulgaris* lives in fields, and *P. cristatus* in woods; *Carabus auratus* is found in open situations, and *C. auronitens* is confined to forest. Species distribution is determined by microclimate, and particularly by temperature and humidity. Climatic fluctuations are more pronounced in open environments, while woodland attenuates the variations (cf. chapter 2.1).

Woodland		Cultivated land	
Abax ater	194	*Pterostichus versicolor*	7,727
Nebria brevicollis	125	*Ophonus sabulicola*	1,913
Pterostichus cristatus	54	*Harpalus rufipes*	1,815
Harpalus rufipes	50	*Trechus quadristriatus*	938
Trechus quadristriatus	34	*Bembidion lampros*	780
Ophonus sabulicola	27	*Platynus dorsalis*	271
Notiophilus biguttatus	18	*Harpalus aeneus*	177
Asaphidion flavipes	11	*Clivina fossor*	78
Clivina fossor	9	*Agonum viduum*	57
Notiophilus rufipes	9	*Bembidion quadrimaculatum*	23

Table 19.2.   *List of the ten most abundant species and numbers of individuals trapped over two years. Note the complete absence in woodland of* Pterostichus versicolor, *which is the commonest species in cultivated land, and the presence of* Abax ater *in woodland but not in the fields. Few species are common to both environments and abundance is very diverse (Dajoz, 1993).*

Most carabids are predators. *Calosoma* eat caterpillars, which they often hunt in trees. *Calosoma sycophanta* devours the caterpillars of defoliating moths such as *Lymantria dispar, Tortrix viridana* and *Operophtera brumata.* An allied species, *Calosoma inquisitor,* often occurs abundantly in oak woods during gradations of the oak processionary caterpillar. Four categories of predatory Carabidae may be defined according to their size in relation to that of their prey and abundance of the latter (Loreau, 1983):

– Species larger than 15 mm such as *Carabus problematicus* and *Abax ater* are polyphagous and attack all kinds of prey except Collembola.

– Species between 10 and 15 mm such as *Nebria brevicollis* are also polyphagous but they also feed on Collembola and ignore molluscs.

– Species between 6 and 10 mm such as *Loricera pilicornis* are specialised feeders on Collembola but also take Nematocera and aphids.

– Species smaller than 6 mm such as *Notiophilus biguttatus* have diets consisting of small mites, Collembola and a great variety of other small, abundant insects.

The part played by ground-dwelling beetles as predators of *Operophtera brumata* was analysed in an oak wood near Oxford in England (Frank, 1967). Predatory soil Coleoptera are represented by 43 species of carabid and 78 species of staphylinid. The main carabid predators of pupae, which were identified by direct observation or by the serological method, are *Pterostichus madidus, Abax ater, Pterostichus cupreus* and *P. melanarius* (cf. figure 5.4). Abundance of three of these species, in individuals per square metre, was as follows: *P. madidus*: 10.8; *A. ater*: 5.8; *P. melanarius*: 2.5. Carabids accounted for 38% of *Operophtera brumata* pupal mortality in this wood.

In Solling beech wood in Germany the biomass of Carabidae is about the same as that of centipedes and spiders. It varies from 100 to nearly 400 g m^{-2} according to the time of year. The metabolism of the two dominant species, *Pterostichus oblongopunctatus* and *P. metallicus* has been evaluated. Their main prey are small Coleoptera as well as aphids, Collembola and various small arthropods. *P. oblongopunctatus* reduces weevil populations, and particularly that of *Phyllobius*, by 20%. The allied species *P. metallicus* is less active and only reduces the *Phyllobius* population by 1 to 2%. These examples, chosen from among many others, show that carabids play an important part in controlling soil arthropod populations in the forest environment.

Carabid species are numerous in forest. Southwood (1978) showed that competition is avoided and ecological niches differentiated by various mechanisms such as differences in size, and in annual rhythms of activity, and that different daily rhythms of activity are also characteristic of each species. Although they are less sensitive to forestry practices than species that live in dead wood, carabids are nevertheless affected by them. The consequences of felling on the carabids of boreal forest have been described in Canada and in Finland (cf. chapter 4.2).

A forest's richness in carabid species depends on many factors; surface area is only one of them. It is known that coniferous forests contain fewer species than broadleaf forests and that the dominant species in each are not the same (Butterfield & Benitez Malvido, 1992).

4.2. Other families

The family Staphylinidae is very rich in species and is well-represented in forest soils, but their biology is as yet poorly known. Various members of the family Elateridae have larvae that live in the soil. They may be pests in nurseries where they attack the roots of seedlings and young trees; cases in point are *Adelocera murina, Dalopius marginatus* and *Athous niger*. The most abundant curculionids in soil are species of the genus *Otiorrhynchus*, whose larvae attack the roots of young conifers in nurseries and whose adults gnaw the needles and bark. Serious damage has been caused to spruce by *Otiorrhynchus niger* and *O. ovatus*.

5. HYMENOPTERA

Ants are the main group of Hymenoptera in the forest soil fauna. The best known species is the wood ant, *Formica rufa*. This ant's colonies are large, each nest being polycalic, in other words, an initial society may build up to around a hundred annex mounds, sometimes extending to an area of half a hectare and up to 50 metres from the central nest. These nests are conical mounds of conifer needles mixed with debris and may reach a diameter and height of 1.3 metres. Under the mounds the ants dig galleries extending one metre below ground level. The microclimate that prevails inside the nest features a temperature of 25 to 30 °C, and an atmosphere saturated with formic acid and 30 to 40 times richer in carbon dioxide than the outside air. Because of their diet, ants of the *Formica rufa* group are regarded as useful insects and sometimes successful attempts have been made at using them in biological control of forest pests (cf. chapter 6).

6. ORTHOPTEROIDS

Few species of this group are particular to the forest environment in temperate regions. Indeed many of these insects tend to colonise open terrain, which is not the case in the tropics. Various species of cockroach belonging to the genus *Ectobius* live in dry leaf litter and among grasses and in shrubs. The crickets are represented by the wood cricket, *Nemobius sylvestris*, which is distributed in Europe and in North Africa. The life cycle of this species lasts two years. Adults, which live on average over one year, are mainly abundant in summer and autumn (Campan, 1965). The wood cricket lives in dry leaf litter, especially in oak woods; it does not dig a gallery. Its living space is sufficiently restricted for the cricket to be familiar with its terrain. Its movements, which have been studied by marking individuals, do not usually extend beyond ten metres. Forest edge affects the wood cricket's distribution, as it is an impassable barrier.

REFERENCES

BACHELIER G., (1978). *La faune des sols. Son écologie et son action.* ORSTOM, Paris.

BIRCH L. C., CLARK D. P., (1953). Forest soil as an ecological community with special reference to the fauna. *Quart. Rev. Biol.*, **28**: 13-36.

BORNEBUSCH C.-H., (1930). The fauna of the forest soil. *Forstl. Fors. Danmark*, **11**: 1-224.

BUTTERFIELD J., BENITEZ MALVIDO J., Effect of mixed-species tree planting on the distribution of soil invertebrates. *In:* M. G. R. CANNEL (1992). *The ecology of mixed-species stands of trees.* Blackwell, Oxford, 255-265.

CAMPAN R., (1965). Étude du cycle biologique du grillon *Nemobius sylvestris* dans la région toulousaine. *Bull. Soc. H. N. Toulouse*, **100**: 1-8.

DAJOZ R., (1987). Le régime alimentaire des Coléoptères Carabidae. *Cahiers des Naturalistes*, **43**: 61-96.

DAJOZ R., (1993). Les Coléoptères Carabidae d'une région cultivée à Mandres-les-Roses. Comparaison de la faune des cultures, des lisières et d'un bosquet. *Cahiers des Naturalistes*, **48**: 67-78.

EDWARD C. A., HEATH G. W., The role of soil animals in breakdown of leaf material. *In:* J. DŒKSEN, J. VAN DER DRIFT (1963). *Soil Organisms.* North Holland Publishing Company, 76-84.

FRANK J. H., (1967). The effect of pupal predators on a population of winter moth *Operophtera brumata* (L.) (Hydriomenidae). *J. Anim. Ecol.*, **36**: 375-389 and 611-621.

FUNKE W., Food and energy turnover of leaf-eating insects and their influence on primary production. *In*: H. ELLENBERG (1971). *Integrated Experimental Ecology.* Springer, 83-93.

GHILAROV M.-S., Abundance, biomass and vertical distribution of soil animals in different zones. *In: Secondary Productivity of terrestrial Ecosystems* (1967), **2**: 611-629.

KARPACHEVSKY L.-O., PEREL T.-S., BARTSEVICH V.-V., (1968). The role of Bibionidae larvae in decomposition of forest litter. *Pedobiologia*, **8**: 146-149.

KŒHLER H., (1984). Zum Nahrungspektrum und Nahrungumsatz von *Pt. oblongopunctatus* und *Pt. metallicus* (Coleoptera, Carabidae) im Ökosystem "Buchenwald". *Pedobiologia*, **27**: 171-183.

KRIZELJ S., (1969). Recherches sur l'écosystème forêt. Diptères récoltés dans des bacs d'eau. *Bull. Rech. Agron. Gembloux*, **4**: 111-120.

LOREAU M., Trophic role of Carabid beetles in a forest. *In*: Ph. LEBRUN (1983). *New Trends in Soil Biology*, 281-285.

NEF L., (1957). État actuel des connaissances sur le rôle des animaux dans la décomposition des litières des forêts. *Agricultura*, **5**: 245-290.

PEREL T.-S., KARPACHEVSKY L.-O., YEGOROVA E.-V., (1971). The role of Tipulidae (Diptera) larvae in decomposition of forest litter-fall. *Pedobiologia*, **11**: 66-70.

SOUTHWOOD T. R. E., The components of biodiversity. *In:* L. A. MOUND, N. WALOFF (1978). *Diversity of insect fauna*. Symposium of the Royal Entomological Society, **9**: 19-40.

STEBAYEVA S.-K.,. Phytogenetic microstructure of Collembola associations in steppes and forests of Siberia. *In:* J. VANEK (1975). *Progress in Soil Zoology*. W. Junk, The Hague, 77-84.

THIEDE U., (1976). Quantitative Untersuchungen an Insektenpopulationen in Fichtenforsten des Solling. *Verhandl. Ges. Okol.*, 139-144.

THIEDE U., (1977). Untersuchungen über die Arthropodenfauna in Fichtenforsten (Populationökologie, Energieumsatz). *Zool. Jb. Syst.*, **104**: 137-202.

THIELE U., (1977). *Carabid beetles in their environments*. Springer, Berlin.

VAN DER DRIFT J., (1951). Analysis of the animal community in a beech forest floor. *Tijdsch. voor Ent.*, **94**: 1-168.

WALLWORK J. A., (1976). *The distribution and diversity of soil fauna*. Academic Press, London.

20

North American forest insects

1. SOME CHARACTERISTICS OF NORTH AMERICAN FORESTS

There is a great variety of forests in North America and these forests are much richer in genera and species of tree than European forests (cf. chapter 1.4). The three main types of forest are the following (Barbour & Billings, 1998).

1.1. Boreal forest

This forest forms a continuous belt across the continent from Alaska to Labrador. Although boreal forest covers a vast area its diversity is low. The main tree species are conifers such as *Picea glauca, Picea mariana, Larix laricina, Abies balsamea* and *Pinus banksiana*. Among the broadleaf trees *Populus tremuloides, P. balsamifera* and *Betula papyrifera* are common.

1.2. Eastern deciduous forest

This forest is highly diversified, and seven associations can be identified. Tree species are numerous; there are many oaks (*Quercus velutina, Q. borealis, Q. alba*), many pines (*Pinus rigida, P. echinata, P. elliottii*) and other conifers (*Picea rubens, Abies fraseri*) as well as broadleaf trees such as *Fagus grandifolia, Liriodendron tulipifera, Magnolia* spp., *Acer saccharum, Castanea dentata, Tilia* spp., etc.

1.3. Pacific northern forest

The coniferous forest of the Pacific Northwest extends from Alaska to northern California and is particularly lush in British Columbia, Washington state and Oregon. The mild climate and high rainfall favour the growth of trees. The dominant species are Douglas-fir (*Pseudotsuga menziesii*), Sitka spruce (*Picea sitchensis*), western hemlock (*Tsuga heterophylla*), silver fir (*Abies amabilis*), western red cedar (*Thuja plicata*) and, in southern Oregon and northern California, coast redwood (*Sequoia sempervirens*). This forest is a temperate rain forest in which the height of the trees allows a vertical stratification and the diversification of many microclimates. *Pseudotsuga menziesii* reaches a height of 70 metres, a diameter of 3 metres and can live for more than 1,000 years. Epiphytes are abundant in this forest.

A remarkable characteristic of this coniferous forest is that large fragments that have been very little or not modified by man, known as "old growth forests", still remain. The trees are very old; standing dead trunks (or snags) and trunks lying on the ground have a volume that may reach 500 m^3 per hectare (table 20.1). This dead wood is a habitat for many saproxylic organisms, mostly insects and fungi. The different stages of decomposition of Douglas-fir wood and the successions of insects and other organisms have been described in chapter 17.1.5.

There are similarities but also differences between European and North American forest insects. This chapter gives further information on the ecology of North American species that have not yet been studied in the previous chapters, and focuses on some of the most harmful or remarkable insects.

	Young forest (40 to 80 years)	Mature forest (80 to 200 years)	Old forest (> 200 years)
Number of trees with diameter > 100 cm	0.50	2.5	19
Volume of wood (m^3)	248	148	313
Biomass of wood (t ha^{-1} dry weight)	43	20	66
Number of snags	171	121	60
Snags with diameter > 50 cm	2.8	2.5	6.4
Biomass of snags (t ha^{-1} dry weight)	35	23	57
Volume of snags (m^3)	175	101	221

Table 20.1. *Number, biomass and volume per hectare of trees and of snags in Douglas-fir forests of Oregon and Washington (Spies et al., 1988).*

2. CANOPY ARTHROPODS OF NORTH AMERICAN FORESTS

In a Sitka spruce forest in British Columbia sampling of the canopy was made by branch clipping at three different heights (33 m, 45 m and 54 m). The guild composition of the canopy arthropods, in percentages, is as follows: phytophages: 41.3; predators: 37.2; parasitoids: 11.8; epiphyte fauna: 8.3; scavengers: 1.0; tourists: 0.4. These percentages are different from those given by Moran & Southwood (1982) and by Stork (1987) but similar to those given by Schowalter & Crossley (1987) for a coniferous forest. The main characteristic of the phytophagous guild is that it is composed of a small number of species but a large number of individuals. The great rarity of ants contrasts with the abundance of these insects in tropical forests. No differences in the structure of the guilds can be detected between the low canopy and the high canopy. Among the epiphytes that are abundant in the canopy, mosses and other plants form mats which can be 28 cm deep and support a well-developed soil. These mats are colonised by microarthropods, particularly by oribatid mites, some of which are strictly arboreal species (Winchester, 1997).

The structure and evolution of the canopy arthropod community have also been described by Martin (1966) and by Schowalter (1988) (cf. chapter 7.3). Martin studied stands of red pine (*Pinus resinosa*) in Ontario. There are three distinct main stages in the growth of the trees: the transition stage when the trees are about 5 years old and 2 feet high, the monoculture stage when the trees are about 15 years old and 15 feet high, and the young forest stage that begins when the trees are about 25 years old and 35 feet high. Abundance of the arthropod fauna during the growth of the trees shows a peak when the trees are about 15 years old, after which there is a decrease in abundance (figure 20.1). The most significant change is the decrease in the number of abundant or very abundant species and the increase in the number of rare species (figure 20.2).

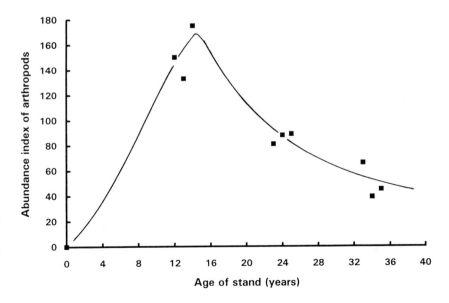

Figure 20.1. *Abundance of arthropods in relation to the age of the stand (and to crown size) during the development of a red pine community (Martin, 1966).*

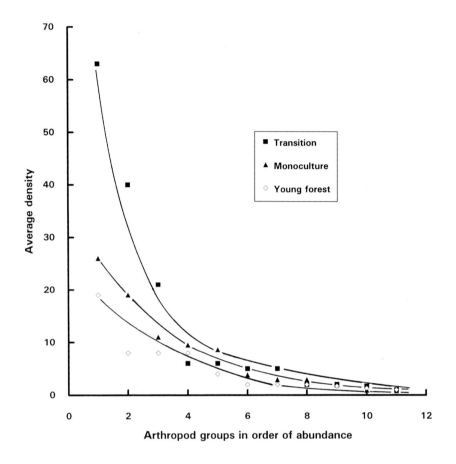

Figure 20.2. *Abundance of different arthropod groups in the tree crown during the three stages in the development of a red pine community. There are more abundant species in the transition stage than in the young forest (Martin, 1966).*

The canopy of an old growth Douglas-fir forest may have more than 1,500 species of invertebrate in a single stand. A minority of species belonging to the Araneida, Acarina, Homoptera, Collembola, Neuroptera, Psocoptera and Thysanoptera spend all their life cycle in the canopy. The majority of the species encountered in the canopy are adults that spend their larval stage on the forest floor. This is the case of the Mycetophilidae (or fungus gnats, Diptera) which occur as larvae in fungi on the forest floor. The most numerous arthropods are predatory spiders that eat the many and various flies. Other arthropods feed on bacteria, fungi or debris (Franklin *et al.*, 1981). Some species have lost the ability to fly, a characteristic that seems to be associated with habitats that have remained stable for a long time (Lattin, 1990; Moldenke & Lattin, 1990). There are successional trends in canopy arthropods, with a high diversity of predators in the old growth forest, whereas there is a high diversity of sap-sucking insects and a low diversity of other groups in the regenerating forest.

3. DEFOLIATING INSECTS

The number of species of insect attacking different species of tree is variable. This is a function of the area occupied by the trees, as is seen in the case of leaf-mining Lepidoptera on North American oaks (figure 20.3), in the case of cynipid gall wasps on Californian species of oak (figure 20.4) or in the case of Scolytidae that attack pines (cf. figure 4.2).

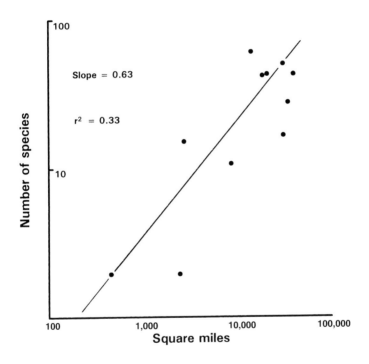

Figure 20.3. *Species-area relationship between the number of species of leaf-mining Lepidoptera of North American oaks and the host plant range (Opler, 1974).*

The number of species is also a function of other characteristics. Many trees intolerant of shade (such as *Pinus ponderosa*) are attacked by many scolytids, and long-lived trees (such as *Sequoia sempervirens*) are attacked by a low number of defoliators. The number of species that attack some conifers of western North America are as follows:

Tree species	Number of defoliators	Number of Scolytidae
Sequoia sempervirens	2	1
Libocedrus decurrens	6	2
Thuja plicata	11	3
Abies lasiocarpa	6	9
Larix occidentalis	12	4
Pinus monticola	6	12
Abies concolor	18	10
Picea sitchensis	14	14
Picea engelmanni	12	16
Tsuga heterophylla	22	7
Pinus contorta	33	25
Pinus ponderosa	42	23
Pseudotsuga menziesii	49	24

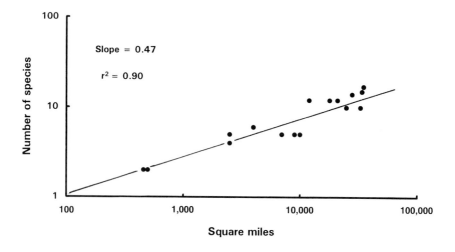

Figure 20.4. *Species-area relationship between the number of species of cynipid gall wasp on Californian oaks and the area occupied by these trees (Cornell & Washburn, 1979).*

The North American fauna is very rich in species (cf. chapters 1 and 4). Many papers deal with the forest insects of this region, mostly with the species that have an economic impact. Some European species arrived in North America where they have become major pests (cf. chapter 6.2). The importance of defoliators in North American forests is obvious (table 20.2).

Species	Defoliated areas (in 10^3 ha)	
	1992	1994
Choristoneura fumiferana	10,104.2	5,089.4
Choristoneura pinus	159.5	419.6
Lambdina fiscellaria fiscellaria	16.0	18.3
Malacosoma disstria	16,266.0	785.1

Table 20.2   *Estimate of areas of moderate to severe defoliation by four major defoliators in Canada.*

3.1. Lepidoptera

The families of Lepidoptera that are economically important are the same as in Europe.

3.1.1.   A case-study: the spruce budworm, **Choristoneura fumiferana**

The tortricid moth *Choristoneura fumiferana* occurs in the east of Canada and of the United States. It is primarily an inhabitant of the boreal forest. It feeds on the new growth of *Abies balsamea, Picea rubens, Picea glauca* and *Picea mariana*, and sometimes also on *Larix laricina, Tsuga canadensis, Picea engelmanni, Abies lasiocarpa* and *Pinus strobus*. The spruce budworm is a threat to more than fifty million hectares of forest. The radial growth of trees declines by 30 to 90% in the year following a heavy defoliation; tree mortality begins in the third year of an outbreak and seed production is almost nil. Old trees are particularly sensitive and they die after a few years of severe defoliation. The percentage of defoliation by *Choristoneura fumiferana* increases with the age and height of the trees (figure 20.5).

Choristoneura fumiferana is a wasteful feeder, eating the needles at their base and leaving the remainder enveloped in a mass of silk. During outbreaks the leaves turn russet and the damage can be seen from afar. *Choristoneura fumiferana* has a one year cycle with six larval instars. The second instar larvae emerge from hibernation in late April or May. The young larvae are yellowish with a dark brown head. They mine needles, unopened buds or male flowers and feed within the dilated shoots. Mature larvae are 18-24 mm long and have a black head and a reddish-brown body. The larvae gather two or three shoots together in a silk web to form a feeding tunnel. Pupation takes place among the foliage in June. Adults emerge in late June or early July. They have a wingspan of about 20 mm; the forewings are dull grey with brown bands and spots; the hind wings are light grey. The eggs are green and laid in clusters of about 20 on the needles of the host tree.

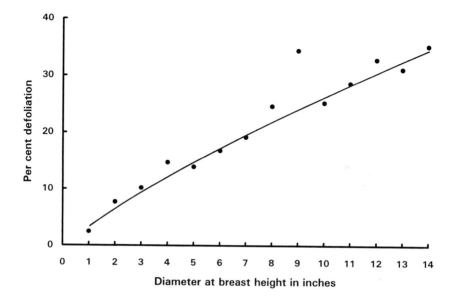

© Tec&Doc – Unauthorized photocopying is an offence

Figure 20.5. *Relationship between percentage of defoliation and diameter (in inches) at breast height of* Abies balsamea. *Black dots are average for many trees of the same diameter (Morris, 1963).*

They hatch in ten days and the larvae disperse among the foliage. The first instar larvae spin a hibernaculum under bark scales and moult to the second instar without feeding. Dispersal of adults is initiated by weather conditions. The moths begin their flight by ascending to a height of 150 to 300 m above the canopy. A large number of moths can invade regions as far as 600 km away from where they started.

Choristoneura fumiferana is attacked by a great number of predators and parasitoids. The parasitoids include more than 100 species of Diptera and Hymenoptera (Morris, 1963; Huber *et al.*, 1996). Only some of these parasitoids are common. *Trichogramma minutum* (Trichogrammatidae) is a parasitoid of eggs; *Apanteles fumiferanae* (Braconidae), *Glypta fumiferanae* and *Synetaeris tenuifemur* (Ichneumonidae) are parasitoids of small larvae; *Meteorus trachynotus* (Braconidae), *Actia interrupta*, *Eumea caesar* and *Phryxe pecosensis* (Tachinidae) are parasitoids of large larvae; *Agria housei* (Sarcophagidae) is a parasitoid of pupae. The parasitoids are not efficient when population densities of the budworm are high (cf. table 20.3). In the case of the parasitoids of pupae the limiting effects are due to a decrease of parasitism as pupal density increases (figure 20.6). To control this pest, insecticides were sprayed from aircraft on expanses of several million hectares.

The intervals between outbreaks range from 17 to 100 years, with an average of about 35 years (cf. figure 6.4). Tree ring analyses show that

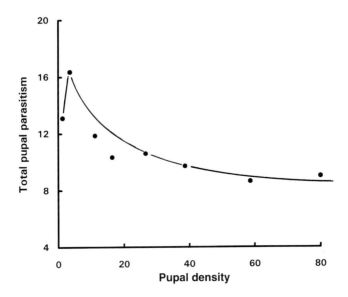

Figure 20.6. *Relationship between total pupal parasitism and pupal density of the spruce budworm (Morris, 1963).*

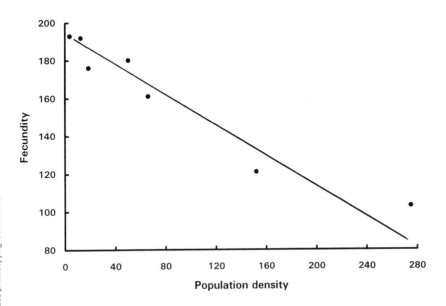

Figure 20.7. *Relationship between fecundity of females of the spruce budworm and the population density of larvae. Each black dot is the mean value for 10 data (Morris, 1963).*

outbreaks have occurred more frequently in the twentieth century than in the previous century, and that they are now of greater size. This is due to modern forestry practices (improved fire protection, increased harvesting, use of pesticides) which favour mature stands of spruce and fir and in turn favour the budworm.

Population studies on *Choristoneura fumiferana* were made by Morris (1963) in New Brunswick, Canada, during the period 1945-1959. In New Brunswick this insect generally remains at densities so low that it can be considered a rare species. Periodically, when conditions are favourable, it breaks out and causes severe defoliation and great mortality over large areas. In Canada *C. fumiferana* is a complex of three forms: the eastern form with one generation a year, a western form also with one generation a year, and another western form with a generation that lasts two years. The fecundity of females decreases when the population density of larvae increases (figure 20.7). Survival S of small larvae is a function of population density N according to the following relationship $S = a + b/N$.

x: age interval	N_x : number alive at beginning of x	Mortality factors and number dying	S_x : survival (%) within interval
Eggs	200	Parasites 18.0 Predators 12.0 Unknown 8.0	81%
Instars I-II	162	Dispersal 132.8	18%
Instars III-VI	29.2	Parasites 11.7 Disease 6.7 Unknown 6.7	14%
Pupae	4.1	Parasites 0.5 Predators 0.2 Unknown 0.7	66%
Moths	2.7		
Generation total	2.7	All factors 97.3	1%

Table 20.3. *Life table for the spruce budworm during an outbreak (Morris, 1963).*

3.1.2. *Some other important Lepidoptera*

Coleophoridae. *Coleophora laricella* (the larch casebearer) is a European species that was found in North America around 1880. This insect occurs in the South of Canada and the North of the United States. It attacks various species of larch and mines the needles. By covering the needles with a silk mesh the larvae build a shelter in which they spend their whole lives. Several enemies of *Coleophora laricella* have been imported from Europe including

Agathis pumila and *Chrysocharis laricinella* (chalcid wasps) which help to reduce the populations and the damage.

Gelechiidae. *Coleotechnites starki* (the lodgepole miner moth) mines the tip of the needles of lodgepole pine (*Pinus contorta*) causing them to turn reddish-brown. The damage may be noticeable, the annual incremental loss ranging from 20 to 75% in heavily infested stands. This moth occurs in British Columbia, Alberta, Saskatchewan and Montana. Prolonged and destructive outbreaks of an allied species, *Coleotechnites milleri*, occurred in Yosemite National Park. Defoliation during such outbreaks may kill mature lodgepole pine, which is the principal host. Other pines, red fir and mountain hemlock are also attacked.

Tortricidae. The genus *Choristoneura* includes many species. *C. occidentalis* (the western spruce budworm) is the most destructive defoliator in western North America. During its numerous outbreaks it attacks *Abies* spp., *Picea* spp., *Larix occidentalis* and *Pseudotsuga*. Another species, *Choristoneura conflictana* (the large aspen tortrix) lives on *Populus tremuloides* in a large part of North America. During outbreaks the foliage of other trees such as *Populus balsamifera*, *Betula papyrifera* and *Salix* may be attacked. This species has many parasitoids including *Omotoma fumiferanae* (Tachinidae) and a species of *Glypta* (Ichneumonidae).

Acleris gloverana (the western blackheaded budworm) whose preferred host is western hemlock also attacks Sitka spruce and white spruce. It is a destructive forest pest characterised by sporadic and spectacular increases in abundance followed two or three years later by rapid decline.

Geometridae. *Lambdina fiscellaria fiscellaria* (the hemlock looper) has an eastern distribution but extends into Alberta. Its hosts are species of *Abies*, *Picea* and *Tsuga*, but during outbreaks it also feeds on many other forest trees. This moth is the main enemy of *Abies balsamea* in Newfoundland where more than 700,000 hectares of forest were devastated. In the late 1960s the main agent of destruction of this moth was a disease caused by two pathogenic fungi, *Entomophthora sphaerosperma* and *E. egrossa*.

Lambdina fiscellaria lugubrosa occurs in the forests of the West. Many severe outbreaks of this insect are known. *Operophtera brumata* comes from Europe. It was first discovered in Canada in 1950; it spread to Nova Scotia and has now reached the Vancouver area. This moth is a pest of oak and of all deciduous trees. Biological control is successful against this pest (cf. chapter 6).

Lymantriidae. Species of the genus *Orgyia* are the tussock moths. *Orgyia antiqua* occurs in southern Canada and the northern United States as well as in Europe. Many trees are attacked, both conifers and hardwood. *Orgyia pseudotsugata* (the Douglas-fir tussock moth) is a major defoliator in forests of *Pseudotsuga menziesii* and of *Abies* spp. in western North America (figure 20.8). Among the most severe infestations were those in Washington State from 1927 to 1930 which caused the loss of over 700,000 m^3 of

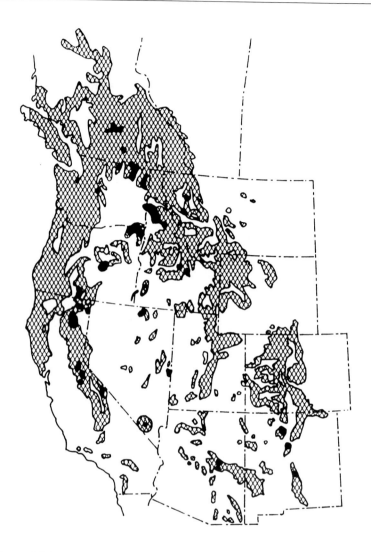

Figure 20.8. *Distribution map of* Orgyia pseudotsugata. *Outbreak areas are in black (Brookes et al., 1978).*

Douglas-fir and grand fir. Population densities of *Orgyia pseudotsugata* fluctuate 1,000 to 10,000-fold in some stands. During the past 70 years a major peak in abundance occurred every 8-9 years. These apparent cycles may be explained by various hypotheses. Severe defoliation by the larvae causes a shortage of food, an increase in larval starvation, higher susceptibility to diseases and enemies, and reduced fecundity. There is also an inverse relationship between the rate of increase and the population density in the two previous generations.

Orgyia pseudotsugata overwinters at the stage of eggs which are in diapause. The larvae hatch in spring; the young larvae feed on young needles which desiccate and turn brown. The older larvae are able to consume old needles. In August the larvae pupate in silk cocoons. Adults emerge in September. The females are wingless and produce a pheromone that attracts the winged males. After mating the females lay a mass of 30 to 300 eggs which overwinter.

A synthetic component, Z-6-heneicosen-11-one, attracts males and is sprayed from aircraft to disrupt insect mating. This method is effective, and no mating was observed when 72 g/ha of the synthetic pheromone were sprayed (Hulme & Gray, 1996, 1997). This method is used in pest management together with some pesticides and a nuclear polyhedrosis virus. The loss of tree growth following severe defoliation by *O. pseudotsugata* may be very important (figure 20.9).

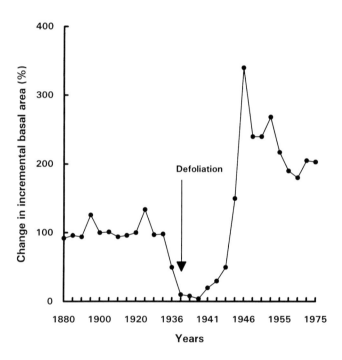

Figure 20.9. *Average annual growth of white fir,* Abies concolor, *defoliated by* O. pseudotsugata *from 1936 until 1938. Growth is expressed as a percentage of the growth of trees that are not defoliated. After the period of defoliation, growth is accelerated (Brookes et al., 1978).*

The gypsy moth, *Lymantria dispar*, was introduced accidentally in New England in 1869. The infestation now extends westward to Ohio, southward to North Carolina and northward to Montreal (Canada) and is likely to spread to most hardwood forests in the eastern United States and Canada in the near future. This major pest has already defoliated at least two million acres of forest and shade trees a year. Although the gypsy moth prefers oaks it feeds on more than five hundred plant species. The larvae migrate from the tree crown to litter at the base of the host tree. This behaviour increases vulnerability to mammals and often results in a high gypsy moth mortality. The vertebrates are essential in maintaining low populations. The main predators are mammals such as the rodent *Peromyscus leucopus* and the insectivores *Blarina brevicauda* and *Sorex cinereus*, and many bird species (Smith & Lautenschlager, 1978). Insect predators include native species of *Calosoma*, known as caterpillar hunters (Coleoptera, Carabidae) which hunt chiefly at night and pursue the caterpillars in trees. *Calosoma sycophanta* was introduced from Europe in the 19th century.

Arctiidae. *Hyphantria cunea*, the fall webworm, is a common defoliator of hardwood trees in the whole of North America.

Saturniidae. Members of this family are called giant silkworm moths. *Coloradia pandora*, the pandora moth, is a defoliator of pines in western forests. Outbreaks occur at intervals of about 20 to 30 years and continue for 6 to 8 years. The pandora moth has a two year life cycle. Mature larvae are about 25 mm long. In periods of abundance they were eaten by the Paiute Indians in California (cf. chapter 2.4).

Hyalophora cecropia, the cecropia moth, has a wingspan of 125 to 150 mm. Its larvae live on various trees including maple, elm and birch. *Hyalophora columbia* is the only species of the family that attacks conifers. It is found in the South of Canada and in the North of the United States. Its attacks on larch are of no economic importance.

Lasiocampidae. *Malacosoma* species are common and destructive insects. The larvae live in colonies in silken nests and are called "tent caterpillars". All species have similar life cycles. Adults are yellowish to reddish, stout, with a wingspan of 25 to 35 mm. They are nocturnal and lay their eggs in clusters of 150 to 250 in bands around twigs. Mature larvae are about 50 mm long and feed on various trees. *Malacosoma disstria*, the forest tent caterpillar, is the most widely distributed and destructive species. It feeds on poplars, willows and oaks. In western Canada outbreaks of this species have persisted up to six years. In 1962 defoliation extended to 200,000 km^2. The effects of forest fragmentation on *Malacosoma disstria* were studied by Rothman & Roland (1998). Outbreaks of this species last longer in fragmented forests than in continuous forests. As forest cover increases fewer moths survive. Increased mortality from a virus and from parasitoids are the main cause of this phenomenon. These effects of fragmentation are attributed to edge effect, that is to say to differences in microclimate between the forest edge and the interior habitat. Conversely, outbreaks of *Choristoneura* species are more severe in continuous than in fragmented forests. Partial protection against *Malacosoma disstria* and *Lymantria dispar* is obtained by transgenic *Populus* plants expressing a *Bacillus thuringiensis* endotoxin (Robison *et al.*, 1994). *Malacosoma californicum*, the western tent caterpillar, has six subspecies distributed in the western United States and

Canada. The host trees are *Quercus, Salix, Populus, Betula* and *Alnus*. Economic damage is minor except in the case of outbreaks of the subspecies *fragile* in the Rocky Mountains.

3.2. Hymenoptera

The suborder Symphyta comprises the sawflies which include major pests. Those that attack conifers belong to the following families.

Tenthredinidae. *Pristiphora erichsonii*, the larch sawfly, is a Holarctic species and the most destructive enemy of *Larix laricina* (cf. chapter 5 and chapter 10.1). Five species of Tenthredinidae which mine the leaves of birch have been introduced into eastern North America from Europe since 1920: *Messa nana, Scolioneura betuleti, Profenusa thomsoni, Fenusa pusilla* and *Heterarthrus nemoratus*. Only two species, *F. pusilla* and *P. thomsoni* have become abundant in Alberta. Female *F. pusilla* emerge in early spring, mate or reproduce parthenogenetically, and oviposit in the mesophyll of young birch leaves. The first four larval instars feed within a blotch mine and the last instar drops to the ground and pupates. This species has one to three generations per year in Europe and three to four in Québec (Digweed *et al.*, 1997).

Diprionidae. Four species were introduced from Europe (cf. chapter 10.1): *Diprion hercyniae* (the European spruce sawfly), *Diprion similis* (the introduced sawfly), *Neodiprion sertifer* (the European pine sawfly) and *Neodiprion abietis* (the balsam fir sawfly).

Some minor species of sawfly attack hardwood trees. *Arge pectoralis* attacks birch. *Cimbex americana* attacks elm, willow and other trees; it is the largest North American sawfly with a body about 25 mm long and a wingspan of about 40 mm.

4. SEED AND CONE INSECTS

The seed and cone insects of North America have been studied by Hedlin *et al.* (1980). Several species of Lepidoptera, Hymenoptera and Diptera and some Coleoptera are major pests of seeds and cones, especially in seed orchards (cf. chapter 13). These insects can destroy a large quantity of the seed production, have an adverse impact on forest regeneration, and reduce the value of seeds produced (table 20.4).

	No insect damage	Insect damage in 1984
Initial number of cones	200	200
Cone mortality %	0	60
Average number of sound seeds per cone	47	0.9
Sound seeds per tree	9,400	72
Pounds of sound seeds per tree	0.82	0.01
Value of seeds per tree	$ 31.60	$ 0.38

Table 20.4. *Seeds produced per* Pinus ponderosa *tree in Arizona. Comparison of trees not damaged by insects with trees damaged by insects in the year 1984 (Blake et al., 1986).*

Cone insects often belong to the same families and genera as in Europe (table 20.5).

Families	Genera			
	Abies	*Picea*	*Pinus*	*Pseudotsuga*
Olethreutidae	*Epinotia* *Laspeyresia* *Barbara* *Eucosma*	*Laspeyresia* *Barbara*	*Eucosma* *Rhyacionia*	*Barbara*
Geometridae	*Eupithecia*		*Eupithecia*	
Pyralidae	*Dioryctria*	*Dioryctria*	*Dioryctria*	
Tortricidae			*Choristoneura*	*Choristoneura*
Cecidomyiidae	*Asynapta* *Dasineura*	*Dasineura* *Mayetiola*	*Asynapta*	*Contarinia*
Chloropidae	*Earomyia*		*Hapleginella*	
Anthomyiidae	*Hylemyia*	*Hylemyia*		
Torymidae	*Megastigmus*	*Megastigmus*	*Megastigmus*	*Megastigmus*
Anobiidae			*Ernobius*	
Scolytidae			*Conophthorus*	
Curculionidae			*Conotrachelus*	

Table 20.5. *Main genera of insects attacking cones and seeds of four genera of conifer.*

4.1. Lepidoptera

The Lepidoptera that attack conifer cones belong to the following three families.

Olethreutidae. *Barbara colfaxiana* is a pest of Douglas-fir. *Laspeyresia youngana* is found in the whole of Canada and in the North of the United States. It feeds on the cones of various species of *Picea* and causes great damage in some areas. Another member of this family, *Melissopus latiferreanus*, attacks hardwood tree seeds such as acorns.

Pyralidae. *Dioryctria abietivorella* is a trans-continental species that attacks the cones of firs, Douglas-fir and other conifers. *Dioryctria auranticella* is a major pest in the West of North America.

Tortricidae. *Cydia strobilella* attacks the seeds of most spruces.

4.2. Hymenoptera

One family, the Torymidae, is of economic importance: members of the genus *Megastigmus*, which has 23 species, spend their whole life cycle in one seed. The only evidence of attack by *Megastigmus* is the adult emergence hole. *Megastigmus laricis* attacks larch and *M. specularis* balsam fir. *Megastigmus spermotrophus* occurs on *Pseudotsuga menziesii* and *P. macrocarpa* in western North America, and is introduced in Europe (cf. chapter 13.1).

4.3. Diptera

The family Cecidomyiidae contains many species that attack cones. The Douglas-fir gall midge, *Contarinia oregonensis*, induces the formation of galls in the seed coat that destroy the seeds. Moreover necrosis of the cone scales follows the infestation. *Mayetiola thujae* damages the seeds of *Thuja plicata*. The family Anthomyiidae includes species of *Strobilomyia* known as cone maggots. The larvae of *Strobilomyia neanthracina* and *S. appalachensis* (the spruce cone maggots) attack spruce cones and in some years may destroy the entire seed crop. *Dioryctria abietivorella* preferentially exploits cones previously damaged by *Strobilomyia* (Fidgen & Sweeney, 1996).

4.4. Coleoptera

The scolytid genus *Conophthorus* contains a large number of species that attack numerous species of conifer. *C. resinosae* lives in the cones of red pine and can be one of its major pests. This small (3 mm) scolytid bores a gallery to the centre of the cone where it lays its eggs. The larvae feed on the seeds and scales of the cone. *Conophthorus coniperda* attacks the cones of white pine (*Pinus monticola*) and is a serious pest in orchards. Pest management is limited to the burning of infested cones while they are on the ground. The pheromones of this insect have recently been identified. The female produces (+)-*trans*-pityol and (2*R*,5*S*)-2-(1-hydroxy-1-methylethyl)-5-methyl-tetrahydrofuran. These pheromones may be an attractant that can be used to lure males and to lower populations (de Groot & DeBarr, 1998).

The Coleoptera also include species of *Ernobius* (family Anobiidae) and *Conotrachelus* (Curculionidae) that attack cones and seeds.

5. INSECTS THAT ATTACK BUDS AND SHOOTS

5.1. A case-study: the white pine weevil, *Pissodes strobi*

The genus *Pissodes* has a Holarctic distribution with 29 species in North and Central America. Most species attack weakened or recently dead conifers and are not therefore of economic importance. Others such as *Pissodes strobi* and *P. terminalis* attack the terminal leaders of living trees and cause considerable damage. *P. schwarzi* is known to attack and kill seedlings. *P. fasciatus* and *P. strobi* are implicated in the dissemination of fungal diseases of conifers.

The most destructive species is *P. strobi*, the white pine weevil, a transcontinental species present from Canada to the North of California that attacks the terminal leaders of 2 to 5 year old *Pinus* and *Picea* saplings. In some areas 30 to 40% of trees may be infested each year. The consequences of these attacks are deformations of trees, defects, and loss of saleable wood. The impact of *P. strobi* has given rise to a vast body of literature on this species (Alfaro *et al.*, 1994; Langor, 1998). The adults of *P. strobi* overwinter at the base of the trees from which they emerged in the previous autumn. In late March or April they fly to the terminal leaders of trees where they feed and mate. Oviposition begins soon after. Eggs are laid under the apical bud in feeding punctures. After hatching, the larvae consume the phloem around the leader, causing its destruction. Pupation takes place in chambers excavated in the xylem. New adults emerge from July to September. Two aggregation pheromones, grandisol and grandisal, were identified from males of *Pissodes strobi* and *P. approximatus* (cf. chapter 15.4 and figure 15.15). A tree that has been attacked can grow a new leader which produces deformed and worthless tree with a bayonet-shaped trunk with a forked appearance.

In natural stands of spruce *P. strobi* is a rare insect. This is because spruce in these forests regenerates under shade. The weevil is negatively affected by shade because the adults require certain conditions of temperature and humidity for oviposition. "The widespread adoption of clearcutting in the

60's and 70's and the planting of single-species stands created optimum conditions for weevil development. Plantation of large areas with vigorous regeneration created an enormous food supply, planted in open stands where heat accumulation is more than appropriate for weevil development. Under these conditions population explosions or outbreaks developed." (Alfaro *et al.*, 1994, p. 9).

Laboratory experiments have shown that myrcene (a monoterpene) is highly repulsive to *P. strobi*. The percentage of myrcene in fast-growing terminals is lower than in slow-growing ones, and the former are more attacked than the latter (figure 20.10). Trees resistant to attack also have a

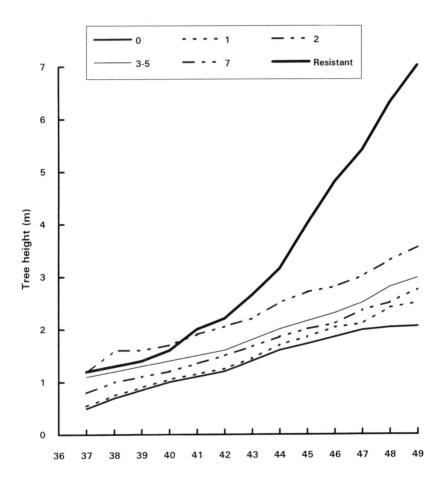

Figure 20.10. *Height growth of Sitka spruce in an experimental plot in British Columbia. The trees were subjected to between zero and seven attacks during the period 1937-1949. Note that attacks are more numerous on fast-growing trees. One fast-growing tree was resistant and was never attacked (Alfaro et al., 1994).*

greater number of vertical resin canals than susceptible trees (figure 20.11). *Pissodes strobi* responds to host kairomones when searching for a suitable host. Two of these kairomones are monoterpenes (isoamyl and isopentenyl isovalerate) which are present in lower amounts in resistant trees than in susceptible trees.

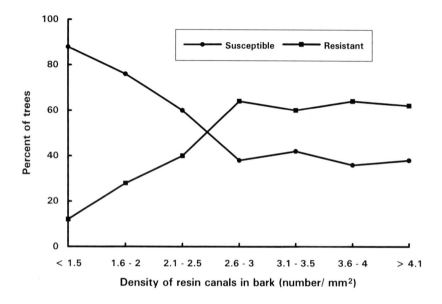

Figure 20.11. *Percentage of trees that are resistant or susceptible to attack by* Pissodes strobi *versus total resin canal density in bark (Alfaro et al., 1996).*

Few natural enemies of *Pissodes strobi* are known. One of these is the facultative predator *Lonchaea corticis* (Diptera). There is a strong negative relationship between the survival of weevils within the terminal shoot and the abundance of *Lonchaea corticis* (Nealis, 1998). Integrated pest management, and particularly the use of resistant trees, seems to be the best method of control of *Pissodes strobi*.

5.2. Other bud and shoot species

The order Lepidoptera includes *Rhyacionia buoliana* (family Olethreutidae), a moth of European origin the larvae of which mine the buds and shoots of young pines and deform them (figure 12.15). The larvae of *Eucosma gloriola*, a moth of the same family, attack pines and girdle the shoots, which wilt or break at the point of girdling. Some species of *Dioryctria* (family Pyralidae) bore into the branches and shoots of pines. The larch shoot moth *Argyresthia laricella* (family Yponomeutidae) attacks tamarack (*Larix laricina*) and western larch (*Larix occidentalis*) and kills the tips of the twigs. The genus *Pityophthorus* (Scolytidae) is the largest genus of North American bark beetle, with about 120 species. The larvae of several species kill the tips of pine shoots, causing the trees to become bushy.

6. SAP-SUCKING AND GALL-FORMING INSECTS

6.1. Sap-sucking insects

These insects belong to the order Homoptera, to the four families Aphididae, Adelgidae, Cercopidae and Eriosomatidae, and to the superfamily Coccoidea (cf. chapter 11.3 for Saratoga spittle bug).

Aphididae. Most of the aphids that attack conifers belong to the genus *Cinara* (cf. chapter 11.2). The spruce aphid, *Elatobium abietinum*, feeds almost exclusively on spruce and is native to Europe (cf. chapter 15.2 and figure 11.8). The woolly aphid, *Eriosoma lanigerum*, feeds on elm and on apple trees, on both of which it forms galls.

Adelgidae. (or Chermesidae). These insects only attack conifers (cf. chapter 12.2) *Dreyfusia piceae* is a European species which is highly destructive to several species of *Abies*. Another species, *Gilletteella* (= *Adelges*) *cooleyi*, attacks various spruces (cf. chapter 12).

Eriosomatidae. *Pamphigus* species form galls on the leaves of poplars.

Coccoidea. The species of this superfamily are the scale insects. They include species of *Matsucoccus* and *Chionaspis* which live on pines, and the beech scale insect, *Cryptococcus fagisuga*, which was introduced from Europe and causes beech bark disease on *Fagus grandifolia* in the North-east of the United States and the South-east of Canada.

6.2. Gall-forming insects

Besides mites (Acarina), gall-forming insects are found among the Homoptera, but above all in the orders Diptera (family Cecidomyiidae) and Hymenoptera (family Cynipidae).

6.2.1. Cecidomyiidae

There are between 1,500 and 2,000 species of gall midge in North America. *Paradiplosis tumifex* causes the formation of galls on needles of balsam fir. The needles become swollen at the base and infestations may cause premature needle drop. Young balsam fir trees are more severely affected than older ones. This species can cause damage in Christmas tree plantations. *Dasineura balsamicola* is an inquiline that lives in the galls formed by *Paradiplosis*.

6.2.2. Cynipidae

The gall wasps form, as they do in Europe, a great variety of galls on trees, and mainly on oaks. Although they are often abundant these galls cause no appreciable damage. The genus *Neuroterus* has more than fifty species in North America. *Neuroterus saltatoris*, the jumping gall wasp, is a western species that produces small round galls 1.0 - 1.5 mm in diameter. The name of this species refers to the jumping behaviour of the galls after they drop to the ground. They cause premature browning of the foliage of oak trees.

7. PHLOEM- AND WOOD-BORING INSECTS

There are many species of economic importance among phloem- and wood-boring insects (cf. chapters 14 to 16). In North America the main species belong to the beetle families Scolytidae and Cerambycidae. The other main families of insect in this category are Coleoptera of the families Buprestidae and Lyctidae, and Lepidoptera, of the families Cossidae and Sesiidae. Termites do not usually play an important role in forest (cf. chapter 16.5).

Some Elmidae are aquatic wood-boring beetles (cf. chapter 16.1.9), as well as some Diptera (cf. chapter 16.2.3).

7.1. Scolytidae (bark beetles)

The three most important genera of Scolytidae are *Dendroctonus, Ips* and *Scolytus*.

7.1.1. *The genus* Dendroctonus

The genus *Dendroctonus* (the name means "tree killer") has 19 species, 17 of which are found in North and Central America. The non-American species are *Dendroctonus micans* which lives in Europe and Asia (cf. chapter 15.3.10) and *D. armandi* in China. These two taxa, as well as other scolytids, migrated from a spruce refugium in Alaska to Siberia during the Wisconsin glaciation. Evidence that some of these migrations occurred recently is shown by the existence of at least twelve species of scolytid that are identical or very similar in America and Eurasia. These include *D. micans* and *D. punctatus* (Furniss, 1996). With the exception of *D. frontalis* the most important *Dendroctonus* are mainly western species (cf. chapter 15.3 and figure 15.20):

(a) *Dendroctonus adjunctus* (the roundheaded pine beetle) is destructive in stands of overmature ponderosa pines.

(b) *Dendroctonus approximatus* (the larger Mexican pine beetle) attacks

Pinus ponderosa, P. engelmanni, P. leiophylla and many Mexican pines. It is a secondary species that attacks the base of trees infested by other *Dendroctonus* and *Ips*.

(c) *Dendroctonus brevicomis* (the western pine beetle) breeds in overmature ponderosa pine and Coulter pine (cf. chapter 15.2.2). A study of the population dynamics of these species was made by Stark & Dahlsten (1970).

(d) *Dendroctonus frontalis* (the southern pine beetle) is a destructive enemy of many species of pine.

(e) *Dendroctonus murrayanae* (the lodgepole pine beetle) breeds in *Pinus contorta* and *Pinus banksiana*.

(f) *Dendroctonus ponderosae* (the mountain pine beetle) has four major hosts in the western United States: *Pinus lambertiana, P. contorta, P. ponderosa* and *P. monticola*. It is the most important enemy of these trees in the West (cf. figures 6.12 and 15.12). A study of the allozymes and of the morphology of *D. ponderosae* suggests that there is a genetic differentiation between beetles living in the various species of pine and that the formation of host races is occurring (Sturgeon & Mitton, 1986). A list of 68 natural enemies and insects associated with *Dendroctonus ponderosae* is given by Dahlsten & Stephen (1974). The influence of forest management practices on populations of this species was described by Amman & Logan (1998). According to these authors the practices "have created conditions that seriously debilitate the capacity of the host to defend itself... Forest management often has attempted to maintain a status quo of current stand conditions that extends beyond the natural cycle of forest regeneration and renewal. The result of both management practices has been the same – large acreages of forests that are highly vulnerable to mountain pine beetle attack." In 1898 an outbreak of *Dendroctonus ponderosae* in the Black Hills severely damaged ponderosa pines. The studies made by the entomologist Andrew D. Hopkins in 1901 following this attack are at the heart of American forest entomology and of the foundation in 1902 of the *Division of Forest Insect Investigations* of the U.S. Department of Agriculture (Furniss, 1997). In Canada a survey of mountain pine beetle infestations using remote sensing techniques was initiated for the identification of stands killed by beetles, mapping their extent and the damage to trees. In a managed area of 4,673 hectares the estimated loss of volume was more than 453,000 m^3 of lodgepole pine, averaging 93 m^3 per hectare (Gimbarzevsky *et al.*, 1992).

(g) *Dendroctonus pseudotsugae* (the Douglas-fir beetle) is the most important bark beetle of Douglas-fir and of western larch (cf. figure 15.13). A list of insects associated with Scolytidae in western Washington, and particularly with *D. pseudotsugae* and *Scolytus unispinosus*, was made by Deyrup & Gara (1978).

(h) *Dendroctonus punctatus* (the boreal spruce beetle) is an enemy of *Picea glauca* and of *Picea sitchensis* in boreal North America, from Newfoundland to Alaska.

(i). *Dendroctonus rufipennis* (the spruce beetle) occurs throughout the range of the genus *Picea* in North America where it attacks mature spruce trees (*Picea engelmanni, P. glauca* and *P. sitchensis*). Bark beetles are associated with blue stain fungi. During an attack by aggressive bark beetles the combined effect of the insect and the fungi is responsible for the tree mortality (cf. chapter 3.1). The two blue stain fungi *Ceratocystis rufipenni* and *Leptographium abietinum* are associated with *Dendroctonus rufipennis* and their pathogenicity has been established experimentally. *Ceratocystis rufipenni* plays the most important role in the death of trees following a mass attack by *Dendroctonus*. This species is destructive to *Picea lutzii* and *Picea glauca* in Alaska. A list of arthropods associated with *D. rufipennis* in Alaska (Gara *et al.*, 1995) mentions more than 30 Coleoptera, 16 Diptera and about 30 Hymenoptera. The Coleoptera include 11 species of Scolytidae. The mean density (number per 1,000 cm^2 of bark) of some scolytids, predators and parasitoids in a locality is as follows:

Scolytidae:

> *Dendroctonus rufipennis*: 50.9
> *Dryocoetes affaber*: 175.7
> All Scolytidae: 460.1

Predators and parasitoids:

> Coleoptera (mainly *Rhizophagus* and *Lasconotus*): 3.8
> Diptera (*Medetera* and *Lonchaea*): 14.1
> Hymenoptera (Braconidae and Chalcidoidea): 1.8
> All predators and parasitoids: 19.7

This study shows that the most important factor in the survival of *Dendroctonus rufipennis* is the interspecific competition with other scolytids, particularly with *Dryocoetes affaber*. The impact of predators and parasitoids is less important.

(j). *Dendroctonus valens* (the red turpentine beetle) occurs in practically all forests and attacks all species of pine.

(k). *Dendroctonus jeffreyi* (the Jeffrey pine beetle) is host-specific on Jeffrey pine.

(l). *Dendroctonus simplex* (the eastern larch beetle) infests *Larix laricina* in Alaska and in the western provinces of Canada.

The biology of *Dendroctonus* has been much studied because of the great economic importance of these insects. Data have already been given on these scolytids in chapter 15. The two species *D. micans* and *D. punctatus* have several unusual biological features (Furniss, 1996) such as mating among siblings in the brood chamber before emergence from their host tree, aggregation by larvae to overcome the defences of the tree rather than aggregation by the adults, and development of a complete generation in a tree without killing it.

7.1.2. A case-study: **Dendroctonus frontalis**

This species, known as the southern pine beetle (SPB), is a very destructive enemy of pines in the south-western United States and in parts of Mexico and Central America. The adults are black-brown and about 3 mm long. The SPB has four to seven generations per year. Females are pioneer beetles that find suitable host trees. "Primary" attraction to volatile compounds emitted by trees exists in several species of scolytid but not in the case of the SPB. Random landing by the dispersing beetles seems to be the means used by the SPB to find its host trees. "Secondary" attraction begins as soon as the females have bored into the tree. This phase of aggregation is triggered by a release of frontalin which is the primary aggregation pheromone, and is found in the hindgut of newly-emerged females. *Trans*-verbenol is also produced by females but its role is not well known. Together with the odour of volatile compounds of the tree (mainly α-pinene), frontalin attracts numerous individuals of the SPB (mainly males). The males release *endo*-brevicomin which is attractive to the females. The males also produce verbenone which is believed to affect the behaviour of these insects by reducing the number of males attracted. When the number of SPB increases, the high concentration of the pheromone has an inhibitory effect and the tree loses its attractiveness (Thatcher *et al.*, 1981).

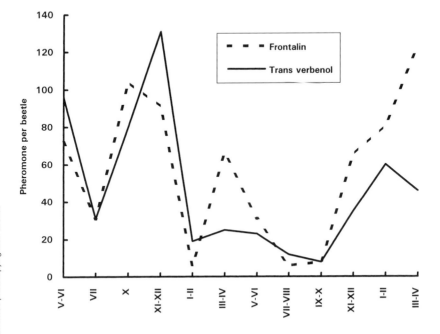

Figure 20.12. *Seasonal variations in pheromone content of the females of* Dendroctonus frontalis. *The amount of pheromone is in μg per beetle x 1,000 for frontalin and x 10 for* trans-*verbenol. The maximum is in autumn (Thatcher* et al., *1981).*

The pheromone content of the SPB reaches a peak in autumn when the beetles are less aggregated and when chemical communication between them requires a greater amount of pheromones than in summer (figure 20.12). The percentage of response of the beetles to the pheromones is lower in summer when colonisation of the trees expands and when the pheromone content is lower than in autumn (figure 20.13).

There are changes in habitat conditions during the development of the bark beetles. In the case of *Dendroctonus frontalis* the beetle influences the moisture in the phloem in an attacked tree (figure 20.14). In general there is a drying of the phloem from the egg stage to the larval stages. When high

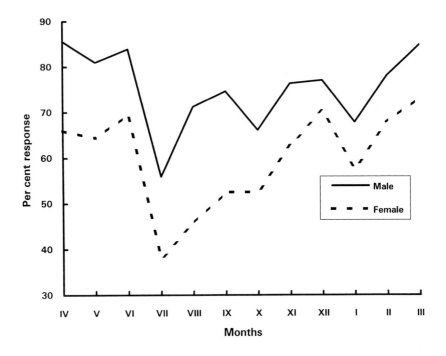

Figure 20.13. *Seasonal variation in response to an attractant (frontalin, verbenone and turpentine) by* Dendroctonus frontalis *in the laboratory. The minimum response is in summer (Thatcher et al., 1981).*

phloem moisture is maintained in bolts in the laboratory, the survival of the SPB is lower than in bolts with lower phloem moisture. Phloem moisture may be a factor in the expansion or decline of the SPB (Webb & Franklin, 1978).

Populations of *Dendroctonus frontalis* from *Pinus taeda* in the United States differ in pheromone production and do not interbreed with beetles from *Pinus leiophylla* in Mexico. The populations from *Pinus tenuifolia* in

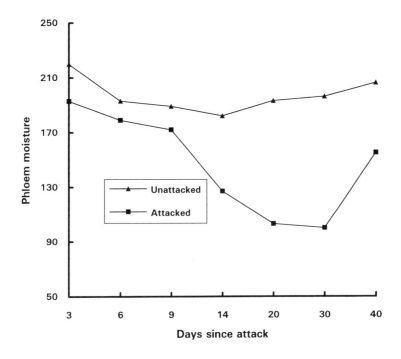

Figure 20.14. *Mean phloem moisture (% of dry weight) in healthy shortleaf pine trees and in trees attacked by the southern pine beetle (Thatcher et al., 1981).*

Guatemala are close to the Mexican populations. These differences support the conclusion that the populations in Mexico and Guatemala differ from those in the United States and constitute a different species.

Like many scolytids, the southern pine beetle has many predators, parasitoids and commensals that live in its galleries. Birds, especially woodpeckers, are important enemies. Woodpeckers may be fifty times more numerous in stands infested by the SPB than in non-infested stands. Insects and acarians are the main enemies of the SPB; they almost all belong to the same families (and often to the same genera) as those that are found with other scolytids. A hundred species are known to predate the SPB (cf. chapter 15.5); the main ones are listed in table 20.6. The clerid beetles *Enoclerus lecontei* and *Thanasimus dubius* are major predators of scolytids and particularly of *Dendroctonus* (figure 20.15).

The number of parasitoids of *Dendroctonus frontalis* is influenced by two factors: bark thickness and the number of scolytid eggs. Most parasitoids are more numerous when bark is thinner. Lowest parasitoid populations are found in autumn and in winter when the SPB populations are also low. A list of the parasitoids is given in table 20.7.

Hemiptera	Anthocoridae	*Lyctocoris campestris*
		Lyctocoris elongatus
		Scoloposcelis flavicornis
		Scoloposcelis mississippensis
Coleoptera	Carabidae	*Tachyta parvicornis*
		Dromius piceus
		Pinacodera limbata
		Stenolophus lineola
	Histeridae	*Cylistix attenuata*
		Cylistix cylindrica
		Epierus pulicarius
		Platysoma parallelum
		Plegaderus pusillus
	Staphylinidae	*Leptacinus parumpunctatus*
	Trogositidae	*Corticotomus parallelus*
		Temnochila virescens
		Tenebroides marginatus
	Cleridae	*Cymatodera undulata*
		Phyllobaenus pallipennis
		Thanasimus dubius
	Silvanidae	*Nausibius clavicornis*
		Cathartosilvanus imbellis
	Colydiidae	*Aulonium ferrugineum*
		Colydium lineola
		Lasconotus referendarius
	Tenebrionidae	*Corticeus glaber*
		Corticeus parallelus
Diptera	Dolichopodidae	*Medetera bistriata*
		Medetera maura
	Lonchaeidae	*Lonchaea auranticornis*
Hymenoptera	Formicidae	*Camponotus* spp.
		Solenopsis picta
		Dorymyrmex pyramicus

Table 20.6.   *Main arthropod predators of the southern pine beetle*
Dendroctonus frontalis *(Thatcher et al., 1981).*

Competition is known to exist between the southern pine beetle and other wood-boring Coleoptera. Foraging by the larvae of *Monochamus titillator* (Cerambycidae) reduces the survival rate of the scolytid larvae, the mortality of which increases up to 70%. Competition also exists with other scolytids. In a loblolly pine occupied by five species of scolytid the distribution from the base of the tree to the upper part is as follows:

Dendroctonus terebrans is almost always near the base; then come *Dendroctonus frontalis, Ips grandicollis* and *Ips calligraphus. Ips avulsus* is found in the upper part of the tree.

Braconidae	*Atanycolus comosifrons*	Bethylidae	*Parasierola* sp.
	Atanycolus ulmicola	Eurytomidae	*Eurytoma cleri*
	Cenocoelius nigrisoma		*Eurytoma tomici*
	Coeloides pissodis		*Eurytoma* sp.
	Compyloneurus movoritus		
	Dendrosoter sulcatus	Scelionidae	*Gyron* sp.
	Meteorus hypophloei		*Idris* sp.
	Spathius canadensis		*Leptoleia* sp.
	Spathius pallidus		*Probaryconus heidemanni*
	Vipio rugator		*Telenomus podisi*
		Pteromalidae	*Dinotiscus dendroctoni*
			Heydenia unica
Torymidae	*Lochites* sp.		*Rhopalicus pulchripennis*
	Liodontomerus sp.	Ichneumonidae	*Cremastus* sp.
	Roptrocerus xylophagorum		
	Roptrocerus eccoptogastri	Eupelmidae	*Arachnophaga* sp.
	Roptrocerus sp.		*Eupelmus cyaniceps*

Table 20.7. *Main species of arthropod parasitoids of the southern pine beetle. All species belong to the Hymenoptera.*

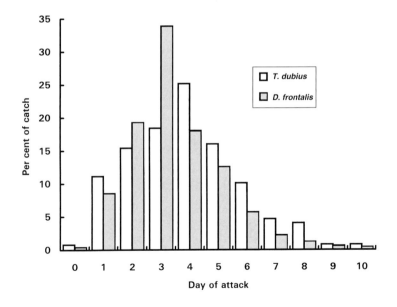

Figure 20.15. *Sequence of arrival of* Thanasimus dubius *and of* Dendroctonus frontalis *on trees under mass attack by the scolytid. The predator arrives almost at the same time as the scolytid (Thatcher* et al., *1981).*

7.1.3. The genus *Ips*

Species of the genus *Ips* come second in destructiveness after *Dendroctonus*. They mainly attack pines and spruces. A characteristic feature of the genus is the concave declivity of the elytra, each elytron being margined with three to six spines. Among the main species the following may be mentioned:

Ips confusus (the piñon ips) lives on *Pinus edulis, P. monophylla* and other pines.

Ips paraconfusus (the California five-spined ips) resembles *I. confusus* and lives on numerous species of pine.

Ips pini (the fir engraver) attacks pines and sometimes also spruces. It is a transcontinental species, one of the commonest bark beetles, and is sometimes a serious pest.

Ips calligraphus (the six-spined ips) attacks pines in the whole of North America.

Like all scolytids, *Ips* have many predators and parasitoids. Some of the most abundant are colydiids belonging to the genus *Lasconotus* that inhabit the galleries of bark beetles; these are omnivorous beetles that feed on phloem but also on the larvae and pupae of *Ips* and other scolytids (Rohlfs & Hyche, 1981). Six species of Colydiidae were collected on *Pinus* attacked by *Ips grandicollis* and *Ips calligraphus*; these species are *Pycnomerus sulcicollis, Colydium nigripenne, Aulonium tuberculatum, A. ferrugineum, Lasconotus pusillus* and *L. referendarius*. The peak arrival of adult *Lasconotus pusillus* is synchronised with the peak arrival of attacking *Ips*.

7.1.4. The genus Scolytus

Many species of *Scolytus* attack conifers. *Scolytus unispinosus*, the Douglas-fir engraver, lives on *Pseudotsuga menziesii* and occasionally on *Abies* and *Tsuga*. It breeds in the phloem of recently-killed Douglas-fir trees. This species is localised in small trees and branches partially exposed to sunlight and is common in western forests. The insects associated with *Scolytus unispinosus* have been studied by Deyrup (1989).

The primary hosts of *Scolytus ventralis*, the fir engraver, are species of *Abies*. This scolytid has caused heavy losses in the western United States. Two other species were introduced from Europe: *Scolytus multistriatus* is the main vector of Dutch elm disease, and *Scolytus rugulosus* breeds mainly in fruit trees.

7.2. Cerambycidae and Buprestidae

Wood-boring insects belong to many families (cf. chapter 16). The family Cerambycidae, or "wood borers", is the richest in species. The larvae of the

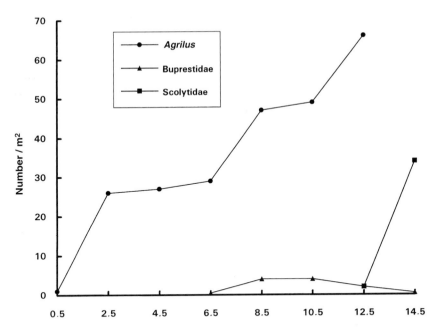

Figure 20.16. *Mean abundance of* Agrilus bilineatus, *of other Buprestidae and of Scolytidae in oaks after one season of attack by* Agrilus bilineatus *(Haack et al., 1983). In abscissa, height above ground (metres).*

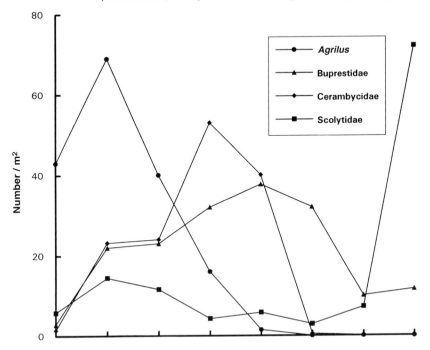

Figure 20.17. *Mean abundance of various Coleoptera in oak after two years of attack by* Agrilus bilineatus *(Haack et al., 1983). Same abcissa as in figure 20.16.*

genus *Monochamus* are known as "sawyers". These larvae cause damage to the wood of dying or recently-dead and felled conifers (*Pinus, Picea, Abies, Pseudotsuga*). Buprestid larvae are known as "flat-headed wood borers" because of their flat shape (cf. figure 16.18), and the adults as "metallic wood borers" or "jewel beetles" because of the brilliant metallic colours of many species. The genera *Chalcophora* and *Chrysobothris* have numerous species. *Melanophila* species are attracted to scorched trees. *Buprestis aurulenta* is a brightly-coloured species that lives in conifers; its larvae can live fifty years or more.

Successions of insects in weakened or dying trees have been described in chapter 17.1.5. Another example is that of oaks (*Quercus* spp.) stressed by drought and by defoliation caused by *Alsophila pometaria*, the fall cankerworm (Lepidoptera, Geometridae). The biology of the buprestid *Agrilus bilineatus* has been studied by Dunbar & Stephen (1976). This species is normally the first wood-borer to appear in declining oaks (cf. chapter 5.1 and 6.3). The attack occurs first in the crown and then downward along the trunk (figure 20.16). The oak tree is then likely to become host to other wood-borers and phloem-borers of the families Scolytidae, Buprestidae and Cerambycidae. After one season of attack by *Agrilus bilineatus* (figure 20.16) the buprestid *Chrysobothris femoratus* and the scolytid *Pseudopityophthorus minutissimus* appear. After two seasons of attack *Agrilus bilineatus* becomes less abundant and various scolytids (*Monarthrum* spp., *Pseudopityophthorus minutissimus, Xyloterus politus*) are abundant, mostly higher up in the tree; the cerambycids *Xylotrechus colonus, Neoclytus acuminatus* and *Sarosesthes fulminans* are abundant, especially at mid height (figure 20.17). After more than two seasons of attack *Agrilus bilineatus* disappears and cerambycids become the most abundant species in the dead tree (Haack *et al.*, 1983).

REFERENCES

ALFARO R. I. *et al.*, (1994). The White Pine Weevil: Biology, Damage and Management. *Canada-British Columbia Partnership Agreement on Forest Resource Development, Victoria, B. C.*, 311 pp.

ALFARO R. I. *et al.*, (1996). White spruce resistance to white pine weevil related to bark resin canal density. *Canad. J. Bot.*, **75**: 568-573.

AMMAN G. D., LOGAN J. A., (1998). Silvicultural control of mountain pine beetle: prescriptions and the influence of microclimate. *American Entomologist*, **44**: 166-177.

BARBOUR M. G., BILLINGS W. D., (1988). North American Terrestrial Vegetation. *Cambridge University Press*.

BERRYMAN A. A., (1966). Studies on the behavior and development of *Enoclerus lecontei* (Wolcott), a predator of the western pine beetle. *Canad. Entomol.,* **98**: 519-526.

BLAKE E. A. *et al.,* (1986). Insects destructive to ponderosa pine cone crops in northern Arizona. Proceedings Conifer Tree Seed in the Inland Mountain West Symposium. *USDA For. Serv. Gen. Tech. Rep. INT-203.*

BROOKES M. *et al.,* (1978). The Douglas-fir tussock moth: A synthesis. *U.S. Dept. Agric. Tech. Bull.* 1,585.

CORNELL H. V., WASHBURN J. O., (1979). Evolution of the richness area correlation for cynipid gall wasps on oak trees: a comparison of two geographic areas. *Evolution,* **33**: 257-274.

DAHLSTEN D. L., STEPHEN F. M., (1974). Natural enemies and insect associates of the mountain pine beetle, *Dendroctonus ponderosae* (Coleoptera: Scolytidae), in sugar pine. *Canad. Entomol.,* **106**: 1,211-1,217.

DE GROOT P., DeBARR G. L., (1998). Factors affecting capture of the white pine cone beetle, *Conophthorus coniperda* (Schwarz) (Col., Scolytidae) in pheromone traps. *J. Appl. Ent.,* **122**: 281-286.

DEYRUP M., (1989). Deadwood decomposers. *Natural History*, 84-91.

DEYRUP M., GARA R. I., (1978). Insects associated with Scolytidae (Coleoptera) in western Washington. *Pan-Pacific Entomologist,* **54**: 270-282.

DIGWEED S. C. *et al.,* (1997). Exotic birch-leafmining sawflies (Hymenoptera: Tenthredinidae) in Alberta: distributions, seasonal activities, and the potential for competition. *Canad. Entomol.,* **129**: 319-33.

FIDGEN L. L., SWEENEY J. D., (1996). Fir coneworm, *Dioryctria abietivorella* (Groté) (Lepidoptera: Pyralidae), prefer cones previously exploited by the spruce cone maggots *Strobilomyia neanthracina* Michelsen and *Strobilomyia appalachiensis* Michelsen (Diptera: Anthomyiidae). *Canad. Entomol.,* **128**: 1221-1224.

FRANKLIN J. F. *et al.,* (1981). Ecological characteristics of old-growth Douglas-fir forests. USDA, *Pacific Northwest Forest and Range Experiment Station, technical report PNW-118*: 49 pp.

FURNISS M. M., (1996). Taxonomic status of *Dendroctonus punctatus* and *D. micans* (Coleoptera: Scolytidae). *Ann. Entomol. Soc. Amer.,* **89**: 328-333.

FURNISS M. M., (1997). American forest entomology comes on stage. Bark beetle depredations in the Black Hills forest reserve 1897-1907. *American Entomologist,* **43**: 40-47.

GARA R. I. *et al.,* (1995). Arthropod associates of the spruce beetle *Dendroctonus rufipennis* (Kirby) (Col., Scolytidae) in spruce stands of south-central and interior Alaska. *J. Appl. Ent.,* **119**: 585-590.

GIMBARZEVSKY P. *et al.,* (1992). Assessment of aerial photographs and multi-spectral scanner imagery for measuring mountain pine beetle damage. *Forestry Canada, Pacific and Yukon Region, Information Report BC-X-333*, 31 pp.

HAACK R. A. *et al.*, (1983). Buprestidae, Cerambycidae, and Scolytidae associated with successive stages of *Agrilus bilineatus* (Coleoptera: Buprestidae) infestation of oaks in Wisconsin. *The Great Lakes Entomologist*, **16**: 47-55.

HALL J. P., (1994). Forest insects and disease conditions in Canada 1992. *Forestry Canada, Ottawa.*

HEDLIN A. F. *et al.*, (1980). Cone and seed insects of North American conifers. *Canadian Forestry Service.* (122 pages, 227 references, 98 colour photos).

HUBER J. T. *et al.*, (1996). The Chalcidoid parasitoids and hyperparasitoids (Hymenoptera: Chalcidoidea) of *Choristoneura* species (Lepidoptera: Tortricidae) in America North of Mexico. *Canad. Entomol.*, **128**: 1,167-1,220.

HULME M., GRAY T., (1996). Effect of pheromone dosage on the mating disruption of Douglas-fir tussock moth. *J. Entomol. Soc. Brit. Columbia*, **93**: 99-103.

HULME M., GRAY T., (1997). Control of Douglas-fir tussock moth *Orgyia pseudotsugata*, using a pheromone and virus treatment. *Technology Transfer in Mating Disruption. IOBC wprs Bulletin*, **20**: 133-137.

IVES W. G. H., WONG H. R., (1988). *Tree and shrub insects of the prairie provinces.* Information *Report NOR-X-292. Canadian Forestry Service* (327 p; 657 references and 117 colours plates).

LANGOR D. W., (1998). Annotated bibliography of North and Central American species of bark weevils, *Pissodes* (Coleoptera: Curculionidae). *Northern Forestry Centre, Information Report NOR-X-355.* 79 pages and 698 references.

LATTIN J. D., (1990). Arthropod diversity in old-growth forests. *Wings,* summer 1990: 7-10.

MARTIN J. L., (1966). The insect ecology of red pine plantations in Central Ontario. IV. The crown fauna. *Canad. Entomol.*, **98**: 10-27.

MOLDENKE A. K., LATTIN J. D., (1990). Dispersal characteristics of old-growth soil arthropods: the potential for loss of diversity and biological function. *Northwest Environ. J.*, **6**: 408-409.

MORAN V. C., SOUTHWOOD T. R. E., (1982). The guild composition of arthropod communities in trees. *Journal of Animal Ecology*, **51**: 289-306.

MORRIS R. F., (1963). The dynamics of epidemic spruce budworm populations. *Memoirs Entomological Society Canada*, **31**:1-332.

NEALIS V. G., (1998). Population dynamics of the white pine weevil *Pissodes strobi*, infesting jack pine, *Pinus banksiana*, in Ontario, Canada. *Ecological Entomology*, **23**: 305-313.

OPLER P. A., (1974). Oaks as evolutionary islands for leaf-mining insects. *Amer. Scient.*, **62**: 67-73.

ROBISON D. J. *et al.*, (1994). Responses of gypsy moth (Lepidoptera: Lymantriidae) and forest tent caterpillar (Lepidoptera: Lasiocampidae) to transgenic poplar, *Populus* spp., containing a *Bacillus thuringiensis* d-endotoxin gene. *Environmental Entomology*, **23**: 1030-1041.

ROHLFS W. M., HYCHE L. L., (1981). Colydiidae associated with *Ips* in southern pines: relative abundance and time of arrival of adults at pines under attack by *Ips* spp. *Journal of Economic Entomology*, **74**: 458-460.

ROTHMAN L. D., ROLAND J., (1998). Forest fragmentation and colony performance of forest tent caterpillar. *Ecography*, **21**: 383-391.

SCHOWALTER T. D., (1998). Canopy arthropod community structure and herbivory in old-growth and regenerating forests in western Oregon. *Canad. J. For. Res.*, **19**: 318-322.

SCHOWALTER T. D., CROSSLEY D. A., Canopy arthropods and their response to forest disturbance. *In:* W. T. Swank & D. A. Crossley (1987). *Forest Hydrology and Ecology at Coveeta*. Springer, Berlin, 207-218.

SMITH H. R., LAUTENSCHLAGER R. A., (1978). Predators of the gypsy moth. *U S Department of Agriculture, Agriculture Handbook*, no. 534.

SOLHEIM H., SAFRANYIK L., (1997). Pathogenicity to Sitka spruce of *Ceratocystis rufipenni* and *Leptographium abietinum*, blue-stain fungi associated with the spruce beetle. *Canad. J. For. Res.* **27**: 1,336-1,341.

SPIES T. A. *et al.*, (1988). Coarse woody debris in Douglas-fir forests of Western Oregon and Washington. *Ecology*, **69**: 1,689-1,702.

STARK R. N., DAHLSTEN D. L., (1970). Studies on the population dynamics of the western pine beetle, *Dendroctonus brevicomis* LeConte (Coleoptera: Scolytidae). *University of California, Berkeley*, 174 pp.

STORK N. E., (1987). Guild structure of arthropods from Bornean rain forest trees. *Ecological Entomology*, **12**: 61-80.

STURGEON K. B., MITTON J. B., (1986). Allozyme and morphological differentiation of mountain pine beetles *Dendroctonus ponderosae* Hopkins (Coleoptera: Scolytidae) associated with host trees. *Evolution*, **40**: 290-302.

THATCHER R. C. *et al.*, (1981). The Southern Pine Beetle. *U.S Dep.Agric., Technical Bulletin* 1631, 266 pages.

VITÉ J.-P. *et al.*, (1974). Biochemical and biological variation of southern pine beetle populations in North and Central America. *Z. ang. Ent.* **75**: 422-435.

WEBB J. W., FRANKLIN R. T., (1978). Influence of phloem moisture on brood development of the southern pine beetle (Coleoptera: Scolytidae). *Environmental Entomology*, **7**: 405-410.

WINCHESTER N. N., Canopy arthropods of coastal Sitka spruce trees on Vancouver Island, British Columbia, Canada. *In:* N. E. Stork *et al.* (1997). *Canopy Arthropods*. Chapman & Hall, London, 151-168.

21

Tropical rain forest insects

Tropical rain forests are found in regions where the macroclimate is characterised by a high rainfall during all months of the year, by a mean monthly temperature of about 25 °C, and daily temperature fluctuations varying from 6 to 12 °C. For example, in Buitenzorg (Java), the mean monthly temperature varies only from 24.3 °C to 25.3 °C; annual precipitation is 4,370 mm; rainfall in the wettest month is 450 mm and 230 mm in the driest month. Within a day, temperature can vary from 23.4 °C to 32.4 °C, and relative humidity between 80% and less than 40%.

Tropical rain forests in South America are in the Amazon basin, along the west (Pacific) coast of Colombia and in part of Central America; in Africa they are in the Congo basin, Cameroon and the Ivory Coast, while in southeast Asia they are in Sri Lanka, Malay peninsular, Sumatra, Java, Borneo, New Guinea and the Philippines.

Tropical forests are characterised by the richness in tree species, the abundance of life forms such as lianas and epiphytes, the height of the trees and a very heterogeneous structure. These forests, when not disturbed by man, have numerous gaps due to the death of old trees, lightning strikes, trees uprooted by wind, and fires. The old gaps are progressively filled by young trees. In French Guyana gaps due to fallen trees may represent 90 to 95% of some forests. This mosaic structure contributes to the great richness in animal species. A heterogeneous spatial structure due to the same natural causes existed in temperate forests before they were modified by man. This structure is still present in the few "primeval forests" of Europe and in the "old growth forests" of North America (cf. chapter 1.6).

The profile diagram of a forest in French Guyana (figure 21.1) shows a succession of more or less well-defined strata. In this case we can see an "emergent" tree about 40 m high. Under this emergent tree, at a height of about 20-30 metres, the canopy begins and it is in this part of the forest that density of the leaf phytomass is the highest. In a forest near Manaus in Brazil 58% of the total phytomass and nearly 40% of the leaf phytomass are in the

(a)

/ʬ/ = Epiphytes or foliage of lianes

Figure 21.1. *Structure of a tropical rain forest. Profile diagram of a plot 20 m x 20 m at Rivière Sinnamary, French Guyana (Oldeman, 1974). The large "emergent" tree (thick dashed outline) is a* Newtonia *which is outside the plot. Fine dotted outlines are immature trees. The dotted line represents the inversion surface. Note the abundance of lianas and epiphytes.*

canopy between 16.7 and 25.9 m above ground level. The lower boundary of the canopy is the "inversion surface", a surface joining the bases of the lowest branches (Oldeman, 1974). The inversion surface separates the tree crowns, which are more or less fully exposed to sunlight, from the lower stratum which is shaded. Richards (1983) called the canopy above the inversion surface the *euphotic zone* and the region below it the *oligophotic zone*. The microclimate of the canopy is very different from the one that prevails within the subjacent zone, as is shown by a recent work by Blanc (1990) and by earlier works such as that of Cachan (1974). Air movements, temperature and humidity are variable in and above the canopy; below the canopy the air is still and the microclimate more constant. Food resources are also different. Leaves, flowers and fruits provide food for numerous and diverse species of herbivorous animals: mammals, birds and insects. In the oligophotic zone, leaves, flowers and fruits are scarcer and the main source of food is wood and decaying plant tissues.

Below the canopy there is a cluster of small trees between 8 and 10 m high, interspersed with palm trees of the genus *Astrocaryum,* and another stratum of taller trees between 18 and 22 m high. According to Oldeman there are three different types of tree: mature trees ("arbres du présent") which are very tall and tower high above the others; immature trees ("arbres du futur") which are smaller and in the shadow of mature trees, and giant dead trees which are still standing ("arbres du passé").

The study made by Cachan (1974) in the forest of Banco in the Ivory Coast provides data on the microclimate inside the forest. The most remarkable feature is the vertical gradient of temperature, illumination, wind and relative humidity. When on a sunny day we climb down from the top of the trees (40 m high) to the ground, temperature falls from 32 °C to 27 °C; relative humidity rises from 30% to 80%; the illumination falls from 100,000 lux to 100-200 lux and the speed of the wind subsides from 7 m per sec at the top of the trees to zero near the ground. These characteristics explain the vertical stratification of many species (cf. below).

1.   THE STUDY OF INSECTS – PAST AND PRESENT

Rain forests were explored in the 19th century by naturalists such as Henry Walter Bates (1863), Alfred Russell Wallace (1878) and Charles Darwin, who noted the great diversity and the originality of the fauna and particularly of the insects. Wallace wrote:

> "Animal life is, on the whole, far more abundant and varied within the tropics than in any other parts of the globe, and a great number of peculiar forms are found there which never extend into temperate regions. Endless eccentricities of form and extreme richness of colour are the most prominent features, and these are manifested in the highest degree in those equatorial lands where the vegetation acquires its greatest beauty and fullest development... The special interest of this country to the naturalist is, that while there appears to be at first so few of the higher forms of life, there is in

reality an inexhaustible variety of almost all animals... There are, for instance, few places in England where during one summer more than thirty different kinds of butterflies can be collected, but here, in about two months, we obtained more than four hundred distinct species, many of extraordinary size, or of the most brilliant colours." (A. R. Wallace. Tropical Nature and other Essays).

Henry Walter Bates studied many insects such as Carabidae in South America and noted the importance of arboreal species and of their adaptations:

"The carnivorous beetles at Caripi were, like those of Para, chiefly arboreal. Some were found under the bark of trees (*Coptodera, Goniotropis, Morio*, etc.), others running over the slender twigs, branches and leaves (*Ctenostoma, Lebia, Calophaena, Lia*, etc.), and many were concealed in the folds of leaves (*Calleida, Agra*, etc.). Most of them exhibited a beautiful contrivance for enabling them to cling and run over smooth or flexible surfaces, such as leaves. Their tarsi or feet are broad, and furnished beneath with a brush of short, stiff hairs; whilst their claws are toothed in the form of a comb, adapting them for clinging to the smooth edges of leaves, the joint of the foot which precedes the claw being cleft so as to allow free play to the claw in grasping." (The Naturalist on the River Amazons).

Bates is also the discoverer of Batesian mimicry, so frequent among tropical butterflies. We must mention the more recent works of Orlando Park (1937, 1938) in tropical America, and the pioneering study of the canopy in the Ivory Coast forest made by Renaud Paulian (1947). Charles Elton (1973) gave a detailed account of his research in the rain forest at Belem (Brazil) and in Barro Colorado Island (Panama). He noted the small size of diurnal insects, the great species diversity, the abundance of rare species and the diversity of flying nocturnal species which are of a larger size. Elton suggests that the low population density of species inside the forest in day time is due to the action of predators. The species that survive are rare or are nocturnal. Janzen is the author of numerous papers on tropical insects which he mostly studied in the forest of Costa Rica.

The tropical forest canopy has been called "the last frontier". Its fauna was studied as early as 1884 by Bates, and in 1917 by William Beebe who wrote about the canopy: "Yet another continent of life remains to be discovered, not upon the earth, but one to two hundred feet above it. Up to now gravitation and tree-trunks swarming with terrible ants have kept us at bay, and of the tree-top life we have obtained only unconnected facts and specimens."

The first techniques used for these studies obliged researchers to climb up the trees, to fell them, or to install platforms similar to those built by Paulian (1947) in the Ivory Coast, or to set up various traps. Later, permanent ladders, walkways and cranes were used. In Barro Colorado a crane was installed by the Smithsonian Institution; this crane is 33 metres high and can revolve and reach to a distance of 39 metres. The canopy raft ("radeau des cimes") is an other device useful to study the canopy (Hallé & Blanc, 1990). Martin (1966) obtained a representative sample of canopy arthropods by spraying trees with a knock-down insecticide and by collecting what fell on trays (cf. chapter 20.2). The fogging technique consists in spraying a cloud of insecticides. This technique was first developed by Roberts (1973) and Gagné (1979) and it is now widely used. These studies were made at different sites in the world, particularly in South and Central America, tropical Asia, Africa (Cameroon), New Caledonia and even in temperate forests in North America and Europe. They were focused on the study of several groups of arthropods, mainly Coleoptera but also Orthoptera, Diptera, Collembola and even spiders.

The study of the insects of tropical forests is now making great progress. This is due to the interest these insects represent for scientists. Moreover the fast and increasing destruction of tropical forests (at a rate of 7 to 12 million hectares per year) urges entomologists to study these insects as quickly as possible since they fear that a great number of species might disappear before they are discovered.

There is today a large amount of information on tropical forest insects. In this chapter their main characteristics are described, with special reference to their originality and what makes them different from those of temperate forests.

2. SPECIES DIVERSITY AND ABUNDANCE

2.1. Biodiversity

There is much controversy about the magnitude of global species-richness on earth. But many authors agree that, apart from micro-organisms, the bulk of this richness is to be found in the arthropods of tropical forests, and perhaps in the canopy of these forests.

2.1.1. Some examples

Tropical forests cover only 7% of terrestrial land, but they contain about 3 million species whereas temperate regions have only 1.5 million species. The richest in biological diversity are the wet tropical forests of south-east Asia and South America. This is true of trees, mammals and insects, with some exceptions such as termites (cf. below). The richness of the fauna is obvious not only in a geographic area (γ diversity), but also in a small community such as a tree (α diversity).

There are fewer quantitative data on the number of species of insect than on the species of vertebrate. Bates collected 550 species of butterfly on the Amazon, and Beebe 400 species at Kartabo, Guyana. Eighty five species of termite were collected at the same locality of Kartabo and 43 from 1.5 ha in the rain forest of Cameroon (cf. figure 4.3). The local diversity of ants is substantial: 172 species belonging to 59 genera in 2.6 km^2 of a lowland rain forest in New Guinea; 219 species in 63 genera in an equal area in a cocoa plantation in Ghana; 272 species belonging to 71 genera in a comparable area in Brazil. In two years an entomologist collected at least 350 species of ant in an area of the western Amazon Basin in Peru. In a *single tree* in the Peruvian Amazon, Wilson collected 43 species of ants; the terrestrial fauna under the tree was not studied. Temperate forests are less rich: there are about 180 species of ant in France (figure 21.2).

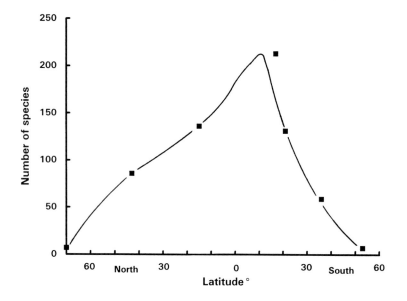

Figure 21.2. *Latitudinal pattern of species-richness of ants (Kusnezov, 1957).*

Among the Lepidoptera, 3,600 species of butterfly are known from the Afrotropical region and 890 are recorded from Ghana. One thousand two hundred and nine species of butterfly were recorded within the 55 km^2 of the Tambopata Reserve in Peru. Nymphalid butterflies of the subfamily Ithomiinae are exclusively neotropical with 310 species belonging to 52 genera. These insects participate in mimicry complexes rich in species. At Jatun Sacha Biological Station in Ecuador these mimetic complexes involve 124 species: 55 Ithomiinae, 34 other butterflies belonging to other families, 34 moth species and 1 species of damselfly (Beccaloni, 1997). In one hectare of Panamanian rain forest, Erwin estimates that over 18,000 species of insect are present (only 24,000 species are known from the whole of the USA and Canada). Many other groups of insects are known to be very rich in species in the tropics. Among the Coleoptera these are the Cerambycidae, Buprestidae, Lucanidae, Brenthidae, Cetoniidae, Scolytidae and Dynastidae; among Orthoptera they include the subfamily Pseudophyllinae; among the Homoptera the Fulgoridae and Membracidae; among the Diptera the Drosophilidae. In French Guyana there are 1,700 species of Cerambycidae, 600 of which were described in the last seventeen years. The insect fauna of French Guyana is estimated at 300,000 species.

This species-richness is found also in aquatic communities. There are 30-60 species of insect in streams in tropical America and only 10-30 species in the temperate USA (Stout & Vandermeer, 1975).

2.1.2. The assessment of species diversity

2.1.2.1. The studies of Erwin

Erwin & Scott (1980) made the first field study of the number of insect species in the canopy of a tropical forest. This study also provides some data on the structure of a community of Coleoptera in the canopy of a tropical forest. By using a fogging technique Erwin gathered data on canopy insects, and particularly on Coleoptera, from 19 trees of the species *Luehea seemannii* in a seasonal lowland forest in Panama. In this experiment more than 1,200 species of beetle were collected. The study of trophic groups allows an estimated percentage of host-specific beetles of 13.5% corresponding to 163 host-specific species (table 21.1).

Trophic groups	Number of species	Number host-specific (estimated)	Mean length (mm)
Herbivores	682	136	3.89
Predators	296	15	3.72
Fungivores	69	7	3.38
Scavengers	96	5	3.28

Table 21.1.   *Number of estimated host-specific species per trophic group among Coleoptera on* Luehea seemannii, *and mean length of the Coleoptera in each trophic group (Erwin & Scott, 1980).*

An average tropical forest has about 70 species of tree per hectare. There could be 163 x 70 = 11,410 host-specific species of beetle, plus 1,038 species that could be found anywhere. There are 11,410 + 1,038 = 12,448 species of beetle per hectare in the canopy alone. Assuming that 40% of all insect species consist of beetles, that the canopy is twice as rich as the forest floor, and that there are 50,000 species of tree in the world, we can deduce that there must be 30 million species of insect in the world.

In the same study Erwin establishes that trophic groups respond differently to seasonal changes (figure 21.3). All groups are at their maximum species-richness in the early rainy season. Species dependent upon plants (herbivores and fungivores) vary in density and in number of species according to the three seasons. Predators and scavengers (which depend on animal food) have a two-phase annual cycle only. In all trophic groups the average length of species decreases in the dry season.

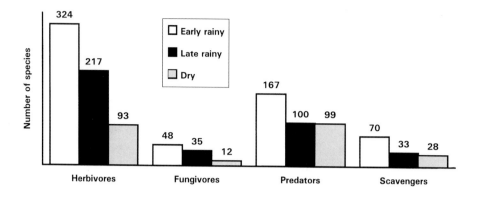

Figure 21.3. *Distribution of Coleoptera collected by fogging on the tree* Luehea seemannii *in a seasonal forest in Panama (Erwin & Scott, 1980).*

A more extensive study was made by Erwin (1983) in four types of forest within 70 km of Manaus (Brazil). One type of forest is in "*terra firma*", and three on inundated land when the river water level begins to subside: *white water forest, mixed water forest* and *black water forest* (see description, of the forests in Adis, 1981). A study of 49% of the trays on which the insects where collected shows that among 24,108 insects, Hymenoptera are the most numerous because of the great abundance of ants which dominate in the canopy. Coleoptera and Homoptera come second and third in abundance (table 21.2). Eighteen orders of insects are present, many represented by very few specimens.

Orders	Number of specimens	Orders	Number of specimens
All Hymenoptera	12,245	Neuroptera	58
Ants	10,311	Odonata	46
Coleoptera	4,845	Trichoptera	38
Homoptera	2,063	Ephemeroptera	9
Diptera	1,755	Embioptera	8
Orthoptera	1,177	Thysanura	7
Psocoptera	939	Dermaptera	2
Hemiptera	557	Strepsiptera	1
Thysanoptera	228	Plecoptera	1
Isoptera	129		

Table 21.2. *Number of insects collected by fogging in four types of forest near Manaus, Brazil (Erwin, 1983). Note the presence of orders rare or absent in the canopy of temperate forests.*

The abundance of insects in each sample is lower in white water forest and higher in mixed water forest. Among the Coleoptera 83% of the species are restricted to one type of forest, 14% to two types, 2% to three types and only 1% to four types. This seems to indicate a high rate of endemism. The percentage of host-specific species is estimated at 16.3%.

The size of the beetles collected is small (figure 21.4), particularly on *terra firma* which has many small species in the 1 mm size class and few large species (20 mm). We may question the reliability of the fogging method in sampling all the insects of the canopy, because large species are notable in tropical forests. It may be that these larger species are not canopy species or it may be that fogging does not bring them down and they are able to escape.

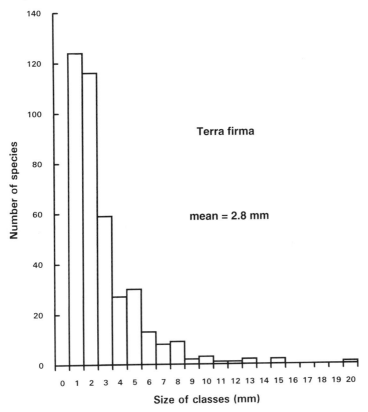

Figure 21.4. *Size of Coleoptera collected by fogging in a forest of* terra firma *type near Manaus, Brazil (Erwin, 1983).*

The Coleoptera collected may be classified according to four trophic groups. Herbivores (including xylophagous and foliage-eaters) are the most numerous (795 species). Their two main families are Curculionidae (337 species) and Chrysomelidae (170 species); the other main families are Cerambycidae (42 species), Mordellidae (37 species), Anobiidae (36 species), Elateridae (35 species), and

Buprestidae (29 species). Predators (145 species) come second in importance. The main families of predators are Coccinellidae (50 species), Staphylinidae (35 species) and Carabidae. There are 100 species of scavenger belonging mainly to Tenebrionidae (57 species), Nitidulidae (16 species) and Anthicidae (13 species). Fungivores fall into 29 species in families such as Anthribidae (6 species), Endomychidae (6 species) and Platypodidae (4 species). The mean length of the species varies according to the trophic levels: herbivores 3.2 mm; scavengers 3.0 mm; predators 2.7 mm; fungivores 2.0 mm. Scarabaeidae are the largest herbivores (mean length 11.2 mm); Lampyridae are the largest predators (mean length 9.57 mm); Tenebrionidae are the largest scavengers (mean length 4.17 mm) and Anthribidae are the largest fungivores (mean length 3.5 mm).

The most important conclusion that Erwin drew from this study is "a substantiation of the prediction that there may be as many as 30,000,000 species of insects in the world" (Erwin, 1983 page 73).

2.1.2.2. *Other hypotheses*

According to many entomologists Erwin's hypothesis on the percentage of host-specific species and on the ratio between the number of species in the canopy and on the ground are too optimistic. The Coleoptera represent about 20% and not 40% of the species of the canopy and the real number of insect species is very likely much lower. These objections seem to have been accepted by Erwin (1991). Hodkinson & Casson (1991) when studying the diversity of Hemiptera in tropical forests gave an estimation of 2 million species only. In their collection only 63% of the species were undescribed. In a study of the previously unknown psyllid fauna of Panama 72% of the species were undescribed. In the neotropical genus *Heteropsylla* only 54% of the species were undescribed and many have a wide range in Central and South America (Muddiman *et al.*, 1992). Hodkinson (1992) thinks that the simplest way of estimating arthropod diversity is to examine the proportion of undescribed species in samples. This author remarks that of the 14,700 species of insect that Bates collected along the river Amazon there was an average value of 68% of undescribed species. Assuming that the number of described species is one million, the total number N of species is:

$$N = 1,000,000 \times [100 / (100 - 68)] = 3,125,000$$

It is possible to estimate the number of species by extrapolation. Britain has 67 species of butterfly and 22,000 species of insect. The world butterfly fauna contains 17,500 species. If the ratio observed in Britain is the same in the whole world, the number of species of insect is: $(22,000 / 17) \times 17,500 = 5.7$ million. According to Gaston (1992) the ratio between the number of insect species and the number of plant species lies between 10 and 20. Since the number of plant species is between 275,000 and 350,000, the number of insect species is between 2.75 and 7.5 million.

In conclusion we may assume that the number of insect species is between 5 and 8 million. Most of these species are found in the tropical forests; they are still undescribed and their biology is as yet unknown. The destruction of tropical forests is now so rapid that this enterprise will, perhaps, never be achieved.

2.2. Latitudinal gradient

Many taxa of plants and animals have more species in tropical than in temperate regions. This pattern was recognised early in conspicuous groups such as birds, trees or butterflies (figure 21.5). It applies also to many groups of insects such as termites, ants, beetles, dragonflies, etc. Some of these patterns have been well-documented (cf. chapter 4.2). For example there are about 75 species of dung beetle of the subfamily Scarabaeinae in the United States whereas in a lowland forest of Costa Rica a single site contains 40 to 60 species. A single pile of horse dung contained as many as 15 species of Scarabaeinae (Howden & Nealis, 1975). The higher biodiversity of Psocoptera in tropical areas was established by Wolda (1983). The α diversity index increases from the temperate climate of England to the tropical areas:

Localities	Mean α	Range of α
England	1.34	0.68 - 2.36
South Australia	3.22	3.01 - 3.58
Jamaica	4.95	3.66 - 6.52
East Africa	5.74	5.74
Trinidad	12.84	11.81 - 14.81
Panama lowland (canal zone)	24.38	22.77 - 25.75

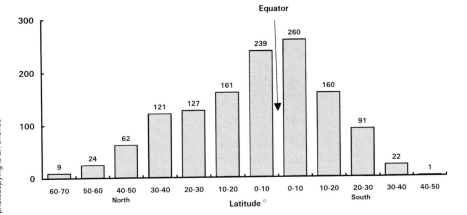

Figure 21.5. *Number of species of Papilionidae (Lepidoptera) at different latitudes. The greatest richness is near the Equator.*

Examples of higher diversity in temperate regions than in tropical regions can however be found, particularly among the following insects:

Ichneumonidae, Psyllidae, Aphididae (figure 21.6), sawflies of the families Diprionidae and Tenthredinidae, Agromyzidae (Diptera that are leaf-miners and gall-formers, Price, 1991). Aphididae are insects that cannot live long without food, are highly specific to host-plant species, but locate their host with difficulty. Therefore they cannot exploit the rare plants that are represented by many different species in the tropics (Dixon *et al.*, 1987). Many Aphididae and Diprionidae are associated with plants adapted to cool temperate latitudes such as conifers, and Tenthredinidae are often associated with *Salix* and *Populus* which are rare or absent in the tropics.

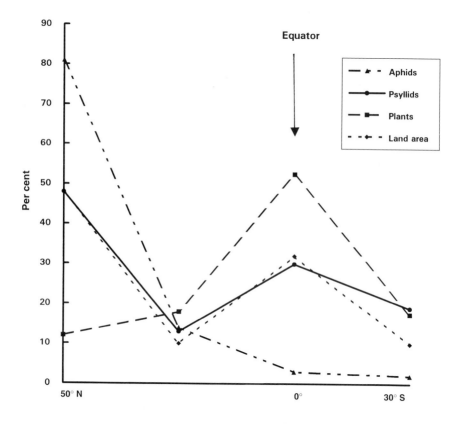

Figure 21.6. *Distribution of aphids, psyllids and plants (in percentage of total number of species) and distribution of land areas (in percentage of total area). Note the preponderance of psyllids and aphids in north temperate regions and the importance of plant species and of land area near the Equator (Eastop, 1978).*

Owen & Owen (1974) first observed that ichneumonids, a large family of Hymenoptera that parasitise other insects – mainly Coleoptera and Lepidoptera, are less rich in species in tropical Africa than in England (table 21.3). This exception was confirmed by Janzen (1981), Noyes (1989) and Askew (1990). Janzen shows that the peak of species-richness in North America lies between 38° and 42° N. There are two different interpretations of the latitudinal pattern of ichneumonid distribution. The former is that of Janzen & Pond (1975) and Janzen (1981) who proposed the *resource fragmentation hypothesis*. Populations of ichneumonid hosts tend to be smaller and more scattered in the tropics. Below a certain host density, parasitoids cannot find their specific hosts and must be polyphagous. The latter hypothesis stresses the fact that tropical plants may be generally more toxic than the plants of temperate regions and that the toxic substances in the tissues of their phytophagous hosts can injure the larvae of parasitoids. This observation led Gauld *et al.* (1992) to propose the *nasty host hypothesis*: on average, in the tropics increased toxicity makes hosts less accessible, which leads to a decline in species-richness. Gauld *et al.* noted that diversity is not lower among tropical parasitoids attacking stages such as eggs or pupae that are not chemically well-defended by the plant chemicals they contain, or in those attacking insects that feed on plant tissues containing few toxic substances. Even among ichneumonids, the Rhyssinae are parasites of wood borers and they are richer in species in the tropics than in temperate areas. The same is true of egg parasitoids belonging to the superfamilies Chalcidoidea, Proctotrupoidea and Ceraphronoidea.

In the order Coleoptera, carabid species of the tribe Agonini are more numerous in the tropics, whereas the Pterostichini are more numerous in temperate areas. The reason for these opposite patterns is to be found in the different origins of the two taxa.

Localities	1	2	3	4	5
Uganda	2,268	293	4.524 ± 0.032	116	10.1
Sierra Leone	1,979	319	4.934 ± 0.059	117	4.9
England	2,495	326 [a]	4.937 ± 0.024	122	3.2
Sweden	10,994	758	5.481 ± 0.014	203	5.5

Table 21.3. *Diversity index and number of individuals and of species collected in tropical and temperate areas. 1: sample size. 2: number of species. 3: diversity index H of Shannon ± s.e. 4: number of rare species with only one specimen collected. 5: abundance of the most common species, in percentage of the total of insects collected. (a): the inventory now includes 529 species in this locality (Owen & Owen, 1974).*

2.3. Altitudinal gradient

A decrease in species-richness with increasing altitude is a common pattern for plants and animals in temperate as well as in tropical regions. In the tropics this pattern is well-documented for arthropods (Janzen *et al.*, 1976; Wolda, 1987; Olmstead & Wood, 1990; Fernandes & Price, 1991;

etc.). In Costa Rica the study of four families of Lepidoptera shows variations according to altitude. Species-richness decreases with altitude and with variations among taxa (figure 21.7). Above 1,200 metres the species-richness of all families decreases. The Riodinidae represent a third of all lowland species and less than 10% at 2,000 m. The wingspan of butterflies, on average, increases from 0 m to 1,600-2,000 m and then decreases. But variations in body size are generally small and each family shows a unique pattern of variations. Janzen *et al.* (1976) also found that arthropods were, on average, larger at mid-altitude in the Andes.

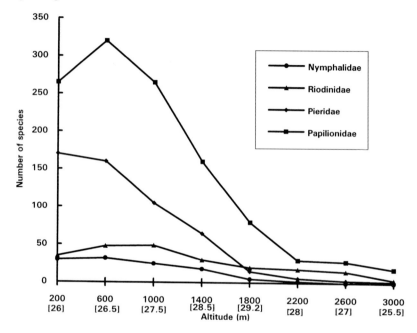

Figure 21.7. *Variations in the species-richness of four families of Lepidoptera in Costa Rica according to altitude. The numbers between square brackets and below the altitude correspond to the mean length (in mm)of the forewing for the four families of Lepidoptera (Hawkins & De Vries, 1996).*

The diversity and seasonality of weevils in tropical Panama was deduced from light-trap catches by Wolda *et al.* (1998). The number of species is high (2,030 species) but the average number of species and of individuals as well as the α diversity index decreases with altitude. In Costa Rica, ants, termites, and wood-boring beetles decrease in density and species-richness above 2,200 m and are almost absent above 3,000 m. Dung beetles also become rare at higher altitude. The cause seems to be the same: the wet soil never warms up at high altitude even when the air may be warm (Janzen, 1983). In Sierra do Cipo (Brazil) the number of species of gall insect on trees and on all plants decreases on an altitudinal gradient (figure 21.8)

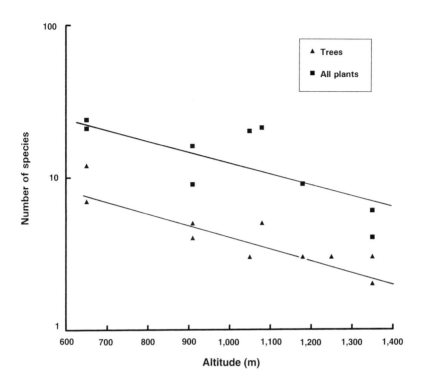

Figure 21.8. *Number of gall insect species on trees and on all plants on an altitudinal gradient in Sierra do Cipo, Brazil (Fernandes & Price, 1991).*

Variations of the insect fauna according to altitude in Dumoga-Bone National Park (Sulawesi) were described by Stork & Brendell (1990). Insect abundance does not appear to vary with the altitude but the percentage of Diptera declines while Coleoptera and spiders increase in abundance. The number of ants is very low at high altitude and they are replaced by other predators such as spiders or Carabidae (table 21.4).

Taxa	Altitude			
	210 m	400 m	1,150 m	1,760 m
Coleoptera	44.0	12.3	49.5	82.5
Diptera	243.2	73.0	70.5	84.5
Formicidae	21.1	6.3	11.0	0.9
Others (mainly spiders)	15.2	3.3	38.0	100.9

Table 21.4. *Mean number of arthropods collected per m² of tray after fogging the canopy in Dumoga Bone National Park (Stork & Brendell, 1990).*

In some cases there is a mid-elevation peak species-richness. This pattern is observed in several taxa. In Sulawesi five groups of Lepidoptera show a peak between 600 m and 1,000 m (Holloway *et al.*, 1990). The sampling of leaf litter invertebrates in western Panama reveals that both species-richness and the number of individuals of the different taxa show a mid-elevation peak near 800 m and then a pronounced decline above that peak (figure 21.9). Both Carabidae and Curculionidae peak at 1,750 m instead of 800 m but they have few species and display great variations of abundance along the transect. Many species are restricted to a narrow altitudinal belt: 73% have an altitudinal range of 500 m or less, 55% occur at only one station and the turnover of species is rapid. If this structure with a peak of diversity at mid-elevation is general (or frequent) it may have implications for the conservation of the biodiversity: it is therefore necessary to preserve tropical forests that span an important range of elevations (Olson, 1994).

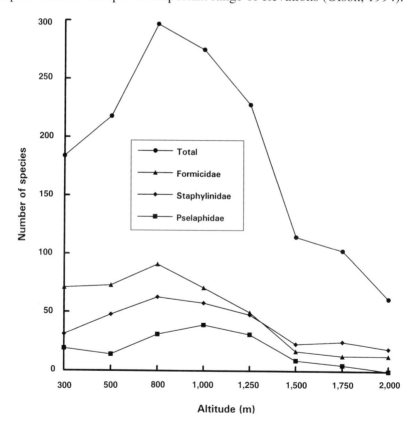

Figure 21.9. *Number of species for all arthropods and for three families of insect collected at various elevations in western Panama (Olson, 1994).*

The study of dung beetles and carrion beetles (genus *Nicrophorus*) between 200 m and 1,750 m in northern Sulawesi reveals that, generally, there is an "individualistic" pattern of distribution, each species having its upper and lower limits independent of the others. But a more detailed observation indicates a mutually exclusive range of altitude in two species of *Nicrophorus* and in some species of *Copris* and *Onthophagus*. Interspecific competition is suggested as the cause of these phenomena (Hanski & Niemela, 1990).

The study by Russell-Smith & Stork (1994) in the rain forest of Sulawesi suggests that, contrary to other taxa of arthropods, the abundance and species-richness of spiders increase with altitude in the tropical rain forests.

To sum up, the insects of tropical forests have many characteristics in common: (a) species-diversity increases from high latitudes to low latitudes; (b) the number of species per unit area is usually greater in tropical rain forests and lower in temperate forests; (c) tropical mountains support a larger number of species than equivalent temperate mountains because the environment ranges from lowland tropical conditions to alpine conditions; (d) species-diversity is not uniform. Areas with a large number of species are interspersed with areas of lower species-diversity.

2.4. Abundance and importance of insects

Many studies come to the conclusion that insects are the animals that have the highest biomass in tropical forests. This biomass reaches several tons per hectare whereas that of mammals and birds is only a few kilograms (cf. chapter 4.5). The species-richness of tropical bees (Apoidea) may not be superior to that of bees of temperate regions but their ecology is more varied and their abundance very high. Six hundred colonies of *Melipona* were counted per km^2 in a forest in Central America. These *Melipona* collect 10^{10} kilojoules of food from the vegetation and their dead bodies and waste material amount to several hundred kg per km^2 (Roubik, 1989). Insects are also important since they are at the beginning of food chains for many vertebrates and play a role in the pollination of many entomophilous plants.

About 50% of insects are herbivorous and flowering plants are their main food resource. Since there are so many species of insect and plant the phenomenon of herbivory must be particularly important and varied in the tropics. The predominant role of insects as foliage consumers has been established for several tropical sites (Leigh & Smythe, 1978; Janzen, 1981). In Barro Colorado Island the three main arboreal leaf-eating vertebrates are the howler monkey (*Alouatta* sp.), the sloth (*Bradypus* sp.) and the iguanas. The howler monkeys eat nearly 30 kg (dry weight) of leaves per hectare per year; the sloths 40 kg and the iguanas 50 kg. The sum total of vertebrates eat between 100 and 300 kg (dry weight) of leaves $ha^{-1} yr^{-1}$. Leaf consumption by insects in the same forest is 680 kg $ha^{-1} yr^{-1}$, that is, twice to six times greater than by vertebrates. These approximate numbers give an idea of the importance of insects in the forest. Dirzo & Miranda (1991) made a study of

1,103 seedlings of plants collected at random in the understorey of a mature forest at Los Tuxtlas Tropical Research Station in the state of Veracruz (Mexico). They found that the damage caused by insects and pathogens was the most frequent type of damage and that nearly 40% of leaves were intact. None of the plants suffered damage caused by vertebrates. This is particularly surprising since several leaf-eating mammals such as peccaries and tapirs are present in the area and might be expected to be important consumers of plants. Is this anomalous or is it true of other tropical forests?

Why do insects not eat more? *First*, other animals eat insects and control their populations. Birds eat 24 kg ha^{-1} yr^{-1} (dry weight) of folivorous insects on Barro Colorado Island, and 34 kg ha^{-1} yr^{-1} in the Liberian moist forests. Other predators reduce populations. In New Guinea, spiders alone catch 50 kg ha^{-1} yr^{-1} (dry weight) of insects. We must also mention predatory insects, parasitoids, frogs, bats, monkeys, etc. *Second*, plants, and particularly leaves, contain chemicals which render them unpalatable, indigestible or even poisonous for insects. An insect dependent on one tree may spend much time searching for its host. Interactions between plants and insects have been much studied in temperate and tropical regions and in the latter they show certain characteristics.

What strikes the naturalist familiar with temperate forests is the large size of some insects, particularly in the Amazon basin which harbours many of the world's largest insects, including *Thysania agrippina*, the moth with the largest wingspan (32 cm), *Titanus giganteus*, the largest beetle (25 cm), *Megasoma acteon* (Coleoptera, Dynastidae) the heaviest insect (200 g). *Pepsis heros*, the largest species of Hymenoptera, is a parasite of the largest spider, the tarantula *Teraphosa leblondii*. On the continental scale the largest South American butterfly is *Morpho hecuba* (25 cm). *Lethocerus annulipes*, the largest aquatic bug (10 cm), is a predator of small fishes. These gigantic insects coexist with numerous minute species.

2.5. Seasonal and daily variations

Seasonal variation is the rule rather than the exception among tropical animals, even in areas where seasonal changes in the weather are minimal. This has been well established for many groups of insect (Dobzansky & Pavan, 1950; Wolda, 1983; Broadhead, 1983; etc.). In eastern Colombia only one out of seven species of mosquito does not show seasonal variations (figure 21.10). The abundance and diversity of Psocoptera vary with the seasons in the forest of Barro Colorado Island. The diversity is slightly higher near the ground than in the canopy; abundance and diversity are highest in the wet season (table 21.5).

Janzen (1983) gave many examples from Costa Rican forests. In the deciduous forest of Santa Rosa National Park the abundance and diversity of

Location of the traps	Individuals per week		α diversity index
	Dry season	Wet season	
Canopy traps	62.5	125.6	22.77
Traps near the ground	85.6	94.9	25.75

Table 21.5. *Abundance and diversity of Psocoptera according to the seasons and the height at which the traps are set up (Wolda, 1983).*

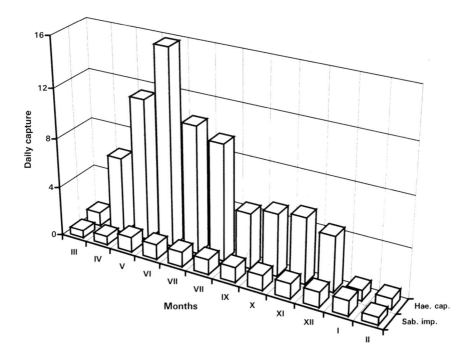

Figure 21.10. *Seasonal changes of two species of mosquito in eastern Columbia.* Haemagogus capricornii *has a peak of abundance in June (wet season) whereas* Sabethoides imperfectus *has about the same abundance all the year round (Bates, 1945).*

dung beetles is maximal in May, shortly after the first rain. By the end of July the density of these insects begins to decrease and by January, early in the dry season, dung beetles are gone except for some small species. For some insects (bees, bruchids, tiger beetles, ant lions) the dry season is the time of abundance, since there are no possible sites of oviposition and of larval development for these species in the rainy season. The wet season is the time when the migrating moth *Urania fulgens* appears and when biting ceratopogonids of the genus *Culicoides* are most abundant. Most butterflies appear at certain times of the year in response to the availability of food, particularly flowers and fruits. Some species have a wet and a dry season colour form. In the African butterfly *Precis octavia* the wet season form is orange with black markings and the dry season form is black with blue markings (Owen, 1966).

Catches in interception traps and Malaise traps in Dumoga-Bone National Park (Sulawesi) show that the peak of abundance of all insects is early and late in the year (Hammond, 1990) and that low abundance is around mid-July until mid-September, which is the driest period of the year (figure 21.11). In the canopy of an Amazonian rain forest, Farrell & Erwin (1988) found no seasonal variation in abundance and species diversity in the family Chrysomelidae. This stability seems to be due to the long life span of these beetles that feed on plants at all stages of their life cycles.

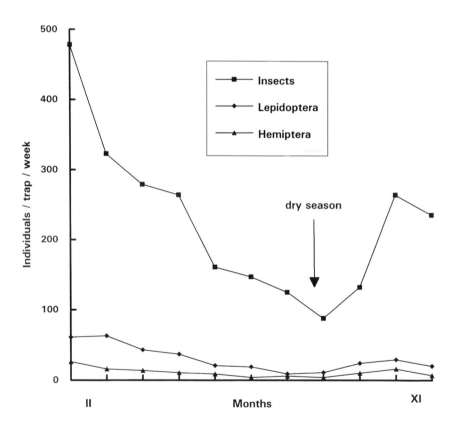

Figure 21.11. *Seasonal variations, in Dumoga-Bone National Park (Sulawesi) in the catches with a Malaise trap for 11 periods from February to November. The period of low abundance in the dry season is marked with an arrow. The term "Insects" applies to all insects except Diptera and Collembola (Hammond, 1990).*

The annual abundance of insects at Cocha Cashu Biological Station in Peru was measured by collecting with an ultraviolet light trap during 3-4 consecutive nights every two weeks. There is a peak of abundance in the period of transition between the rainy season and the dry season i.e. September-November. The insects collected are dominated by Orthoptera, Coleoptera and Lepidoptera. The seasonal patterns of abundance of those three orders are the same (Janson & Emmons, 1990).

Some species undergo diapause in the event of adverse climatic conditions. Diapause has been detected in a number of species in the Sarcophagidae (Diptera), Carabidae (Coleoptera), Pyralidae (Lepidoptera), and parasitic Hymenoptera. For tropical insects diapause is generally of short duration and, when it occurs in adults, is restricted to an arrest of gonad development while the individuals remain active and continue feeding. In *Stenotarsus rotundus* (Coleoptera, Endomychidae) diapause lasts 6 months of the wet season and 4 months of the dry season. The beetles in diapause cluster in aggregations of up to 70,000 individuals on the stem of a palm tree (Wolda & Denlinger, 1984).

Some species of Carabidae such as *Colpodes* spp. collected by fogging in the canopy are in a state of reproductive dormancy, while specimens collected at the same time on the ground are mature. These results suggest that these carabids use the canopy as a refuge during a stage of low reproductivity (Paarmann & Stork, 1987).

In the Amazon basin the water level of some rivers varies by an average of 10 metres throughout the year. In this region there is a rainy season (December to May) with an average precipitation of 1,550 mm, and a period of low precipitation (June to November) when average precipitation is 550 mm. During the high water period, areas of forest are flooded for 5 to 7 months. This seasonality is reflected by adaptations of invertebrates that Adis (1997) named "survival strategies". *Megacephala sobrina punctata* is a nocturnal tiger beetle that inhabits sandy beaches. During the high water phase it lives on driftwood and hides submerged during daytime (Adis & Messner, 1997). *Pentacomia egregia* is a diurnal and soil-dwelling tiger beetle that during the high water period climbs on the lower area of the tree trunks above the water level. Experiments (figure 21.12) showed that the availability of soil is the factor that induces a rapid maturation of the gonads and oviposition in females (Amorim *et al.*, 1997).

There are few data on the daily activity of insects. Hammond (1990) used Malaise traps in North Sulawesi and showed that all major insect orders have a low level of nocturnal flight activity. Only 6.3% of insects were trapped in the period between 20.00 h and 04.00 h. The peak of flight activity for Coleoptera, Diptera and Hemiptera is between 15.00 h and 16.00 h; for Hymenoptera it is between 11.00 h and 12.00 h but remains high until 16.00 h, and for Lepidoptera it is between 18.00 h and 19.00 h. At the species level, flight activity can be restricted to a period of time of less than an hour.

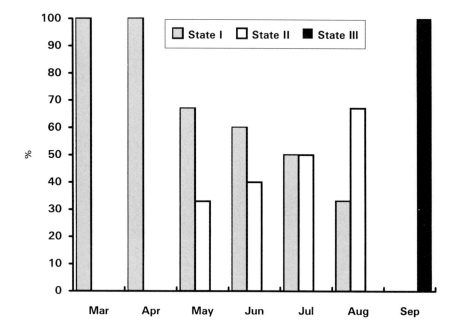

Figure 21.12. *State of the gonads of females of* Pentacomia egregia *according to the period of the year. June to November is the period of low precipitation. State I: immature state; ovarioles not differentiated. State II: maturing state; oocytes clearly differentiated. State III: mature state; eggs in the ovary (Amorim et al., 1997).*

2.6. Stratification of the fauna

Like many other animals such as mammals or birds, insects are unequally distributed in the different strata of the forest. This distribution is often due to variations of the microclimate, as Cachan (1974) showed for twelve species of bark beetle in the forest of Banco, Ivory Coast (cf. chapter 2.1.5).

The importance of the vertical variation of the microclimate is illustrated by the behaviour of two mosquitoes in the rain forest of Trinidad. *Anopheles homonculus*, which requires a humid atmosphere, settles closer to the ground than *Anopheles billater*, which requires a less humid atmosphere (Ewusie, 1980). Around midday when the humidity decreases, the two species move higher in the forest; after midday when the humidity increases they go down, but they always maintain their relative positions in order to remain in a constant (or almost constant) microclimate (cf. also chapter 2.1.5.4).

Besides the litter that is on the forest floor, many tropical forests (and some wet temperate forests) support a pool of organic matter on the trunks and tree branches. The biomass of organic matter in the canopy ranges between 1.4 to 14.0 t ha^{-1} and reaches a maximum in montane tropical cloud forests. The *live component* of this biomass is made up of epiphytes,

parasitic plants and vines and their fauna such as the invertebrates living in Bromeliaceae and other phytotelmata (cf. Frank & Lounibos, 1983). The *dead component* is derived from intercepted nutrients in the rain, litter from the overstorey, decaying epiphytes and dead canopy animals. This dead organic matter supports an invertebrate fauna which was sampled by Nadkarni & Longino (1990) and compared with the fauna of the ground litter. The dominant taxa in the canopy and on the ground were Acarina, adult Coleoptera, larvae of holometabolous insects, Formicidae, Collembola and crustaceans (amphipods and isopods). These taxa are the major agents of the fragmentation of the dead organic matter. The percentage of these different taxa is the same both in the canopy and in the ground litter which share a similar invertebrate community. But the mean abundance of invertebrates is 2.6 times greater on the ground than in the canopy, except for ants. The lower abundance of invertebrates in the organic matter of the canopy could be explained by the microclimate and its characteristic features of which are greater insolation, more wind and frequent cycles of wetting / drying.

In the rain forest of Sulawesi, traps were installed at a height of 1 metre, 9 m, and 19 m in the canopy, and 28 m above the canopy (Sutton, 1983). The insects were caught each day between 19.00 h and 22.00 h. Among the 51,798 insects collected, the three most numerous orders were Diptera (28.7%), Hymenoptera (26.6%) and Coleoptera (25.02%). The majority of Hymenoptera and Homoptera were found above the canopy, whereas the majority of Coleoptera and Diptera were found in the canopy. Only the Ephemeroptera were more numerous at the lower altitude of 9 metres.

Among the butterflies of Barro Colorado Island some species have a flying height which depends on their colours. The orange and yellow species are associated with the upper forest (they fly preferentially about 30 m high); the clearwing species have a preference for the forest floor (they fly around 2 m above the ground); other colour groups (blue, red and tiger) have no preferences and fly within a wide range of heights. It is difficult to explain these patterns. One theory suggests that butterflies choose a flying height where the luminance makes them look more similar to their unpalatable models and therefore less attractive prey for birds. This theory has been contested (Burd, 1994) since the visual capacities of insectivorous birds are not well-known.

The last study we will mention is that of Rees (1983) who described the vertical distribution of Homoptera and Hemiptera in Sulawesi (figure 21.13). The number of individuals and of species increases with height and is maximal at 30 metres within the canopy – see also the vertical distribution of ants (figure 21.18) and of termites (figure 21.19).

Seasonal vertical movements of soil arthropods are known in tropical areas. They move downwards in the dry season and upwards in the wet season. In temperate areas temperature is the factor that determines the migration (Strickland, 1947).

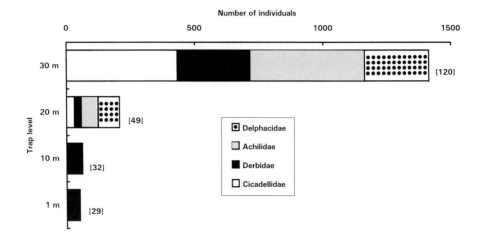

Figure 21.13. *Vertical stratification of Homoptera belonging to four families in a rain forest of Sulawesi. The numbers in square brackets are the numbers of species (Rees, 1983).*

2.7. Why so many species?

It is difficult to explain how high diversity originated and is maintained in tropical regions (Lugo, 1988). Many hypotheses have already been put forward yet none of them seems to be quite satisfactory. We will mention these hypotheses only briefly because most are based on the study of organisms (plants, mammals, birds) other than insects.

2.7.1. *The hypothesis of environmental stability*

High species diversity occurs in the tropics because the environmental stability over geological time favoured the evolution of species and a low rate of extinction. In tropical forests the climatic variations (such as temperature and precipitation) are low and predictable. This stable environment offers opportunities for specialised species (such as frugivorous species) that cannot persist where climate or resources fluctuate. This view was challenged with observations of high endemism and diversity in areas such as the Andes where disruptions entail a rapid evolution of plants and animals. In regions such as Amazonia, environmental stability has not always been the rule, since there were important climatic changes during the glacial times (cf. Pleistocene refugia theory, 2.8, below).

2.7.2. *The hypothesis of greater specialisation*

Another possible explanation for the high biodiversity in tropical forests is that, compared with the forests of the temperate zone, organisms have narrower niches, and are more specialised, which allows more species to

coexist in a given habitat. The niche breadth of herbivorous insects may be measured by their degree of host-specificity. The estimate of species-richness made by Erwin is based on untested estimates of insect specificity. There are some studies of the diet breadth of herbivores (Wood and Olmstead, 1984; Basset, 1992; etc.). An experimental study (Barone, 1998) on 46 herbivorous insect species shows that 26% specialise on a single plant species, 22% are limited to a single genus, 37% are able to feed on several genera within a family and 15% are generalists able to feed on plants from different families. Janzen (1983) estimated that about half the species of Lepidoptera in a forest in Costa Rica feed on a single plant species and that generalist species are few. Papilionidae, Pieridae and Nymphalidae have a narrower host-range in the wet forest of Costa Rica than in temperate sites, and the same latitudinal pattern exists for grasshoppers. Ithomiinae and *Heliconius* (Nymphalidae) generally feed on one to three plant species (Brown, 1987). Most Hemiptera studied in Indonesia by Hodkinson & Casson (1991) are restricted to hosts belonging to a single plant family. The different species of *Acanthoscelides* (Coleoptera, Bruchidae), the larvae of which live in seeds, show no trend in specificity on a latitudinal gradient and the number of host plants they parasitise does not differ significantly in a tropical region and in a temperate one (figure 21.14).

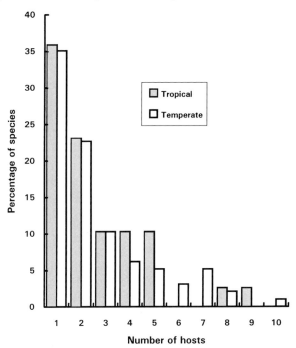

Figure 21.14. *Percentage of the species number of Bruchidae of the genus* Acanthoscelides *in a tropical region (Central America) and in a temperate region (West of the USA) according to their number of host plants. (Johnson, 1983).*

These data contradict others. Beaver (1979) found that Scolytidae and Platypodidae are less host-specific in the tropics (cf. table 3.1). Stork (1987) also mentions that his data on the arthropod fauna of Bornean rain forest trees support the view that for many insect groups there is a lower host-specificity in the tropics than in temperate regions. According to Wood & Olmstead (1984), 46% of the tree hopper species (Membracidae) in Costa Rica are polyphagous as opposed to only 16% in Ohio. Among the non-phytophagous insects of the family Ichneumonidae the groups with species that parasitise many hosts have a higher species-diversity in the tropics than in temperate regions.

The host-specificity of tropical canopy beetles is discussed by Mawdsley & Stork (1997). There are numerous tree species in tropical forests, but many of these species belong to the same family. For example in England there are 20 species of tree belonging to 15 genera and 10 families; in Sabah there are 511 species of tree belonging to 164 genera and 59 families. According to these authors specificity of tropical beetles is primarily at the family level. This is in accordance with studies that suggest that host-specificity in tropical forests is lower than expected.

All these data make it difficult to assert whether or not the high species diversity of tropical faunas is due to host-specificity which might be higher in the tropics than in temperate regions.

2.7.3. The hypothesis of the role of parasitism and competition

According to MacArthur, competition and parasitism are more intense in the tropics. This allows the coexistence of many species that are less abundant and that can therefore share the resources provided by the environment. Yet competition seems to be rare among phytophagous insects. Strong (1982) studied a group of 13 species of hispine beetle (Chryso-melidae) that coexist as adults in the rolled leaves of *Heliconia*. These neo-tropical beetles are long-lived, closely-related, eat the same food and occupy the same habitat. The ecological niches of these species partially overlap and 8 of the species can be found on a single plant and 5 species in one rolled leaf. In spite of this, the beetles lack any aggressive behaviour within or between species and their host-specificity does not change when there are other species that might be competitors. Competition, abundance of food and availability of habitat are not the limiting factors for these beetles. They do however suffer parasitism and predation, particularly by Hymenoptera Trichogrammatidae which parasitise 30 to 50% of individuals. Experiments carried out on 3 species of hispine at different densities on *Heliconia imbricata* showed no effect caused by the presence of these species on subsequent emigration or immigration by co-occuring species (Seifert, 1982). In conclusion, this example shows that competition does not structure phytophagous insect communities and this may be true of other communities as Lawton & Strong (1981) proposed.

2.7.4. The hypothesis of the role of surface area

The surface of tropical regions is larger than that of temperate regions and this may be the reason why the biodiversity in the tropics is greater (Rosenzweig, 1992). One theory predicts that tropical regions, because of their greater surface area, have a higher speciation rate and a lower extinction rate than temperate regions. According to the theory of island biogeography, the relationship between the number of species S and the area A is $S = cA^z$. This species-area relationship is not restricted to islands. It can also be applied to continental areas, and to trees and other plants and to their insect fauna (cf. chapter 4.2 and figure 21.15). The evidence of this species-area relationship is given by the loss of species following fragmentation of tropical forests (cf. 3.1, below). In the formula, z is a parameter which ranges between 0.15 and 0.35. If $z = 0.25$ a tenfold increase in area doubles the number of species.

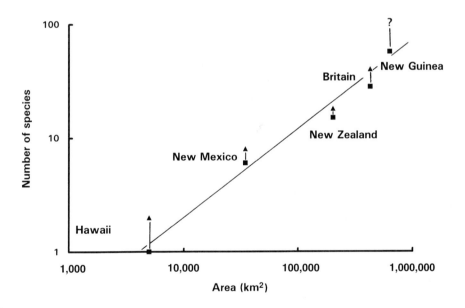

Figure 21.15. *Species-area relationships for phytophagous insect species living on bracken (*Pteridium aquilinum). *The areas where bracken covers a large area have more species. Triangles include infrequent or uncertain species (Lawton, 1976).*

2.7.5. The energy diversity theory

Some geographic areas have shown significant correlations between Net Primary Productivity (NPP) or a related measure, and species-richness. This is the *energy-diversity theory* (see Currie, 1991). There are few data supporting this theory which was tested for termites by Eggleton *et al.* (1994). These authors have shown: (1) that termite generic richness for various sites across the globe is correlated with species-richness and is higher in forested areas than in non-forested areas; (2) that the Ethiopian region has a low NPP and a high generic richness, whereas the neotropical and Indo-Malayan regions have high NPP and low generic richness. The highest generic richness is found in Cameroon with 65 genera; West Amazonia has 54 genera and North Borneo 46 genera; (3) that the generic richness per given NPP value is highest in the Ethiopian region, lowest in the Indo-Malayan region and is intermediate in the neotropical region. These data show that variations in NPP do not explain the patterns observed. A historical hypothesis is suggested. The Quaternary climatic disturbances have been more intense in Africa, which experienced a series of drier and wetter periods. In response there has been recent speciation in termites, particularly among "higher termites" (Termitidae) which constitute a young clade. It is a fact that there is a higher proportion of higher termites in the Ethiopian region than in other tropical regions.

2.8. Diversity, endemism and indicator taxa

Tropical forest species are not evenly distributed. Some regions have more species and a greater percentage of endemism. Tropical America contains more species than tropical Africa. The study of plants in South America has revealed a number of localities – 26 or more – where there is an unusual concentration of species and of endemism. The regions delimited by studies of plants are approximately the same when insects are studied; it is the case for Lepidoptera Heliconiini and Ithomiinae of the family Nymphalidae (cf. Brown, 1987). Many of these areas coincide with the "Pleistocene refugia" (Prance, 1982; Myers, 1986). According to the "Pleistocene refugia theory" endemism and diversity must be concentrated in the areas where forests subsisted during dry periods, which corresponds to the glaciations. This theory has been criticised (see Bush, 1994). Some think that these regions are simply the ones that have been best studied. The centres of endemism used to support the theory of unchanged refugia were not areas of maximum stability but of maximum disturbance. In some high-diversity areas such as the Andes or New Guinea, diversity seems due to recent speciation, which contrasts with the model of Pleistocene refugia. Moreover, areas of high diversity and of endemism frequently do not coincide, and rare species are often found outside areas of high diversity.

Other areas of high diversity and endemism are located in Africa, south-east Asia and New Caledonia (table 21.6). These centres of diversity are "hotspot areas" for tropical forests. The original surface of these hotspots was $2{,}204 \times 10^3$ km^2. It is now only 293×10^3 km^2. It is likely that these hotspots, which were established with the help of surveys of plants, are also very rich in insect species.

Regions	Original area (x 10^3 km^2)	Present area (x 10^3 km^2)
Madagascar	62	10
Atlantic forest of Brazil	1,000	20
West Ecuador	27	2.5
Choco, Colombia	100	72
Mountains of West Amazonia	100	35
East Himalaya	340	53
Malaysia (peninsular)	120	26
North Borneo	190	64
Philippines	250	8
New Caledonia	15	1.5

Table 21.6. *Hotspot areas of tropical forests (Myers, 1988).*

Taxa indicators of biodiversity are found in insects. Two of these are tiger beetles (Cicindelidae) and swallowtail or birdwing butterflies (Papilionidae). Among the ten countries that have the greatest number of species in these two families, seven are common to both the Cicindelidae and the Papilionidae (table 21.7). In the Papilionidae, the majority of species are associated with forest or woodland; some 54% of the 570 species of papilionid are concentrated in five countries: Indonesia, the Philippines, China, Brazil and Madagascar. Some species of swallowtail have a very restricted distribution so that 50 countries have endemic species, some of which are known only from a small area. The world's largest birdwing *Ornithoptera alexandrae* from New Guinea and *Papilio homerus* from

Papilionidae	Number of species	Number of endemics	Cicindelidae	Number of species	Number of endemics
Indonesia	121	53	Madagascar	176	174
Philippines	49	21	Indonesia	217	103
China	104	15	Brazil	184	97
Brazil	74	11	India	193	82
Madagascar	13	10	Philippines	94	74
India	77	6	Australia	81	72
Mexico	52	5	Mexico	116	57
Taiwan	32	5	USA	111	45
Malaysia	56	4	New Guinea	72	45
New Guinea	37	4	South Africa	94	40

Table 21.7. *Number of species of Papilionidae and of Cicindelidae in the ten areas where these two families are richest in species (Pearson & Cassola, 1992; New & Collins, 1991).*

Jamaica are examples of these highly localised species that are endangered by deforestation and trade by collectors who are ready to pay several thousand dollars for a specimen. Rodriguez *et al.* (1998) provided evidence that tiger beetles may be used as bio-indicators of the degree of disturbance of tropical forests. These data are useful to determine and to protect the most interesting areas.

3. FRAGMENTATION AND DISRUPTION OF TROPICAL FORESTS

3.1. How many species are lost?

The first studies applied the theory of island biogeography to fragmented forests to determine the number of species that are lost when the forest surface is reduced. According to this theory (cf. 2.7.4, above) the number of species decreases with the area and, if the area of a forest is reduced to 10% of its original area, half the species disappear. Today tropical forests are being depleted at an appalling rate. In Brazilian Amazonia, the annual loss of forest rose from between 1 and 2 million hectares in 1991, to an average of 19.2 million hectares during the years 1992-1994. In 1995 more than 2.8 million hectares were destroyed, which represents the area of Belgium. In western Ecuador more than 90% of the forests have been destroyed during the past four decades. Lugo (1988) gives the following estimates for the number of species:

Regions	Species present (x 10^3)	Projected deforestation 1980-2000 (%)	Extinctions (x 10^3)
Latin America	300 to 1,000	17.1	30 to 100
Africa	150 to 500	8.9	6 to 20
Asia	300 to 1,000	15.1	30 to 100
Total	750 to 2,500	12.3	66 to 220

Assuming that insects constitute half the number of all living species, the number of species of tropical forest insect destined to become extinct by the year 2010 is between 33,000 and 110,000.

Following some studies it has become evident that the isolation and fragmentation of the continuous forest also induces changes in the structure of the communities, affects many kinds of organisms and causes changes in biological processes such as pollination, seed dispersal and nutrient recycling. Insects are important actors in these processes, and abiotic factors, particularly edge effect, play a role in the modification of insect communities. The consequences of tropical forest fragmentation have now been studied for more than 20 years near Manaus (Brazil) thanks to the *Biological Dynamics of Forest Fragments Project* (see chapter 2.5). In this field of research, the most difficult problem lies in determining which aspects of the changes observed in the communities are due to surface effect, to edge effect or to the isolation of forest fragments.

One of the earliest and most unexpected effects of forest fragmentation is an increase in insect diversity and abundance in disturbed areas. The insects living on the edge of forests that are adjacent to pastures or in small forest fragments, are two to three times more common than in undisturbed forests. But this increase is caused by the invasion of common and widely spread insects called "supertrampers". When the Amazonian forest was intact these supertrampers survived in the gaps where trees had fallen down and in other naturally disturbed areas. In recent years they have invaded the forest fragments caused by man and the insects that normally live in the deep forest, far from the edge, have decreased in abundance and number of species.

3.2. Effects of fragmentation on some insects

Malcolm (1997) evaluated the biomass of insects living in five habitats of the Amazonian forest: continuous forest CF, the edge of the continuous forest, two forest fragments of 1 and 10 ha, and secondary forest. He showed that there is an increase of insect biomass in the understorey of the forest fragments and a decrease of insect biomass in the overstorey (figure 21.16). These changes could be attributed to edge effect. The changes in the vertical distribution of insects have important implications for vertebrate insectivores. In general the insect biomass in secondary forest is similar to that of the fragments of forest, and lower than that of the edge and of the CF. The edge has the highest biomass. The following numbers give the dry biomass (± s.d.) of all captures (mg per trap per night):

Secondary forest:	269.8 ± 35.2	Edge: 724.7 ± 183.5
1 hectare fragment:	232.2 ± 70.7	CF: 590.1 ± 105.8
10 hectare fragment:	329.8 ± 42.1	

The impact of habitat fragmentation varies greatly according to species (table 21.8). Here are a few examples:

Species	Effects of fragmentation	General habitat and vertical strata	Distance insects will not cross (m)
Apidae			
Euglossa augaspis	Positive	OPEN ALL	-
Euglossa iopyrrha	Negative	PF ALL	100
Eulaema bombiformis	Unaffected	PF ALL	100
Formicidae			
Atta sexdens	Negative	PF GRO	-
Eciton burchelli	Negative	- GRO	250
Scarabaeinae			
Canthon triangularis	Negative	PF GRO UND	15
Glaphyrocanthon (4 spp.)	Positive	OPEN	-
Ithomiinae	Negative	EDGE PF	-
Morphinae	Positive	ALL	-
Nymphalinae	Positive	ALL	-

Table 21.8. *Effects of forest fragmentation on some neotropical forest insects. OPEN: open, non-forest habitat; EDGE: forest edge and gaps due to treefall; PF: primary forest; ALL: all habitats; GRO: ground; UND: understorey.*

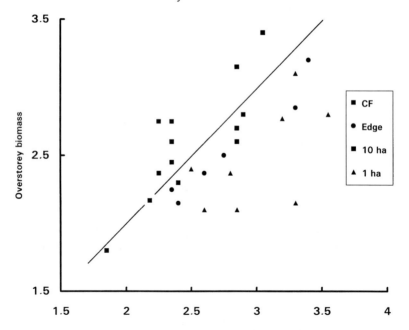

Figure 21.16. *Log-transformed overstorey biomass versus log-transformed understorey biomass (in abscissa) in continuous forest (CF), edge, 10-ha and 1-ha plots. In the continuous forest the overstorey biomass is generally greater than the understorey biomass (Malcolm, 1997).*

3.2.1 Euglossine bees

Euglossine bees (Hymenoptera Apidae of the genera *Euglossa* and *Eulaema*) which pollinate orchids are a typical example of the effects of fragmentation (Lovejoy *et al.*,1986; Powell & Powell, 1987). Some of their species are more abundant in small fragments of forest or even in cleared areas, but many others decline and disappear in fragments of less than 100

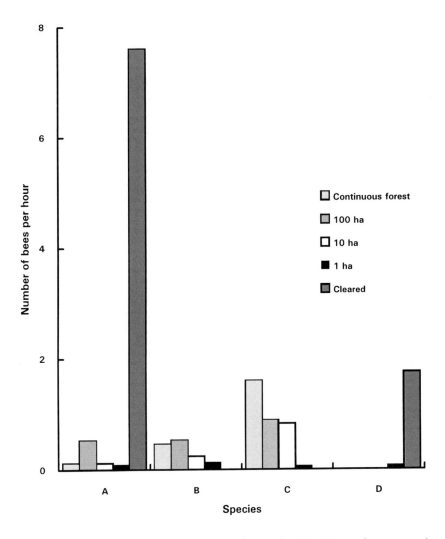

Figure 21.17. *Effects of forest fragmentation on the abundance of some species of euglossine bee. Number of male bees observed per hour at chemical baits in five different habitats. A:* Eulaema mocsaryi*; B:* Euglossa iopyrrha*; C:* Euglossa chalybeate*; D:* Euglossa prasina *and* E. augaspis *(Lovejoy* et al., *1986).*

hectares (figure 21.17). This results in reduced flower pollination. Euglossine bees are long-distance pollinators which can fly 23 km a day. Females visit flowers for resin, nectar and pollen, to build their nests. These bees have specific foraging routes known as "traplines" (Janzen, 1971) and they visit the same flowers over a long period. The plants these females visit produce few flowers each day and flower over a long period of time. Male euglossine bees fail to cross a narrow (100 m wide) cleared area, possibly because of changes in microclimate conditions.

Loss of habitat is one of the reasons for loss of species diversity and decline in abundance of other pollinating insects such as various species of bee (of the genera *Apis, Melipona, Ceratina, Megachile*, etc.). In 1972, in a site in the seasonal dry forest of Costa Rica, about 70 species of bee were collected on the flowers of the tree *Andira inermis* (family Fabaceae). In 1996 only 28 species were collected and their abundance had decreased by 90%. This is attributed to the partial destruction of the forest for agricultural development and to fire (Frankie *et al.*, 1997).

3.2.2   Ants and termites

Ants of the genus *Eciton*, or army ants, are characteristic of Amazonian forests. A single colony of *Eciton burchelli* needs about 30 hectares of *continuous* forest so it is not surprising that this ant should disappear in fragments of only 10 hectares. The birds known as 'ant-followers' that follow the ant swarms to eat the insects they drive out of the litter disappear too (Lovejoy *et al.*, 1986). The influence of different levels of disturbance on species-richness of termites in Cameroon was described by Eggleton *et al.* (1995). The clearance of forest by logging reduces termite species-richness. There are 46 species in a primary forest and only 16 in a cleared area. Variations in species-richness depend on the taxa. Termitinae are dominant in the primary forest and absent in cleared areas, whereas Apicotermitinae are richer in species in cleared areas than in the primary forest. Among functional groups, soil-feeders predominate in the primary forest and are greatly reduced in cleared areas; wood-feeders are less disturbed and have about the same number of species in all the plots. Species-richness in this area of Cameroon is greatest in the old secondary plantation forest, followed closely by the primary forest and by the least disturbed habitat, whereas species-richness is lowest in severely disturbed plots. This pattern confirms the theory that predicts that the maximum species-richness prevails at an intermediate level of disturbance.

These observations contradict those made on termites in Sarawak (Collins, 1980), on moths in Malaysia (Holloway *et al.*, 1992) and on dung beetles in Brazil (cf. figure 2.11). In all these cases diversity decreases when the forest habitat becomes more fragmented or more disturbed. Some differences exist between communities of Amazonian termites located in the continuous forests and those located in the fragmented forests. In the fragments of forest the proportion of rare species of termite is higher than in the continuous forest and the number of species is lower (Fonseca da Souza & Brown, 1994). This is in agreement with observations that show that population size is lower on the islands than on the continents and that there are fewer species on an island than on a continental area of equal size. Isolated populations become extinct either by loss of fitness due to inbreeding or by an increase of niche overlap due to the lower amount of resources that are available.

3.2.3. Dung beetles

Dung beetles are good indicators of habitat disturbance. Klein (1989) shows the loss of species diversity with the reduction of forest area. Population density declines seven-fold in a 1 ha forest compared with continuous forest. This may be more important for the ecosystem processes than the loss of species-richness. In the state of Veracruz (Mexico) dung beetles show characteristic variations of species. The pasture area is the poorest habitat. The areas of the forest fragments and the distance between them are the two main factors that affect species diversity. Small or isolated forests have a poor mammal fauna and thus few species of dung beetle (Estrada *et al.*, 1998). In Mexico the percentage of species of dung beetle belonging to the "dung rollers" is higher in open areas than in forests. Necrophagous species of this family are less numerous, and diurnal species more abundant in open areas following deforestation (Howden & Nealis, 1975; Halffter *et al.*, 1992). There is a decrease of the number of species and of the mean length, caused by the disappearance of large species linked to the forest (table 21.9). In order to conserve a high biodiversity of dung beetles it is necessary to maintain large enough forests. However the minimum surface necessary for this conservation is not yet known.

	Forest	Edge	Open area
Diurnal species	6	4	4
Nocturnal species	21	7	1
Diversity H	2.5	2.12	1.01
Equitability	0.76	0.88	0.62
Mean length (mm)	13.63	12.38	11.76

Table 21.9. *Characteristics of the dung beetles (Scarabaeidae) in a forested area of south Mexico. There are more (mainly nocturnal) larger species than in open areas (Halffter et al., 1992).*

3.2.4. Butterflies

According to Brown & Hutchings (1997) the effects of area on butterfly communities are less important than those predicted by the theory of island biogeography. Their observations suggest that environmental heterogeneity rather than the size of the forest fragment is the most important factor in determining the species-richness of butterflies. They conclude that "to sustain a large number of species, an isolated reserve should be large enough (100-1,000 ha) both to include a number of internal tree-fall clearings of reasonable size (0.02-0.1 ha) and varying ages, and to resist negative edge effects from destructive land uses in the surrounding area".

The number of species of butterfly is often greater in fragmented forests than in continuous forests. Edge effect is characterised by increased wind, low humidity and mainly by an increase in the light level, and is felt up to 200-300 metres inside the forest. It disturbs the butterfly species of the interior and allows the arrival of butterfly species of the clearings and secondary growth vegetation. In Trinidad a comparison was made between an undisturbed forest and a recently selectively-logged forest where the gaps caused by tree felling interrupted the continuity of the canopy. This comparison showed that butterfly species-richness was greater in the disturbed area, the most disturbed habitats having more species than the undisturbed habitats (Wood & Gillman, 1998). However the number of species and of individuals caught in fruit traps was lower, the fruit-feeding species being specific to a closed canopy. Entomologists have also found a greater butterfly species-richness in disturbed areas in Amazonia (Lovejoy *et al.*, 1986). This is due to the invasion by light-loving species which are characteristic of secondary growth vegetation because of the high light level caused by clearings. These results are similar to those obtained in temperate forests when management of rides increases the variety of habitats and of the number of butterfly species (cf. chapter 2.5.1).

Chey *et al.* (1997) have compared the diversity of moths in forest plantations, in secondary forests and in natural forests in Sabah. They give the following numbers: secondary forest, 1,048 species and α diversity index 315; plantation sites, 675 to 872 and α diversity index varying from 208 to 331 according to the tree species. The species diversity is lowest in forest plantations, but these plantations can play a role in the conservation of important faunal groups such as moths when compared with total clearance of forest for agriculture.

> Log-normal distribution of species abundance is often associated with communities in equilibrium. According to Hill *et al.* (1995) only the species distribution of unlogged tropical forest butterflies fits in with the log-normal distribution, but not the species distribution of butterflies in forests that have been selectively logged. This theory needs to be confirmed by other observations since it has not been observed by Nummelin (1998) in his study of certain beetles (Coccinellidae, Cassidinae and dung beetles) from Kibale forest in western Uganda.

3.2.5. *Influence on soil fauna*

More than 1,000 species of beetle live in leaf litter in any one patch of undisturbed Amazonian forest. Among them there are 300 species of

Carabidae, Staphylinidae and Scarabaeidae. Of these 300 species 70% are rarely found and are sometimes represented by a single individual in the samples. Six species of beetle represent 30% of the total number of individuals. These common species are the first to disappear from fragments of forest. The insects of the soil fauna play a major role in the decomposition of dead organic matter. In small forest fragments the rate of the breaking down of the leaves is one third of that in undisturbed forests, partly because of the loss of insect biodiversity. Together with many beetles, ants and litter-feeding termites disappear and leaf litter accumulates on the soil.

Didham (1997) describes the influence of edge effect and forest fragmentation on leaf litter invertebrates, particularly on three families of Coleoptera (Carabidae, Staphylinidae and Scarabaeidae) and on ants. He found that the abundance of these insects is higher at distances of 26 to 105 metres away from the edge, which is at a mid-distance from the deep forest. The most likely explanation is that this zone is an ecotone where there is an overlap of edge species and of interior species. For all the beetles of the litter, the proportion of predatory species is higher near the edge and the proportion of xylophagous species is higher at the core of the fragment of forest. This seems to be in contradiction with the theory that species at higher trophic levels are more vulnerable to disruptions, particularly those caused by fragmentation of the habitat. In conclusion, it is evident that the loss of biodiversity, the reduction in abundance of some species and the changes in the structure of forest insect communities greatly affect the functioning of the ecosystem.

4. THE MAIN INSECTS OF TROPICAL FORESTS

4.1. Ants

4.1.1. *The importance of ants*

The local diversity of ants is frequently high. Because of the role they play, termites and ants are the most important insects and even the most important animals in rain forests. In the Amazonian *terra firma* rain forest they constitute about one third of the entire animal biomass, each hectare containing more than 8 million ants and 1 million termites. With bees and wasps these insects represent 75% of the insect biomass (Fittkau & Klinge, 1973). Ants and termites also dominate in the forests and savannah of Zaire (Dejean *et al.*, 1986).

Ants are found in all strata of the forest where they exploit many types of food resources. Stork (1987) has established that the ants collected by fogging in the Bornean rain forest may be separated into different guilds. The two largest guilds are the opportunists (51% of individuals and 58% of species) which eat all that is available, and the predators / scavengers

(42% of individuals and 11% of species). Few ants are specialised predators (3% of individuals and 2% of species), or feed on honeydew from Homoptera, whereas Pseudomyrmeciinae feed on extra floral nectaries.

Many species of ant are predators that play a part in structuring the arthropod communities in tropical forests (Hölldobler & Wilson, 1990; see also Basset, 6.4, below). The impact of ants on insects in the arboreal habitat is considerable. In many cases they affect the arthropod guild structure, with an increase in the proportions of the sucking phytophagous species and a reduction of the proportion of the chewers and other guilds. Ants also protect some plants such as Acacia from the attacks of herbivorous species (cf. 4.3, below). Reduced herbivory as a consequence of ant foraging has been observed many times (Grant & Moran, 1986; etc.). Other ant species are herbivorous. Some species live in symbiosis with plants.

4.1.2. Stratification of ants in forests

Like many other insects, different ant species are unequally distributed in the different strata of the forest. In a rain forest of Sabah, Borneo, 524 'morphospecies' could be distinguished; they belong to 7 subfamilies and 73 genera. The leaf litter harbours as many species as the lower vegetation or the canopy, and has the highest generic diversity (figure 21.18).

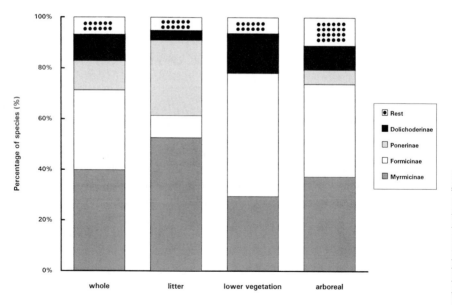

Figure 21.18. *Percentage of ant species belonging to various subfamilies in different habitats in a rain forest, Kinabalu National Park, Borneo. "Whole" stands for all the habitats (Bruhl et al., 1998).*

The stratification of ants seems to be very strict, with 75% of the species associated with only one stratum and with 17% in the canopy (Bruhl *et al.*, 1998). This contrasts with the stratification of beetles in a rain forest of Sulawesi. In this forest only 10% of the beetle species are restricted to a specific stratum and only 10% to the canopy (Hammond, 1990). The most numerous ants in the litter layer of tropical forests are Myrmicinae and Ponerinae; the percentage of Myrmicinae varies from 53 to 63% and the percentage of Ponerinae from 20 to 30%.

4.1.3. Some ant species

In the neotropical region, leaf-cutting or gardening ants (tribe Attini, genera *Atta* and *Acromyrmex*) are perhaps the most important herbivores. Species of the genus *Atta* are found between latitudes 44° S and 40° N. Their nests are abundant in many lowland tropical forests and each nest may cover an area of 250 m^2 or more. A nest of 35 m^2 in surface reached a depth of 5 m and contained 3,000 chambers, the total volume of which was 5 m^3. In lowland rain forests the density of the nests varies from 0.6 ha^{-1} in Trinidad to 1-18 ha^{-1} in Brazil. In some forests leaf-cutting ants consume 12 to 17% of the leaf production. The quantity of leaves carried to the nest was estimated at 0.253 g m^{-2} day^{-1} (Lugo *et al.*, 1973). Columns of workers carrying leaves are spectacular: sometimes the column is 15 cm wide and moves at a speed of 1.3 m min^{-1}.

The location of the nests is very variable and many are abandoned. The ant nests could occupy the equivalent of the entire forest area within 200 to 300 years. When the nest is active there is, in and around it, a reduced diversity and abundance of understorey plants. But in abandoned nests there is a greater diversity and abundance of small understorey plants compared with the nearby forest. *Atta* nests could therefore play a role in the maintenance of plant species diversity (Garrettson *et al.*, 1998). The plants collected by *Atta* are varied, and whole bushes and trees may be defoliated in a short time. In Guyana, the leaves of 61 species of tree are collected but those of *Terminalia amazonica* and *Emmotum fagifolium* are preferred, whereas trees such as *Marlierea montana* are avoided because their leaves contain tannins or latex (Cherrett, 1968, 1972). *Atta* cultivate a fungus on a substrate made of fresh leaves, flowers, fruits and other vegetable matter. This fungus is a basidiomycete similar to the species used by some termites of the subfamily Macrotermitinae. These ants eat special nutrient bodies of the fungus. The fresh leaves that are carried to the nest are chewed and divested of the surface waxes, which have antibiotic properties. The fungal hyphae grow on this mixture of plants, to which workers add faeces that contain nitrogen wastes and fungal enzymes that degrade the constituents of the plants collected.

Atta sexdens has four types of worker which can be identified according to the width of their heads. The largest carry leaves and defend the nest and the smallest are fungal gardeners and nurses. *Atta* are tolerant ants which, in Guyana, coexist with other insects such as termites or the ant *Eciton burchelli*. Many species of ant build their nests on the large nests of *Atta* (Weber, 1972).

Ants of the genus *Azteca* are easily identified by their odour which is produced by various methylated heptanones. These neotropical ants are very aggressive. They live in trees, particularly in several species of *Cecropia*. The trees that are colonised have hollow trunks in which the ants build their nests. *Azteca* ants protect their host plants against defoliators such as *Atta* and eliminate lianas growing around their trees. Another advantage for the plant is the nitrogen supplied by the excreta and the dead bodies accumulated in the hollow stems of the host plants.

Neotropical ants of the genus *Eciton* are known as army ants. They are important as predators. In Panama the density of *Eciton burchelli* is of three colonies per km^2. Their raids progress on a 20 metre wide front at 15 metres per hour. Fifty "professional" species of bird follow these raids to eat the insects flushed out by the ants. Large workers or soldiers with enlarged mandibles defend the columns of workers and carry large prey. Because they drastically reduce the abundance of insects and other small animals, *Eciton* may be considered as key species in their habitat.

African *Dorylus*, or driver ants, have queens which vary in length from 39 to 50 mm, making them the largest ants. Their colonies contain as many as 22 million workers with a total weight of 20 kg; there may be 3 colonies per 10 ha of forest in the Ivory Coast. Their swarms extend like many metre long pseudopods. These ants kill all animals too slow to get out of their way.

4.2. Termites

Termites were originally polyphagous, feeding on a variety of organic matter. The most primitive species alive, *Mastotermes darwiniensis*, eats dung and plant litter as well as wood. Termites play a major role in the decomposition of dead trees in the tropics, compared with the temperate forests where they are rare or absent (cf. chapter 16). Primitive families such as the Kalotermitidae and Rhinotermitidae digest wood with the aid of flagellate Protozoa, whereas Termitidae digest cellulose thanks to the bacteria present in their gut. The stages in the decomposition of logs were described by Abe (1980). Termites may choose the wood they eat; when colonised by fungi the wood may be an attractant or a repellent, depending upon the species. Termites can digest 70 to 90% of the cellulose of the wood but no more than a third or a half of the lignin (Brian, 1978).

The three subfamilies of Termitidae include soil-feeding species which eat the humus in the soil. The hindgut of these species contains bacteria (spirochetes and actinomycetes) that break down the components of the humus. Thanks to these symbionts 54 to 93% of the humus are degraded, which is much higher than for other soil arthropods (mites, cockroaches, isopods, diplopods) for which degradation of the humus ranges between 14 and 40%. In the subfamily Macrotermitinae the termites make a comb in which a fungus of the genus *Termitomyces* grows. The fungus breaks down the lignin and the cellulose of the litter collected by the termites, which feed on the fungus.

In an African savanna, termites eat 24% of the wood and litter production (Collins, 1981). The abundance and biomass of termites in different forests are compared in table 21.10. In Africa the Macrotermitinae construct mounds that may be 3 to 4 metres high and contain 1.10^5 to 24.10^5 kg ha^{-1} of soil and cover 30% of the surface. Maximum abundance and biomass are of 4,000 m^{-2} and 10 g m^{-2} but with great differences between different types of forest. In Sarawak there is a marked decrease in abundance with increasing altitude. In Pasoh Forest, West Malaysia, there are 57 species of termite belonging to 23 genera (cf. figure 21.19). One hectare may have as many as 52 species. The biomass ranges from 3 to 4 kg m^{-2} live weight (or 2.2 to 2.6 kg m^{-2} dry weight). Thirty seven species feed on dead wood on the ground, 24 species on leaf litter, 17 species eat humus and two species eat lichens in the canopy as high as 40 m above the ground (Abe & Matsumoto, 1979). Termite nests may be classified under five types: nests within wood, subterranean nests, epigeal nests, arboreal nests and nests of intermediate type. The sites of the nests can be located in all strata of the forest, from more than 30 m high to a depth of 30 cm in the soil (table 21.11).

CLASSIFICATION OF ANTS

Ants constitute the family Formicidae, which is divided into 11 subfamilies with 297 genera and 8,800 species. Many parts of the tropics are poorly collected, particularly the moist forests where ants are abundant and diverse. The subfamilies are:

– **Myrmicinae** have a world-wide distribution. Genera such as Atta (leaf-cutting ants), *Cardiocondyla, Crematogaster, Eurhopalothrix* and *Tetramorium* have species in tropical forests.

– **Ponerinae** are present in the Old and New World tropics. Examples: *Cerapachys, Odontomachus, Myrtrium, Prionopelta*.

– **Leptanillinae** are species of the Old World temperate and tropical regions.

– **Dorylinae** are ants of the Old World tropics and subtropics. Example: *Dorylus* (driver ants).

– **Ecitoninae** are New World species that range from the tropics to the warm temperate zone. They are known as "army ants".

– **Myrmeciinae** have species in Australia and New Caledonia.

– **Pseudomyrmeciinae** have species in the tropics and subtropics. The genus *Tetraponeura* has abundant arboreal species.

– **Aneuretinae** has only one species of the genus *Aneuretes* from the Sri Lankan rain forest.

– **Dolichoderinae** have a world-wide distribution. *Azteca is* arboreal in the New World.

– **Nothomyrmeciinae** are confined to southern Australia.

– **Formicinae** are distributed world-wide. The species of *Oecophylla* are the "weaver ants". *Formica, Cataglyphis, Lasius, Camponotus* have species in the tropics.

Localities	Numbers per m^2	Biomass g/m^2 (w.w.)
Malaysia, dipterocarp forest	3,160 to 3,810	8.69 to 10.13[b]
Sarawak, dipterocarp forest	1,527	2.4
Sarawak, lower montane forest	38	0.09
Sarawak, upper montane forest (1,310-1,860 m)	99 to 95	0.01 to 0.78
Sarawak, upper montane forest (1,970-2,376 m)	0	0
Sarawak, Kerangas forest	2,271	4.1
Trinidad	4,450	n.d.
Surinam forest	0 to 1,420	n.d.
Nigeria, riparian forest	2,646	6.9
Zaire, riparian forest	1,000	11.0[a]

Table 21.10. *Abundance and biomass of termites in some tropical forests. [a] this biomass is certainly an exaggeration. [b] these data are underestimated (after several authors).*

Subfamilies	Nests						Number of species
	In tree	Arboreal	Intermediate	Epigeal	Subterranean	Unknown	
Coptotermitinae			1			1	2
Rhinotermitinae	2		3				4
Amitermitinae		4			1		5
Termitinae			8	5	7		14
Macrotermitinae				2	10		12
Nasutitermitinae	1	5	7		7	1	19
Unidentified					1		1
Total species	3	9	19	7	26	2	57

Table 21.11. *Spatial distribution and abundance of nests in the diverse subfamilies of termites in a 1-ha plot in Pasoh Forest, Malaysia (Abe & Matsumoto, 1979).*

Localities	Consumption (g m^{-2} yr^{-1})	% of annual litter production
Sarawak, three forests	7.4 to 35.3	0.6 to 3.4
Malaysia, Pasoh forest	155.4 to 173.2	14.7 to 16.3
Venezuela, three forests	21 to 59	3 to 5

Table 21.12. *Consumption of organic matter by rain forest termites in three regions (Collins 1983).*

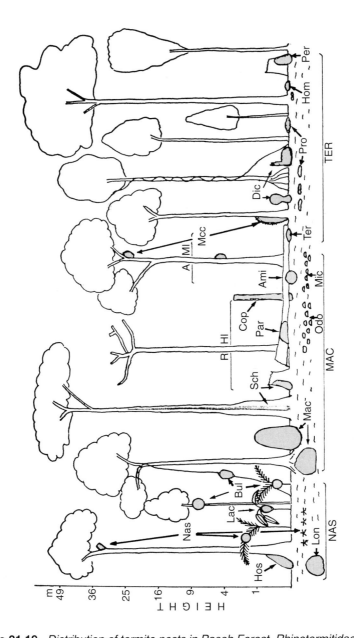

Figure 21.19. *Distribution of termite nests in Pasoh Forest. Rhinotermitidae (RHI).*
Sch: Schedorhinotermes; *Par:* Parrhinotermes; *Cop:* Coptotermes.
Amitermitinae (AMI). Ami: Amitermes; *Mcc:* Microcerotermes.
Termitinae (TER). Ter: Termes; *Per:* Pericapritermes;
Dic: Dicuspiditermes; *Pro:* Procapritermes; *Hom:* Homallotermes.
Macrotermitinae (MAC). Mac: Macrotermes; *Odo:* Odontotermes;
Mic: Microtermes. *Nasutitermitinae (NAS). Hos:* Hospitalitermes;
Lon: Longipeditermes; *Nas:* Nasutitermes; *Lac:* Lacessitermes;
Bul: Bulbitermes. *Nests are located, according to species, from*
underground to the canopy (Abe & Matsumoto, 1979).

Consumption of organic matter by termites is difficult to evaluate. In different rain forests consumption varies from 7.4 to 173.2 g m^{-2} yr^{-1}, which is between 0.6% and 16.3% of annual litter production (table 21.12).

Detailed studies of the role played by termites in turning over soils in forests and savannas in tropical Africa was made by Boyer (1971) and Grassé (1986). Soil termites belong mainly to the family Termitidae. In rain forests the most abundant termites are those that feed on humus. These humivorous species are unable to digest wood and leaves, and only eat humus, in other words vegetable debris already decayed by bacteria. Humivorous termites play a role in the conservation of the humus of forest soils comparable with that of earthworms. Lignivorous forest termites are numerous in the soil, while species that build epigeal nests are less abundant. Each species builds a network of galleries completely independent from the galleries of the other species. Some species, such as those of the genus *Pseudacanthotermes*, gnaw at fragments of wood lying on the ground and replace the fragments they carry away with clay. By so doing they bring a large quantity of material up to the surface of the soil. All these changes are an important part of the cycle of forest soils. The importance of termites is even higher in savannas where their action compensates the degradation of the soils caused by the farming methods. In conclusion, termites are the main component of the soil microfauna in tropical forests and the dominant decomposer insects in tropical ecosystems. They act in maintaining soil fertility, decomposition processes, and the carbon and nitrogen cycles. Wilson (1992) went so far as to write "termites are premier decomposers of wood, rivals of earthworms as turners of the soil, owners of 10 percent of the animal biomass in the tropics".

Termite nests attract other invertebrates. Some species of termite colonise the nests of other species. Many termitophilous insects (particularly Coleoptera of the family Staphylinidae) live with termites in their nests (cf. Grassé, 1986). Termites as well as ants are eaten by specialised mammals such as neotropical anteaters (*Tamandua* spp.) which have strong claws with which they rip open wood and ant and termite nests. In the forests of Gabon some mammals that are consumers of termites and ants avoid competition by their behaviour. Of the two species of pangolin, *Manis tetradactyla* is nocturnal and feeds from arboreal ant nests; *Manis tricuspidata* is also nocturnal but feeds on termites and ants and part of its activity takes place on the ground (Pagès, 1970).

4.3. Coleoptera

4.3.1. Curculionidae *and* Chrysomelidae

These two phytophagous families are the most numerous in species in tropical forests. There are more than 50,000 species of Curculionidae in the world fauna but little is known of the ecology of tropical species. Many Curculionidae exhibit a high degree of host-specificity and certain groups of species are associated with certain families of plants such as Palmae (genus *Rhynchophorus*), Musaceae (genus *Cosmopolites*) and Bromeliaceae (genus *Paradiaphorus*). Some subfamilies (Cossoninae, Magdalinae) live in dead or decaying wood like the species of temperate areas. The large weevil *Gymnopholus lichenifer* (3 cm) lives in the montane rain forest of New Guinea. It carries micro-algae, liverworts, lichens and mosses on its back. This miniature garden supports several species of Collembola, Psocoptera, rotifers, nematodes and minute oribatid mites of the genus *Symbioribates* (0.2 mm long). This phenomenon, which seems to be restricted to New Guinea, was discovered by Gressitt (1969) and is called *epizoic symbiosis*.

The family Chrysomelidae is the second largest family of phytophagous insects after the Curculionidae. The majority of species are foliage-eaters with a high degree of specificity. There are about 35,000 described species and the estimated number of 60,000 species seems reasonable. (Jolivet *et al.*, 1988). Some species of Chrysomelidae in tropical and temperate areas are threatened by excessive use of insecticides or the extinction of their host plants. *Elytrosphaera* species in South America are rapidly disappearing, since their food plant (genus *Adenostemma*, family Compositae) lives only in the clearings of primary forests and only at higher altitudes. Food selection in Chrysomelidae has been extensively reviewed.

CLASSIFICATION OF TERMITES

Termites constitute an insect order, Isoptera, with 7 families (Grassé, 1986).

– **Mastotermitidae.** Contains only one species that has many archaic characteristics, *Mastotermes darwiniensis*, from N W Australia.

– **Termopsidae.** These termites, known as "damp wood termites" are primitive, and their habitat is damp wood. They have no workers, live with symbiotic flagellates and are world-wide in distribution.

– **Hodotermitidae.** These termites live in the soil, forage for food and displace a great quantity of earth. They are mainly present in the arid areas of the Old World and are known as "harvester termites".

– **Kalotermitidae.** These termites are called "dry wood termites" because they live in wood that is not in contact with the soil. Their only source of water is the moisture of the wood. They are distributed world-wide in the tropics and subtropics.

– **Rhinotermitidae.** These termites live in soil or in wood and are distributed world-wide.

– **Serritermitidae.** Contains a single species, *Serritermes serrifer*, from Amazonia.

– **Termitidae.** This is the family with the largest number (80%) of species. These termites do not have symbiotic flagellates in their hindgut. They are mostly wood-eating and either subterranean or mound-builders. The most remarkable are the Macrotermitinae that cultivate a fungus of the genus *Termitomyces* (Basidiomycetes). The genera *Bellicositermes* and *Macrotermes* are African. The towering nest mound of *Bellicositermes bellicosus*, known in French as "termitière cathédrale", may be more than 3 metres high. *Pseudocanthotermes* species live in the rain forests of Gabon and Zaire.

Hispine beetles that roll leaves are almost exclusively restricted to plants of the order Zingiberales in the New World. These beetles live a long time, at least 18 months according to Janzen who called them "elephants of phytophagous insects". The most interesting specialisation is the association of genera such as *Aulacoscelis* or *Carpophagus* with cycads, which are very primitive plants. The toxicity of the leaves and flowers of cycads is so high that the beetles that eat these plants are also toxic and have very few predators. Some weevils are associated with cycads: species of Cossoninae belonging to the genera *Porthetes* and *Amorphocerus* in South Africa attack the male flowers of these plants.

Erwin (1983) identified 650 species of leaf beetle in the Amazonian forest at Tambopata in Peru, (which is about the number of species present in eastern North America). This extraordinary diversity is difficult to explain. Many leaf beetles produce poisonous or repellent secretions that function as defence mechanisms. Species with defence mechanisms that make them unpalatable are numerous in the tropics. Such is the case, for example, of the Oedionychina (subfamily Alticinae) a group of neotropical flea beetles that has 2,000 species in South America.

4.3.2. Wood-borers

In the order Coleoptera the main families of wood-borers are the Cerambycidae, Buprestidae, Scolytidae and Platypodidae, Lucanidae, Scarabaeidae, Dynastidae and Passalidae, and minor families such as the Brentidae and Tenebrionidae. As is the case in temperate forests, there are also some other wood-boring insects in the orders Lepidoptera, Diptera and Hymenoptera, and the most important role in this habitat is often played by termites. Some wood-boring Coleoptera are polyphagous: the cosmopolitan scolytid *Xyleborus ferrugineus* occurs in 74 host plants representing 29 families.

The two most important subfamilies of Cerambycidae in the tropics are the Prioninae and Lamiinae. The largest species, such as *Titanus giganteus* in Amazonia, belong to the Prioninae. The Brentidae are a family of beetle (superfamily Curculionoidea) found mainly in tropical and subtropical forests. Some of its species are myrmecophilous but most are wood-boring, their larvae developing inside the wood of dead or dying trees. Many species are long, slender beetles with the mandibles borne at the end of an elongate rostrum. The larvae of many species of Dynastidae live in decaying wood. The larva of *Megasoma elephas* (a beetle 54 to 82 mm long and 18 to 28 g in weight) develops in large logs and lives 3 to 4 years. This South American species survives only in lowland rain forests that are not greatly disturbed. It has become a rarity; an additional cause of its disappearance is the collecting of adults for collections and decorative purposes. Scarabaeid larvae of the subfamily Rutelinae are associated with rotting logs and decaying roots. Species of the genera *Pelidnota* and *Plusiotis* are numerous in the New

World, where, because of their green or golden colour, they are called "gold bugs". The Passalidae are beetles which live in family groups in the decaying wood of tree trunks. They have a sub-social behaviour; a family group consists of a male, a female and their offspring. These beetles are all saproxylic insects. Their digestive tract includes fermentation chambers containing micro-organisms similar to those found in the digestive tracts of Scarabaeidae (cf. figure 16.17). The larvae, which are unable to break wood into small pieces, are fed by the adults. Passalids are equipped with a stridulatory apparatus. The sounds they emit ensure cohesion of the social group; those emitted by the female seem to induce the larvae to stay together (Reyes-Castillo & Halffter, 1983). Castillo & Moron (1992) have shown that consumption of wood by different species of passalid beetle is a function of their weight.

Scolytidae and Platypodidae are abundant in tropical forests. As is the case in temperate forests, many predators and parasites live under of the bark of trees attacked by these insects. The best known group is the family Colydiidae. Mayné (1960, 1962) collected 70 species of Colydiidae under the bark of different species of tree in Congo. In Africa, many larvae of the genus *Sosylus* parasitise the nymphs of Platypodidae, and the adults of *Sosylus* attack the adults of the Platypodidae. *Sosylus spectabilis* parasitises 13 species of Platypodidae (Browne, 1962; Hywel, 1969).

Dead wood is an important constituent of tropical forests. In Central Amazonian forests the biomass of dead wood on the forest floor varies from 1.8 to 33.0 ton (dry weight) ha^{-1}. To this biomass we must add 1.1 to 7.6 ton ha^{-1} for the standing dead trees and 1.1 to 45.6 ton ha^{-1} for the attached dead branches (Martius & Bandeira, 1998). The processes of wood decomposition in tropical forests are less well known than in temperate forests (some are described in chapter 17.2). Another study of the role of Coleoptera in the decomposition of dead wood was made by Rivera-Cervantes & Moron (1991) in the cloud forest of the Sierra de Manantlan in Mexico. The saproxylic beetles of rotten wood in this forest belong to 29 families. One melolonthid (*Macraspis rufonitida*) is the dominant species; it is followed by three tenebrionids (*Uloma mexicana, Nosoderma morbillosum* and *Eleodes* sp.), one curculionid (*Cossonus* sp.) and three cerambycids (*Aplagiognathus spinosus, Trichoderes pini* and *Derobrachus* sp.).

Insect pests in tropical forests have been reviewed by Gray (1972) and more recently by Löyttniemi & Heliövaara (1996). Wood-boring insects, and particularly wood-boring Coleoptera, are among the most important pests.

4.3.3. Carabidae, Cicindelidae, Staphylinidae

The family Staphylinidae or "rove beetles" is perhaps the richest in species, both in the soil and in the canopy, but little is known of the biology of these species. In contrast with temperate regions where Carabidae live (with very few exceptions) in the litter and in the soil (which explains why they are called ground beetles) there are many arboreal species in tropical forests. The biology of Carabidae in the neotropical region was thoroughly studied by Erwin (1975). Their apparent rarity in South America was noted by Bates (1884). Erwin showed that the apparent rarity of forest floor carabids is due to a patchy distribution, many living only under the crowns of trees, in litter covered with fermenting fallen fruits and blossoms. Some genera of arboreal Carabidae have numerous species. On Barro Colorado Island there are 298 species of Carabidae, 34 of which belong to the genus *Agra*. Adult arboreal carabids have developed certain characters relating to their habitat. These include (1) an elongate and relatively narrow body form such as in *Agra* and *Collyris*; (2) metallic coloration; (3) palpi with the broadest surface of the last segment oriented anteriorly, which Erwin supposes is an adaptation to searching for and eating arboreal snails which are particularly abundant in the neotropical region; (4) tarsal pads of long and curved setae (similar to those of chrysomelids and some cerambycids); these pads together with the pectinate claws give the beetle a secure foothold on the leaves; (5) spinose elytra are present in groups such as the Lebiini and Agonina; the function of the spines seems to be a protective device against predatory frogs which are abundant in the tropical arboreal community.

> Darlington (1971) made a very detailed study of the ecology of the Carabidae of New Guinea. He suggested that in New Guinea, ants (which are dominant insects) may affect the size distribution of Carabidae. Size distribution of the New Guinea lowland Carabidae is bimodal. Two modes occur at 2.0 - 2.95 mm and at 6.0 - 6.95 mm with a deficiency especially at 4.0 - 4.95 mm. Very small Carabidae may hide away from ants and large ones are able to protect themselves, whereas Carabidae in the 4.0 - 4.95 size class may be especially vulnerable to competition or predation by ants. Montane species of Carabidae have a unimodal distribution: ants are relatively few at a high altitude in New Guinea.

Cicindelidae, or tiger beetles, are represented by numerous arboreal species. The South East Asian genus *Collyris* includes 230 described species. They are diurnal insects that live on low, broad-leaved bushes; they fly from one leaf to another and feed mostly on small ants. Perching and roosting are behaviours observed in several species of tropical forest tiger beetle. In a tropical forest in Peru, tiger beetles may be arboreal (*Ctenostoma, Iresia*), or live on the forest floor (*Odontochila* and *Pentacomia*) or are riparian (*Megacephala, Chiloxia, Oxychila, Cicindela*). All these species are predators of small arthropods. When they are disturbed they perch for some time in the vegetation. At night they roost together for 10-12 hours at a height which depends on their length. Perching is attributed to protection against predators such as large tiger beetles of the genus *Megacephala* that forage in large numbers throughout the night (Pearson & Anderson, 1985). Perching is also known in tropical dung beetles: this behaviour is attributed to their need to search for a limited resource (Peck & Forsyth, 1982).

Nocturnal perching and roosting of butterflies is attributed to protection against predators, to thermoregulation, and to protection from unfavourable abiotic factors (Young & Thomason, 1975).

4.3.4. Dung beetles

Dung beetles (Coleoptera: Scarabaeidae) are conspicuous in rain forests. They exploit the dung of forest vertebrates, mainly herbivores such as primates, and occasionally the dung of birds and reptiles, as food and a substrate for oviposition and feeding their larvae (Hanski, 1989; Hanski & Cambefort, 1991). Carrion and decaying fungi are also exploited, especially in the neotropical region. Dung is a limiting resource for dung beetles. Some species are active by day, others are nocturnal. There are also differences in foraging methods including "ball-rolling" species and "non ball-rolling" species. This reduces the intense interspecific competition usually observed (Estrada *et al.*, 1993). Dung beetles may act as dispersers of the seeds present in large numbers in the dung of frugivorous mammals. This dispersal may be an advantage to the plant since it reduces competition among seedlings and decreases the probability of discovery by seed predators (Feer, 1999; Shepherd & Chapman, 1998).

Dung beetles generally forage on the ground where the resources they exploit are to be found. In some tropical forests where arboreal mammals are numerous (particularly monkeys), dung and carrion in the canopy have led some dung beetles to become arboreal, and some species never descend to the forest floor. Arboreal dung beetles are known to occur in the forest canopy in Gabon (Walter, 1984), in Borneo (Davis *et al.*, 1997), in South America (Howden & Young, 1981) and in Madagascar (Vadon, 1947). The main arboreal dung beetles are species of *Onthophagus*, a large genus with many hundreds of species. In Borneo there is one group of species that have curved and elongated hind metatarsi and a second group of species that have a tibial spur on their legs. The former group is more abundant than the latter and lives higher in the canopy (figure 21.20). The abundance between 5 and 20 metres is due to the concentration of dung and of the monkey troops. Arboreal dung beetles are diurnal, with maximum activity around 14.00 h. Some species of arboreal dung beetle belong to genera other than *Onthophagus*. One of these genera is *Canthon* in south America. *Canthon angustatus* rolls balls made of howler monkey dung and then falls to the ground holding the ball between its hind legs. Other species belonging to the genera *Trichillum* and *Uroxys* live in the fur of sloths (Halffter & Matthews, 1966). In Gabon there are four species of *Onthophagus* and one of the genus *Sisyphus*. In Madagascar one ball-rolling species of the genus *Arachnodes* lives 50 cm above the ground on small trees.

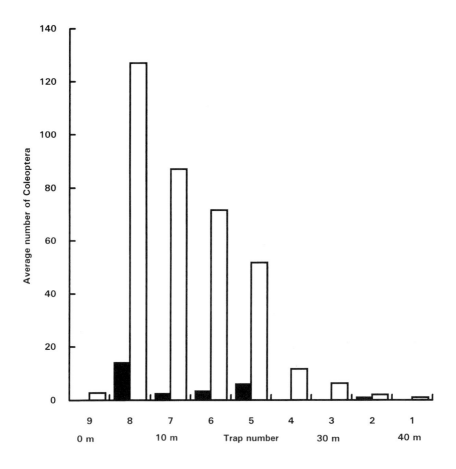

Figure 21.20. *Mean number of dung beetles per pitfall trap at heights between 0 and 40 metres above the ground (cf. text). The former group of species is in white and the latter group in black (Davis et al., 1997).*

4.4. Collembola

Collembola constitute an important group in edaphic environments because of their abundance and because of their role in nutrient cycling in the soil. They can also be abundant in environments such as soil and litter accumulated in epiphytic plants, and in "hanging soils" ("sols suspendus") (Delamare Deboutteville, 1948, 1951; Palacios-Vargas, 1981). Collembola are often abundant and form a large percentage of the rain forest arthropods. In one hectare of forest in Seram, Collembola constitute almost half the 12.3 million arthropods (Stork, 1988). The role of Collembola can be important, especially where there is an accumulation of organic matter in the canopy. Palacios-Vargas *et al.* (1998) studied the canopy of a tropical forest in Mexico by fogging (table 21.13). Collembola are most numerous in the canopy. In the shrub layer arthropods were collected with a Malaise trap.

This trap is not suitable for sampling Collembola but nevertheless the results show that Collembola in this stratum rank fourth in relative abundance.

Taxa	Canopy fogging			Malaise trap	
	Abundance	Percentage	Specimens/m^2	Abundance	Percentage
Collembola	1,044,032	95.06	15,196.97	2,633	6.35
Acari	13,691	1.25	199.29	3,411	8.23
Hymenoptera	9,850	0.90	143.38	5,401	13.03
Araneae	7,333	0.67	106.74	1,162	2.80
Diptera	4,920	0.45	71.62	14,600	35.23
Coleoptera	3,872	0.35	56.36	5,566	13.43
Homoptera	3,598	0.33	52.37	2,144	5.17
Psocoptera	2,314	0.21	33.68	795	1.92
Thysanoptera	2,310	0.21	33.62	139	0.34
Larvae	2,114	0.19	30.77	393	0.95
Orthoptera	919	0.08	13.38	518	1.25
Heteroptera	838	0.08	12.20	301	0.73
Isopoda	688	0.06	10.01	4	0.01
Dictyoptera	477	0.04	6.94	85	0.21
Lepidoptera	408	0.04	5.94	3,863	9.32
Pseudoscorpionidae	359	0.03	5.23	2	<0.01
Isoptera	355	0.03	5.17	129	0.31
Chilopoda	69	0.01	1.00	10	0.02
Thysanura	35	<0.01	0.51	33	0.08
Neuroptera	29	<0.01	0.42	92	0.22
Embioptera	26	<0.01	0.38	20	0.05
Scorpionida	8	<0.01	0.12	2	<0.01
Mecoptera	1	<0.01	0.01	3	0.01
Solifuga	1	<0.01	0.01	5	0.01
Odonata	1	<0.01	0.01	-	-
Total	1,098,248	100	15,986.14	41,443	100

Table 21.13. *Density, abundance and percentage of arthropods from canopy collected by fogging, and from shrub layer collected by a Malaise trap in a Mexican tropical deciduous forest (Palacios-Vargas et al., 1998).*

4.5. Lepidoptera

The species diversity of moths is high but our knowledge of these insects in tropical forests is superficial. The main families are the same as in temperate forests. The biology of the caterpillars is varied. Many are defoliators and in some cases can completely defoliate certain trees; others are leaf-miners (Nepticulidae), stem-borers (some Noctuidae and Pyralidae), flower-feeders (Noctuidae Nolinae and Geometridae Larentinae), fruit- and seed-predators (such as Tortricidae), or wood-borers (Cossidae, Hepialidae). Some pyralids live in the fur of sloths, and noctuids of the genus *Eublemma* are predators of Heteroptera. Moth larvae have a considerable impact on plants in the rain forest. Janzen showed that 80% of defoliation in the rain forest of Costa Rica is caused by caterpillars. One of the best known families is the Sphingidae, or hawk moths, which are pollinators adapted to many plants, especially among the Mimosaceae, Rubiaceae, Cactaceae and Orchidaceae. Janzen has recorded at least 121 species of Sphingidae in Costa Rica (the whole of the United States and Canada have 115 species and the world fauna about one thousand).

Butterflies are mostly diurnal. They visit flowers for food and are frequently major pollinators. Some species, such as nymphalids of the sub-family Charaxinae, are almost never seen at flowers but feed on rotting fruits. The main families are the Papilionidae, Pieridae, Ithomiidae, Danaidae, Satyridae, Nymphalidae and Lycaenidae. Most satyrid species of the subfamily Brassolinae have ocelli on the underside of their hind wings. These ocelli seem to be vertebrate eye mimics. *Caligo memmon* is a spectacular Brassolinae known as the "owl butterfly". Its larva lives on Musaceae and *Heliconia* spp. This moth has a wingspan of about 13 cm and the underside of its hind wings displays two large round spots which look like the eyes of an owl. Species of subfamily Morphinae (including the genus *Morpho*) are large and fast-flying blue butterflies. The richness of the butterfly fauna is attested by the provisional check-list made by Janzen which lists more than 1,000 species from Costa Rica. Tropical cyclones can have important effects on insect abundance. Fifteen species of Lepidoptera occurred in large numbers after the passage of a cyclone over Puerto Rico. This is due to the new foliage that appears on the trees following the cyclone (Torres, 1992). In Ghana more than 60 species of Lepidoptera are major defoliators of trees and can occur in outbreaks. This is the case of the noto-dontids *Epicerura pulverulenta*, which lives on *Terminalia ivorensis*, and *Anaphe venata*, which lives on *Triplochiton scleroxylon*.

Mimicry is common among tropical butterflies and provided Bates with what Darwin considered to be some of the most convincing evidence for his theory of evolution by natural selection. There are two forms of mimicry. Distasteful butterflies frequently advertise their unpalatability by an aposematic coloration. In Batesian mimicry a palatable species imitates an unpalatable one. In Muellerian mimicry several unpalatable species converge in appearance, each species gaining protection from its similarity to the others. A good example of Muellerian mimicry is that of the two species *Heliconius melpomene* and *Heliconius erato*. Each species has more than twenty easily recognisable geographical races. In each area resemblance between the races of the two species is close enough to confuse an entomologist. These butterflies are protected against predators by the cyanoglucosides they synthesise. In *Heliconius melpomene* and *H. erato* the colours of the warning pattern are predominantly black, red and yellow. In a second group of species the pattern is black-blue and yellow; in a third group it is striped; in the fourth group (species that fly over the canopy rather than in the shade) it is orange. This variety of patterns is a widespread phenomenon. Differentiation of the races of *Heliconius melpomene* and of *H. erato* arose in the Pleistocene refugia about 18,000 years ago when the neotropical forest was restricted to a few separate areas (Turner, 1984). An example of Batesian mimicry is that of the African butterfly *Papilio dardanus* which has many geographical races. These races mimic seven unpalatable species belonging to two different families and to three different genera (*Danaus, Amauris, Bematistes*). Each mimetic morph can only be found where its model lives.

New Guinea is covered with a tropical rain forest in which some extraordinary insects are found. They include the world's two largest butterflies, *Ornithoptera alexandrae* and *Ornithoptera goliath*, the largest of all stick insects, the largest cricket, belonging to the family Tettigoniidae, the largest dipterous fly, belonging to the family Diopsidae, and a large weevil which is remarkable because of the epizoic symbiosis phenomenon mentioned above. The *Ornithoptera* are much sought after by collectors and laws have been passed for their protection. New Guinea is the only country whose constitution includes insects as a natural renewable resource. Only the *Insect Farming and Trading Agency* is allowed to sell insects. These are provided by country people who collect only a few specimens in order not to endanger the populations. Other commercial *Ornithoptera* must be raised. The profit made in this way amounts to 400,000 dollars a year. "Butterfly ranching" provides country people engaged in it with several hundred dollars a year. It also helps the inhabitants to understand that the forest is the source of their livelihood, and encourages them to respect it.

4.6. Orthoptera

Among Orthoptera, the Acridoidea or grasshoppers are an important group of phytophagous insects. They are abundant in the canopy of the forests in the neotropical region and are represented by numerous species. They are also known in all continents, particularly Madagascar and South East Asia. The arboreal species constitute about 40% of all the Acridoidea species in the neotropical region. All these species have characteristics that are considered as adaptations to their way of life: elongation of the metatarsi, concavity of the sternum, modifications of tibiae and femorae, and egg pods laid either on or in the leaves. In the neotropical region these Acridoidea belong mainly to the following families: Romaleidae, Acrididae, and Eumastacidae. Among the Romaleidae there is a particular feeding behaviour: some species exploit the trunk and the branches and feed on bark or on small epiphytes (Amedegnato, 1997).

5. INSECT / PLANT RELATIONSHIPS IN TROPICAL FOREST

5.1. Insects as pollinators

Pollination by wind is rare in tropical forests where insects are the main pollinators (figure 21.21). Pollination has been studied mainly in the

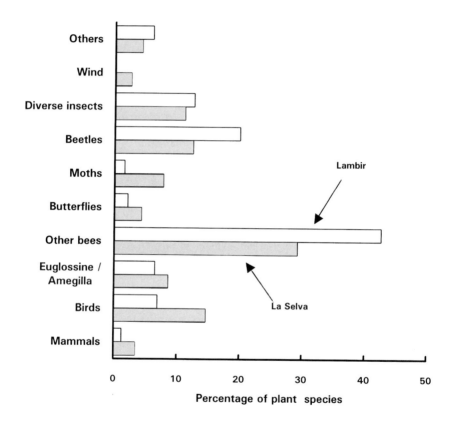

Figure 21.21. *Percentage of plant species pollinated by wind and by animals in a locality in Costa Rica (La Selva) and a locality in Sarawak (Lambir). Pollination by wind is rare in both localities (Momose* et al., *1998).*

neotropical region but also in South East Asia (Momose *et al.*, 1998). The characters that adapt flowers to insect pollinators are comparable to those of the temperate regions, but there are also some peculiarities. More than 300 species of insect visit *Cordia inermis* (Boraginaceae), a plant with very small flowers (Opler *et al.*, 1976). Large solitary bees such as *Xylocopa* are important pollinators. Female euglossine bees visit and pollinate orchids. Fig wasps (family Agaonidae), which are specialised pollinators of the many species of *Ficus*, are an example of extreme mutualism between a plant and an insect. Among the Lepidoptera, Sphingidae are frequent pollinators. Tropical flowers pollinated by hawk moths have long corolla tubes or spurs. The record length for an insect proboscis is held by *Xanthopan morgani praedicta*. In the forests of Madagascar, this sphingid pollinates the orchid *Angraecum sesquipedale* which has a 30 cm long spur. This relationship between the sphingid and the orchid is usually considered to be an example of coevolution, but recently Wasserthal (1997) explained that this elongation of the proboscis could be a device for avoiding predators that lie in ambush on flowers. Sphingids and euglossine bees are long-distance flyers which can transport pollen 10 to 15 km away from the plant they have visited, and they have a longevity exceeding two months. These insects are efficient pollinators of plants that are widely scattered in the tropical forest. Female *Heliconius* butterflies suck nectar from flowers and then collect a ball of pollen from the flowers of *Anguria* or *Gurania* (Cucurbitaceae). They then regurgitate the nectar on the pollen so that amino-acids diffuse out of the pollen into the nectar which they absorb (Gilbert, 1975).

Coleoptera are more important as pollinators in tropical forests than in temperate forests. For example some plants of the family Araceae attract beetles by their odours. In Ghana *Amorphophallus johnsonii* is pollinated by the carrion beetle *Phaechrous amplus* (and by the occasional dung flies *Hemigymnochaeta unicolor* and *Paryphodes tigrinus*). The beetles are trapped in the spathe overnight and escape the following evening (Beath, 1996). *Erioscelis emarginata* (Dynastidae) is the sole pollinator of *Philodendron selloum* in Brazil. This beetle is attracted by the volatilisation of odoriferous substances from the spadix which can maintain a temperature reaching 46 °C for 20 to 40 minutes (Gottsberger & Silberbauer-Gottsberger, 1991). Palm trees of the genus *Bactris* are pollinated by beetles of the genera *Phyllotrox* (Curculionidae) and *Epuraea* (Nitidulidae) (Listabarth, 1996). In Colombia, the palm *Phytelephas seemannii* is pollinated principally by pollen-eating staphylinid beetles of the genus *Amazoncharis* and by their predators, other staphylinids of the genus *Xanthopygus* (Bernal & Ervik, 1996). Even Diptera may be pollinators in tropical forests, particularly for primitive plants (Thien, 1980). Cycads, a group of primitive plants, were long thought to be pollinated by wind, but recently pollination by insects has been established. A weevil, *Rhopalotria mollis*, completes its life cycle upon and within the male cones of the cycad *Zamia furfuracea*. Its larvae feed on the tissues of the cones and when adults emerge they are coated with pollen. They visit female cones and are capable of effecting the pollination (Norstog *et al.*, 1986). There are other examples of pollination of cycads by insects, mainly Coleoptera.

One of the consequences of coevolution between plants and pollinating insects is that flowering phenologies are different for trees that have the same pollinators. Flowering at different times avoids competition between plants for pollinators and provides insects with food for a longer period of time. In Costa Rica there is a vertical stratification among pollinating insects. The greatest number of bees, mostly Anthophoridae and Xylocopidae, is found at the greatest height; Megachilidae and Halictidae predominate at a lower height (Frankie, 1975).

5.2. Two examples of mutualism and of coevolution

5.2.1. Heliconius *and* Passiflora

The relationships between *Heliconius* butterflies, their larval food plants *Passiflora* and their adult food plant *Anguria* which provides pollen and nectar are an example of coevolution between plants and insects (Gilbert, 1975). The *Passiflora* plants are widely scattered within the forest. To find the young plants that are the necessary food for the larvae, females must fly long distances in the forest and have a long life span (several months, which is a long time for a butterfly). They have keen eyesight which enables them to detect the right plants. *Heliconius* larvae are cannibalistic and females seek host plants carefully before ovipositing and avoid plants on which there is already an egg. A remarkable adaptation is the development on the stipules of structures that mimic *Heliconius* eggs and give the impression that the host plant is occupied. These structures have been described as "most striking coevolved adaptations".

5.2.2. Ants and plants

Many plants have specialised structures called domatia (or myrmecodomatia) for the housing of ants. Domatia are found in numerous plant families such as the Leguminosae (genus *Acacia*), Moraceae (genus *Cecropia*), Piperaceae (genus *Piper*), Rubiaceae (genera *Hydnophytum* and *Myrmecodia*) etc. Some domatia (i.e. in the genus *Myrmecodia*) have a complex structure. The detritus accumulated by the ants may be a source of nutrients for the plant. A striking example of this mutualism is that of some species of *Acacia* with ants of the genus *Pseudomyrmex* (Janzen, 1966, 1967). More than 90% of the American species of *Acacia* are protected against herbivores by cyanogenetic compounds in the leaves but some species lack this chemical defence. *Acacia cornigera* has swollen and hollow stipular thorns. The ant *Pseudomyrmex ferruginea* lives in these thorns, which provide it with a measure of protection. The *Acacia* also supplies the ants with food: sugar which is secreted by nectaries located on the petioles of the leaves, and proteins secreted by Beltian bodies which are at the tips of the leaves. The aggressive ants patrol the plant and drive the herbivores away.

An ant garden is an aggregate of epiphytic plants assembled by ants. Ant gardens are round and may measure more than 60 cm in diameter. The ants build their nest in the tangled roots by collecting humus, dead leaves and other debris. Thus the roots of the epiphyte are provided with more nutrients than when ants are not present. Sometimes the ants also bring seeds that germinate. They protect the ant gardens and spread small seeds.

5.3. Plant defences and phytophagous insects

Defensive chemical substances play an important part in the relationships between phytophagous insects and plants. Are there differences in anti-herbivore defences between temperate and tropical forest plants? The annual rate of herbivory (i.e. percentage of leaf area damaged per year) is 10.9% in tropical forests and only 7.5% in temperate forests. This may be due either to higher pressure by phytophagous insects, or to a lower level of defensive substances. In the tropics the frequency of plants with alkaloids is higher (35%) than in temperate regions (16%) (Levin, 1976; Levin & York, 1978). Tannins (or polyphenols) act as defences against herbivores (cf. chapter 3.1). The leaves of temperate forest plants have a lower concentration of tannins (1.9% dry weight) than the leaves of tropical plants (6.0% dry weight). Among the non-chemical defences many tropical plants possess extra-floral nectaries which attract ants. In return the ants provide the plant with partial protection against herbivores (Bentley, 1977). The families of plants with extra-floral nectaries are mainly found in the tropics and are few in temperate regions. An adaptation against phytophagous species is either the synchronous production of leaves that satiates herbivores, or the production of leaves in a season when herbivores are rare. In Barro Colorado Island 44% of the production of leaves occurs in the dry season when herbivores are rare.

To conclude, the comparison between tropical and temperate forest plants allows us to think that tropical forest plants have been exposed to high herbivore pressure which triggered the production of a greater number of defensive reactions in them than are found in the plants of temperate regions.

6. SOME EXAMPLES OF TROPICAL FOREST INSECT COMMUNITIES

Knowledge of the structure of arboreal arthropod communities in tropical forests is still fragmentary because studies generally describe only species composition and guild structure. Few data are available on the vertical and spatial distribution of arthropods within rain forest trees. This is due to the difficulties entomologists have in reaching the tree-crowns and in the technique of sampling, since the fogging method does not allow them to distinguish which insects come from each of the different strata of vegetation.

6.1. An arboreal community of beetles in Brunei

Morse *et al.* (1988) described the relationships between the number of species, their abundance and their body size for a community of 859 species and of 3,919 individuals of Coleoptera. These insects were collected by fogging the canopy of ten trees in Brunei (Borneo). The 859 species belong to 4 guilds: herbivores (384 species), predators (200 species), scavengers (120 species) and fungivores (141 species).

The study of the distribution of species abundance reveals that 58% of the species are represented by only one individual, and that few species are very abundant. The rank / abundance relationship of the species can be fitted to the equation

$$\text{abundance} = 470 \times \text{rank}^{-0.96} \ (P < 0.001)$$

The distribution of species body length shows that the maximum number of species occurs in a size class of about 2 mm. This proves that the beetles of the canopy are small in size, a characteristic already mentioned by Erwin (cf. figure 21.4). The same histogram for the different guilds shows that the size-classes that are the richest in species are different. This size-class is 5 for herbivores, 0 for fungivores and 2 for predators and scavengers. The distribution of the total number of individuals in the different body length classes is characterised by a modal length classes 6 (that is about 3.5-4 mm).

This community of arboreal beetles is characterised by large number of species. Most of these species are not very abundant and are of small size (cf. chapter 4.3). A comparison with temperate forests shows that a higher percentage of larger species can be found in the latter (Lawton, 1991).

These conclusions may not be true in all cases since the size structure of arboreal insect communities in tropical and temperate areas is influenced by factors such as season and moisture. For example the size structure of samples of beetles in Panama varies with the seasons, the beetles of the dry season being somewhat smaller than those of the wet season (cf. 2.1, above). The modal size of Homoptera and Heteroptera in a rain forest in Sulawesi increases with rainfall. During the dry period more than 60% of the Homoptera are less than 3.6 mm in body length, but in the wet season more than 50% are over 6.4 mm in body length (Rees, 1983).

The small size of the insects is confirmed by Basset (1997) who, among all the chewing insects collected on New Guinea trees, found a modal size class of 3.4 to 4.1 mm, the largest species being > 54.6 mm. The abundance of rare species is also confirmed: among 994 species of Coleoptera, 489 species are rare.

The distribution of beetles among feeding groups varies in different regions. The following figures allow a comparison between three tropical forests and a temperate forest:

Countries	Percentages					Number of species
	Herbivores	Xylophages	Fungivores	Saprophages	Predators	
Sulawesi	25.1	16.1	27.5	23.8	17.4	1,355
Brunei	34.6	7.7	18.5	15.5	23.7	875
Australia	19.8	21.1	23.1	8.4	27.3	454
UK	23.5	10.0	26.5	11.0	29.0	200

The taxonomic distribution also varies. In Sulawesi (Hammond *et al.*, 1997) fogging samples contain species belonging to 17 superfamilies, the best represented being the Curculionoidea, followed by Cucujoidea, Tenebrionoidea, Staphylinoidea and Chrysomeloidea. There are few species of Caraboidea, Hydrophiloidea and Scarabaeoidea. The families Eucnemidae, Mordellidae, Aderidae and Anthribidae are particularly well represented compared with other tropical regions. The families Elateridae, Tenebrionidae, Chrysomelidae, Pselaphidae and Dryopidae are poorly represented.

6.2. Communities of Coleoptera in a lowland forest in Borneo

A recent study by Floren & Linsenmair (1998) provides interesting data on the structure of a community of arboreal Coleoptera in a rain forest in Borneo. The percentage of Coleoptera is always around 5% of all the insects collected. The estimated number of species of Coleoptera from only 19 trees is about 2,000 (that is equivalent to a quarter of all species found in central Europe) belonging to 77 families. Chrysomelids are the largest group represented (485 species and 23.9% of the specimens) followed by curculionids (257 species and 11.1% of the specimens) and staphylinids (species number not determined and 10.9% of the specimens). Most of the species collected are highly mobile; less mobile species and larvae were almost totally absent. This is the consequence of the high predatory pressure by ants which dominate in all trees. These ants prevent the herbivorous species from laying their eggs in the crowns of the trees. This is in contrast with temperate forests in which ants are rare and the larvae of many phytophagous insects are abundant in the canopy. The number of rare species is high: 96% of the species are represented by fewer than 10 individuals and 59% by only one specimen (cf. table 21.14). This contrasts with temperate forests where there are few rare species. In Central Europe a third of the most common species usually accounts for 85% of individuals. Moreover, no tree-specific beetles were found, contrary to what is observed in temperate regions. The abundance of rare species and the rarity of tree-associated species allow the authors to conclude that the arboreal Coleoptera of lowland rain forests resemble a non-interactive system (i.e. having no measurable impact on their food plants) and that they constitute a non-equilibrium community.

Regions	Main families	Number of species	% species < 10 individuals	% species with one individual
Panama	C, S, P, M	954	at least 60	
Peru	C, Cu, S		+ 500 species with fewer than 5 individuals	
Brunei, Borneo	C, S, A, Cu	859	at least 92	58
Sabah, Borneo	C, S, Cu	1,183	96	59
Sulawesi		1,355		46
New Guinea	S, Cu, C, A	633	at least 90	50.7
New Guinea		418	85.2	47.6
Rwanda	A, An	84		67.5
Rwanda	C, Ap, Cu	230		56.2
Congo	C, S, Cu	137		86

Table 21.14. *Characteristics of some arboreal communities of Coleoptera in tropical lowland rain forests. Note the high percentage of rare species represented by only one specimen (Erwin & Scott, 1980; Farrell & Erwin, 1988; Morse et al., 1988; Floren & Linsenmair, 1998; A: Attelabidae; An: Anthicidae; Ap: Apionidae; C: Chrysomelidae; Cu: Curculionidae; M: Mordellidae; P: Ptilodactylidae; S: Staphylinidae.*

6.3. Beetle fauna of a rain forest in Sulawesi

Perhaps the most extensive study of the taxonomy, abundance, and diversity of insects is that of Hammond (1990) in the lowland rain forest in the Toraut region of Sulawesi. The material collected at various elevations amounts to several million insects, including about 1,172,000 Coleoptera. Diptera are unusually abundant (about 800 specimens per trap per week in ground-level Malaise traps); the mean number of all other arthropods (except Diptera and Collembola) is 325 specimens per trap per week. Termites are poorly represented and ants less abundant than in other comparable forests. A study of the Coleoptera reveals the presence of 5,649 species in the 500 hectares sampled, but the real number may be as high as 6,200.

The insects were collected with light traps, flight-interception traps, pitfall traps, Malaise traps and by canopy fogging. For all the samples taken together, the superfamily with the greatest number of species is the Staphylinoidea, followed by Curculionoidea and Tenebrionoidea (table 21.15). The best represented trophic guild is that of predators, followed by fungivores. The fungivores include 2.1% of all species that feed on 'ambrosia' fungi carried by ambrosia beetles (Scolytidae and Platypodidae) and 1.1% that feed on slime moulds. Among habitat-groups the richest in species is wood; followed by litter and non-woody (herbaceous) plants.

Categories	Percentage of all species
I. Superfamilies	
Staphylinoidea	26.6
Curculionoidea	18.7
Tenebrionoidea	12.9
Cucujoidea	10.9
Chrysomeloidea	7.8
All other superfamilies	23.1
II. Trophic guilds	
Predators	29.0
Fungivores	23.4
Herbivores	17.4
Xylophages	16.0
Saprophages	12.7
Parasitoids	0.9
III. Habitat groups	
Wood	32.7
Litter	24.8
Non-woody plants	23.3

Table 21.15. *Structure of the community of Coleoptera in a rain forest in Sulawesi: superfamilies, trophic guilds and habitat groups (Hammond, 1990).*

The proportions of herbivorous species are low. Even if the Coleoptera are not entirely representative of all the insects, we may conclude with Hammond that "assumptions that the majority of animal species are more or less directly associated with living plants may prove unjustified when tropical insect faunas have been more fully investigated".

Catches of Coleoptera with Malaise traps in the canopy amounted to 322 species. The most important groups in term of species and individuals were Anthribidae, Cerambycidae of the subfamily Lamiinae, Cleridae, Chrysomelidae of the subfamily Eumolpinae, and Curculionidae. Xylophagous species are better represented at canopy level and saprophagous species less well represented. The litter samples contained 181 species belonging to 32 families. The best represented, belonging to the superfamily Staphylinoidea, are Pselaphidae (38 species), Staphylinidae of the subfamilies Aleocharinae (18 species) and Paederinae (17 species), and Ptiliidae (16 species). There is no very common species, the most abundant, *Oxytelopsis* sp. (Staphylinidae) accounting for only 8.6% of the individuals. Flight-interception trap catches are rich in Coleoptera and Diptera. The Coleoptera are mainly Staphylinidae (Aleocharinae, Oxytelinae and Staphylininae) and Ptiliidae.

6.4. Arthropods in a lowland forest in Cameroon

A study in different strata (canopy and shrub layer) with the aid of the canopy raft has shown that most arthropods of the canopy are ants (44%) and herbivorous species (31%). Scavengers amount only to 11%, spiders to 8%, parasitoids to 1% and others to 5%. A total of 2,271 arthropods distributed in 88 families were collected in samples corresponding to about 117 m^2 of foliage (Basset *et al.*, 1992). The mean density of the fauna is three times higher in the canopy than in the shrub layer, and species diversity is also higher in the canopy (table 21.16). However, the grazing pressure of herbivores in the shrub layer, is twice as high. This may be due to a lower

Variables	Shrub layer	Canopy
All arthropods *	7.00 ± 0.91	24.13 ± 4.91
Ants *	1.31 ± 0.35	12.12 ± 2.84
Spiders	1.75 ± 0.57	0.88 ± 0.11
Parasitoids	0.18 ± 0.07	0.17 ± 0.05
Herbivores *	1.48 ± 0.38	7.97 ± 2.30
Scavengers *	1.49 ± 0.17	2.22 ± 0.23
Others	0.77 ± 0.17	0.75 ± 0.12
OTU *	4.80 ± 0.28	6.68 ± 0.33

Table 21.16. *Mean number of individuals from samples of the shrub layer and from the canopy in a lowland forest of Cameroon (Basset* et al.*, 1992). The asterisk indicates significant differences between the shrub layer and the canopy. OTU : number of Operational Taxonomic Unit or "morphospecies".*

amount of defensive substances in the shade-leaves of the shrub layer than in the sun-leaves of the canopy (Lowman, 1985; Coley, 1988). The greater life-span of shade-leaves may also increase the estimate of damage (Greenwood, 1990). The high number of predatory ants in the canopy may depress population levels of herbivorous insects such as caterpillars which are frequently responsible for high leaf damage (Basset, 1991).

How can we explain the greater number of insects in the canopy? Four hypotheses are possible: (1) primary productivity in the canopy is higher than in the shrub layer; (2) there are more young leaves which are more nutritious; (3) the high illumination and temperature enhance foraging and oviposition; (4) the smaller leaves may protect herbivores from predators.

6.5. Canopy arthropod diversity in a New Caledonian primary forest

Because of its richness in endemic species New Caledonia is one of the world's "hotspots". In the Rivière Bleue reserve the canopy of the primary rain forest towers 20-25 metres above the ground and its understorey is relatively dense. The mean annual precipitation is 3,181 mm. The flora is highly diversified with 350 species of Phanerogams within little more than 5 hectares. Fogging the canopy produced 9,608 specimens; they belong to 107 families and to 20 orders (table 21.17). This corresponds to

Taxa	%	Taxa	%
Collembola	18.6	Acarina	5.5
Diptera	18.0	Homoptera	3.5
(Nematocera)	12.8	Thysanoptera	2.5
(Brachycera)	5.2	Heteroptera	1.8
Coleoptera	13.6	Orthoptera	0.8
Hymenoptera	12.7	Lepidoptera	0.6
(Formicidae)	7.2	Amphipoda	0.4
(Other Hymenoptera)	5.5	Pseudoscorpionida	0.5
Psocoptera	11.4	Dermaptera	0.2
Araneae	6.7	Others	1.7

Table 21.17. *Relative frequencies of individuals belonging to major taxa collected by fogging at Rivière Bleue (New Caledonia).*

106.76 specimens per m^2. This is low compared with other samples: 35 to 161 specimens per m^2 in Amazonia (Adis *et al.*, 1984), 51 to 218 specimens per m^2 in Borneo (Stork, 1987). These arthropods belong to the following guilds (Guilbert, 1994; Guilbert *et al.*, 1995):

– Epiphytic grazers: these are Collembola and Psocoptera.

– Predators: these are mostly Acarina, and spiders Clubionidae and Theridiidae.

– Parasitoids: all these are Hymenoptera, particularly Scelionidae and Chalcidoidea such as Trichogrammatidae and Aphelinidae.

– Tourists: these belong to Diptera Nematocera of the families Chironomidae, Ceratopogonidae and Cecidomyiidae

– Chewers: these are all phytophagous Coleoptera of the families Chrysomelidae and Curculionidae.

– Sap-feeders: these are Thysanoptera as well as Cicadellidae and Psyllidae.

– Scavengers: these belong to Dictyoptera, to Sciaridae (Diptera) and to Corylophoridae (Diptera).

– Predators: the main predators are Staphylinidae (Coleoptera), Dolichopodidae (Diptera) and Sphecidae (Hymenoptera).

The percentages of the various taxa vary with the seasons. The dominant order is Collembola, followed by Diptera, Coleoptera and Psocoptera. Coleoptera are represented by 31 families, the three most numerous being the Curculionidae, Chrysomelidae and Corylophidae, followed by Staphylinidae and Pselaphidae. The most abundant Hymenoptera are ants, followed by Aphelinidae and Chalcidoidea. There are few spiders. We may conclude that the canopy fauna of this New Caledonian forest is characterised by the importance of Collembola, Psocoptera and Diptera. This may be related to the composition of the canopy and particularly to the presence of "hanging soils".

REFERENCES

ABE T., MATSUMOTO T., (1979). Studies on the distribution and ecological role of termites in a lowland rain forest of West Malaysia. (3) Distribution and abundance of termites in Pasoh Forest Reserve. *Jap. J. Ecol.,* **29**: 337-351.

ADIS J., (1981). Comparative ecological studies of the terrestrial arthropod faunas in Central Amazonian inundation-forests. *Amazoniana,* **7**: 87-173.

ADIS J., (1997). Terrestrial invertebrates: survival strategies, group spectrum, dominance and activity patterns. *In*: W.-J. Junk (ed.), *The central Amazon floodplain, Ecology of a pulsing system*, 287-299. Ecological Studies, 126. Springer, Berlin.

ADIS J., MESSNER B., (1997). Adaptations to life under water: tiger beetles and millipeds. *In*: W.-J. Junk (ed.), *The central Amazon floodplain, Ecology of a pulsing system*, 299-317. Ecological Studies, 126. Springer, Berlin.

AMEDEGNATO C., Diversity of an Amazonian canopy grasshopper community in relation to resource partitioning and phylogeny. *In*: N. E. Stork *et al.* (1997). Canopy Arthropods. *Chapman & Hall, London*, 281-319.

AMORIM M.-A. *et al.*, (1997). Ecology and adaptations of the tiger beetle *Pentacomia egregia* (Chaudoir) (Cicindelinae: Carabidae) to Central Amazonian floodplains. *Ecotropica*, **3**: 71-82.

ASKEW R. R., Species diversity of hymenopteran taxa in Sulawesi. *In*: W. J. Knight, J. D. Holloway (1990). *Insects and the Rain Forests of South East Asia (Wallacea)*. Royal Entomological Society, London, 255-260.

BARONE J. A., (1998). Host-specificity of folivorous insects in a moist tropical forest. *Journal of Animal Ecology*, **67**: 400-409.

BASSET Y., (1991). Influence of leaf traits on the spatial distribution of insect herbivores associated with an overstorey rainforest tree. *Oecologia*, **87**: 388-393.

BASSET Y., (1992). Influence of leaf traits on the spatial distribution of arboreal arthropods within an overstorey rainforest tree. *Ecological Entomology*, **17**: 8-16.

BASSET Y., (1992). Host-specificity of arboreal and free-living insect herbivores in rain forests. *Biological Journal of the Linnean Society*, **47**: 115-133.

BASSET Y., Species abundance and body size relationships in insect herbivores associated with New Guinea forest trees, with particular reference to insect host-specificity. *In*: N. E. Stork *et al.* (1997). *Canopy Arthropods*. Chapman & Hall, London, 237-264.

BASSET Y. *et al.*, (1992). Abundance and stratification of foliage arthropods in a lowland rain forest of Cameroon. *Ecological Entomology*, **17**: 310-318.

BATES H. W., (1863). *The Naturalist on the River Amazons*. John Murray, London.

BATES M., (1945). Observations on climate and seasonal distribution of mosquitoes in eastern Columbia. *J. Anim. Ecol.*, **14**: 17-25.

BEATH D. D. N., (1996). Pollination of *Amorphophallus johnsonii* (Araceae) by carrion beetles (*Phaeochrous amplus*) in a Ghanaian rain forest. *Journal of Tropical Ecology*, **12**: 409-418.

BEAVER R. A., (1979). Host-specificity of temperate and tropical animals. *Nature*, **281**: 139-141.

BECCALONI G. W., (1997). Ecology, natural history and behaviour of Ithomiinae butterflies and their mimics in Ecuador (Lepidoptera: Nymphalidae: Ithomiinae). *Tropical Lepidoptera*, **8**: 103-124.

BEEBE W., (1917). *Tropical wildlife in British Guiana*. New York Zoological Society.

BENTLEY B. L., (1977). Extrafloral nectaries and protection by pugnacious bodyguards. *Ann. Rev. Ecol. Syst.,* **8**: 407-427.

BERNAL R., ERVIK F., (1996). Floral biology and pollination of the dioecious palm *Phytelephas seemannii* in Columbia: an adaptation to Staphylinid beetles. *Biotropica,* **28**: 682-696.

BLANC P., Bioclimatologie comparée de la canopée et du sous-bois. *In:* F. Hallé & P. Blanc (1990). *Biologie d'une canopée de forêt équatoriale. Rapport de Mission Radeau des Cimes Octobre-Novembre 1989, Petit Saut - Guyane Française,* 42-43. Montpellier/Paris.

BOYER P., Les différents aspects de l'action des termites sur les sols tropicaux. *In:* P. Pesson (1971). *La vie dans les sols,* 281-334. Gauthier-Villars, Paris.

BRIAN M. V., (1978). *Production ecology of ants and termites.* Cambridge University Press.

BROADHEAD E., The assessment of faunal diversity and guild size in tropical forests with particular reference to the Psocoptera. *In:* S. L. Sutton *et al.* (1983) *Tropical Forest: Ecology and Management.* Blackwell, Oxford, 107-119.

BROWN K. S., Biogeography and evolution of neotropical butterflies. *In:* T. C. Whitmore., G. T. Prance (1987). *Biogeography and Quaternary History in Tropical America.* Oxford Science Publications, 66-104.

BROWN K. S., HUTCHINGS R. W., Disturbance, fragmentation, and the dynamics of diversity in Amazonian forest butterflies. *In:* W. F. Laurence & R. O. Bierregaard (1997). *Tropical Forest Remnant. Ecology, Management, and Conservation of Fragmented Communities.* Chicago University Press, 91-1110.

BROWNE F. G., (1962). *Sosylus spectabilis* Grouvelle (Coleoptera, Colydiidae) a predator and parasite of African ambrosia beetles. *Fifth WATBRU Report:* 91-96.

BRUHL C. A. *et al.*, (1998). Stratification of ants (Hymenoptera, Formicidae) in a primary rain forest in Sabah, Borneo. *Journal of Tropical Ecology,* **14**: 285-297.

BURD M., (1994). Butterfly wing colour patterns and flying heights in the seasonally wet forest of Barro Colorado Island, Panama. *Journal of Tropical Ecology,* **10**: 601-610.

BUSH M. B., (1994). Amazonian speciation: a necessarily complex model. *Journal of Biogeography,* **21**: 5-17.

CACHAN P., Importance écologique des variations microclimatiques du sol à la canopée dans la forêt tropicale humide. *In:* P. Pesson (1974). *Écologie Forestière.* Gauthier-Villars, Paris, 21-42.

CASTILLO M. L., MORON M. A., (1992). Observaciones sobre la degradacion de madera por algunas especies de pasalidos (Coleoptera, Lamellicornia). *Folia Entomologia Mexicana,* **84**: 35-44.

CHERRETT J. M., (1968). The foraging behaviour of *Atta cephalotes* (L.) (Hymenoptera, Formicidae). I. Foraging pattern and plant species attacked in tropical rain forest. *Journal of Animal Ecology,* **37**: 387-403.

CHERRETT J. M., (1972). Some factors involved in the selection of vegetable substrate by *Atta cephalotes* (L.) (Hymenoptera, Formicidae) in tropical rain forest. *Journal of Animal Ecology,* **41**: 647-660.

CHERRETT J. M., Chemical aspects of plant attack by leaf-cutting ants. *In:* J. B. Harborne (1972). *Phytochemical ecology.* Academic Press, London, 13-24.

CHEY V. K. *et al.,* (1997). Diversity of moths in forest plantations and natural forests in Sabah. *Bulletin of Entomological Research,* **87**: 371-385.

COLEY P. D., (1988). Effects of plant growth rate and leaf lifetime on the amount and type of anti-herbivore defence. *Oecologia,* **74**: 531-536.

COLLINS N. M., (1980). The effect of logging on termite diversity and decomposition processes in lowland dipterocarp forests. *Tropical Ecology and Development,* **198**: 113-121.

COLLINS N. M., (1981). The role of termites in the decomposition of wood and leaf litter in the southern Guinea savanna of Nigeria. *Oecologia* **51**: 389-399.

COLLINS N. M., (1983). Termites. Ecosystems of the World, 14 B, Tropical rain forest Ecosystems, 455-471. *Elsevier, Amsterdam.*

CURRIE D. J., (1991). Energy and large scale patterns of animal and plant species richness. *Amer. Nat.,* **137**: 27-49.

DARLINGTON P. J., (1971). The Carabid beetles of New Guinea. Part IV. General considerations; analysis and history of fauna; taxonomic supplement. *Bull. Mus. Comparative Zoology,* **142**: 129-337.

DAVIS A. J. *et al.,* The ecology and behaviour of arboreal dung beetles in Borneo. *In:* N. E. Stork *et al.* (1997). *Canopy arthropods.* Chapman & Hall, London, 417-432.

DEJEAN A. *et al.,* (1986). Les termites et les fourmis, animaux dominants de la faune du sol de plusieurs formations forestières et herbeuses du Zaïre. *Actes du Colloque Insectes Sociaux,* **3**: 273-283.

DELAMARE DEBOUTTEVILLE C., (1948). Étude quantitative du peuplement animal des sols suspendus et des épiphytes en forêt tropicale. *C. R. Ac. Sc.,* **226**: 1544-1546.

DELAMARE DEBOUTTEVILLE C., (1951). *Microfaune du sol des pays tempérés et tropicaux.* Hermann, Paris.

DIDHAM R., The influence of edge effects and forest fragmentation on leaf litter invertebrates in central Amazonia. *In:* W. F. Laurance & R. O. Bierregaard (1997). *Tropical Forest Remnants: Ecology, Management, and Conservation of Fragmented Communities.* University of Chicago Press, 55-70.

DIRZO R., MIRANDA A., Altered patterns of herbivory and diversity in the forest understorey: a case study of the possible consequences of contemporary defaunation. *In*: P. W. Price *et al.*, (1991) *Plant-Animal Interactions*. Wiley, New York, 273-287.

DIXON A. F. G. *et al.*, (1987). Why are there so few species of aphids, especially in the tropics? *Amer. Nat.*, **129**: 580-592.

DOBZHANSKY T., PAVAN C., (1950). Local and seasonal variation in relative frequency of species of *Drosophila* in Brazil. *J. Anim. Ecol.*, **19**: 1-14.

EASTOP V. F., (1978). Diversity of the Sternorrhyncha within major climatic zones. *Symposium of the Royal Entomological Society of London*, **9**: 71-88.

EGGLETON P. *et al.*, (1994). Explaining global termite diversity: productivity or history? *Biodiversity and Conservation*, **3**: 318-330.

EGGLETON P. *et al.*, (1995). The species richness of termites (Isoptera) under differing levels of forest disturbance in the Mbalmayo Forest Reserve, southern Cameroon. *Journal of Tropical Ecology*, **11**: 85-98.

ELTON C. S., (1973). The structure of invertebrate populations inside Neotropical rain forest. *J. Anim. Ecol.*, **42**: 55-104.

EMMEL T. C., LARSEN T. B., (1997). Butterfly diversity in Ghana, West Africa. *Tropical butterfly*, **8**: 1-13.

ERWIN T. L., Thoughts on the evolutionary history of ground beetles: hypotheses generated from comparative faunal analyses of lowland forest sites in temperate and tropical regions. *In:* T. L. Erwin *et al.*, (1975). *Carabid beetles*. W. Junk, The Hague, 539-592.

ERWIN T. L., (1983). Tropical forest canopies: the last biotic frontier. *Bull. Entom. Soc. Amer.*, **29**: 14-19.

ERWIN T. L., Beetles and other insects of tropical forest canopies at Manaus, Brazil, sampled by insecticidal fogging. *In:* S. L. Sutton *et al.*, (1983). *Tropical Rain Forest: Ecology and Management*. Blackwell, Oxford, 59-75.

ERWIN T. L., (1991). How many species are there - revisited. *Conservation Biology*, **5**: 330-335.

ERWIN T. L., SCOTT J. C., (1980). Seasonal and size patterns, trophic structure and richness of Coleoptera in the tropical arboreal ecosystem: the fauna of the tree *Luehea seemannii* Triana and Planch in the Canal Zone of Panama. *Coleopterists Bulletin*, **34**: 305-322.

ESTRADA A. *et al.*, (1993). Dung beetles attracted to mammalian herbivore (*Alouatta palliata*) and omnivore (*Nasua narica*) dung in the tropical rain forest of Los Tuxtlas, Mexico. *Journal of Tropical Ecology*, **9**: 45-54.

ESTRADA A. *et al.*, (1998). Dung and carrion beetles in tropical rain forest of Los Tuxtlas, Mexico. *Journal of Tropical Ecology*, **14**: 577-593.

EWUSIE J. Y., (1980). *Elements of tropical ecology*. Heinemann, London.

FARRELL B. D., ERWIN T. L., Leaf-beetle community structure in an Amazonian rainforest canopy. *In:* P. Jolivet *et al.*, (1988). *Biology of Chrysomelidae*. Kluwer Academic Publishers, Dordrecht, 73-90.

Feer F., (1999). Effects of dung beetles (Scarabaeidae) on seeds dispersed by howler monkeys (*Alouatta seniculus*) in the French Guianan rain forest. *J. Tropical Ecol.*, **15**: 129-142.

Fernandes G. W., Price P. W., Comparison of tropical and temperate insect galling species richness: the roles of environmental harshness and plant nutrient status. *In:* P. W. Price *et al.*, (1991). *Plant-animal interactions. Evolutionary ecology in tropical and temperate regions.* Wiley & Sons, New York, 91-115.

Fittkau E. J., Klinge H., (1973). On biomass and trophic structure of the central Amazonian rain forest ecosystem. *Biotropica*, **5**: 2-14.

Floren A., Linsenmair K. E., (1998). Non-equilibrium communities of Coleoptera in trees in a lowland rain forest of Borneo. *Ecotropica* **4**: 55-67.

Fonseca de Souza O. F., Brown V. K., (1994). Effects of habitat fragmentation on Amazonian termite communities. *Journal of Tropical Ecology*, **10**: 197-206.

Frank J. H., Lounibos L. P., (1983). *Phytotelmata: terrestrial plants as hosts for aquatic insect communities.* Plexus, Medford, New Jersey.

Frankie G. W., Tropical forest phenology and pollinator plant coevolution. *In:* L. E. Gilbert & P. H. Raven (1975). *Coevolution of Animals and Plants.* Texas University Press, Austin, 192-209.

Frankie G. W. *et al.*, (1997). Diversity and abundance of bees visiting a mass flowering tree species in disturbed seasonal dry forest, Costa Rica. *J. Kansas Ent. Soc.*, **70**: 281-296.

Gagné W.-C., (1979). Canopy-associated arthropods in *Acacia koa* and *Metrosideros* tree communities along an altitudinal transect on Hawaii island. *Pacific Insects*, **21**: 56-82.

Garrettson M. *et al.*, (1998). Diversity and abundance of understorey plants on active and abandoned nests of leaf-cutting ants (*Atta cephalotes*) in a Costa Rica rain forest. *Journal of Tropical Ecology*, **14**: 17-26.

Gaston K. J., (1992). Regional number of insects and plant species. *Functional Ecology*, **6**: 243-247.

Gauld I. D. *et al.*, (1992). Plant allelochemical, tritrophic interactions and the anomalous diversity of tropical parasitoids: the 'nasty' host hypothesis. *Oikos*, **65**: 353-357.

Gilbert L. E., (1975). Ecological consequences of a coevolved mutualism between butterflies and plants. *In*: L. E. Gilbert, P. H. Raven (1975) *Coevolution of animals and plants.* University of Texas Press, Austin, 210-240.

Gottsberger G., Silberbauer-Gottsberger I., (1991). Olfactory and visual attraction of *Erioscelis emarginata* (Cyclocephalini, Dynastinae) to the inflorescences of *Philodendron selloum* (Araceae). *Biotropica*, **23**: 23-28.

Grant S., Moran V. C., (1986). The effects of foraging ants on arboreal insect herbivores in an undisturbed woodland savanna. *Ecological Entomology,* **11**: 83-93.

GRASSÉ P. P., (1986). *Termitologia. Tome III. Comportement, socialité, écologie, évolution, systématique.* Masson, Paris.

GRAY B., (1972). Economic tropical forest entomology. *Ann. Rev. Ent.,* **17**: 313-354.

GREENWOOD S. R., Patterns of herbivory on the edge of the tropical forest in North Sulawesi. *In:* W. J. Knight, J. D. Holloway (1990). *Insects of the rain forests of South East Asia (Wallacea).* The Royal Entomological Society, London, 309-312.

GRESSITT J. L., (1969). Epizoic symbiosis. *Entomological News,* **80**: 1-5.

GUILBERT E., (1994). Biodiversité des arthropodes de la canopée dans deux forêts primaires en Nouvelle Calédonie. Thesis, Paris, Muséum National d'Histoire Naturelle.

GUILBERT E. *et al.,* (1995). Canopy arthropod diversity in a New Caledonian primary forest sampled by fogging. *Pan-Pacific Entomologist,* **71**: 3-12.

HALFFTER G. *et al.,* (1992).A comparative study of the structure of the scarab guild in Mexican tropical rain forests and derived ecosystems. *Folia Entomologia Mexicana,* **84**: 131-156.

HALFFTER G., MATTHEWS E. G., (1966). The natural history of dung beetles of the subfamily Scarabaeinae (Coleoptera: Scarabaeidae). *Folia Entomologica Mexicana,* **12-14**: 1-312.

HALLÉ F., BLANC P., (1990). *Biologie d'une canopée de forêt équatoriale. Rapport de Mission Radeau des Cimes Octobre-Novembre 1989, Petit Saut-Guyane française.* Montpellier / Paris.

HAMMOND P. M., Insect abundance and diversity in the Dumoga-Bone National Park, North Sulawesi, with special reference to the beetle fauna of lowland forest in the Toraut region. *In:* W. J. Knight, J. D. Holloway (1990), *Insects and the Rain Forests of South East Asia (Wallacea).* Royal Entomological Society of London, 197-254.

HAMMOND P. M. *et al.,* Tree-crown beetles in context: a comparison of canopy and other ecotone assemblages in a lowland tropical forest in Sulawesi. *In:* N. E. Stork *et al.* (1997). *Canopy Arthropods.* Chapman & Hall, London, 184-223.

HANSKI I., (1989). Dung beetles. *Tropical Rain Forest Ecosystems,* **14**: 5-21.

HANSKI I., CAMBEFORT Y., (1991). *Dung Beetle Ecology.* Princeton University Press.

HANSKI I., NIEMELA J., (1990). Elevational distribution of dung and carrion beetles in northern Sulawesi. *In:* W. J. Knight, J. D. Holloway (1990). *Insects and the Rain Forests of South East Asia (Wallacea).* Royal Entomological Society of London, 145-152.

HAWKINS B. A., DEVRIES P. J., (1996). Altitudinal gradients in the body size of Costa Rica butterflies. *Acta Oecologia,* **17**: 185-194.

HILL J. K. *et al.,* (1995). Effects of selective logging on tropical forest butterflies in Buru, Indonesia. *Journal of Applied Ecology,* **32**: 754-760.

HODKINSON I. D., (1992). Global insect diversity revisited. *Journal of Tropical Ecology,* **8**: 505-508.

HODKINSON I. D., CASSON D., (1991). A lesser prediction for bugs: Hemiptera (Insecta) diversity. *Biological J. Linn. Society,* **43**: 101-109.

HÖLLDOBLER B., WILSON E. O., (1990). *The Ants.* Springer, Berlin.

HOLLOWAY J. D. *et al.,* Zonation in the Lepidoptera of northern Sulawesi. *In:* W. J. Knight, J. D. Holloway (1990). *Insects and the Rain Forests of South East Asia (Wallacea).* The Royal Entomological Society of London, 153-166.

HOLLOWAY J. D. *et al.,* (1992). The response of some Rain forest insect groups to logging and conversion to plantation. *Phil. Trans. Royal Soc., ser. B,* **335**: 425-436.

HOWDEN H. F., NEALIS V. G., (1975). Effects of clearing in a tropical rain forest on the composition of coprophagous scarab beetle fauna (Coleoptera). *Biotropica,* **7**: 77-83.

HOWDEN H. F., YOUNG O. P., (1981). Panamanian Scarabaeinae: taxonomy, distribution, and habits. *Contributions to the American Entomological Institute,* **18**: 1-204.

HYWEL R., (1969). A note on the Nigerian species of the genus *Sosylus* Er. (Col. Colydiidae) parasites and predators of ambrosia beetles. *J. Nat. Hist.,* **3**: 85-91.

JANSON C. H., EMMONS L. H., Ecological structure of the nonflying mammal community at Cocha Cashu Biological Station, Manu National Park, Peru. *In:* A. H. Gentry (1990). *Four neotropical rainforests.* Yale University Press, New Haven, 314-338.

JANZEN D. H., (1966). Coevolution of mutualism between ants and acacias in Central America. *Evolution,* **20**: 249-275.

JANZEN D. H., (1967). Interaction of the bull's horn acacia (*Acacia cornigera* L.) with an ant inhabitant (*Pseudomyrmex ferruginea* F. Smith) in eastern Mexico. *Univ. Kansas Sci. Bull.,* **47**: 315-558.

JANZEN D. H., (1971). Euglossine bees as long-distance pollinators of tropical plants. *Science,* **171**: 203-205.

JANZEN D. H., (1981). The peak in North American ichneumonid species richness lies between 38° and 42° N. *Ecology,* **62**: 532-537.

JANZEN D. H., (1981). Patterns of herbivory in a tropical deciduous forest. *Biotropica,* **13**: 59-63.

JANZEN D. H., (1983). *Costa Rican Natural History.* Chicago University Press.

JANZEN D. H., POND C. M., (1975). A comparison by sweep sampling of the arthropod fauna of secondary vegetation in Michigan, England and Costa Rica. *Trans. Roy. Ent. Soc. London,* **127**: 33-50.

JANZEN D. H. et al., (1976). Changes in the arthropod community along an elevational transect in the Venezuelan Andes. *Biotropica,* **8**: 193-203.

JOHNSON C. D., (1983). Ecosystematics of *Acanthoscelides* (Coleoptera: Bruchidae) of southern Mexico and Central America. *Entomol. Soc. Amer. Misc. Publ.,* **56**: 1-370.

JOLIVET P. et al., (1988). *Biology of Chrysomelidae.* Kluwer Academic Publishers, Dordrecht.

KLEIN B. C., (1989). Effects of forest fragmentation on dung and carrion beetle communities in central Amazonia. *Ecology,* **70**: 1,715-1,725.

KUSNEZOV M., (1957). Number of species of ants in faunas of different latitudes. *Evolution,* **11**: 298-299.

LAWTON J. H., (1976). The structure of the arthropod community on bracken. *Bot. J. Lin. Soc.,* **73**: 167-216.

LAWTON J. H., Species richness, population abundances, and body size in insect communities: tropical versus temperate comparisons. *In:* P. W. Price *et al.* (1991). *Plant-Animal Interactions: Evolutionary Ecology in Tropical and Temperate Region.* Wiley & Sons, New York, 71-89.

LAWTON J. H., STRONG D. R., (1981). Communities patterns and competition in folivorous insects. *Amer. Nat.,* **118**: 317-338.

LEIGH E. G., SMYTHE N., Leaf production, leaf consumption, and the regulation of folivory on Barro Colorado Island. *In:* G. G. Montgomery (1978). *The Ecology of Arboreal Folivores.* Smithsonian Institution Press, Washington, 33-50.

LEVIN D. A., (1976). Alkaloid-bearing plants: an ecogeographical perspective. *American Naturalist,* **110**: 261-284.

LEVIN D. A., YORK B. M., (1978). The toxicity of plant alkaloids: an eco-geographical perspective. *Biochemical Systematics and Ecology,* **6**: 61-76.

LISTABARTH C., (1996). Pollination of *Bactris* by *Phyllotrox* and *Epuraea.* Implications of the palm breeding beetles on pollination at the community level. *Biotropica,* **28**: 69-81.

LOVEJOY T. E. et al., Edge and other effects of isolation on Amazon forest fragments. *In:* M. E. Soulé (1986). *Conservation Biology.* Sinauer Associates, Sunderland, 257-285.

LOWMAN M. D., (1985). Temporal and spatial variability in insect grazing of the canopies of five Australian rain forest tree species. *Australian Journal of Ecology,* **10**: 264-268.

LÖYTTNIEMI K., HELIÖVAARA., (1996). Tropical Forest Insects. An intro-duction to tropical forest entomology. *University of Helsinki, Department of applied zoology, Report* **24**, 106 pages. (in Finnish).

Lugo A. E., (1988). Diversity of tropical species. Questions that elude answers. *Biology International,* **19**: 1-37.

Lugo A. E. *et al.,* (1973). The impact of leaf cutter ant *Atta columbia* in the energy flow of a tropical forest. *Ecology,* **54**: 1,292-1,301.

Malcolm J. R., Insect biomass in Amazonian forest fragments. *In:* N. E. Stork *et al.* (1997). *Canopy arthropods.* Chapman & Hall, London, 510-533.

Martin J. L., (1966). The insect ecology of red pine plantations in central Ontario. IV. The crown fauna. *Canadian Entomologist.* **98**: 10-27.

Martius C., Bandeira A. G., (1998). Wood litter stocks in tropical moist forests in central Amazonia. *Ecotropica.* **4**: 115-118.

Mawdsley N. A., Stork N. E., Host-specificity and the effective specialization of tropical canopy beetles. *In:* N. E. Stork *et al.* (1997). *Canopy Arthropods.* Chapman & Hall, London, 104-130.

Mayné R., (1960). Hôtes entomologiques du bois. I. Espèces relevées à Yangambi. *Publications de l'I.N.E.A.C., Bruxelles, sér. scientifique,* no. 83, 116 pp.

Mayné R., (1962). Hôtes entomologiques du bois. II. Distribution au Congo, au Rwanda et au Burundi. Observations éthologiques. *Publications de l'I.N.E.A.C., Bruxelles,* no. 100, 514 pp.

Momose K. *et al.,* (1998). Pollination biology in a lowland dipterocarp forest in Sarawak, Malaysia. I. Characteristics of the plant-pollinator community in a lowland dipterocarp forest. *Amer. J. Botany,* **85**: 1,477-1,501.

Morse D. R., Stork N. E., Lawton J. H., (1988). Species number, species abundance and body length relationships of arboreal beetles in Bornean lowland rain forest trees. *Ecological Entomology,* **13**: 25-37.

Muddiman S. B. *et al.,* (1992). Legume feeding psyllids of the genus *Heteropsylla* (Homoptera: Psylloidea). *Bull. Entom. Res.,* **82**: 73-117.

Myers N., (1986). Tropical deforestation and a mega-extinction spasm. *In*: M.-E. Soulé, *Conservation Biology.* Sinauer Associates, Sunderland, 394-409.

Myers N., (1988). Threatened biotas: "hotspots" in tropical forests. *Environmentalist,* **8**: 1-20.

Nadkarni N. M., Longino J. T., (1990). Invertebrates in canopy and ground organic matter in a Neotropical montane forest, Costa Rica. *Biotropica,* **22**: 286-289.

New T. R., Collins N. M., (1991). *Swallowtail butterflies: an action plan for their conservation.* UICN, Gland, Suisse.

Norstog K. J. *et al.,* (1986). The role of beetles in the pollination of *Zamia furfuracea* L. fil. (Zamiaceae). *Biotropica,* **18**: 300-306.

Noyes J. S., (1989). The diversity of Hymenoptera in the tropics with special reference to Parasitica in Sulawesi. *Ecological Entomology,* **14**: 197-207.

NUMMELIN M., (1998). Log-normal distribution of species abundances is not a universal indicator of rain forest disturbance. *Journal of Applied Ecology,* **35**: 454-457.

OLDEMAN R .A. A., (1974). L'architecture de la forêt guyanaise. *Mémoires ORSTOM, no. 73.*

OLMSTEAD K. L., WOOD T. K., (1990). Altitudinal patterns in species richness of neotropical treehoppers (Homoptera: Membracidae): the role of ants. *Proc. Ent. Soc. Washington,* **92**: 552-560.

OLSON D. M., (1994). The distribution of leaf litter invertebrates along a Neotropical altitudinal gradient. *Journal of Tropical Ecology,* **10**: 129-150.

OPLER P. A. *et al.,* (1976). Reproductive biology of some Costa Rican *Codia* species (Boraginaceae). *Biotropica,* **7**: 234-247.

OWEN D. F., (1966). *Animal Ecology in tropical Africa.* Oliver & Boyd, Edinburgh.

OWEN D. F. & OWEN J., (1974). Species diversity in temperate and tropical Ichneumonidae. *Nature,* **249**: 583-584.

PAARMANN W., STORK N. E., (1987). Seasonality of ground beetles (Coleoptera: Carabidae) in the rain forest of N. Sulawesi (Indonesia). *Insect Science and Application,* **8**: 483-487.

PAGES E., (1970). Sur l'écologie et les adaptations de l'oryctérope et des pangolins sympatriques du Gabon. *Biologia Gabonica,* **6**: 27-92.

PALACIOS-VARGAS J. G., (1981). Collembola asociados a *Tillandsia* (Bromeliacea) en el derrame lavico del Chichinautzin, Morelos, Mexico. *Southwestern Entomol.,* **6**: 87-98.

PALACIOS-VARGAS J. G. *et al.,* (1998). Collembola from the canopy of a Mexican tropical deciduous forest. *Pan Pacific Entomologist,* **74**: 47-54.

PARK O., (1937). Studies in nocturnal ecology VI. Further analysis of activity in the beetle, *Passalus cornutus* and description of audio-frequency recording apparatus. *J. Anim. Ecol.,* **6**: 208-223.

PARK O., (1938). Studies in nocturnal ecology. VII. Preliminary observations on Panama rain forest animals. *Ecology,* **19**: 208-223.

PAULIAN R., (1947). *Observations écologiques en forêt de basse Côte d'Ivoire.* Lechevalier, Paris.

PEARSON D., ANDERSON J. J., (1985). Perching heights and nocturnal communal roosts of some tiger beetles (Coleoptera: Cicindelidae) in Southeastern Peru. *Biotropica,* **17**: 126-129.

PEARSON D. L., CASSOLA F., (1992). Worldwide species richness patterns of tiger beetles (Coleoptera: Cicindelidae): indicator taxon for biodiversity and conservation. *Conservation Biology,* **6**: 376-391.

PECK S. B., FORSYTH A., (1982). Composition, structure, and competitive behaviour in a guild of Ecuadorian rain forest dung beetles (Coleoptera: Scarabaeidae). *Canad. J. Zool.,* **60**: 1,624-1,634.

POWELL A. H., POWELL G. V. N., (1987). Population dynamics of male euglossine bees in Amazonian forest fragments. *Biotropica,* **19**: 176-179.

PRANCE G., (1982). *Biological diversification in the tropics.* Columbia University Press, New York.

PRICE P. W., Patterns in communities along latitudinal gradients. *In:* P. W. Price *et al.* (1991). *Plant-animal interactions. Evolutionary ecology in tropical and temperate regions.* John Wiley & Sons, New York, 51-69.

REES C. J. C., Microclimate and the flying Hemiptera fauna of a primary lowland rain forest in Sulawesi. *In:* S. L. Sutton *et al.* (1983) *Tropical Rain Forest: Ecology and Management.* Blackwell, Oxford, 121-136.

REYES-CASTILLO P., HALFFTER G., (1983). La structure sociale chez les Passalidae. *Bull. Soc.Ent. Fr.,* **88**: 619-635.

RICHARDS P. W., The three dimensional structure of tropical rain forest. *In:* S. L. Sutton, T. C. Whitmore, A. C. Chadwick, (1983). *Tropical Rain Forest: Ecology and Management.* Blackwell, Oxford, 3-10.

RIVERA-CERVANTES L. E., MORON M. A., (1991). La comunidad de Coleopteros asociados al arbolado muerto en un bosque mesofilo de montana de la Sierra de Manantlan, Jalisco, Mexico. *Folia Entomologica Mexicana,* **85**: 65-76.

ROBERTS H. F., (1973). Arboreal Orthoptera in the rain forest of Costa Rica collected with insecticide: a report of the grasshoppers (Acrididae) including new species. *Proceedings of the Academy of Natural Science of Philadelphia,* **125**: 46-66.

RODRIGUEZ J. P. et al., (1998). A tests for the adequacy of bioindicator taxa: are tiger beetles (Coleoptera: Cicindelidae) appropriate indicators for monitoring the degradation of tropical forests in Venezuela? *Biological Conservation,* **83**: 69-76.

ROSENZWEIG M. L., (1992). Species diversity gradients: we know more and less than we thought. *Journal of Mammalogy,* **73**: 715-730.

ROUBIK D. W., (1989). *Ecology and Natural History of Tropical Bees.* Cambridge Univ. Press.

RUSSELL-SMITH A., STORK N. E., (1994). Abundance and diversity of spiders from the canopy of tropical rainforests with particular reference to Sulawesi, Indonesia. *Journal of Tropical Ecology,* **10**: 545-558.

SEIFERT R. P., (1982). Neotropical *Heliconia* insect communities. *Quart. Rev. Biol.,* **57**: 1-28.

SHEPHERD V. E., CHAPMAN C. A., (1998). Dung beetles as secondary seed dispersers: impact on seed predation and germination. *Journal of Tropical Ecology,* **14**: 199-215.

STORK N. E., (1987). Guild structure of arthropods from Bornean rain forest trees. *Ecological Entomology,* **12**: 69-80.

STORK N. E., (1988). Insect diversity: facts, fiction and speculation. *Biological Journal of the Linnean Society,* **35**: 321-337.

STORK N. E., BRENDELL M. J. D., Variation in the insect fauna of Sulawesi trees with season, altitude and forest type. *In:* W. J. Knight, J. D. Holloway (1990). *Insects and the Rain Forests of South East Asia (Wallacea)*, The Royal Entomological Society, London, 173-190.

STOUT J., VANDERMEER J., (1975). Comparison of species richness for stream-inhabiting insects in tropical and mid-latitude streams. *Amer. Nat.,* **109**: 263-280.

STRICKLAND A .H., (1947). The soil fauna of two contrasted plots of land in Trinidad, British West Indies. *Journal of Animal Ecology,* **16**: 1-10.

STRONG D. R., (1982). Harmonious coexistence of hispine beetles on *Heliconia* in experimental and natural communities. *Ecology,* **63**: 1,039-1,049.

SUTTON S. L., The spatial distribution of flying insects in tropical rain forests. *In:* S. L. Sutton *et al.,* (1983), *Tropical Rain Forest: Ecology and Management.* Blackwell, Oxford, 77-91.

THIEN L. B., (1980). Patterns of pollination in the primitive angiosperms. *Biotropica,* **12**: 1-13.

TORRES J. A., (1992). Lepidoptera outbreaks in response to successional changes after the passage of Hurricane Hugo in Puerto Rico. *Journal of Tropical Ecology,* **8**: 285-298.

TURNER J. R. G., Extinction as a creative force: the butterflies of the rain-forest. *In:* A. C. Chadwick & S. L. Sutton (1984). *Tropical Rain Forest.* Leeds Philosophical and Literary Society, 195-204.

VADON J., (1947). Les epilissiens de Madagascar (Coleoptera, Scarabaeidae, Canthoniini). II. Biologie. *Bulletin de l'Académie Malgache,* **26**: 173-174.

WALLACE A. R., (1878). *Tropical Nature and other Essays.*

WALTER P., (1984). Contribution à la connaissance des Scarabéides coprophages du Gabon (Col.). 2. Présence de populations dans la canopée de la forêt gabonaise. *Bull. Soc. ent. Fr.,* **88**: 514-521.

WASSERTHAL L. T., (1996). Of hawkmoth species with long 'tongues'. *Rept. DFG,* **1**: 22-25.

WEBER N. A., (1972). Gardening ants-the attines. *Memoirs of the American Philosophal Society, Philadelphia,* **92**

WILSON E. O., (1992). *The diversity of life.* W. W. Norton & Co., New York.

WOLDA H., Spatial and temporal variation in abundance in tropical animals. *In:* S. L. Sutton *et al.* (1983). *Tropical Forest: Ecology and Management.* Blackwell, Oxford, 93-105.

WOLDA H., DENLINGER D. L., (1984). Diapause in a large aggregation of a tropical beetle. *Ecological Entomology,* **9**: 217-230.

WOLDA H. *et al.,* (1998). Weevil diversity and seasonality in tropical Panama as deduced from light-trap catches (Coleoptera: Curculionidae). *Smithsonian Contributions to Zoology,* **590**: 1-79.

Wolda J. R., (1987). Altitude, habitat and tropical insect diversity. *Biological Journal Linn. Soc.*, **30**: 313-323.

Wood B., Gillman M. P., (1998). The effects of disturbance on forest butterflies using two methods of sampling in Trinidad. *Biodiversity and Conservation*, **7**: 597-616.

Wood T. K., Olmstead K. L., (1984). Latitudinal effects on treehopper species richness (Homoptera: Membracidae). *Ecological Entomology*, **9**: 109-115.

Young A. M., Thomason J. H., (1975). Notes on communal roosting of *Heliconius charitonius* (Nymphalidae) in Costa Rica. *J. Lep. Soc.*, **29**: 243-255.

Conclusion

Insect fauna and forest management

The foregoing chapters have shown that biodiversity of forest insect fauna is high, even in temperate regions, and that fauna plays a major part in the functioning of the ecosystem. We have also seen that present methods of forest management have created woodland considerably different in the way it functions and in structure from primeval forests untouched or little exploited by man. Of the main features of primary forest that have been lost in managed forest, the most notable may be summarised as:

– Heterogeneity in the horizontal plane, due to juxtaposition of stands of different age and structure.

– Complexity of structure in the vertical plane (sometimes almost as complex as that of tropical forests), due to the presence of old or very old trees of great height and girth.

This heterogeneity ensures "forest continuity" for insects, and the maintenance of ecological niches that otherwise vanish, together with their associated fauna.

A few examples among others may be recalled. Systematic removal of large dead old trees, often riddled with cavities, causes saproxylic species dependent on this habitat to vanish. The clearing of forest rides, which destroys the characteristic structure of forest edge, leads to the disappearance of many species, especially Lepidoptera (Noblecourt, 1996). Fragmentation of forest stands reduces them to small islands surrounded by inhospitable cultivated land treated with insecticides. Removal of old trees from avenues in urban parks destroys the last refuges of many insects. Bad forest management practices, such as excessive planting of conifers, makes trees more vulnerable and predisposes them to mass attacks by pests whose numbers usually remain very low in natural forest.

Loss of biodiversity has become evident in the last few decades, particularly in the forest environment. "Many species (of colydiid beetles) live in dead wood, both as larvae and as adults, and they are often localised in old forests that are untouched or little changed by man. Many of these species are becoming rare in the whole of Europe and their ranges of distribution are more and more fragmented due to destruction of their habitats" (Dajoz, 1977). Rational management of the forest heritage must conserve what is left, and should aim to provide endangered species with the means of reclaiming areas from which they have vanished owing to interference by man. Some of the concepts which were accepted in the past must be revised. Management must become aware that temperate forest ecosystems naturally include large quantities of dead wood, and that this dead wood reflects a normal functioning of the ecosystem rather than a dysfunction.

Forest should be managed with three aims in view: to produce timber, to provide recreational space, and to protect the biodiversity of the plants and animals that inhabit it. To achieve the aim of maintaining biodiversity it is necessary to establish a network of sufficiently large nature reserves, and where possible to link them with corridors.

We know today that insects, like other groups of living organisms, are "indicators of environmental quality". Biodiversity conservation is a subject that concerns responsible organisations such as the Office National des Forêts and the European Council. Together with other measures, biodiversty conservation is essential for sustainable development.

REFERENCES

DAJOZ R., (1977). *Coléoptères Colydiidae et Anommatidae paléarctiques.* Faune de l'Europe. Masson, Paris, 277 pages.

NOBLECOURT T., (1996). La protection de l'entomofaune et la gestion forestière. *Rev. For. Fr.*, **48**: 31-38.

Office National des Forêts, (1993). *Prise en compte de la diversité biologique dans l'aménagement et la gestion forestière.* Two fascicles 37 and 18 pages.

SPEIGHT M.-D., (1989). *Les invertébrés saproxyliques et leur protection.* Conseil de l'Europe, Strasbourg, 75 pages.

Index of insect taxa

Impression : EUROPE MEDIA DUPLICATION S.A.
F 53110 Lassay-les-Châteaux
N° 7734 - Dépôt légal : Septembre 2000
N° 331 - LK 80°